The Bacteria

VOLUME I: STRUCTURE

THE BACTERIA
A TREATISE

Volume I: Structure
Volume II: Metabolism
Volume III: Biosynthesis
Volume IV: Growth
Volume V: Heredity

The Bacteria

A TREATISE ON STRUCTURE AND FUNCTION

edited by

I. C. Gunsalus
Department of Chemistry
University of Illinois
Urbana, Illinois

Roger Y. Stanier
Department of Bacteriology
University of California
Berkeley, California

VOLUME I: STRUCTURE

1960

ACADEMIC PRESS • NEW YORK AND LONDON

QR41
.G78
v.1

Copyright © 1960, by Academic Press Inc.

ALL RIGHTS RESERVED

NO PART OF THIS BOOK MAY BE REPRODUCED IN ANY FORM
BY PHOTOSTAT, MICROFILM, OR ANY OTHER MEANS,
WITHOUT WRITTEN PERMISSION FROM THE PUBLISHERS.

ACADEMIC PRESS INC.
111 Fifth Avenue
New York 3, N. Y.

United Kingdom Edition
Published by
ACADEMIC PRESS INC. (London) Ltd.
Berkeley Square House, London, W. 1

Library of Congress Catalog Card Number 59-13831
First Printing 1960
Second Printing 1962

PRINTED IN THE UNITED STATES OF AMERICA

CONTRIBUTORS TO VOLUME I

Thomas F. Anderson, *Biology Division, University of Pennsylvania, and Institute for Cancer Research, Philadelphia, Pennsylvania*

E. Klieneberger-Nobel, *Lister Institute of Preventive Medicine, London, England*

E. S. Lennox, *Division of Biochemistry, Noyes Laboratory of Chemistry, University of Illinois, Urbana, Illinois*

S. E. Luria, *Department of Biology, Massachusetts Institute of Technology, Cambridge, Massachusetts*

Allen G. Marr, *Department of Bacteriology, University of California, Davis, California*

Kenneth McQuillen,* *Department of Biochemistry, University of Cambridge, Cambridge, England*

R. G. E. Murray, *Department of Bacteriology and Immunology, University of Western Ontario, London, Ontario, Canada*

C. F. Robinow, *Department of Bacteriology and Immunology, University of Western Ontario, London, Ontario, Canada*

M. R. J. Salton, *Department of Bacteriology, University of Manchester, Manchester, England*

Claes Weibull, *Central Bacteriological Laboratory of Stockholm City, Stockholm, Sweden*

*Also Research Associate of the Carnegie Institution of Washington, Washington, D.C.

PREFACE

Bacteriology developed in the late nineteenth and early twentieth centuries very largely as an applied science, with momentous consequences for human health and welfare. Its outlook was largely empirical, and its growth was influenced only to a minor extent by general biological theory. During the past twenty-five years, a radical change has occurred. Although the applied aspects of this science have continued to grow, there has emerged a vigorous and widespread interest in the bacteria as biological entities. This originated historically from the realization of the fundamental unity of biochemical processes in living organisms, coupled with the recognition that bacteria provide ideal experimental systems for the investigation of many biochemical problems. Somewhat later, the experimental analysis of bacterial variation by scientists aware of the concepts of classical genetics led to the emergence of bacterial genetics as an important branch of genetic science. Here again, it was quickly realized that the biological properties of bacteria make them particularly suitable for the study of certain genetic problems; in addition the unique mechanisms of gene transfer in bacteria enriched genetic theory with new insights. Lastly, the study of bacterial cell structure has made great strides, aided particularly by the development of electron microscopy and of modern cytochemical techniques.

Heretofore, there has been no comprehensive source of information about modern knowledge of the general biological properties of bacteria. The present treatise will, it is hoped, help to fill this gap in the bacteriological literature. When the editors first discussed plans for the treatise some five years ago, they recognized the desirability of covering also the special properties of the various bacterial groups. However, it soon became apparent that a treatise which dealt with both the general and special aspects of bacteriology would prove an enormous undertaking, and it was accordingly decided to confine the work to general aspects. These will be treated in five volumes, concerned, respectively, with: structure; energy-yielding metabolism; biosynthesis; growth and general physiology; and heredity.

An undertaking of this sort, which involves the collaboration of many different specialists, inevitably lacks the unity and cohesiveness of a work produced by one or two authors. This is unfortunate, but the tremendous expansion of science in our time allows comprehensive coverage of a broad field only at this price. The editors have refrained from attempts to impose a single format, or to delineate in detail the area assigned to individual authors, beyond attempting to eliminate obvious large overlaps. The willingness with which the authors have responded to our appeals for contributions and have accepted suggestions for revisions has made the task of the editors a pleasant one. We are also greatly appreciative of the understanding, enthusiasm and practical assistance extended by the publishers and their staff. We trust that this undertaking will prove to be of value to the community of microbiologists.

I. C. GUNSALUS
R. Y. STANIER

December, 1959

CONTENTS

Contributors to Volume I... v
Preface... vii
Contents of Volume II, III, IV..................................... xiii

1. The Bacterial Protoplasm: Composition and Organization............ 1
S. E. Luria

    I. Living Matter, Cell Theory, and the Unity of Biochemistry...... 1
    II. The Bacterial Cell.. 5
    III. The Materials of Bacteriology................................... 8
    IV. Chemical Analysis of Bacteria................................... 13
    V. Isolation of Functional Constituents............................ 22
    VI. Isolation of Organized Bacterial Constituents................... 25
    VII. Specialized Differentiations of Bacterial Cells.................. 31
    References... 32

2. The Internal Structure of the Cell................................. 35
R. G. E. Murray

    I. Introduction... 35
    II. The Cytoplasm and Its Surface.................................. 36
    III. Chromatin Bodies... 64
    References... 93

3. Surface Layers of the Bacterial Cell................................ 97
M. R. J. Salton

    I. Introduction... 97
    II. Anatomy of the Bacterial Surface................................ 98
    III. Extracellular Surface Components, Slime, and Capsular Materials. 99
    IV. Cell Walls.. 115
    References... 144

4. Movement... 153
Claes Weibull

    I. Introduction... 153
    II. Theoretical Aspects of the Movements of Bacteria................ 154
    III. Flagellar Movement... 158
    IV. Movements of the Spirochetes................................... 174
    V. Gliding Movement.. 180

CONTENTS

VI. Bacterial Movements Considered as Tactic Responses to
    External Stimuli 188
    References 198

5. Morphology of Bacterial Spores, Their Development and Germination 207
C. F. ROBINOW

I. Introduction 207
II. Distribution of the Ability to Form Spores 209
III. General Observations on the Development of Spores 211
IV. The Brightness of Spores 216
V. The Interior of Spores 216
VI. The Skin of Spores 216
VII. The Imperviousness of Spores to Stains 229
VIII. The Chromatin of the Spore 230
IX. Germination 237
X. The Chromatin of Germinating Spores 243
XI. Parasporal Bodies 243
XII. Conclusion 245
    References 246

6. Bacterial Protoplasts 249
KENNETH McQUILLEN

I. Concepts and Definitions 250
II. Formation of Protoplasts 261
III. Morphology and Structure 282
IV. Physicochemical Properties of Protoplasts 288
V. Composition of Protoplasts 295
VI. Physiology and Biochemistry of Protoplasts 311
    References 353

7. L-Forms of Bacteria 361
E. KLIENEBERGER-NOBEL

I. Introduction 361
II. The Discovery of the L-Form 362
III. Definition of L-Form 363
IV. Appearance of Growth on Solid and in Liquid Media 363
V. Production of L-Form 365
VI. Microscopic Demonstration of L-Form 367
VII. Morphology of L-Form 368
VIII. Properties of L-Form 371
IX. The Similarities of L-Forms and Pleuropneumonia-like
    Organisms 374
X. Electron Microscopic Demonstration of L-Forms of
    Bacteria and of PPLO 377
XI. L-Forms and Protoplasts 381
XII. Summary and Conclusions 382
    References 383

8. Bacterial Viruses—Structure and Function 387
Thomas F. Anderson

  I. Introduction 387
  II. Structure of Bacteriophage Particles 394
  III. Relation of Structure to Function—Mechanism of Infection 408
  IV. Importance of Bacteriophages in Bacteriology 411
       References 412

9. Antigenic Analysis of Cell Structure 415
E. S. Lennox

  I. Introduction 415
  II. Preparation of Antisera 417
  III. Quantitative Methods of Using Antisera 423
  IV. Applications of Serological Techniques to Problems of
       Bacteriology 426
       References 439

10. Localization of Enzymes in Bacteria 443
Allen G. Marr

  I. Introduction 443
  II. Direct Cytochemistry 444
  III. Analytical Morphology 446
  IV. Pigments of Photosynthetic Bacteria 461
  V. Endospores 464
       References 465

Author Index 469

Subject Index 489

The Bacteria

A TREATISE ON STRUCTURE AND FUNCTION

VOLUME II: METABOLISM

Energy-Yielding Metabolism in Bacteria
    I. C. Gunsalus and C. W. Shuster

Fermentation of Carbohydrates and Related Compounds
    W. A. Wood

Fermentations of Nitrogenous Organic Compounds
    H. A. Barker

Cyclic Mechanisms of Terminal Oxidation
    L. O. Krampitz

The Dissimilation of High Molecular Weight Substances
    H. J. Rogers

Survey of Microbial Electron Transport Mechanisms
    M. I. Dolin

Cytochrome Systems in Aerobic Electron Transport
    Lucile Smith

Cytochrome Systems in Anaerobic Electron Transport
    Jack W. Newton and Martin D. Kamen

Cytochrome-Independent Electron Transport Enzymes of Bacteria
    M. I. Dolin

Bacterial Photosynthesis
    David M. Geller

Bacterial Luminescence
    W. D. McElroy

VOLUME III: BIOSYNTHESIS

Photosynthesis and Lithotrophic Carbon Dioxide Fixation
    S. R. Elsden

Assimilation of Carbon Dioxide by Heterotrophic Organisms
    Harland G. Wood and Rune L. Stjernholm

Inorganic Nitrogen Assimilation and Ammonia Incorporation
  L. E. Mortenson
Pathways of Amino Acid Biosynthesis
  Edwin Umbarger and Bernard D. Davis
The Synthesis of Vitamins and Coenzymes
  J. G. Morris
Biosynthesis of Purine and Pyrimidine Nucleotides
  B. Magasanik
Tetrapyrrole Synthesis in Microorganisms
  June Lascelles
Synthesis of Polymeric Homosaccharides
  Schlomo Hestrin
The Biosynthesis of Homopolymeric Peptides
  Riley D. Housewright
Biosynthesis of Bacterial Cell Walls
  Jack L. Strominger
The Synthesis of Proteins and Nucleic Acids
  Ernest Gale
The Synthesis of Enzymes
  Arthur B. Pardee
Author Index—Subject Index

VOLUME IV: GROWTH*

Synchronous Growth
  O. Maaløe
Nutritional Requirements of Microorganisms
  Beverly M. Guirard and Esmond E. Snell
Ecology of Bacteria
  R. E. Hungate
Growth Inhibitors
  B. D. Davis
Permeability
  G. N. Cohen and A. Kepes
Halophilism
  Helge Larsen
Temperature Relationships
  John Ingraham
Physiology of Sporulation
  Harlyn O. Halvorson
Exoenzymes
  M. R. Pollock
Author Index—Subject Index

* Tentative contents.

CHAPTER 1

The Bacterial Protoplasm: Composition and Organization

S. E. LURIA

I. Living Matter, Cell Theory, and the Unity of Biochemistry............... 1
   A. Life and Organization.. 1
   B. Cell Theory and the Unity of Biochemistry........................ 2
   C. Biochemical Cytology... 5
II. The Bacterial Cell... 5
   A. Bacteria as "Unicellular" Organisms.................................. 5
   B. The Cell as a Factory... 7
III. The Materials of Bacteriology.. 8
   A. Size of Bacterial Cells.. 8
   B. Amounts of Bacterial Cells.. 10
   C. Water Content.. 13
IV. Chemical Analysis of Bacteria.. 13
   A. Cellular and Extracellular Materials................................ 13
   B. Elementary Composition.. 14
   C. Organic Constituents... 15
V. Isolation of Functional Constituents.................................... 22
   A. Tests of Functional Activity.. 22
VI. Isolation of Organized Bacterial Constituents.......................... 25
   A. An Outline of Bacterial Cytology.................................... 25
   B. Cytologically Identifiable Fractions................................. 25
   C. Submicroscopic Structure and Macromolecular Organization......... 28
VII. Specialized Differentiations of Bacterial Cells......................... 31
   A. Endospores; Bacteriophage Particles................................ 31
   References... 32

I. Living Matter, Cell Theory, and the Unity of Biochemistry

A. LIFE AND ORGANIZATION

When we examine living beings, the feature that immediately strikes us is organization. The organization of the world of life expresses itself at various levels. Some are taxonomic levels, which reflect the historical aspects of life and define the common descent and degree of relationship among living beings. Other levels of organization are detected within the individual units of living matter, which are in fact called *organisms*. The organized features of living beings reach all the way from the macroscopic to the molecular level, and one of the major tasks of biology is to understand and describe this organization, its interdependent levels, its stability, and its variability. In fact, life can be defined as the set of processes by which the organization of organic matter is maintained and reproduced, that is, imposed upon other matter which becomes assimilated into living organisms. We may define as *living matter* or *protoplasm* that portion of matter endowed at any one time with this self-maintaining organization.

The progressive evolution and diversification of living forms result from the flexibility of the organization of living matter, which, by reproducing its own pattern with accurate but not always absolute fidelity, makes it possible for new forms of life to arise, to be tried out in the ever-changing material environment, and to persist or disappear, depending on their fitness to carry out the processes of maintenance and reproduction.

B. Cell Theory and the Unity of Biochemistry

The unique features and historical continuity of the organization of living matter or protoplasm are embodied in two basic generalizations: the cell theory and the theory of the unity of biochemistry.

The *cell theory*[1] recognizes that all living matter consists of units, called *cells*, which have a common basic pattern of organization and which represent the simplest elements that can carry out all processes of life. The *theory of the unity of biochemistry*[2] recognizes that these life processes involve the same basic chemical reactions in all cells, because the common pattern of organization is embodied in a chemically common material substrate.

The cell theory was founded on a morphological basis. The microscope showed that all tissues and organs of animals and plants consisted of cells and of intercellular materials. The essential steps in the formulation of the cell theory were, on the one hand, the realization that in any organism only the cells are living, that is, are capable of assimilation and reproduction, while the extracellular materials, no matter how specific or complex, are nonliving, incapable of assimilation, metabolically inert products of cell activity; and, on the other hand, the recognition that all cells have a common basic structure, and that the almost infinite variety of cells represent differentiations and modifications of the common pattern—modulations and variations on a basic theme.

1. Cell Theory and Cellular Structure

The common pattern of organization of all cells is embodied in the idealized cytological picture of a *generalized cell*. The generalized cell has a nucleus, which, within a nuclear membrane, contains one or more nucleoli and a characteristic number of chromosomes. These are identified by cytogenetic experiments as the seat of the discrete genetic determinants that obey Mendelian heredity rules in sexual reproduction. Outside the nucleus is the cytoplasm, consisting of a basic substance or *hyaloplasm*, in which are immersed a variety of organelles and granules, including functional structures—*mitochondria, centrosomes, plastids, kinetosomes*—and reserve materials—starch granules, fat globules, and vacuoles.

The electron microscope recognizes further levels of morphological organization, which reflect more and more closely the patterns of molecular organization. For example, in the hyaloplasm we recognize the *ergastoplasm* or *endoplasmic reticulum*[3] with its constituent granules or *microsomes*; in the mitochondria, elaborate systems of lamellae. On the outer edge the cell is bounded by a *cytoplasmic membrane*, which contains lipids, acts as a semipermeable barrier, and is the site of active transport mechanisms, retaining the larger cellular constituents and regulating the exchanges of smaller molecules between the cell and its environment.

The actual cell types may deviate morphologically from the generalized cell in a number of ways. They may acquire an outer rigid *cell wall*, as many plant cells do. They may develop complex surface structures, such as cilia, flagella, and even "mouth parts," as in certain protozoa, probably by differentiated functions of kinetosomes.[4] They may acquire contractile functions by synthesizing fibers of contractile proteins. They may lose irreversibly their reproductive ability, as in differentiated nerve cells, or even lose their nuclei, as in mammalian erythrocytes. Supracellular or apparently acellular complexes may arise by failure of cellular division to accompany nuclear division, as in fungi with nonseptate hyphae, or by actual fusion of preexistent cells.

2. Unity of Biochemistry and Cellular Function

In the same way as the cell theory recognizes a general cell pattern and interprets the multitude of actual cell types as modifications of the general pattern, so does the theory of the unity of biochemistry recognize a common chemical substrate of all protoplasm and a common set of chemical reactions providing the energy and the building blocks for the construction of protoplasm. The functionally diverse types of cells represent modifications of the basic chemical pattern, specifically adapted to certain functions and environments.

The unity of protoplasmic chemistry reveals itself in two ways. The first and most basic aspect is the composition of the macromolecules—proteins and nucleic acids—that carry out the specific chemical activities of protoplasm, *enzyme action* and *genetic action*; that is, chemical catalysis and control of the specific patternization of new macromolecules—the so-called "template action." All proteins and all nucleic acids consist of a limited number of common ingredients, in the same way as all words of a language consist of a limited number of letters and syllables: the proteins, of amino acids; the nucleic acids, of nucleotides. As the meaning of language results from combinations and permutations of letters and syllables and from the arrangements of words into sentences, thus all functional specificity or "information" in cells results from the combinations and permutations of about 20 amino acids in proteins or 4 nucleotides in nucleic acids, respectively, and from the folding and juxtaposition of the macromolecules. Folding and mutual relations of the macromolecular chains are probably themselves fully determined by the inner sequence of digits[4a].

Such a system of chemical alphabets makes available innumerable functional specializations embodied in a common substrate; but the actual chemical processes by which proteins and nucleic acids carry out their catalytic or template functions are still largely not understood.

The second way in which the unity of biochemistry expresses itself is that a given chemical result is accomplished in different cells and organisms, not by a great variety of mechanisms, but in only a few alternative ways, sometimes in a unique way. This limited number of alternative pathways reflects the opportunism of evolution, which tends to select and preserve any successful device among the chemically, thermodynamically and ecologically possible ones. Clear examples are observed in energy-yielding and biosynthetic pathways. There are four known prototypes of energy-yielding mechanisms:[5] anaerobic fermentation, pentose phosphate cycle, tricarboxylic acid cycle* and photosynthesis. One or more of these are found with minor variations in all cells and organisms, with similar or identical substrates, key intermediates, and cofactors. Even the enzymes have enough common features to bespeak a common molecular structure. Likewise, each of the necessary chemical ingredients of protoplasm, or *essential metabolites*, is synthesized in different organisms by one of a few alternative pathways. Even the processes involved in all forms of motion of living organisms seem to share a common basic mechanism: stimulation of contractile proteins with ATPase activity by the common substrate adenosine triphosphate (ATP).[6]

All quantitative and qualitative differences among cells and organisms can be traced biochemically to variations of the basic pattern, which fall into three main categories: (1) failures to synthesize some essential metabolite, which thereby becomes a *required nutrilite* (or "growth factor," or "vitamin") and must be supplied from the outside, either by other cells or from the external milieu; (2) specific stimulation or inhibition of chemical processes by endogenous or exogenous substances that regulate enzyme function, like hormones, or enzyme synthesis, like "inducers"[7] or "repressors,"[7a] or agents which alter cell permeability; (3) production of specific functional differentiations, such as contractile proteins for motion,

---

* The glyoxylate bypass is here considered as an essential variant of the tricarboxylic acid cycle permitting net synthesis of four-carbon compounds for cellular biosynthesis during growth on two-carbon compounds exclusively.[5a]

or rigid materials for structural support, or permeability barriers for selective retention and exclusion of enzymes, substrates, and products. For example, mechanically rigid structures may be achieved either by excretion of intercellular substances, such as collagen and its derivatives in connective and bone tissues of animals, or by production of rigid cell walls, as in plants. On the other hand, a rigid cell wall may be used to provide rigid support for a delicate membrane subject to high osmotic pressures, as in molds and bacteria.

C. Biochemical Cytology

The morphological, genetic, and biochemical approaches to cellular organization have converged in recent years to create a "biochemical cytology"[8] which endeavors to give a complete picture of cellular structure and function in terms of the location and mutual relation of functional constituents. Progress has been made, on the one hand, in locating the sites of specific biochemical functions within cells or sections of cells by cytochemical reagents or by microautoradiography using radioactive substrates;[9] and, on the other hand, in fractionating cells, detecting chemical constituents or specific catalysts in certain fractions, and identifying these fractions with cellular layers or organelles. Thus, for example, we recognize the mitochondria as the sites of oxidation, electron transport, and oxidative phosphorylation; the microsomal fraction as the site of cytoplasmic protein synthesis; the chloroplasts as the seat of the primary reactions in photosynthesis; the deoxyribonucleic acid as the basic component of genetic matter, present in constant amount in all cells of a given genetic constitution.

II. The Bacterial Cell

A. Bacteria as "Unicellular" Organisms

Both for technical and for historical reasons, morphology played a relatively less important role in the study of the bacterial cells than in that of other cells. The small size of bacteria long delayed the recognition of inner structures whose specific functions had to be explained. The roots of bacteriology in the study of disease and fermentations caused the attention of bacteriologists to center on chemical function; and the rise of bacteriology as a science coincided with the flourishing of cellular physiology and with the rise of modern biochemistry. Hence, it was possible for some time to ignore the inner organization of bacterial cells or even to deny it and to consider bacteria simply as "bags of enzymes."

The formulation of the theory of the unity of biochemistry led to the recognition of bacterial cells as a superb material for the study of cellular functions. The pure culture methods provided a variety of cell types, with characteristic metabolic patterns, which could be grown and prepared under chemically controlled conditions. In turn, the similarity of metabolic functions in bacteria and in other organisms required a careful comparison between the organization of bacterial cells and other cells. Biochemical cytology revitalized the study of bacterial cytology, bringing to it the tools of enzymology, electron microscopy, chemical fractionation, and tracer methodology.

Thus, there emerged a picture of a *generalized bacterial cell*, comparable to the picture of the generalized cell, as an abstraction embodying the basic feature of organized bacterial protoplasm. In this field, most variations on the common theme represent different organisms rather than different types of cells of the same organism. This is what we mean when we say that bacteria are unicellular organisms. More correctly, we assume that in a pure culture of bacteria we can equate the properties of any one cell with those of all other cells; that is, in a pure culture or *clone* derived from a single ancestor there is no regular morphogenetic differentiation of structure and function, barring genetic mutation, and no irreversible loss of totipotency distinguishable from sterility or death.

Functional coordination within clones derives from metabolic interactions, including competition for substrates, accumulation of by-products, and production of extracellular enzymes, within a population of equivalent individuals. Numerous reports of bacterial "life cycles" have been explained as resulting from an interplay of two types of processes: physiological response of bacterial cells to changes in environment, and genetic variation.[10, 11] Acting together, these two processes lead to changes in growth rates and growth patterns and, secondarily, to a selection of mutant types preadapted or more readily adaptable to the altered environment. Thus, deviations from the prototype or generalized bacterial cell represent either genetic differences or physiological adaptations to a changing environment. These adaptations do not give rise to permanently differentiated cell types with specialized functions within a clone.

Phenomena such as the sporadic appearance and transient persistence of individual motile cells within nonmotile clones[12, 13] indicate, however, that some intraclonal specialization can occur in common bacteria. Under favorable circumstances, such specializations might become stabilized and provide a basis for a true organismic differentiation. Another type of differentiation is observed in cells that initiate the production of an adaptive enzyme.[7, 14] Enzyme adaptation is an "all-or-none" phenomenon and, during the transition from the unadapted to the adapted state, a culture contains a mixture of adapted and nonadapted cells.[15]

Still another intraclonal differentiation in bacteria results from mating phenomena.[16, 17] The only well-analyzed instance, in *Escherichia coli*, reveals that the mating polarities, which differentiate the members of a potentially interbreeding population, are mutational or infectious in origin and do not represent intraclonal differentiations. All cells of a given polarity appear to be potential gametes. The only differentiated individuals are the direct products of mating or "zygotes" and their occasional unsegregated descendants[18] (see note added in proof, page 34).

Differentiation might arise in some bacteria such as *Streptomyces* by segregation from heterocaryons, that is, from organisms containing nuclei of two or more genetic types in a nonseptate mycelium.[19] Differentiation takes place in Myxobacteria with the formation of resting cells, generally within fruiting bodies produced by aggregation and transformation of vegetative cells; the details of these processes are still obscure.

By and large, however, all cells of a pure clone in a common environment can generally be considered as equivalent. Even an irregularly multinucleated condition or the truly multicellular condition observed in some rod-shaped bacteria may be overlooked without harm except in studies on genetic and mutational phenomena.

A complete description of bacteria requires an analysis of the basic composition, structure, and organization of bacterial protoplasm, as well as an analysis of the peculiar properties and functions of specific bacteria, and of the evolutionary relationship among bacteria and between bacteria and other organisms.

We classify as bacteria a great variety of organisms. These include groups whose basic similarities are rather obvious, and also very deviant groups. The latter may have derived from the common bacterial types by physiological evolution reflected in morphological changes. Thus, nutritionally fastidious bacteria may have evolved into obligate parasites like the rickettsiae. Very fragile forms, like the pleuropneumonia organisms, may represent bacteria that have lost more or less permanently the ability to form rigid cell walls. Other forms, like the Spirochetales and the Myxobacteriales, probably represent more distant lines of descent, but little is known yet of their chemistry and physiology.

The detailed presentation of our knowledge about bacterial structure and organization will be the subject matter of later chapters. In this introductory chapter we shall limit our discussion to the over-all composition of bacterial protoplasm, as exemplified in the commonest bacteria, the Eubacteriales, and outline the operational nature and limitations of the analysis of cellular organization.

B. The Cell as a Factory

As already stated, a living cell is a dynamic system, whose specific organization is perpetuated through assimilation and replication. An instructive analogy is to compare the cell to a factory. Like a cell, a factory is a system with a specific pattern of organization resulting in a set of specialized functions. Like a cell, a factory has both maintenance and production problems. The special features of living cells—growth and reproduction—can be found in an expanding factory, which grows by building copies of itself according to its own intrinsic blueprint. How do we analyze such a factory? What do we want to learn about it and how do we proceed to learn it?

We may simply examine its layout from the outside or in various sectional views; this corresponds to a morphological study. Or we may tear the factory apart, reduce it to fragments, and describe the various ingredients—bricks, mortar, metals, glass—of which it is built; this corresponds to a chemical analysis of cellular matter. Or we may extract from the factory individual pieces of machinery and study their functions in isolation—the equivalent of enzymology. Next, we may inquire into the factory's power plant: the fuels it uses, the power yield, the residues or smokes—in other words, the energy-yielding mechanisms. We can then examine the raw materials used by the factory—its nutrition—and the changes these materials undergo in the course of production—the intermediary metabolism. We ask about the organization and interrelation of production lines—that is, about the integration of metabolic pathways. As we observe production, we become aware of the existence in the factory of a hierarchy of control mechanisms: board of directors, planning staff, blueprint selectors, and production checkers—the genetic mechanisms. We observe changes in the functions and products of a factory, reflecting changes in raw materials or in demand—adaptation—or changes in the internal controlling systems—mutation and hereditary variation.

When all these observations are brought together, we recognize the factory or the organism as an integrated whole, which persists, functions, and grows by the maintenance and replication of its material pattern of organization.

III. The Materials of Bacteriology

A. Size of Bacterial Cells

The bacteriologist works with bacterial cells and bacterial cultures. All chemical analyses of bacterial protoplasm are carried out on pure cultures. Some data on the size of bacteria define the scale of the materials under study. Table I gives the size and shapes of some representative organisms. The values provide a basis for a rapid approximate calculation of volume and mass. Thus, a spherical coccus 1.2 μ in diameter is about 10^{-12} cm.³ in volume. Since the density of wet bacteria, measured by centrifugation,[20] is close to 1.1, the mass of such cell would be 1.1×10^{-12} g. A rod 0.8μ in

TABLE I
Sizes of Typical Bacterial Cells[a]

Order	Species	Shape	Average width (μ)	Average length (μ)	Average diameter (μ)	Conditions
Eubacteriales	Pasteurella tularensis	Rod	0.2	0.3–0.7		Dry, stained
	Brucella melitensis	Rod	0.3	0.3–1.0		Dry, stained
	Escherichia coli	Rod	0.4–0.7	1.0–3.0		Dry, stained
	Escherichia coli	Rod	1.1–1.5	2.0–6.0		Living
	Proteus vulgaris	Rod	0.5–1.0	1.0–3.0		Dry, stained
	Proteus vulgaris	Rod	1.0–1.4	1.4–3.1		Living
	Pseudomonas fluorescens	Rod	0.3–0.5	1.0–1.8		Dry, stained
	Pseudomonas aeruginosa	Rod	0.5–0.6	1.5–3.0		Dry, stained
	Bacillus subtilis	Rod	0.5–0.8	1.6–4.0		Dry, stained
	Bacillus subtilis	Rod	0.9–1.05	1.8–4.8		Electron micrograph
	Bacillus subtilis spores	Oval	0.5–1.0	0.9–1.6		Dry, stained
	Bacillus anthracis	Rod	1.0–1.3	3.0–10.0		Dry, stained
	Bacillus megaterium	Rod	0.9–1.7	2.4–5.0		Dry, stained
	Bacillus megaterium	Rod	1.6–2.0	3.7–9.7		Living
	Bacillus megaterium	Rod	1.0–1.4	2.0–3.9		Electron microscope
	Bacillus megaterium spores	Oval	0.6–1.2	0.9–1.7		Dry, stained
	Sarcina lutea	Sphere			1.0–1.5	Dry, stained
	Sarcina ventriculi	Sphere			2.5	Dry, stained
	Staphylococcus aureus	Sphere			0.8–1.0	Dry, stained
	Streptococcus lactis	Sphere to rod	0.5–1.0	0.5–1.0		Dry, stained
	Streptococcus faecalis	Sphere to rod	1.4–1.7	1.4–2.1		Living
	Spirillum volutans	Spiral	2.0–3.0	5–20		Dry, stained
Actinomycetales	Actinomyces bovis	Mycelial	1.0–1.5			Dry, stained
Rickettsiales	Bartonella bacilliformis	Rod	0.25–0.5	1–3		Dry, stained
	Rickettsia prowazekii	Rod	0.25–0.3	0.4–0.7		Electron microscope

TABLE I—*Continued*

Order	Species	Shape	Average width (μ)	Average length (μ)	Average diameter (μ)	Conditions
Caryopha- nales	*Lineola*	Rod	1.5	10–50		Living
	Caryophanon latum	Rod	2.5–3.2	6–30		Living

[a] Data from R. S. Breed, E. G. D. Murray, and N. R. Smith, "Bergey's Manual of Determinative Bacteriology," 7th ed. Williams & Wilkins, Baltimore, Maryland, 1957; and G. Knaysi, "Elements of Bacterial Cytology," 2nd ed. Comstock (Cornell Univ. Press), Ithaca, New York, 1951.

cross diameter and 2μ in length (an average cell of *Escherichia coli*) also has a volume of 10^{-12} cm.³. A very large bacillus, with cells 1.6 μ thick and 5 μ long, would be 10 times as large. An ellipsoidal yeast cell, 8×5 μ, is another 10 times as large; 1 gram would contain about 10^{10} such cells.

B. Amounts of Bacterial Cells

The amounts of bacterial cells available for analysis depend on the concentration of cells and on the volume of cultures. Bacterial growth ceases either when the substrate becomes exhausted or unavailable or when the accumulation of metabolic products reaches an inhibitory level. The two principles involved in obtaining maximal growth are the maximal exploitation of the nutritional and metabolic capacity of the organism, especially for utilization of energy sources, and the removal or neutralization of metabolic products. The application of these principles to any specific instance depends on the nature of the organism; two main sets of problems are encountered.

With organisms capable of complete oxidation of organic substrates, the main problems are the choice and amount of oxidizable substrate and the provision of adequate O_2 as hydrogen acceptor. Good aeration greatly increases the yield of aerobic bacteria; for example, in standard nutrient broth, which contains less than 8 g./liter of organic materials (5 g. of peptone and 3 g. of meat extract) *E. coli* and other aerobes of the same size reach titers of about 2×10^8 cells/ml. in stationary cultures and 5×10^9 cells/ml. with shaking or aeration, or about 1 g. dry weight per liter (a yield of at least 15%). With adequate aeration, much more substrate can be utilized and higher cell concentrations obtained. With *Brucella suis*, whose cells are one-half to one-quarter of the size of those of *E. coli*, concentrations of 8×10^{10} cells/ml., that is, 5 to 10 g./liter dry weight, are reached in a medium with 2% casein hydrolyzate +1% glucose.[21] At higher bacterial concentrations, the supply and circulation of O_2 by proper spargers and by agitation, and the prevention of foaming by nontoxic antifoam agents, become important.

Even higher yields can often be obtained in chemically defined media, where better buffering and more complete utilization of organic substrates can be achieved (up to 10^{11} cells/ml. for *B. suis*).[22] Of course, all nutritional requirements for nitrogen, sulfur, phosphorus, and essential growth factors must be satisfied in utilizable form and in amounts sufficient for the maximal crop.

In a synthetic medium it is often possible to utilize more effectively multiple energy-yielding pathways by providing a mixture of substrates. Thus, with *Serratia marcescens*, which can oxidize both glucose and citrate, yields of 29 g./liter of dry weight of cells (2×10^{11} cells/ml.) were obtained from 57 g./liter glucose plus 20 g./liter citric acid.[23] The w/w yield is 37.5%; the yield of assimilated carbon 40%, the rest being mainly oxidized to CO_2. Similar results can be obtained with organisms like *E. coli*, which cannot use citrate, by replacing it with malate or other utilizable Krebs cycle intermediates. With yeasts, amounts up to 44 g./liter dry weight of cells can be obtained, with practically complete utilization of 100 g./liter glucose and yields of 42–45%.[24]

The oxidizable substrates may have to be added gradually to prevent growth inhibition by high osmotic pressure and by accumulation of intermediates in the medium. Well-regulated continuous flow systems may give equally high yields of cells if the replacement of medium is not too rapid. Continuous flow culture can yield 13 g. dry weight/liter/hour of yeast cells, with complete sugar utilization. When large volumes are needed, the construction of huge fermenters presents complex engineering problems, especially with regard to supplying oxygen, mixing the culture, and collecting the cells.[24, 25]

Some of the yields reported above for aerobic bacteria and yeasts correspond to between 10 and 40% partial cell volume in the culture. At this point the mechanical problems of substrate feeding, aeration, and cell collection become overwhelming. Physical crowding, apart from substrate limitations, does not in itself limit growth. In fact, an agar-solidified medium often yields as much surface growth per gram of nutrient as the same medium used in liquid form.

With organisms that carry out incomplete oxidations or fermentations the problems are generally more complex. We must obtain maximal utilization of substrate and, in addition, prevent toxic effects of the products. A variety of examples may be given. With acid producers, pH control throughout growth is essential and the nature and capacity of the buffers are often critical. With lactobacilli, both homolactic and heterolactic types, buffering with citrate or acetate at the optimum pH gives faster growth and better yields than buffering with phosphate or amino acids (as peptone).[26] Here some advantage may derive from utilization of the acetate. The principle of double substrate is well exemplified with *Lactobacillus arabinosus*, which by producing large amounts of the adaptive malic enzyme can utilize malate, producing lactate and CO_2. Addition of 1% malate to a medium with 2% glucose raises the yield by over 50% to 0.5 g./liter cell nitrogen (about 5 g. dry weight).[27]

When all nutritional requirements are satisfied, the yield of cells in fermentative, anaerobic growth is strictly proportional to the amount of fermentable substrate and remarkably constant for different organisms. It is also quite constant for different substrates, provided the amounts of substrate are normalized to the relative amounts of ATP which the organism can generate from them.[27a] Thus, the bacterium *Streptococcus faecalis* and the yeast *Saccharomyces cerevisiae*, growing anaerobically on glucose, produce 22 and 21 μg. dry weight of cells per μmole of glucose; supplementation with L-arginine for *S. faecalis* gives an additional 10 μg. dry weight per μmole of arginine. These "molar growth yields"[27a] correspond to yields of about 12% g. dry weight per gram of glucose, and 6% for arginine. Since both *S. faecalis* and *S. cerevisiae* obtain 2 moles of ATP for each mole of glucose fermented, and *S. faecalis* obtains 1 mole of ATP per mole of arginine fermented to ornithine, the respective yields, in μmoles dry weight per μmoles of ATP produced, are 11, 10.5, and 10, respectively. *Propionibacterium pentosaceum*, which produces acetate and propionate from glucose, yields 37 μg. dry weight of cells per μmole of glucose, suggesting that it can obtain about 3.4 μmoles of ATP per μmole of glucose fermented, instead of 2 μmoles as in the lactic or alcoholic fermentation.

With fermentative, but not strictly anaerobic organisms, the yield often can actually be increased by aeration. With *Streptococcus faecalis* and other streptococci, which do not possess cytochromes but can use O_2 as hydrogen acceptor for a pair of hydrogens from pyruvate without accumulating peroxide, aeration during growth on glucose increases the amount of growth by 40%.[28] Even with organisms that are generally difficult to grow, the cell yield can be improved by proper choice of conditions. For example, yields of 150 mg./liter dry weight can be obtained with *Clostridium kluyveri*, fermenting ethanol and acetate to butyrate, by using bicarbonate buffer at low pH.

It seems reasonable to assume that there is no limit to the amount of cells of any one organism that we could obtain if we knew how to provide optimal growth conditions. The most difficult cases may be those of organisms caught in peculiar metabolic impasses, for example, an iron bacterium

TABLE II
The Water Content of Bacteria[a]

Organism	Water loss upon drying (% weight)	Intercellular volume (%)
Escherichia coli	73; 78	10
Serratia marcescens	78	
Klebsiella pneumoniae (capsulated)	85	
Pseudomonas aeruginosa	75	
Vibrio comma	73	
Malleomyces mallei	76	
Bacillus anthracis	80	
Corynebacterium diphtheriae	84	
Mycobacterium tuberculosis	86	

[a] Data from J. R. Porter, "Bacterial Chemistry and Physiology." Wiley, New York, 1946, and from Roberts et al.[29]

requiring both oxygen and reduced iron for growth. Such organisms have become adapted by natural selection to an almost marginal life in complex biotic environments.

C. Water Content

The values in Section III, A for the weight of bacteria are wet weights. Drying (at 105–110°C. for several hours in air or in a vacuum oven) removes between 70 and 90 % of the mass of washed, packed wet cells. The water content may vary, depending on the organism, the growth conditions, and the accumulation of certain hydrophilic (polysaccharides) or hydrophobic (lipidic) by-products; a reasonable estimate for the average dry weight is about 20 %. Representative data are given in Table II.

Part of the water in packed wet cells is in intercellular spaces. These can be measured by means of isotopically labeled proteins, which do not penetrate the cells. Packed *E. coli* cells include an intercellular space about 10–20 % of the total volume.[29]

IV. Chemical Analysis of Bacteria

A. Cellular and Extracellular Materials

Chemical analyses can be made on entire cultures, on cells collected from cultures and washed, or on the culture medium and the wash fluids. The washed cells contain the living protoplasm. The external media often contain important products of cellular function. These include unutilized metabolites, such as the end-products of fermentation, incomplete oxidation, or incomplete biosynthesis (some of which escape as gases); metabolic by-products, including polypeptides and other substances whose role in the metabolism of the cells that produce them is still unknown but some of which have important biological properties, like the antibiotics chloromycetin, bacitracin, subtilin, streptomycin, and the tetracyclines; polysaccharides, including dextrans, levans, galactans, cellulose, and more complex ones, which may form gums, slime layers, or true pericellular capsules; extracellular enzymes and other proteins, such as the exotoxins. It is not always easy to decide whether an extracellular product is excreted by living cells, or synthesized by secreted enzymes, or released by disintegrated cells.

Once we have secured a batch of clean washed cells, we can analyze their protoplasm in three general ways. We may proceed to a variety of chemical analyses on the whole cells; or we may attempt to isolate cellular constituents such as enzymes in their native, unaltered state, test them for functional integrity, and analyze their chemical and biological functions; or we may isolate fractions corresponding to specific cellular structures, analyze their composition and functional activities, and reconstruct their place, role, and mutual relations in the intact cell. These types of analysis will be outlined in the following sections.

B. Elementary Composition

The elementary composition of bacterial cells reflects both the basic composition of bacterial protoplasm and that of accessory materials, such as stored polysaccharides and lipids. Unfortunately, the available data, especially for the less abundant elements, are few and not very reliable. Representative values for *E. coli* are shown in Table III.

The *carbon* content, determined as CO_2 by combustion or isotopically, is about $50 \pm 5\%$ of the dry weight. Higher values are found in mycobacteria, presumably because of their high lipidic content.

The *nitrogen* content has been reported variously between 8 and 15%. The higher figures are closer to the value of 154 mg. N/g. dry weight found by isotopic analysis of *E. coli* cells grown in a synthetic medium.[29] Lower values are indicative of stored carbohydrates or fats. It is useful to remember that the widely employed micro-Kjeldhal method reveals only about 80–90% of the cellular nitrogen, since it measures completely only the nitrogen in amino, imino and amide groups, not in nitro or azo groups nor in the rings of purines and pyrimidines.

Hydrogen, measured as water, accounts for about 10% of the cell dry weight of bacterial protoplasm. *Oxygen*, generally measured by difference, accounts for another 20%. The remaining 5% of the dry weight consists of elements which after incineration are found in the *ash* as oxides. The ash content varies rather widely, reflecting more the nature of the medium and the growth conditions than actual variations in the composition of protoplasm. Typical figures are given in Table III.

TABLE III
Elementary Composition of *Escherichia coli*

Element	Per cent of dry weight
Carbon	50[a]
Nitrogen	15,[a] 10.3
Phosphorus	3.2[a]
Sulfur	1.1[a]
Ash (total)	12.75%
Fixed salts (nonextractable by water after heat killing)	7.25
Free salts (extractable)	5.5

Element	Per cent of fixed salt fraction	Per cent of free salt fraction
Sodium	2.6	19.8
Potassium	12.9	9.9
Calcium (as CaO)	9.1	13.8
Magnesium (as MgO)	5.9	2.0
Phosphorus (as P_2O_5)	45.8	41.3
Sulfur (as $SO_4^=$)	1.8	4.4
Chlorine	0.0	7.4
Iron (as F_2SO_3)	3.4	tr
Manganese		(20 p.p.m.)
Copper		(80 p.p.m.)
Aluminum		(100 p.p.m.)

[a] Values from isotope analysis of growing cells (Roberts et al.[29]). The values on ash composition are from M. Guillemin and W. P. Larson, *J. Infectious Diseases* **31**, 349 (1922). The values in parentheses are data for other aerobic bacteria (in parts of element per million parts of dry weight).

The *phosphorus* content is variously reported between 2 and 6% of the dry weight; about 50% of the ash consist of P_2O_5. In growing *E. coli* cells the total phosphorus, measured with P^{32} as tracer, was 3.2% of the dry weight; of this, 65% was in nucleic acids, 15% in phospholipids, and 20% in the acid-soluble fraction, mainly as phosphate esters.[29]

Sulfur, about 1% of the *E. coli* dry weight, is about 75% in proteins, as cysteine and methionine, the rest as acid-soluble compounds, mainly glutathione. The amount of sulfur in the many important cofactors (coenzyme A, thiamine, lipoic acid, biotin) is only a minute amount of the total cell sulfur. Sulfur can be present in much larger amounts in sulfur bacteria. Some of these accumulate sulfur while oxidizing H_2S; others take in elementary sulfur, like *Thiobacillus thiooxidans*, or thiosulfate, like *T. thioparus*, and use them as oxidizable substrates.[30]

Most metallic elements, including *sodium, potassium, calcium, magnesium*, and *manganese*, are present either as ions or in readily dissociated organic salts. The sodium content is the most variable, depending on the concentration in the medium. Magnesium and manganese are specific activators of various enzyme activities. Magnesium is also part of the chlorophyll of photosynthetic bacteria. Some *iron* is bound to macromolecular constituents, such as catalase and cytochromes, as part of the heme group; its amount is lower in cells without such enzymes, and also in cells grown in iron-deficient media.[31] Little is known of the cellular iron content in the iron bacteria, which deposit ferric hydroxide outside the cells.

Numerous elements are found in traces in bacterial cells, and the reasons for considering them as constituents of protoplasm rather than as contaminants is our knowledge of their physiological activities. Thus, we know that some *zinc* is bound in the cell wall, as revealed in the interaction between bacteriophage and *E. coli* cells.[32] *Cobalt* is part of the vitamin B_{12} molecule, which is synthesized by many bacteria and serves as activator of some enzymes. *Molybdenum*, replaceable by *vanadium*, stimulates nitrogen fixation by *Azotobacter* and other bacteria and is certainly present in these organisms.[33] In addition, both zinc and molybdenum are cofactors for some dehydrogenases.

C. Organic Constituents

1. Acid-Soluble Fraction

It is customary to start by separating two major fractions: an acid-soluble fraction, containing low-molecular weight constituents, and an acid-insoluble fraction. These are separated by extraction with 5% trichloroacetic acid or 10% perchloric acid in the cold to minimize hydrolysis.

The acid-soluble fraction contains, besides inorganic cations and anions, a pool of organic nutrilites and metabolic intermediates. The content of the pool differs in different nutritional and metabolic types of bacteria and is influenced by the composition of the external medium and by the washing procedures. Although no complete analyses are available, we know that the pool contains sugars, organic acids, amino acids, nucleotides and other nucleoside phosphates such as ATP, a variety of other phosphate esters, vitamins, coenzymes, polyamines such as spermidine and putrescine, and oligopeptides such as glutathione. The vitamin content of whole cells, as estimated by microbiological assay, is given in Table IV. Some of the minor organic constituents of cells are often revealed only by specific nutritional requirements and not by chemical detection in cellular extracts.

The total amount of the acid-soluble organic constituents is low, repre-

TABLE IV

THE AMOUNTS OF B VITAMINS IN BACTERIA[a]

(All figures are in parts per million of dry weight of cells)

Vitamin	Aerobacter aerogenes		Pseudomonas fluorescens		Clostridium butylicum	
	Cells	Medium	Cells	Medium	Cells	Medium
Thiamine	11	8.9	26	48	9.3	30
Riboflavin	44	110	67	310	55	180
Nicotinic acid	240	390	210	350	250	1680
Pantothenic acid	140	640	91	220	93	225
Pyridoxin	6.8	20	5.7	70	6.2	17
Biotin	3.9	44	7.1	61	—	—
Folic acid	14	91	8.8	66	2.8	16
Inositol	1400	—	1700	—	870	—

[a] The data (from R. C. Thompson, *Texas Univ. Publ.* **4237**, 87, 1942) are based on microbiological assays on cells and medium from cultures in vitamin-free media (except for *C. butylicum*, which requires biotin).

senting in growing *E. coli* cells 8% of the total carbon and 20% of the total phosphorus. This fraction contains less than 10^7 organic molecules per cell, 3% of which may be glutathione.[29] Yet the acid-soluble fraction constitutes a very characteristic internal milieu, in which, due to active transport systems or permeases located in the cytoplasmic membrane,[34] compounds such as amino acids may be hundreds of times more concentrated than in the external medium. The concentration of solutes in the internal milieu confers to it a high osmotic pressure, which has been estimated at 20–25 atmospheres for *Staphylococcus aureus*, 5–6 atmospheres for *E. coli*.[35]

Notable in the composition of the acid-soluble fraction is the absence of highly polymerized intermediates of protein and nucleic acid synthesis, in line with current ideas of the biosynthesis of these macromolecules directly from amino acids and nucleoside phosphate esters.[36] The polymerized intermediates are probably always associated with macromolecular constituents; they may be unstable if released in "soluble" form.

2. LIPIDS

The acid-insoluble or "macromolecular" fraction can be fractionated further to yield different classes of substances. Alcohol and alcohol-ether, at 40 to 50°C., extract some of the lipids; others, especially from the cell wall, are extracted by lipid solvents only after hydrolysis with concentrated HCl (6 N at 100°C. for several hours). For example, ether extraction, before or after methanol treatment, extracts only about one-third of the lipids of isolated *E. coli* cell walls and 10% of those of *Bacillus subtilis* cell walls.[37] Clearly, some of these lipids are constituent parts of stable macromolecular structures.

The total lipids in growing *E. coli* represent about 10–15% of the cell carbon and constitute about the same proportion of the dry weight.[29] Lipids are found not only in the cell wall, but also in the cytoplasmic membrane, which is a lipoprotein layer.[38] Lipids accumulate in cells grown with high concentrations of acetate. In *E. coli* the alcohol-ether soluble lipids may be shifted from below 10 to over 20% of the dry weight.[39] In bacilli and other organisms, lipids are stored as granules of poly-β-hydroxybutyrate, which may constitute 10 or even 25% of the dry weight.[40] This is an interesting storing device, which makes available a metabolite less rich in energy, but more readily utilizable than the fully reduced chains of saturated fatty acids.

Special lipid components are found in mycobacteria, in which waxes may constitute 10% or more of the dry weight. Reports of steroids in bacterial cells[41] have remained unconfirmed, but growth requirements for steroids have been encountered.[42]

3. Nucleic Acids

Extraction of the alcohol-ether insoluble residue with hot trichloroacetic acid (5%, 30 minutes at 90°C.) removes the *nucleic acids*. The two components, ribonucleic acid (RNA) and deoxyribonucleic acid (DNA) can be estimated colorimetrically after hydrolysis without further fractionation. If necessary, they can be separated by selective solubility in salt solutions, by selective alkaline hydrolysis of RNA, or by digestion with RNase or DNase followed by alcohol precipitation of the undigested fractions.[43]

The RNA of bacteria represents on the average about 10% of the dry weight. Values as high as 20% are found in growing *E. coli* cells.[29] In this organism, the amount may vary over a factor of 6 depending on the growth phase, from 5×10^{-15} g./cell in old cells to 3×10^{-14} in young growing cells.[44] Similar variations are found in *Salmonella*, where fast growing cells contain over 10^{-13} g. RNA per cell.[44a] The composition of bacterial RNA is the same as that of RNA from other sources, including the usual 5′-ribonucleotides—adenylic, guanylic, cytidylic, and uridylic acid.

The DNA of bacteria represents about 3–4% of the dry weight, around 10^{-14} g./cell in *E. coli*.[44, 45] This represents about 1000 molecules of DNA, if the molecular weight is taken as 8×10^6. In *Salmonella typhimurium* a value of 4×10^{-15} g. DNA per nuclear body has been reported.[44a] Values as high as 7×10^{-14} g. per nuclear body of *Bacillus megaterium* have been reported.[46] Bacterial DNA consists of the usual 5′-deoxyribonucleotides corresponding to adenine, guanine, cytosine, and thymine. 5-Methyl cytosine has not been found in bacterial DNA,[47] but 5-hydroxymethylcytosine is present in a group of bacteriophages.[48] 6-Methylaminopurine has been found in small amounts, which increase under abnormal conditions.[49] The usual 1:1 ratios between adenine and thymine and between guanine and cytosine are observed in bacterial DNA, and the molecular structure of (undegraded) bacterial DNA conforms to the general DNA structure.[50] In different bacteria, the molar ratio of adenine to guanine varies between 0.5 and 2.5;[51] in *E. coli* it is about 1.0.[51a]

4. PROTEINS AND POLYPEPTIDES

Proteins are obtained in the hot acid-insoluble fraction. They represent about 50% of the dry weight of the cells. In growing *E. coli* they contain 60% of the cellular carbon and 70% of the sulfur.[29] About 15–20% of the protein can be extracted directly from washed cells with 75% ethanol (30 minutes at 40–50°C.) along with lipids, from which it can then be separated by ether precipitation. This protein fraction is probably present as lipoprotein; in Gram-negative bacteria a lipoprotein layer is found in isolated cell walls[52] (see Section VI, B, 2).

As an order of magnitude, we may take a figure of 1 to 3×10^6 protein molecules per *E. coli* cell. The relative amounts of protein and RNA in *E. coli* gave estimates of 15 and 8 amino acid residues per nucleotide.[29,53] These ratios are of interest because of the known interrelations between protein synthesis and RNA synthesis.[36]

Upon hydrolysis, the bacterial proteins yield the usual amino acids. Few studies have been made of the composition of special protein fractions. There is evidence for the existence of proteins with low aromatic amino acid content, possibly protamine- or histone-like, in the nuclear bodies of *B. megaterium*[46] (see Section VI,B,5).

The total amino acid content of representative bacterial cells is given in Table V. Note, on the one hand, the over-all similarity to the amino acid composition of complete proteins from other sources; on the other hand, the presence of a unique constituent, α-ϵ-diaminopimelic acid (DAP), which as a protoplasmic constituent has been found only in bacteria and in blue-green algae.[54] DAP can be synthesized and utilized by some organisms, including *E. coli*, as an immediate precursor of lysine by decarboxylation.[55] The unexpected finding of DAP as a protoplasmic ingredient is clarified by the finding[56] that it occurs not in proteins, but in a group of peculiar polypeptides, containing D- and L-alanine, D-glutamic acid, and either DAP or L-lysine, depending on the bacterial species. These polypeptides are present in the cell wall of all bacteria in macromolecular association with the amino-sugar "muramic acid" (3-0-carboxyethyl-N-

TABLE V
The Amino Acid Content of Some Bacteria

Amino acid	*Escherichia coli,*[a] (moles per cent)	Mycobacteria[b] (range for 11 strains, moles per cent)
Alanine	12.7	8.6–11.7
Arginine	5.3	6.4–8.6
Aspartic acid	9.9	6.2–7.3
Cysteine	1.7	—
Diaminopimelic acid	0.44	—
Glutamic acid	10.5	9.6–16.2
Glycine	7.8	—
Histidine	0.97	1.6–2.1
Isoleucine	4.6	4.1–4.5
Leucine	7.9	8.0–9.5
Lysine	7.0	2.9–4.6
Methionine	3.4	1.4–1.7
Phenylalanine	3.3	2.7–3.6
Proline	4.6	—
Serine	6.1	2.9–3.7
Threonine	4.7	—
Tryptophan	1.04	—
Tyrosine	2.1	2.0–2.5
Valine	5.5	5.9–7.0

[a] From Roberts et al.,[29] p. 28. The data were obtained by chromatography and radioisotope measurements. Comparable values, in the same order of abundance, were found for 10 amino acids in *E. coli*, *B. subtilis*, and *S. aureus* by J. L. Stokes and M. Gunness (*J. Bacteriol.* **52,** 195, 1946) using microbiological assays.

[b] From B. Ginsburg, S. L. Lovett, and M. S. Dunn, *Arch. Biochem. Biophys.* **60,** 164 (1956). The data were obtained by microbiological and chromatographic measurements.

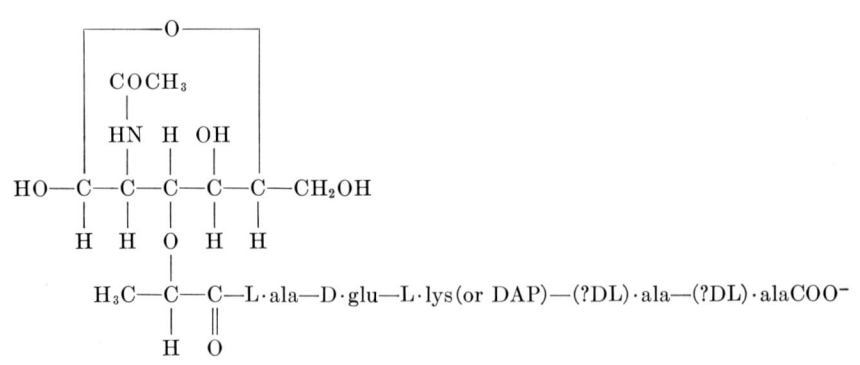

Fig. 1. The probable structure of the muramic acid-peptide component of the bacterial cell wall, according to Park and Strominger[56] and to Dr. J. L. Strominger (personal communication). In most bacteria, L-lysine is replaced by diaminopimelic acid (DAP).

acetyl-hexosamine). The macromolecular constituent (see Fig. 1) is essential for the rigid properties of the bacterial cell walls and, therefore, for the maintenance of the characteristic shapes of the bacterial cells.*

Polypeptides, especially poly-D-glutamic acid, may form true capsules around bacterial cells in the genus *Bacillus*.[57]

5. Carbohydrates

Carbohydrates are found in bacterial cells, not only in the nucleic acids and as oligosaccharides in the acid-soluble fraction, but also in a variety of polysaccharides. Some form capsules, which can easily be removed by washing before analyzing the cells proper. The great variety of capsular polysaccharides, especially well studied in pneumococci, is of particular interest in immunochemistry. True cellulose has been reported only in *Acetobacter xylinum*.[58]

Other polysaccharides are found in the bacterial cell wall, of which they may constitute up to 45% in Gram-positive and 15 to 20% in Gram-negative bacteria. To the cell wall they confer some of its characteristic antigenic properties, including the major specificity of the somatic antigens of the Enterobacteriaceae. These cell wall polysaccharides contain mainly glucose, galactose, arabinose, rhamnose, mannose, a heptose, and hexosamine.[59]

Polysaccharides are also present in bacterial cells as reserve materials, often as discrete granules, which with iodine give either the blue-black reaction of starch (or granulose) as in *Clostridium*, or the reddish color of glycogen as in Enterobacteriaceae. These polysaccharides are extracted in the acid-soluble fraction. The total amount of glycogen and other polysaccharides varies, for example, in *Aerobacter* and *Salmonella*, from less than 10% to over 30% of the dry weight, depending on the nature and concentration of sugars in the medium and on the growth phase.[60] The maximum is reached in the early part of the stationary phase.[61]

6. Pigments

An important group of organic constituents of bacterial protoplasm is that of pigments. Some of these, like prodigiosin, pyocyanin, and violacein, are metabolic by-products formed only under special conditions. Some are deposited within the cells, to which they give their characteristic colors; others are released into the medium. Some pigments—bacterial chlorophyll and some carotenoids—are functional constituents of the protoplasm of photosynthetic bacteria; chlorophyll may be absent when bacteria capable

---

* A new amino acid, probably ϵ-N-methyllysine (which has not been reported elsewhere in nature) has been found in the protein of the flagella of some strains of *Salmonella* (Ambler, R. P., and Rees, M. W., 1959, to be published).

of anaerobic photosynthesis are grown aerobically in the dark as chemo-organotrophs.[62]

7. Polyphosphates

Highly polymerized phosphoric acid as polyphosphate (or metaphosphate) accumulates in many bacteria. Polyphosphate is formed by enzymic catalysis from ATP.[63] The polyphosphates are deposited in granules known as *volutin* or *metachromatic granules*, because of the characteristic metachromasy (color change) they induce in certain dyes.[64]

8. Di-picolinic Acid

A unique constituent of bacteria is di-picolinic acid (pyridine-2-6-dicarboxylic acid), found in the spores of bacteria of the genera *Bacillus* and *Clostridium*. As the calcium salt, it may represent up to 10–15% of the dry weight of the spores.[65] Its function is unknown; upon spore germination, it is largely released into the medium.[66]

V. Isolation of Functional Constituents

A. Tests of Functional Activity

The procedures used for chemical analysis are too brutal to preserve the functional activities of cellular constituents. This can be accomplished by using milder fractionation methods, which isolate materials in their "native" state. At best, the preservation of native cellular elements is a compromise goal since in the cell a functional component, such as an enzyme, may be associated, sterically and functionally, with other cellular constituents. The art of the biochemist[67] in taking apart undamaged the individual cogs of cell machinery aims both at the preservation of the active structures and at the reconstruction of the chemical environment that their function requires. When function is predicated on complex relations among macromolecular constituents, the isolated functional constituents are often complex, "particulate" elements, which can carry out sets of biochemical functions but from which it may prove difficult to separate the individual components that carry out single-step processes. It is customary to characterize as *soluble enzymes* those found in that fraction of broken bacterial cells that is not sedimented in a few hours by ultracentrifugal fields around 10^5 g, and as *particulate enzymes* those found in the sedimentable fraction.

The functional integrity of isolated macromolecular or particulate constituents can be subjected to tests of two kinds: tests for activity *in vitro*, and tests for ability to resume function if reintroduced into a living cell.

1. FUNCTIONAL ACTIVITY *in Vitro*

The first type of test is a less demanding one. As applied to enzymes, it provides an indication that both the chemical structure and the environmental conditions required for the specific catalytic function have been preserved or reconstructed. A large number of bacterial enzymes have been extracted in active forms, some in crude extracts, some highly purified, and a few in crystalline form, like the catalase and the oxaloacetic decarboxylase of *Micrococcus lysodeikticus*[68, 69] and the β-D-galactosidase of *E. coli*.[70] The amount of individual enzymes per cell varies greatly; the values range from 6.6% of the total protein for β-D-galactosidase in fully adapted cells of *E. coli*,[14] to about 1–2% of the dry weight (10^4 molecules per cell) for catalase in *M. lysodeikticus*,[68] to less than 0.01% of the dry weight for the oxaloacetic decarboxylase of the same organism, to only a few molecules per cell for some other enzymes.[71] The relative amounts of various enzymes in an organism can vary, not only as a result of induced enzyme biosyntheses, but also depending on the available nitrogen.[72] The number of different enzymes in an organism like *E. coli*, capable of synthesizing its whole protoplasm from glucose and ammonia, may be of the order of 1000, probably not higher than 2000.

The second test for functional activity, by reintroduction of an extract into a living cell, has not yet been applied successfully to enzymes. In general, there is no evidence that any protein, once extracted from a cell, can ever be reintegrated in functional form into the machinery of another living cell. The reverse situation holds for nucleic acids. Here there are as yet few tests available for *in vitro* functions. If the function of nucleic acids is primarily genetic, that is, if they act as templates embodying the detailed information for the patternization of specific macromolecules, then *in vitro* tests of function require systems capable of performing specific macromolecular biosyntheses in the presence of a specific template. One such system has been reported for DNA.[73] An extract of *E. coli* can produce, apparently by enzyme action, a DNA-like polydeoxyribonucleotide from a mixture of deoxyriboside triphosphates in the presence of a DNA primer. The primer must be a highly polymerized, presumably undegraded DNA; the specificity relations between primer and newly formed polynucleotide are under investigation. Whether the new polymer is truly functional DNA can be decided by means of transformation analysis, which is the prototype of the functional tests by reintroduction of isolated cellular constituents into living cells.

2. Reintroduction of Cellular Constituents into Living Cells.

Transformation[74, 75] is the phenomenon in which highly polymerized DNA extracted from "donor" cells and purified by mild methods, including deproteinization by chloroform or detergents and repeated alcohol precipitations (but not sonic disintegration), can be taken up by intact "recipient" cells and confer upon them, singly or in specifically linked groups, hereditary traits that characterized the donor cells. Transformation phenomena (discussed in Vol. V, Chapter 7) have been analyzed in only few bacterial groups—*Pneumococcus, Streptococcus, Haemophilus*. There is no reason to believe that these groups are in any way unique, except possibly in a more ready permeability of their cells at some stage of their growth cycle to intact DNA molecules. Both transformation phenomena, and the allied "transduction" phenomena, in which genetic traits are transferred from a donor to a receptor cell by the vehicle of bacteriophage particles,[76, 77] indicate that all elements of the bacterial genome are potentially transferable from cell to cell. Transformation proves that DNA is the material substrate of the genetic traits, a role for which its chemical structure[50] appears eminently suitable.

Transformation tests permit definition of functional DNA, but have not yet led to effective fractionation of specific functional elements controlling different genetic traits of a cell.

No satisfactory system is available yet for testing the functional integrity of extracted bacterial RNA. Work on viruses proves that RNA can penetrate cells and act as carrier of genetic specificity. The isolated RNA moiety of the particles of plant viruses of the tobacco mosaic group can penetrate virus-sensitive cells, multiply within them, and produce virus.[78] Hence, we may surmise that bacterial RNA could be tested similarly if we knew what specific activities to expect from the transferred RNA. The known interrelations between RNA and protein synthesis suggest a test for synthesis of specific proteins, especially adaptive enzymes, initiated by externally introduced RNA. Despite some positive reports, such tests with whole cells have generally been unsuccessful. The use as test systems of subcellular fractions capable of synthesizing proteins and nucleic acid is under active investigation.[79, 36]

RNA from tobacco mosaic virus can also be reconstituted with undenatured viral protein to form viruslike particles.[80] Since, however, the reconstitution can also occur when RNA is replaced with enzymically synthesized polyribonucleotides[81] the test for reconstitution does not provide information on the biological integrity of RNA. As for the polynucleotide-synthesizing enzyme system itself,[81] if a template were found to affect the specificity of the products, the system could then be used to test RNA for functional activity.

VI. Isolation of Organized Bacterial Constituents

A. An Outline of Bacterial Cytology

A major contribution to the understanding of the organization of bacterial cells comes from the mechanical and chemical dissection of the cells, the analysis of the isolated fractions, and the recognition of their structural and functional role in the cell. A discussion of this cytochemical work presumes a knowledge of the morphological cytology of bacterial cells, which will be described in later chapters. The detailed study of subcellular fractions will also be the subject of specific chapters. Here we shall simply outline the main findings derived from the cytochemical approach to bacterial organization.

The morphologist recognizes in the typical cells of Eubacteriales a number of essential structures, revealed by appropriate microscopic and cytological methods: a rigid cell wall, a cytoplasmic membrane, and one or more nuclear bodies, in which the cellular DNA is concentrated. The cell wall itself may be absent, as in organisms of the pleuropneumonia group. All motile cells, except the gliding myxobacteria, have flagella, whose distribution on the cell surface is characteristically different in major bacterial groups. Electron microscopy reveals that flagella are attached to basal granules or *blepharoplasts*, located within the cytoplasmic membrane. Many bacteria form extracellular capsules, which differ from simple slime layers by a certain degree of organization related to the cell wall.[82] The cytoplasm may contain, besides the nuclear bodies, a variety of microscopic granules, none of which, except the chromatophores of photosynthetic bacteria, are permanent constituents of the bacterial protoplasm.

The electron microscope adds some essential features to the above cytological picture.[83] The cell wall and the cytoplasmic membrane appear to be the only true membranes recognizable as continuous layers in sections. In the cytoplasm proper no inner membrane is recognizable comparable to the endoplasmic reticulum or the membrane and inner cristae of mammalian mitochondria, nor even a nuclear membrane. The cytoplasmic material is granular, with electron-absorbent granules, mostly around 100–300 A., in a less electron-dense background. The nuclear bodies appear in electron micrographs as convoluted areas with zones of low electron density and areas of greater electron scattering power.

B. Cytologically Identifiable Fractions

After separating the bacterial cells from the medium and removing easily extractable capsular materials, it is possible to isolate fractions, which correspond to one or more of the microscopically recognizable structures.

1. FLAGELLA

Flagella are easily removed by vigorous shaking, without destroying the viability of the cells nor their ability to produce new flagella.[84] The removed material, which does not include the basal granules, consists of single fibers of a protein (flagellin) resembling the contractile proteins. No report of ATPase activity in flagellar protein has been made. The bacterial flagella do not show the complex structure, with two central and nine periferal elements, typical of the cilia and flagella of protozoa.

2. CELL WALL

The cell wall can be removed and isolated by several methods, variously successful with different bacteria, including cell breakage, autolysis, enzyme digestion, or sudden immersion in hot water. The inner protoplasm disintegrates readily and the more resistant cell wall may be washed and purified, more or less intact, and identified morphologically by its shape and rigidity. Its thickness is about 10–15 mμ in Gram-negative bacteria, 15–20 mμ in Gram-positive ones.[59] The chemical composition is characteristic. In Gram-positive bacteria the isolated cell walls consist of a main lipopolysaccharide layer containing not more than 2% lipids, and also the characteristic muramic acid-polypeptide complex shown in Fig. 1. The exact composition of the lipopolysaccharide layer probably varies among different species. To this layer there may be superimposed some characteristic but structurally dispensable materials. For example, the cell walls of group A streptococci have an external protein layer, the M antigen, which can be digested off the living cells by trypsin without impairing viability.[85]

The so-called cell walls of Gram-negative bacteria as isolated by the available procedures are more complex, containing complete proteins and about 20% lipids. The difference between Gram-positive and Gram-negative bacteria is explained as follows:[52] The isolated cell wall of Gram-negative bacteria contains, in addition to the rigid lipopolysaccharide layer similar in composition to the cell wall of Gram-positive bacteria, a lipoprotein layer, removable from isolated cell walls by 90% phenol. A phage enzyme can destroy part of the lipopolysaccharide layer, making it collapse. The presence of the lipoprotein layer in Gram-negative bacteria may explain the greater resistance of their cells to treatments that remove (or prevent synthesis of) the rigid layer, and the ability of some to give relatively stable nonrigid L-forms.[86] In the absence of any definite reports of isolation of the cytoplasmic membrane from Gram-negative bacteria, we may ask whether the lipoprotein layer isolated with the cell wall may be part of a semipermeable cytoplasmic membrane. By analogy with Gram-positive bacteria (see below) this membrane would be expected to possess certain specific enzyme activities.

The somatic surface antigens of Enterobacteriaceae, which can be extracted as macromolecular lipoprotein-polysaccharide complexes, are probably fragments of the bacterial cell walls.

In actinomycetes, the cell walls have the same composition as in Gram-positive Eubacteriales. The cell walls of fungi consist of polysaccharide only, without muramic acid.[86a]

3. Protoplast

The portion of the cell within the cell wall is the *protoplast*. This can be isolated more or less intact by enzymic removal of the cell wall, for example, by lysozyme;[38] or by treatment with antibiotics of the penicillin-bacitracin group, which probably prevent the synthesis of some essential cell wall constituents[56, 87]; or, finally, by growth of bacterial strains requiring external diaminopimelic acid in DAP-deficient media.[88] Protoplasts are only stable in media of high osmotic pressure controlled by empirically selected solutes; otherwise, they undergo lysis. In a stabilizing medium, protoplasts of *B. megaterium* have a rate of oxidative metabolism comparable to that of intact cells.[89]

In Gram-negative bacteria, what is obtained by various treatments that eliminate the rigid cell wall elements may be protoplasts still surrounded by the less rigid lipoprotein layer and also by portions of the lipopolysaccharide layer.*

4. Cytoplasmic Membrane

Gentle disruption of the protoplasts by moderate osmotic shock permits, at least in *B. megaterium*, the isolation of a "ghost" fraction, which is still sedimentable in low centrifugal fields and which can be identified with the cytoplasmic membrane.[38] Its composition indicates that it is the seat of the semipermeable, lipoprotein layer of the cell. The isolated fraction represents about 20% of the bacterial cell weight and contains 15% of the cellular RNA. Most of the cytochromes of the cell remain in this ghost fraction, together with a number of enzyme activities, especially dehydrogenases, but not the full set of Krebs cycle enzymes. When granules of poly-β-hydroxybutyrate are present, some may remain within the ghosts, but most are removed by the osmotic shock and can be separated readily by low speed centrifugation as a chloroform-soluble pellet.[90]

* The osmotically labile forms obtained by various treatments that remove cell wall components from bacterial cells generally contain some residual cell wall materials. It has been proposed (C. Weibull, *Ann. Rev. Microbiol.* **12**, 1, 1958) that these forms be referred to as "protoplasts" in quotes. The word *spheroplasts* has also been suggested (C. Hurwitz, J. M. Reiner, and J. V. Landau, *J. Bacteriol.* **76**, 612, 1958).

5. Nuclear Bodies

Lipase treatment of *B. megaterium* protoplasts in hypertonic medium releases round bodies, which contain, besides protein and RNA, all the cellular DNA (about 7×10^{-14} g. per body).[46] The protein seems to contain some characteristic antigen and includes a protamine or histone-like fraction. These presumed nuclear bodies lack a recognizable nuclear membrane; the DNA is released readily and may be located at the surface of the isolated bodies.

C. Submicroscopic Structure and Macromolecular Organization

1. Macromolecular and Particulate Fractions.

Another aspect of cellular organization emerges if macromolecular components are isolated by physical means and tested for composition and functional activity after disintegration of bacterial cells or of protoplasts. Disintegration of bacteria requires their mechanical breakage,[91] while protoplasts can be disrupted more gently by osmotic shock. A major difference, which warns us of possible pitfalls, appears to be the fate of DNA, which after osmotic disruption remains highly polymerized and viscous, whereas it is more or less completely disintegrated by prolonged grinding or even brief sonication.

The disintegrated bacterial protoplasm, when fractionated by ultracentrifugation, yields a series of fractions of fairly characteristic sedimentation constant and composition.[92, 93] Typical results are shown in Table VI.

The most abundant fractions consist of particles, which were originally reported as the "50 S" and "30 S" components (see Table VI). The 50 S component corresponds to spheres about 150 A. in diameter, with a molecular weight about 2×10^6. The 30 S particles weigh about one-half as much. These fractions constitute 30–40 % of the cell mass and contain protein and RNA in about equal amounts, comprising 90 % of the cellular RNA (at least in Gram-negative bacteria) and 25 % of the cellular proteins, with little or no DNA or lipids. If extraction is carried out in the presence of magnesium, at concentrations around 5×10^{-3} M, most bacteria yield, instead of the 50 S and 30 S fractions, a homogeneous 80 S fraction consisting of nearly spherical particles. Upon lowering the magnesium concentration, the 80 S particles split (reversibly) to give 50 S and 30 S particles.[93a]

The 80 S particles, about 3×10^6 in molecular weight, appear to be the bacterial equivalent of the microsomal particles of animals and plant cells. The collective name "ribosomes" has been proposed for all these RNA-containing particles.[94]

The ribosomal fraction of bacteria, in more or less crude state, contains

TABLE VI
The Sedimentation Constants of Constituents of Broken Bacterial Cells[a]

Organism	Method of rupture	Large particles	About 40 S[c]	About 20–29 S[c]	About 8 S	About 5 S
Aerobacter aerogenes	Glass	—	40	22	9	—
Aerobacter aerogenes	Press	—	40	29, 20	12, 8	5
Azotobacter vinelandii	Alumina	—	39	25	9.0[e]	3.8
Bacillus megaterium	Alumina	—	42	31	—	3.9
Bacillus megaterium	Lysozyme	—	41	26	15–16[d, e]	2.8–3.9
Bacillus megaterium	Sonic	—	43	27	—	4.4
Blue-green alga	Alumina	300	41	30	—	4.7
Clostridium kluyverii	Drying	—	40	26	9.7[d]	6.0
Escherichia coli	Alumina	—	40	29, 20	8[d]	5
Escherichia coli	Alumina	—	32–38	20–25	8.5[d, e]	5
Escherichia coli	Mill	—	39	25, 21	8.7[d, e]	4.9
Escherichia coli	Phage	—	45	29	—	4.4
Escherichia coli	Sonic	—	36–45	23–29	—	4.6–5.8
Micrococcus pyogenes	Alumina	—	39	27	10.0[d]	3.2
Proteus vulgaris	Alumina	—	35	23	—	4.1
Pseudomonas fluorescens	Alumina	—	34–38	22–25	7.9[d]	3.5–5.3
Pseudomonas fluorescens	Decompression	58	44	29	—	4.0
Pseudomonas fluorescens	Sonic	58	44	29	—	6.5
Rhodospirillum rubrum						
Dark-grown	Alumina	55	42	29	—	4.6
Light-grown	Alumina	190	38	29	9.4[d]	6.4
Saccharomyces cerevisiae	Alumina	70	42	—	—	6.3

[a] Data from M. Alexander,[93] p. 74.
[b] 1 S = 1 Svedberg unit = 10^{-13} cm./sec./unit field.
[c] These fractions derive from the breakdown of the 80 S fraction (see text).
[d] Reported as a spiked peak.
[e] Contains deoxypentose nucleic acid.

a variety of oxidative enzymes and, in aerobic organisms, a complete hydrogen transport system; but the full operation of the tricarboxylic acid cycle and the performance of oxidative phosphorylation have been observed only with combinations of the particulate and the soluble, low molecular weight fractions.[93, 95, 96] The size of the particles permits a maximum estimate of the number of enzymes that each particles can contain. An 80 S particle, is 10^6 molecular weight units of protein, could have 20 protein molecules with average weight 50.000. Thus, if the particles were the site of enzymes, each particle could carry a relatively small group of enzymes. There is evidence[96a] that the essential proteins of the bacterial ribosomes have little or no enzyme activity and may instead, as nucleoproteins, be operative in the synthesis of enzymes and other proteins.

The soluble proteins are in the 4–6 S fraction. The DNA, when undegraded, sediments in the ultracentrifuge as a "spike" with a sedimentation constant of about 8 S.

In disrupted photosynthetic bacteria, a characteristic fraction (190 S) contains the *chromatophores*, 40–60 mμ in diameter, with a molecular weight of 3×10^7 in the wet state.[92] In a cell of *Rhodospirillum rubrum* grown in light there may be 5000 such particles, which disappear after aerobic growth in darkness (see Table V). In cell extracts of photosynthetic bacteria, chromatophores and even chromatophore fragments can perform photophosphorylation, synthesizing adenosine triphosphate in the presence of light.[97, 98]

2. Functional and Intramolecular Organization.

This brief description of the main fractions indicates both the potentialities and some of the operational pitfalls of fractionation methods in the study of cellular organization. The manipulations must be evaluated carefully to prevent artifacts. Some components of the particulate and soluble fractions of bacterial protoplasm may in the cell possess a higher degree of functional organization than we can recognize after separation. A full understanding of all the regulatory and inhibitory processes that give the cell factory its exquisite functional coordination and efficiency may require an elucidation of the spatial relations of the bits of machinery to one another within the cell. In addition to the inner structure of the coded macromolecules that catalyze chemical reactions and direct the patternization of other macromolecules, the regulation of cellular functions may require other elements of structural organization: spatial coordination between synthetic and energy-yielding reactions; devices for maintaining local concentrations of substrates and inhibitors; and suppression of the synthesis of unrequired enzymes. Some of the regulatory processes are probably "molar," such as feedback mechanisms, in which a metabolic product (or an external nutrilite) inhibits the function or synthesis of an enzyme in a chain of metabolic reactions.[99] Other coordinating processes may involve arrangements of functional elements into specific patterns, for example, the coupling of two enzymes with a common coenzyme inbetween them. An example of a different sort has been provided by genetic studies. Successive steps of some sequences of biosynthetic reactions are controlled by groups of genetic factors or "gene loci" closely linked to one another in the genetic map, in the same order as the chemical reaction steps they control.[100] This situation, without being general, has been observed in enough instances to suggest some subtle complementariness between the chemical specificities of the genetic material and of the enzymes which it controls.

At the intramolecular level, the coupling of studies of DNA structure, of amino acid sequence in proteins, and of genetic fine structure as revealed by recombination within gene loci promises a rapidly growing insight into the chemical basis of life processes. [101]

VII. Specialized Differentiations of Bacterial Cells

A. Endospores; Bacteriophage Particles

The rapid description of the generalized bacterial cell given in this chapter does not include certain differentiated structures that appear only in some bacteria and under special circumstances. The structures in question include endospores, mating apparatus and bacteriophage particles. These, as well as other specialized forms such as the gonidial elements of *Rhizobium* and other bacteria, will be discussed in appropriate chapters. Here we will briefly consider only the relation of these structures to bacterial protoplasm.

Too little is known yet about the mating apparatus or the gonidial forms of bacteria to justify any description in terms of chemical and functional organization. Endospores and bacteriophage, better known in their chemistry and biosynthesis, have some important features in common.

Both endospore formation and bacteriophage production involve a sudden onset of biosynthesis of specific devices for protecting portions of bacterial protoplasm. The endospore contains, besides the full genetic complement of a bacterium, certain unique chemical constituents, especially 2-6-di-picolinic acid, within a coat made up of novel proteins unrelated to those of the vegetative cell.[102, 103] Other proteins may be formed during sporulation and, in some bacilli, these may crystallize to form *parasporal bodies*.[104] Likewise, the bacteriophage particles contain, within a coat of specific proteins, fragments of genetic material, almost pure DNA, whose production is initiated either by a recently introduced phage DNA element, or by the "induction" of a prophage, that is, a long persistent element of the bacterial genome.[105]

Formation of the endospore coat preserves the bacterial genome and allows it to weather conditions destructive for the vegetative cells. Formation of the bacteriophage coat arrests the replication of phage genomes and provides a vehicle for the transfer of these bits of genetic material from one bacterial cell to another. Not only can the genetic material proper to the phage itself be transferred in this way, but, in the transduction phenomenon,[76] the particles become vehicles for transfer of random bits of bacterial genome that are not "viruses," that is, are themselves incapable of controlling the biosynthesis of a specific transmission apparatus.[106] Both in endospore formation and in bacteriophage formation we observe true differentiations of the bacterial cell; but both spore and phage are incapable of reproduction in the differentiated form. To reproduce, the spore must germinate into a vegetative cell; the phage particle must attack another cell and inject into it its content of genetic material. Thus, the cell remains the sole protagonist of the life process.

REFERENCES

[1] E. B. Wilson, "The Cell in Development and Heredity." Macmillan, New York, 1925.

[2] A. J. Kluyver, "The Chemical Activities of Microorganisms." Univ. London Press, London, 1931.

[3] K. R. Porter, *J. Histochem. and Cytochem.* **2**, 346 (1954).

[4] A. Lwoff, "Problems of Morphogenesis in Ciliates." Wiley, New York, 1950.

[4a] C. B. Anfinsen and R. R. Redfield, *Advances in Protein Chem.* **11**, 1 (1956).

[5] H. A. Krebs and H. L. Kornberg, *Ergeb. Physiol., biol. Chem. u. exptl. Pharmakol.* **49**, 212 (1957).

[5a] H. L. Kornberg and H. A. Krebs, *Nature* **179**, 988 (1957).

[6] W. A. Engelhardt, in "International Symposium on Enzyme Chemistry, Tokyo and Kyoto, 1957," p. 34. Maruzen, Tokyo and Academic Press, New York, 1958.

[7] J. Monod and M. Cohn, *Advances in Enzymol.* **13**, 67 (1952).

[7a] H. J. Vogel, in "The Chemical Basis of Heredity" (W. D. McElroy and B. Glass, eds.), p. 276. Johns Hopkins Press, Baltimore, Maryland, 1957.

[8] J. Brachet, "Biochemical Cytology." Academic Press, New York, 1957.

[9] J. H. Taylor, in "Physical Techniques in Biological Research" (G. Oster and A. W. Pollister, eds.), Vol. 3, p. 546. Academic Press, New York, 1956.

[10] S. E. Luria, *Bacteriol. Revs.* **11**, 1 (1947).

[11] W. Braun, *Bacteriol. Revs.* **11**, 75 (1947).

[12] C. Quadling and B. A. D. Stocker, *J. Gen. Microbiol.* **17**, 424 (1957).

[13] C. Quadling, *J. Gen. Microbiol.* **18**, 227 (1958).

[14] M. Cohn, *Bacteriol. Revs.* **21**, 140 (1957).

[15] A. Novick and M. Weiner, *Proc. Natl. Acad. Sci. U. S.* **43**, 553 (1957).

[16] J. Lederberg and E. L. Tatum, *Science* **118**, 169 (1953).

[17] E. L. Wollman, F. Jacob, and W. Hayes, *Cold Spring Harbor Symposia Quant. Biol.* **21**, 141 (1956).

[18] J. Lederberg, *Proc. Natl. Acad. Sci. U. S.* **35**, 178 (1949).

[19] G. Sermonti and I. Spada-Sermonti, *J. Gen. Microbiol.* **15**, 609 (1956).

[20] D. Ruffilli, *Biochem. Z.* **263**, 63 (1933).

[21] P. Gerhardt, *J. Bacteriol.* **52**, 283 (1946).

[22] W. G. McCullogh, R. C. Mills, E. J. Herbst, W. G. Roessler, and C. R. Brewer, *J. Bacteriol.* **53**, 5 (1947).

[23] C. G. Smith and M. J. Johnson, *J. Bacteriol.* **68**, 346 (1954).

[24] W. D. Maxon and M. J. Johnson, *Ind. Eng. Chem.* **45**, 2554 (1953).

[25] "Bactogen," American Sterilizer Co., Erie, Pennsylvania.

[26] E. E. Snell, *Wallerstein Labs. Communs.* **11**, 81 (1948).

[27] M. L. Blanchard, S. Korkes, A. DelCampillo, and S. Ochoa, *J. Biol. Chem.* **187**, 871 (1950).

[27a] T. Bauchop, *J. Gen. Microbiol.* **18**, VII P (1958).

[28] M. Dolin, *Arch. Biochem. Biophys.* **55**, 415 (1955).

[29] R. B. Roberts, P. H. Abelson, D. B. Cowie, E. T. Bolton and R. J. Britten, *Carnegie Inst. Wash. Publ.* **607** (1955).

[30] W. Vishniac and M. Santer, *Bacteriol. Revs.* **21,** 195 (1957).
[31] A. M. Pappenheimer, Jr. and E. Shaskan, *J. Biol. Chem.* **155,** 265 (1944).
[32] L. M. Kozloff and M. Lute, *J. Biol. Chem.* **228,** 529 (1957).
[33] E. G. Mulder, *Plant and Soil* **1,** 94 (1948).
[34] G. N. Cohen and J. Monod, *Bacteriol. Revs.* **21,** 169 (1957).
[35] P. Mitchell and J. Moyle, *Symposium Soc. Gen. Microbiol.* **6,** 150 (1956).
[36] E. F. Gale and K. McQuillen, *Ann. Rev. Microbiol.* **11,** 283 (1957).
[37] M. R. J. Salton, *Biochim. et Biophys. Acta* **10,** 512 (1953).
[38] C. Weibull, *Symposium Soc. Gen. Microbiol.* **6,** 111 (1956).
[39] S. Dagley and A. R. Johnson, *Biochim. et Biophys. Acta* **11,** 158 (1953).
[40] M. Lemoigne, B. Delaporte, and M. Croson, *Ann. inst. Pasteur* **70,** 224 (1944).
[41] E. Hecht, *Z. physiol. Chem.* **231,** 29, 279 (1935).
[42] H. S. Vishniac, *J. Gen. Microbiol.* **12,** 464 (1955).
[43] J. N. Davidson, "The Biochemistry of the Nucleic Acids," 2nd ed., Chapter 8. Wiley, New York, 1953.
[44] M. L. Morse and C. E. Carter, *J. Bacteriol.* **58,** 317 (1949).
[44a] M. Schaechter, O. Maaløe, and N. O. Kjeldgaard, *J. Gen. Microbiol.* **19,** 592 (1958).
[45] H. Barner and S. S. Cohen, *J. Bacteriol.* **72,** 115 (1956).
[46] S. Spiegelman, A. I. Aronson, and P. C. Fitz-James, *J. Bacteriol.* **75,** 102 (1958).
[47] S. S. Cohen, in "Cellular Metabolism and Infections" (E. Racker, ed.), p. 84. Academic Press, New York, 1954.
[48] G. R. Wyatt and S. S. Cohen, *Nature* **170,** 1072 (1952).
[49] D. B. Dunn and J. D. Smith, *Biochem. J.* **68,** 627 (1958).
[50] J. D. Watson and F. H. C. Crick, *Cold Spring Harbor Symposia Quant. Biol.* **18,** 123 (1953).
[51] K. Y. Lee, R. Wahl, and E. Barbu, *Ann. inst. Pasteur* **91,** 212 (1956).
[51a] A. D. Hershey, J. Dixon, and M. Chase, *J. Gen. Physiol.* **36,** 777 (1953).
[52] W. Weidel and J. Primosigh, *Z. Naturforsch.* **12b,** 421 (1957).
[53] A. B. Pardee and L. S. Prestidge, *J. Bacteriol.* **71,** 677 (1956).
[54] E. Work, *Biochem. J.* **49,** 17 (1951).
[55] B. D. Davis, *Nature* **169,** 534 (1952).
[56] J. T. Park and J. L. Strominger, *Science* **125,** 99 (1957).
[57] C. B. Thorne, *Symposium Soc. Gen. Microbiol.* **6,** 68 (1956).
[58] K. Mühlethaler, *Biochim. et Biophys. Acta* **3,** 15 (1949).
[59] M. R. J. Salton, *Symposium Soc. Gen. Microbiol.* **6,** 81 (1956).
[60] S. Levine, H. J. R. Stevenson, E. C. Tabor, R. H. Bordner, and L. A. Chambers, *J. Bacteriol.* **66,** 664 (1953).
[61] S. Dagley and E. A. Dawes, *Biochem. J.* **45,** 331 (1949).
[62] M. Griffiths, W. R. Sistrom, G. Cohen-Bazire, R. Y. Stanier, and M. Calvin, *Nature* **176,** 1211 (1955).
[63] S. R. Kornberg, *Biochim. et Biophys. Acta* **26,** 294 (1957).
[64] J. M. Wiame, *J. Am. Chem. Soc.* **69,** 3146 (1947).
[65] J. F. Powell, *Biochem. J.* **54,** 210 (1953).
[66] J. F. Powell and R. E. Strange, *Biochem. J.* **63,** 661 (1956).
[67] S. P. Colowick and N. O. Kaplan, "Methods in Enzymology," Vols. 1–4. Academic Press, New York, 1955–1957.
[68] D. Herbert and J. Pinsent, *Biochem. J.* **43,** 193 (1948).
[69] D. Herbert, *Symposia Soc. Exptl. Biol.* **5,** 52 (1951).
[70] K. Wallenfels and M. L. Zarnitz, *Angew. Chem.* **69,** 482 (1957).
[71] H. McIlwain, *Nature* **158,** 898 (1946).
[72] J. DeLey, *Nature* **164,** 618 (1949).

[73] U. Z. Littauer and A. Kornberg, *J. Biol. Chem.* **226**, 1077 (1957).
[74] O. T. Avery, C. M. MacLeod, and M. McCarty, *J. Exptl. Med.* **79**, 137 (1944).
[75] R. D. Hotchkiss, *J. Cellular Comp. Physiol.* **45** *Suppl. 2*, 1 (1955).
[76] N. D. Zinder and J. Lederberg, *J. Bacteriol.* **64**, 679 (1952).
[77] P. E. Hartman, *in* "The Chemical Basis of Heredity" (W. D. McElroy and B. Glass, eds.), p. 408. Johns Hopkins Press, Baltimore, Maryland, 1957.
[78] A. Gierer and G. Schramm, *Z. Naturforsch.* **11b**, 138 (1956).
[79] S. Spiegelman, *in* "The Chemical Basis of Heredity" (W. D. McElroy and B. Glass, eds.), p. 232. Johns Hopkins Press, Baltimore, Maryland, 1957.
[80] H. Fraenkel-Conrat and R. C. Williams, *Proc. Natl. Acad. Sci. U. S.* **41**, 690 (1955).
[81] S. Ochoa and L. A. Heppel, *in* "The Chemical Basis of Heredity" (W. D. McElroy and B. Glass, eds.), p. 615. Johns Hopkins Press, Baltimore, Maryland, 1957.
[82] J. Tomcsik, *Symposium Soc. Gen. Microbiol.* **6**, 41 (1956).
[83] J. R. G. Bradfield, *Symposium Soc. Gen. Microbiol.* **6**, 296 (1956).
[84] B. A. D. Stocker, *Symposium Soc. Gen. Microbiol.* **6**, 19 (1956).
[85] R. C. Lancefield, *J. Exptl. Med.* **78**, 465 (1943).
[86] L. Dienes and H. J. Weinberger, *Bacteriol. Revs.* **15**, 245 (1951).
[86a] C. S. Cummins and H. Harris, *J. Gen. Microbiol.* **18**, 173 (1958).
[87] E. P. Abraham, "Biochemistry of Some Peptide and Steroid Antibiotics." Wiley, New York, 1957.
[88] L. E. Rhuland, *J. Bacteriol.* **73**, 778 (1957).
[89] K. McQuillen, *Symposium Soc. Gen. Microbiol.* **6**, 127 (1956).
[90] C. Weibull, *J. Bacteriol.* **66**, 696 (1953).
[91] W. B. Hugo, *Bacteriol. Revs.* **18**, 87 (1954).
[92] H. K. Schachman, A. B. Pardee, and R. Y. Stanier, *Arch. Biochem. Biophys.* **38**, 245 (1952).
[93] M. Alexander, *Bacteriol. Revs.* **20**, 67 (1956).
[93a] W. G. Gillchriest and R. M. Bock, in: Ref. 94, p. 1 (1958).
[94] R. B. Roberts, ed. "Microsomal Particles and Protein Synthesis." Pergamon, New York, 1958.
[95] A. F. Brodie and C. T. Gray, *Science* **125**, 534 (1957).
[96] R. Storck and J. T. Wachsman, *J. Bacteriol.* **73**, 784 (1957).
[96a] R. B. Roberts, R. J. Britten, and E. T. Bolton, in: Ref. 94, p. 84 (1958).
[97] A. W. Frenkel, *J. Biol. Chem.* **222**, 823 (1956).
[98] J. W. Newton and M. D. Kamen, *Biochim. et Biophys. Acta* **25**, 462 (1957).
[99] B. Magasanik, *Ann. Rev. Microbiol.* **11**, 221 (1957).
[100] M. Demerec, *Cold Spring Harbor Symposia Quant. Biol.* **21**, 113 (1956).
[101] S. Benzer, *in* "The Chemical Basis of Heredity" (W. D. McElroy and B. Glass, eds.), p. 70. Johns Hopkins Press, Baltimore, Maryland, 1957.
[102] B. W. Doak and C. Lamanna, *J. Bacteriol.* **55**, 373 (1948).
[103] J. W. Foster and J. J. Perry, *J. Bacteriol.* **67**, 295 (1954).
[104] C. L. Hannay, *Symposium Soc. Gen. Microbiol.* **6**, 318 (1956).
[105] S. E. Luria, *Protoplasmatologia* **4**(3), 1 (1958).
[106] S. E. Luria, *Symposium Soc. Gen. Microbiol.* **9**, 1 (1959).

Note Added in Proof:

Recent work has attributed the determination of mating polarities, as well as some other properties of *E. coli*, to a group of genetic elements, called *episomes*, which can shift from a chromosomal to an extrachromosomal state (F. Jacob and E. L. Wollman, *Compt. rend.* **247**, 154, 1958). The change in state of an episome may be considered a form of differentiation.

Chapter 2

The Internal Structure of the Cell

R. G. E. Murray

Page

I. Introduction... 35
II. The Cytoplasm and Its Surface................................. 36
   A. Observations on Living Bacteria.......................... 36
   B. Cytoplasmic Membrane.................................... 39
   C. Cytoplasm... 44
III. Chromatin Bodies... 64
   A. General Characteristics................................. 64
   B. Arrangement and Effects of Environment.................. 71
   C. Behavior during Division................................ 76
   D. Structure of Chromatin Bodies........................... 80
   References... 93

I. Introduction

The past twenty years have provided a wealth of pertinent observations and new understanding of the internal structure of bacteria. A wider interest in structure, not confined to the few dedicated individuals of former years, has been encouraged by a combination of three main factors. *First*, a growing realization not only that bacteria have a discernible internal structure but that selected bacterial species are suitable objects for cytological observations and experiment. *Second*, new and powerful auxiliary means of study have been added to the long-practiced techniques of staining. The availability and usefulness of phase-contrast microscopy in recent years have been most important for the study of living organisms; even more important in the long run has been the devising of techniques for utilizing the high resolution of the electron microscope. A new order of structure is now unfolded to our view that is stimulating to new understanding and to a revision of information obtained by the older approaches and of the standards of excellence or validity that have gone before. *Third*, the spurt of effort directed toward the physiology and the genetics of bacteria has brought to light many questions that require answers in terms of structure. More than ever before the fashion and the requirements of biological science focus the attention not merely upon the systematic phenomena but also upon the site, nature, and function of components.

Cytological studies of sufficient breadth and detail to be useful in a general description of internal structure are few and are largely confined to a small number of bacterial species. This is, perhaps, not surprising, but it is tantalizing because we can now glimpse the bare outline of what can be done to bring our knowledge of bacterial structure to a point where it can be incorporated into the whole picture that we should like to have of the kingdoms of living things. Cytological observations, other than shape and form and habit, are only just becoming convincing to the bacteriologist, let alone the general biologist, and this has been largely stimulated by biochemical probing into the nature of some of the tougher and most durable elements of functional structure. All the approaches to the unraveling of structure are bringing to light unique properties of bacteria that will stimulate more effective research.

This chapter attempts to present a general view of what has been learned of the internal structure of bacteria, and a general look at what is on the horizon. It is not a replacement for the general reviews that are available in the literature for those who wish to survey the early literature and also savor the variant and changing shades of opinion; the monographs of Delaporte,[1] Knaysi,[2] and Bisset[3] should be consulted, as also the reviews of Piekarski,[4] Robinow,[5] Delaporte,[6] and Winkler.[7] The symposium in 1956 on "Bacterial Anatomy"[8] provides a stimulating insight into current approaches and arguments, which can be compared and augmented with a similar symposium held in 1953.[9] Peshkoff[10] has written a detailed general review, which is particularly valuable for its survey of the contributions of Russian microbiologists.

I. The Cytoplasm and its Surface

A. Observations on Living Bacteria

From the earliest days of microscopy observers have remarked upon the peculiarly homogeneous and "glassy" appearance of the protoplasm of living bacteria, but allowance must be made for a confusing variety of cytoplasmic inclusions of lipid, volutin, or sulfur in some species. Although vacuoles and areas with the appearance of vacuoles are sometimes present they are not a prominent feature of most bacteria in the phase of active growth and reproduction; they more generally appear when the period of most rapid growth is passed. The homogeneous density of much of bacterial protoplasm is further evident in phase-contrast microscopy or interference microscopy.

It is possible to gain some impression of the organization in the living cell, aside from the more obvious inclusions which have been a matter of remark and a clue to identification since the days of Cohn, Ehrenberg, and DeBary. Even the most homogeneous of bacteria usually show central areas of irregular shape and of lesser density, which become more obvious if the condenser is lowered and the iris is drawn a little. The phase microscope enhances this contrast and they correspond to the areas in which chromatin may be demonstrated in the fixed and appropriately treated cell (see Plate V).

Some idea of the consistency of the protoplasm is needed. The common statement is that bacteria do not show the streaming of protoplasm that is so evident in the cells of plants and animals. However, this belief is open to some doubt. Puncture of the enveloping membranes usually shows a rapid outpouring of protoplasm, as is also the case when cells burst at the end of the latent period of phage infection,[11] indicating the fluidity of contents. Despite the relatively small size of bacteria and a seemingly short distance for diffusion of nutrilites and metabolites, it does not seem reasonable to exclude the possibility of constant protoplasmic churning. In fact the author has on several occasions watched the movement of single, small and defined cytoplasmic particles within bacterial protoplasm. In each case the particle followed a course from one pole of the cell toward the chromatin areas, and even alongside them, and back again, several times; each cycle varied in time from one to five minutes but never traversed the whole length of the cell. Dr. C. F. Robinow and the author have also watched the movements of the lipid inclusions in *Bacillus megaterium* and concluded that there was a most definite, slow and continuous motion. One might not choose to call this "streaming" of the same order as that so readily observed in, let us say, *Spirogyra* cells or the extraordinary motion of protoplasm in fungal hyphae that carries even nuclei in the stream through septal pores in fungal mycelium.[12] Nevertheless, there is indication that the protoplasmic contents of bacteria are in continuous circulation.[12a]

Phase microscopy provides a means of observing the form, habit, and behavior of cells. Cell division, the behavior of chromatin-containing areas and of various inclusions can be followed in all the detail that the nature of the material and resolution will allow. Little effort is required to make phase microscopy an even more effective tool (see Plates VIII, Fig. 42, and IX, Fig. 43) by adjusting the refractive index of the surrounding medium with, usually, a solution of protein, such as gelatin or serum.[13] The special advantage here is that experimental variation of the refractive index of the mountant fluid can select to some degree the inclusion or structure being demonstrated. The same is true for dark-ground illumination.

It is unfortunate that most of the means of observing the *living* cell fall short on specificity of the information obtained. Experience may show some quantitative usefulness for interference microscopy. Even with ultraviolet light microscopy there have been few examples of useful specific data,[14] although handsome photomicrographs may be obtained of the best possible resolution. At the least, this approach demonstrates that the glassy cytoplasm of actively growing cells has a strong ultraviolet (UV) absorption corresponding to that characteristic of nucleic acids. It adds to the evidence that ribose nucleic acid is both great in amount and rather evenly distributed in the cytoplasm. Curiously, the chromatinic areas (which are Feulgen-positive and undoubtedly contain deoxypentose nucleic acid) usually show as an area of relatively low absorption in UV micrographs at the appropriate wavelength. However, this is a technical difficulty and is more than likely due to the relative concentration of the nucleic acids in the two areas. Unfortunately there have been few if any studies of bacteria in other phases of growth when the ribose nucleic acid would be less obtrusive. Even the excellent pictures of Wyckoff and Ter Louw[15] of sporulation show a disappointing lack of detail of the chromatin of the "mother" cell, perhaps due to the unsuitable wavelength used. But excellent pictures of a variety of bacterial species in a natural environment[16] do indicate possibilities that nobody seems to have exploited. Perhaps it is the difficulty of special equipment and the technical problem of focusing and recording the images on film that have discouraged worthy experiments. The application of a television monitoring device,[17] although ingenious and providing an expensively convenient focusing control, has not yet produced any results convincing of usefulness.

Vital staining has not been particularly helpful in revealing organization or structure in bacteria, partly because few dyes are without toxicity at useful concentration and because few have taken the trouble to exploit the possibilities. Delaporte has made the greatest use of "vital" staining with brilliant cresyl blue and methylene blue. She points out[6] that they are slightly toxic and has usually referred to the procedure as "coloration subléthale." The value is that, aside from demonstrating vacuoles, both cytoplasm and nucleus are slightly stained. Neutral red will accumulate at the sites of metachromatic granules and in certain vacuoles of the living cell, according to Delaporte;[6] but Knaysi[2] (p. 283) considers that this occurs only above a critical concentration of dye, affecting both the permeability and the viability of the cell with consequent effects on internal structures. However this may be, the level of information obtained is not of great use in the interpretation of cellular organization except that more must be learned of the nature and significance of the vacuoles involved. Whether they are or are not equivalent to the cell sap vacuoles is not certain. Knaysi maintains that they are bounded by a membrane that is stainable with basic dyes, iodine, and lipid stains; that environmental conditions during active growth determine their appearance, and that they shrink and disappear when the cell is placed in a hypertonic environment. Further,[18] he considers that these vacuoles are the site of deposition of lipid droplets and of concentration of other important inclusion materials, such as polymetaphosphate (volutin) and glycogen. This merits further experiment and a variety of approaches.

Another technique of promise has been applied by Anderson[19] to the study of normal and phage-infected *Escherichia coli*; in principle it has possibilities of some degree of specificity. It involves treating the cells with a selective fluorescent reagent and observing the distribution of fluorescing areas. The acridines, acridine orange in particular, have been used for the demonstration of deoxyribose nucleic acid-containing structures. The results seem to accord reasonably well to those obtained with the conventional methods. There is hope for some specific linkages with other cellular components of significance in structure.

B. Cytoplasmic Membrane

All that lies within the cell wall of the living cell is the protoplast, and this useful morphological term defines the area under consideration in this chapter. The term has gained an added significance with the discovery that the cell wall can be removed from certain bacterial species and the naked cell will survive if the surrounding medium is of suitable tonicity and composition. The formation and properties of these "protoplasts" will be considered in a later chapter, but certain attributes underline or bring to light some important ideas on the organization of the cell. Until recently, the presence of a cytoplasmic or plasma membrane enclosing the protoplast could only be inferred from the behavior of the protoplast within its cell wall during changes in the osmotic environment; there is now the means for more direct study.

The pioneer observations of Alfred Fischer showed that the bacterial cell, of Gram-negative organisms in particular, is sensitive to the osmotic environment. The cell shrinks or plasmolyzes in strong salt solutions and swells (plasmoptysis) when transferred to distilled water. The essential observation is that the protoplast retracts from the cell wall when it is placed in sucrose or salt solutions of sufficient tonicity, but this does not occur with equivalent solutions of glycerol, urea, and some other substances of low molecular weight or high lipid solubility.[20] Thus it appeared that the surface of the bacterial protoplast, like the surface of other cells, is bounded by a semipermeable and selective plasma membrane.

We are not concerned here with the physiological properties and phenomena attributable to this specialized surface. Osmotic function in relation to bacterial structure has been dealt with in detail with a critical review of approaches by Mitchell and Moyle,[21] who have been outstanding contributors in this field. It is certainly a surface of complex function that is dependent upon the metabolic capability of the cell as a whole. In the obligate halophiles, for instance, it has to cope with an environment containing as much as 25 % sodium chloride.

In the normal cell the surface of the protoplast is in the most intimate contact with the enveloping cell wall. When plasmolysis occurs the protoplast does not retract evenly from the walls and the areas of retraction present a concave surface, which can be considered an index of adhesion between the two. The cell wall retains its shape. When the protoplast is liberated from the corsetlike cell wall it takes on a spherical shape. This indicates, as does much other evidence, that the structural basis of cell shape does not lie in the cytoplasm or its surface, but is a characteristic built in to the enveloping integument.

It was realized early that Gram-negative organisms were the more easily plasmolyzed to show separation of protoplast and cell wall; under the same conditions Gram-positive organisms did not show the reaction. More recently separation of the cell wall and the protoplast has been induced in Gram-positive species, *Bacillus subtilis*[22] and *B. megaterium*[23] in very strong salt solution (approaching 25 % sodium chloride or potassium nitrate). Mitchell and Moyle have questioned whether this is, in fact, true plasmolysis, and put it in the same class as the relaxing effect of ether[23] upon the cell wall of some of the same organisms, leading to a relaxation and a stretching away from the protoplast (Plate I, Figs. 1 and 2). It seems to be plasmolysis but, true or not, it serves the purpose of separating the cell wall and the protoplast so that the individual parts may be selectively stained. Perhaps the adherence of the the protoplasmic surface to the cell wall is stronger in Gram-positive organisms, such as *B. megaterium*, than in Gram-negative organisms, such as *Spirillum undula*. It has also been suggested[24] that the cell wall of some Gram-positive bacteria is not merely a

---

PLATE I. FIG. 1. *Bacillus* M fixed with Bouin's fluid, treated with mercuric chloride, and then stained with Victoria blue B to show selective staining of the cell membranes.

FIG. 2. *Bacillus* M exposed to ether vapor for six minutes before fixation with Bouin's fluid followed by Victoria blue B staining. Ether treatment causes separation of the cell wall and the protoplast. The cytoplasmic membrane is more deeply stained than the cell wall. (Figs. 1 and 2 are reprinted[23] with the permission of the publishers of *Experimental Cell Research*.)

FIG. 3. Micrograph of a section of a *Spirillum* sp. showing a dense membrane at the surface of the protoplast. The cell wall is separated and partly destroyed. This is an enlargement of the right end of the cell in Fig. 4. (It is reprinted[31] with the permission of the *Canadian Journal of Microbiology*.) ($\times 41,000$)

FIG. 4. A longitudinal section of a *Spirillum* showing dense cytoplasmic inclusions, which are polymetaphosphate granules. ($\times 8,000$)

FIGS. 5 and 6. *Spirillum* sp. stained with toluidine blue to show metachromatic granules. (Photomicrographs by Dr. J. P. Truant.)

THE INTERNAL STRUCTURE OF THE CELL

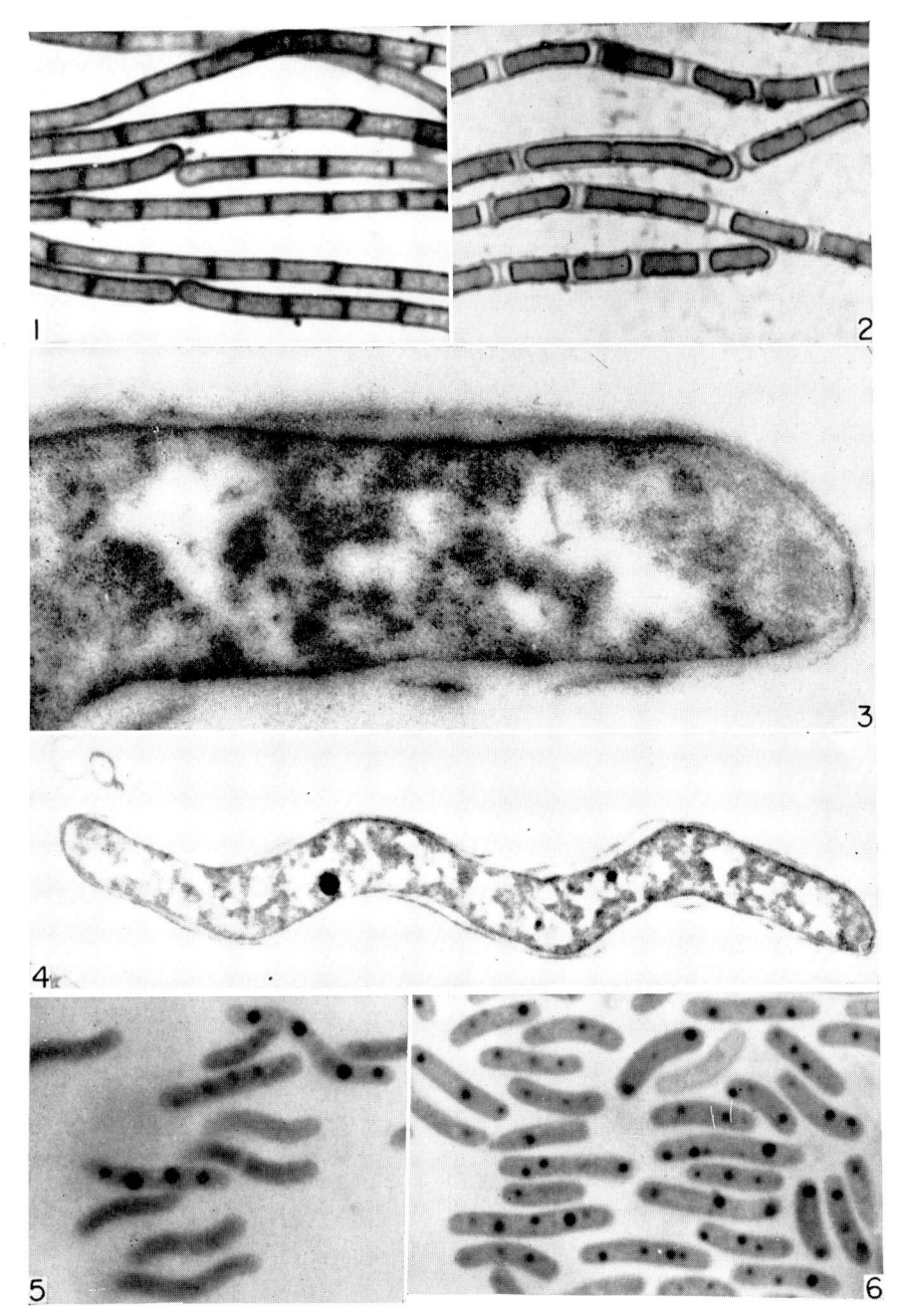

Plate I.

rigid structure conferring shape to the cell but also has elastic properties, which would make the analogy of the cell wall and a "corset" even more apt.

The spherical protoplasts released from *Bacillus megaterium*, treated with lysozyme in a stabilizing environment, behave as osmometers and have provided the most convincing and direct evidence of the presence of a plasma or cytoplasmic membrane at the surface of the protoplasm. When these spheres are lysed by dilution of the stabilizing environment the most prominent structures remaining in suspension are empty vesicles or "ghosts," which are quantitatively released. Thus there is now available a technique for the separation and study of this membranous structure which is considered to be identical to the cytoplasmic membrane. This relationship has been discussed in detail by Weibull.[25] It has always been assumed that these membranes, for all types of cells, are rich in lipids and probably represent a highly organized lipoprotein layer. Weibull[26] has shown that the membranes from protoplasts of *Bacillus* M are indeed rich in lipids and contains 55–75% of the bacterial lipids, despite the fact that they represent only about 15% of the total dry cell weight (but see also[27]).

But despite the very evident demonstration of a membranous surface to isolated protoplasts, the differentiation of such a specialized surface in intact cells, either by the more conventional techniques of staining cytological preparations or by embedding and thin sectioning for electron microscopy, has been difficult indeed, as has been pointed out by Bradfield.[28] In part the difficulty is due to the normal intimate relation of the cell wall and the cytoplasmic surface.

The common basic dyes are taken up so avidly by the cytoplasm that it is impossible to distinguish a special surface. However, a low concentration of Victoria blue selectively stains the superficial layers of fixed Gram-positive cells under the same conditions. Achieving a slight separation of the protoplast and the cell wall by plasmolysis, drying, or treatment with ether before fixation, it can then be demonstrated that the cell wall is only lightly stained and that most of the dye is arrested at or in the cytoplasmic membrane (Plate I, Figs. 1 and 2). The experiments with this staining method[23] indicate that the cell must be intact to attain this degree of differentiation and that, given time, Victoria blue eventually permeates the whole cytoplasm and stains it as do the more commonly used basic dyes. Victoria blue will partition into lipids and this might be invoked as an explanation of the useful properties of the dye. However, both crystal violet and fuchsin will, after suitable pretreatment with tannic acid or mercuric chloride (the useful methods of staining cell surfaces which we owe primarily to Eisenberg), achieve a similar but lesser differentiation of the same order. The Victoria blues are markedly surface active and this may provide a more likely explanation of the phenomena. The high lipid content of cytoplasmic membrane suggests that it should be demonstrable with lipid soluble dyes such as Sudan black and a convincing photograph showing this has been published by Knaysi. However, this has not proved to be reproducible in our hands.

The difficulty of demonstrating a cytoplasmic membrane in electron micrographs of thin sections of intact bacteria requires explanation. Bradfield[28] has observed that "the outermost surfaces of the cytoplasm, lying against the cell wall, sometimes appear more smooth, continuous and uniformly dense over a thickness of about 50A...." He also points out that it would be hard to define a membrane of that order of thickness unless the cut is taken approximately normal to the surface of the membrane. Even where cell wall and cytoplasm have separated in the embedding there is usually no sign of a specific membrane on the surface. However, since this separation is likely to have taken place after fixation it is not too unlikely that the cytoplasmic membrane remains attached to the cell wall in that situation. On the other hand, when the protoplast separates from the cell wall the cytoplasmic membrane remains with the cytoplasmic surface and it has been possible to demonstrate this as a positive layer in sections of the protoplast of *Bacillus* M.[29]

General experience in the electron microscopy of tissues and the study of membranes in particular would indicate that lipid membranes can usually be expected to be of low electron scattering and show little or no particular density on the micrograph. Thus one might expect to find a thin nonscattering layer close to the cell wall rather than a thin scattering layer. And indeed this has been common experience in this laboratory (Plate XII, Fig. 52). Such a layer would be even harder to find or to observe with the electron microscope if it were an intermediate scattering or density and although it would be or should be easily resolved, the difference in density of the surround would not be sufficient to demonstrate it. There have been only a few unequivocal demonstrations of a cytoplasmic membrane in sections. One has been mentioned above, the others have been in sections of *Spirillum*[30, 31] and in *Escherichia coli*[32] and show a dense, well-differentiated layer intimately connected with the cytoplasm (Plate I, Fig. 3). In both these cases it should be noted that the cytoplasm was somewhat dilute so that the effective contrast between cytoplasm and cytoplasmic membrane is increased to its limits. The demonstration in *E. coli* was made possible by phage infection, which caused a marked decrease in cytoplasmic density; normal cells did not show the membrane. In *Bacillus* species the micrographs are less convincing. It is certain that the conditions necessary for the fixation and best demonstration of the cytoplasmic membrane have not yet been found for electron microscopy; some method of specific "staining" would be most desirable. The exact thickness of the observed membrane is hard to gauge but it would seem likely that it is close to and probably a little more than 50 A.

It should be noted that when the cytoplasmic membrane has been clearly distinguished, the membrane forms a smooth envelope for the cytoplasm and that intrusions or membrane systems deriving from the cytoplasmic membrane have not been seen (cf. Section II, C, 4).

C. Cytoplasm

1. General Features

When a cell is suitably fixed and stained the protoplast can be differentiated into various areas that have distinct staining characteristics and cytochemical reactions, which define regions for special study. The bacterial protoplast can be differentiated into a cytoplasmic region, which is rich in ribonucleic acid, and a chromatin region containing the deoxyribonucleic acid of the cell as shown by the Feulgen reaction.

It has been a matter of general observation, from the very earliest experiments with bacteria, that the cytoplasm of bacteria may be intensely stained with both acid and basic dyes. The marked affinity for stains shown by growing bacteria has been of the very greatest use in diagnostic bacteriology. And even with the procedures of air-drying and flame-fixing it has been obvious that the affinity for staining depends upon the physiological state of the organism. The immensely useful Gram procedure illustrates this fact; Gram-positive organisms tend to become Gram-negative during the phase of decline in growth and if the integrity of the cell is broken. Almost any of the common basic dyes are taken up avidly during the logarithmic phase of growth, but by the time the culture has reached rest the affinity is of a low order. Basophilia is the most marked characteristic during the logarithmic phase of growth and it was at this stage of the cycle of growth that the early observer could make out little or no anatomical detail in the bacterial cell. There was then more apparent reward in the study of physiologically "older" cells where cytoplasmic inclusions were more prominent and where the classic staining methods for chromosomes were more regularly successful in showing chromatin bodies. The differentiation obtained by Douglas and Distaso,[33] and particularly by Dobell,[34] taking advantage of the extraordinarily useful differentiating properties of the Romanovsky stains, did not make a general impression on bacteriologists. Their usefulness has now been revived[35] but only after an interval during which the simple application and specificity of the Feulgen reaction and modifications thereof (cf. Section III, A) had allowed a great many bacteriologists to experience for themselves that it is possible to differentiate cytoplasmic and chromatin areas in most bacteria at all stages in the growth cycle.

The current methods of displaying bacterial chromatin in "young" cultures depend upon the abolishment or reduction of the extreme basophilia of the cytoplasm. Classic Feulgen hydrolysis with normal HCl at 60° C. is effective because, as Vendrely and Lipardy[36] showed, it removes all or nearly all of the ribonucleic acid from the cytoplasm; it has the added advantage that the normally refractory chromatin has an increased avidity for basic stains, as will be discussed in a later section. Similarly effective reduction of cytoplasmic basophilia, but not necessarily the concomitant effect on chromatin affinity, may be attained by treatment with ribonuclease, cold perchloric acid and even neutral salts among other treatments (Plate V, Figs. 21–23).

This may seem a rather negative approach to the understanding of the anatomy of the bacterial cell. However, it is of the greatest importance because it allowed a change in the philosophical approach to bacterial structure and provided a beginning point for comparative studies of the internal arrangement and physiological attributes of cytoplasmic and chromatin elements. It became apparent that the cytoplasm of bacteria, particularly in the early phase of growth, shows the tinctorial and histochemical properties of the basophilic component of the cytoplasm of cells in general, which is now ascribed to the Palade or ergastoplasmic granules so well known from the electron microscopy of tissue cells.

The most characteristic general feature of cytoplasm is then a considerable content of ribonucleic acid and it is important to note how this is distributed. Staining with a simple basic dye, such as thionine, shows that this is distributed throughout the whole cytoplasm. It is not apparently contained in the cell wall, as shown by the clear separation between cells in a filament. Some ribonucleic acid is contained within the "chromatinic" nuclear area but this is not obvious in preparations of this sort and the evidence will be discussed in the section on chromatin bodies. One of the many attempts to elucidate the Gram reaction[37] ascribed the reaction to the formation of a complex with magnesium ribonucleate which was assumed to be contained in the cytoplasmic membrane. There is no real supporting evidence for the presence of any considerable amount of ribonucleic acid in the cytoplasmic membrane and much that is against it. Isolated membranes do not stain with this procedure, the cell is Gram-positive only as long as it is intact, and the stained cell does not seem to be stained at its surface alone. There is no doubt that ribonucleic acid may have an important part to play in the formation of the complex, since Gram-positivity is abolished by pretreatment with ribonuclease, but it would seem likely that the retention of the stain is concerned more with the inability of the complex to escape through the cell wall.

2. Cytoplasmic Inclusions

In both the living and the stained cell many types of inclusions may be observed within the cytoplasm. Some seem to be an essential part of the physiology of certain groups and species, e.g., the sulfur granules and the complex lipid inclusions of certain types of sulfur bacteria, but many of the common inclusions (metachromatic granules, lipid droplets, and some polysaccharide accumulations) seem to be absent or inconspicuous during the most active physiological phases of growth and become more prominent during decline. Yet they may be a feature of certain groups and species, reflecting special cytoplasmic characteristics and attributes. And, together with the over-all morphology, the inclusions can be so characteristic that the organism may almost be identified on sight.

The remarkably protean metabolic capabilities of bacteria have stimulated the search for cytoplasmic organelles that correspond to those that are now so well known in the cytoplasm of plant and animal cells. To some of us it is neither surprising nor discouraging to find that things are not quite the same. The truth is, and this should be readily apparent, that we do not yet know quite enough about the architecture and nature of cytoplasm to be sure of the differences and similarities. Certain it is that we know little enough of the nature, structure, physiological and biochemical attributes of even the most prominent bacterial inclusions.

The main means of revealing and identifying the various inclusions is by differential staining. Few have attempted and fewer have accomplished a reasonably complete cytochemical study of any one organism, taking into account at the same time the physiological state and cultural age. One of the few is the study of *B. megaterium* and *E. coli* by Widra.[38] More limited but of great usefulness are the classic studies of Delaporte,[1] but in this case the organisms were for the most part in the later stages of growth. The difficulty lies in providing a synthesis of the individual observations to give an intelligible picture of the cytoplasmic architecture throughout the growth cycle. It is likely that we are not yet sufficiently informed to do so.

The terminology is confused because Meyer[39] used the terms "vacuole" and "inclusion" synonymously. The distinction, if there is one, is a matter of convenience and Knaysi[18] has used the term *vacuole* for an inclusion consisting of substances in aqueous solution, and *inclusion* for droplets or particles of what appear to be pure substances.

a. Vacuoles. The cells of many living bacteria have a vacuolated appearance. The main argument that there are indeed such structures depends on the observation that they retain or concentrate neutral red, as occurs with the cell sap vacuoles of plant cells. Specific staining methods can often show that they also contain identifiable materials, often volutin and sometimes lipid and sulfur.

It seems unlikely from present observations that vacuoles are a consistent feature of the bacterial cell. In most cases vacuoles or inclusions are conditioned by the environment and the physiological state of the cell. Knaysi[2] observes that the cell sap vacuoles act or behave as osmometers, shrinking and often disappearing when the cell is placed in a hypertonic environment, which would argue that there should be a vacuolar membrane. He suggests that this membrane consists of material similar in structure and properties to the cytoplasmic membrane. Certainly some bacteria always contain vacuolar inclusions that are of physiological importance, such as the large terminal vacuole in *Thiobacillus thiooxidans*. These contain sulfur if the organism is grown on a synthetic medium with sulfur as the source of energy; on a thiosulfate agar the content of the vacuole seems to be pure volutin. Mycobacteria contain a complex of vacuole-like structures that are only appreciable by electron microscopy.[40] Here and, indeed, in most cases there is remarkably little really known about the nature and function of vacuoles.

Many cells seem to contain vacuoles even during the active phase of growth in which no definite contents can be identified and they may be assumed for the moment to be fluid or watery. For instance, in some of the *Azotobacter* species that do not contain lipids these can be very prominent and are not stained by any of the usual methods. It is obvious that such structures in these and other bacteria[41] require a more systematic investigation in order to discover their nature and importance.

A study of such vacuolar systems, particularly with thin sectioning and electron microscopy, would be of the greatest importance to form a basis for comparative consideration of cytoplasmic features of all kinds of cells. What is particularly needed is a study of some of the larger bacteria belonging in groups or orders that are not usually studied or have not usually been studied in bacterial cytology. A beginning has been made, using vital staining, by Delaporte on *Achromatium oxaliferum* which can have enormous vacuoles crowding the cytoplasm and which take up neutral red. A study of these after the fashion of Fauré-Fremiet and Rouiller, who have done an excellent electron microscope study of *Thiovulum* (see Section II, C, 4) would undoubtedly produce good returns. The latter also showed clearly that the sulfur droplets are contained within a vacuole having a distinct fine membranous surface (Plate IV, Fig. 15). These do not seem to be continuations of the intrusions of the cytoplasmic membrane into the substance of the cytoplasm. Interesting as these isolated observations are, it is not yet possible to draw any general conclusions because there are not enough examples from diverse groups of bacteria.

b. Metachromatic Granules. Metachromatic granules are perhaps the best known of the bacterial inclusions and are so named because they turn red or reddish when stained with some of the blue dyes, such as toluidine blue and methylene blue (Plate I, Figs. 4–6). They are also called volutin granules from the name given by Meyer[42] and they are found in many fungi, algae, and protozoa, as well as in the bacteria. The main chemical constituent is a polymetaphosphate of very high molecular weight, as was shown first for the metachromatic granules of yeast, but because of strong absorption of ultraviolet light in the region of 2650 A. it was once considered that they consisted chiefly of free ribonucleic acid. Their cytological characteristics have been studied by Winkler,[43] who has done much to characterize these granules and who has reviewed the observations upon them with a critical eye.[7]

Several authors consider that the polyphosphate is precipitated in the interior of vacuoles in most of the protists, but in bacteria polyphosphate accumulations are found directly in the cytoplasm. There is no doubt that the number and size of metachromatic granules increase markedly during the latter part of the growth cycle and that on transfer to fresh medium these are reduced in size and even disappear.

In electron microscopy the metachromatic granules are easily recognizable, not only because they tend to sublime in the electron beam, but also because they are of an extreme density (Plate I, Fig. 4). With sufficient electron bombardment they seem to disappear entirely, leaving a dense wall around the edge of the place that they occupied, which has been used as evidence that the metachromatic granules are formed in a membranous vacuole. This is by no means certain.

c. *Lipid Droplets.* In living bacteria lipid droplets show up as extremely refractile globules, which are usually multiple and pepper the cytoplasm. They may be intensely colored with lipid-soluble dyes, of which the most useful seems to be Sudan black (Plate II, Figs. 8 and 9), and they may take up a considerable portion of the cytoplasmic volume. In *Bacillus megaterium*, for instance, a polymer of beta-hydroxybutyric acid accumulates in very large amounts and can constitute as much as 25% of the total dry weight.[44] The amount or number of lipid globules in the cytoplasm, as in the case of metachromatic granules, increases with the physiological age of the cell and is determined to some extent by the growth medium. Not only the amount but the nature of the lipids that may be present in bac-

---

PLATE II. FIG. 7. Micrograph of a section of *Bacillus cereus* (18-hour culture) showing spaces left in cytoplasm by lipid inclusions. There is no evidence in a membranous border to the inclusions. (×14,000)

FIG. 8. *Bacillus cereus* (10-hour culture) stained with Sudan black to show lipid inclusions.

FIG. 9. Lipid inclusions in *Azotobacter agile* stained with Sudan black. (Photomicrograph by Dr. J. P. Truant.)

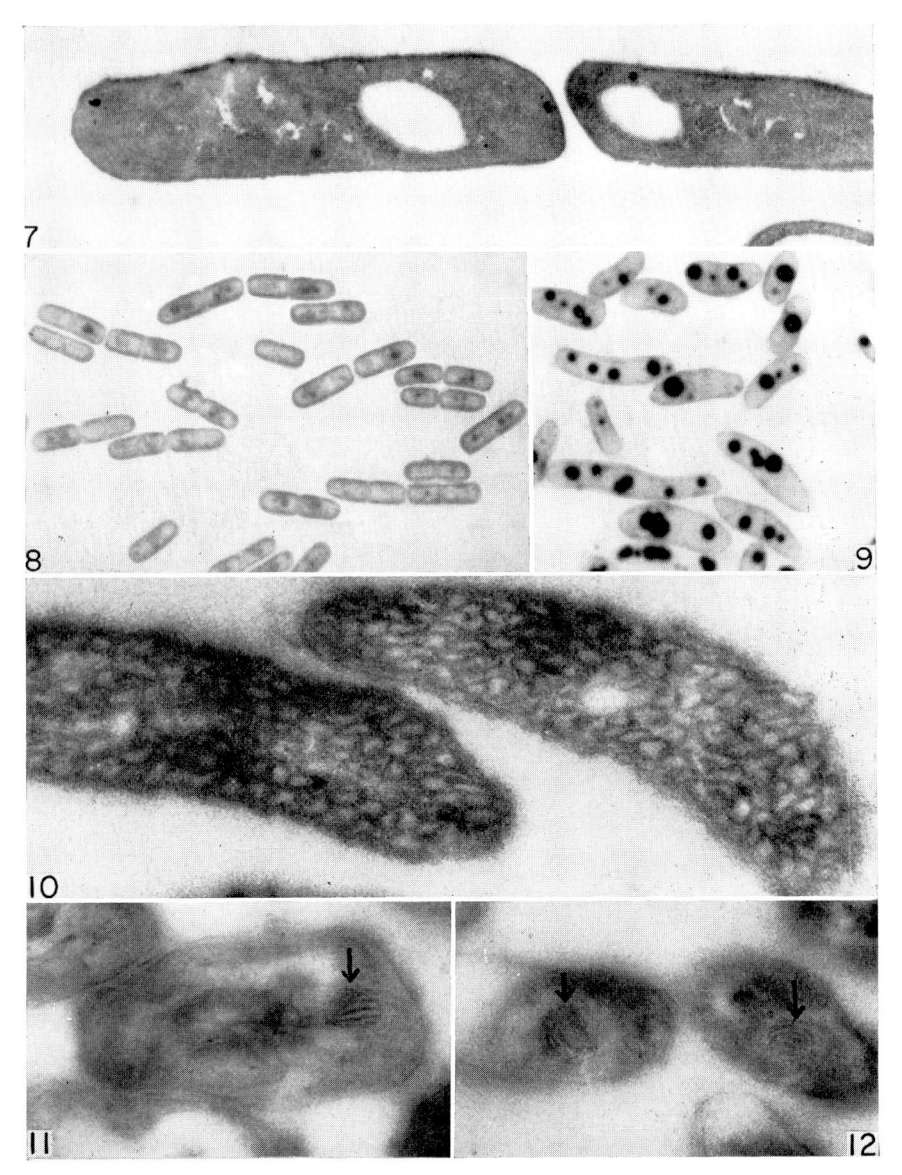

PLATE II.

Fig. 10. Section of *Rhodospirillum rubrum* showing the profiles of the chromatophores, which pack the cytoplasm. (×47,000)

Figs. 11 and 12. Sections of a methacrylate embedding of avian tubercle bacilli (17-hour culture) showing laminated bodies in the cytoplasm. These appear in profusion in some preparations. (Micrographs kindly provided by Dr. Audrey M. Glauert, Strangeways Laboratory, Cambridge, England.) (×62,000 and 61,000, respectively) terial cells is characteristic of the metabolic pattern of the species and even of the strain, as in the case of *B. megaterium* and related strains.[45]

Little is known really of the anatomical location of most of the lipid fractions that may be extracted and the methods that are available to us are not very appropriate for this purpose. There are a number of specific situations in which the location of particular fractions would be of the greatest interest and even practical importance, e.g., the complex lipid fractions extractable from *Mycobacterium tuberculosis* and the peculiar lipid of *Listeria monocytogenes*, which are undoubtedly responsible for characteristic features of natural infections.

Micrographs of thin sections are not particularly helpful for the study of lipids, even though one might expect that some would have an affinity for osmium and be easily apparent and others have a moderate phosphorus content. Both the reagents for the dehydration procedure and the monomers of the plastics used for embedding are lipid solvents and extraction must be expected in some degree. Furthermore, the atoms that make up the lipid molecules are for the most part of low atomic weight and unlikely to produce the considerable scattering effect necessary to appreciate their presence. Thus it is not surprising to see vacuolar areas with no appreciable contents in sections of bacteria that contain lipid inclusions (Plate II, Figs. 7 and 8) or are in a stage of growth (such as approaching sporulation in *Bacillus* species) where one expects to find large lipid droplets.[46]

Even in those cells that do not normally accumulate lipid droplets in the cytoplasm it has been our experience that a few small lipid-staining granules appear in the chromatin areas. The significance of these is very uncertain.

d. Glycogen and Other Polysaccharide Inclusions. Although the majority of bacterial species probably do not accumulate glycogen, those that do so include a tremendous amount in the cells; apparently this is entirely in the cytoplasm.[6] It tends to displace the chromatin bodies to a minimum possible area and commonly this occurs at the end of the growth cycle. Certain species of *Clostridia* accumulate a related substance which gives a red-purple color reaction with iodine and is commonly referred to as granulose. This accumulates at the time of sporulation within the mother cell, which swells exceedingly during the process (see Chapter 5).

As is the case with lipids, accumulations of glycogen or granulose in bacterial cells do not show positively on micrographs of thin sections. The appearance is one of an empty area; indeed, the density of the area may be even less than the density of the surrounding methacrylate of the section, thus indicating dilution of the embedding medium. The same seems to be true of glycogen in sections of liver and starch in sections of plant cells. Specific means of altering this density for identification in electron microscopy are being developed.[46a]

Other polysaccharide inclusions have been described, in particular with the help of the periodic acid-Schiff reaction, in the cytoplasm of many organisms. As has been pointed out by Widra, this cytochemical reaction, although specific for the 1, 2-glycol grouping, is not sufficient for the identification of a polysaccharide inclusion without a considerable range of other tests[38] to define the material. However, both *E. coli* and *B. megaterium* show polysaccharide accumulations in the cytoplasm and particularly in the polar region (see also Section II, C, 5).

 e. Mitochondria or chondrioids. Mudd and his co-workers[47–49] have described a multiplicity of activities in certain cytoplasmic inclusions. They show one or several of a variety of properties: reduction of tetrazoles, color reactions with Janus green, a positive oxidase test with the Nadi reagent, reactions for phospholipids, and often, metachromasia. These centers of extreme oxidation and reduction, together with the somewhat similar reaction given by mitochondria of higher cells, led them to consider the inclusions as the bacterial equivalents of mitochondria. There is no doubt that punctate cytoplasmic staining of the types described does occur and that the sites often appear to be coincident, but the equation of these sites with mitochondria is not at all certain.

The term "mitochondrion" is used to describe a morphological entity in the cytoplasm of higher cells, including both fungi and protozoa.[50-52] It is identifiable in both the living and the fixed cell. It has a definite fine structure, bounded by a double-contoured membrane which sends intrusive cristae or villi into the substance of the organelle. No such structure has appeared in the micrographs of sections of the *majority* of bacteria studied, and these cells have been fixed and prepared in ways that are substantially the same as those effective for the preservation of the mitochondria of tissue cells. However, the recent finding of lamellated structures in the cytoplasm of *Mycobacterium* (Plate II, Figs. 11 and 12), *Streptomyces*, and, in one instance, *B. subtilis* cells is indicative that suggestive structures may be present in some orders or genera (cf. Section II, C, 4).

We now realize that the cytoplasmic granules of bacteria often do contain multiple components, including one or more of the attributes that have provoked the analogy. Difficulties beset the histochemical approach, apart from the problem of resolving small structures and detecting the coincidence of sites of reaction within a small cytoplasmic area. These difficulties include the lipid solubility of some tetrazoles and the Nadi reagent. The actual site of reduction is problematic because Weibull and others[53] have observed that the reduction of triphenyltetrazolium occurs diffusely in the cytoplasm, and the countless small particles aggregate to form the few gross accumulations of formazan that had been taken as the site of mitochondrial activity. It has been shown[54] in *Mycobacterium tham-*

nopheos that tellurite is reduced in the presence of suitable substrates to give deposits of tellurium microcrystals, usually on or around cytoplasmic inclusions which may be identical to those giving the histochemical reactions. Some of the variance in observations will probably be found to be due to distinct differences in the attributes of the cytoplasm in organisms from different groups, which is indicated in the study by Widra.[38]

A biochemical definition of mitochondrial activity is possible although many, if not most of the reactions usually ascribed to mitochondria also take place elsewhere in the tissue cell. One truly characteristic component is cytochrome oxidase. Yet Mudd himself notes,[55] while attempting a new definition of bacterial mitochondria: "In particular the cytochrome system may be absent or deficient." Weibull[56] found that virtually all the cytochrome system of the protoplast of *B. megaterium* is contained in the cytoplasmic membrane ("ghost") fraction. A somewhat similar conclusion was reached in the case of *Azotobacter agilis*.[57] This brings to mind the comment[58] that the whole bacterial cell might be considered as a mitochondrion with the cytoplasmic membrane acting as its active boundary, and one must remember that the order of size of most mitochondria is that of a small bacterial cell. However, the conclusion that cytochrome activity resides only in the cytoplasmic membrane must be accepted cautiously because preparations of ghosts from protoplasts often have minute adherent and pigmented granules[59] which have yet to be examined for enzyme activities.

The nature of the lamellated structures in the cytoplasm of Actinomycetales requires further definition in terms of high resolution micrographs, relation of number and form to metabolic activity, and correlation of staining properties. The attempts at isolation after cell disruption and biochemical characterization[60] will have to aim at the greatest possible purity of the fractions. One must conclude that, although there is a morphological entity in a few bacteria having the semblance of mitochondrial structure, this has not yet been shown to be a common or regular feature of the Eubacteriales. The critical work is yet to come. For this reason it does not seem wise to apply the definitive term "mitochondrion" to the entities here discussed. If a term is to be used it would be better to adopt the "chondrioid" that was coined by Kellenberger for the polar granules of *E. coli*,[61] which represents a region that is certainly replete with the sort of reactivity under discussion.[38]

3. The Fine Structure of Cytoplasm

The information so far obtained with light microscopy, both of living cells and of cells stained by methods designed to give some specific information, has so far given us only rather limited concepts of the organization of bacterial cytoplasm. The difficulty is not merely the problem of determining organization within an extremely small cell or determining the nature of a multiplicity of inclusions that may be visible within that cell, but also to know enough of the intimate organization to be able to undertake reasonable comparisons with cells from other kingdoms. The approach that is particularly suitable to biochemistry involves mechanical or chemical disintegration followed by differential centrifugation and is comparable in scope to the analysis of particle fractions that have been undertaken on tissue cells. The development of methods of embedding and thin sectioning suitable to a study of biological material in the electron microscope has provided a completely new dimension to the study of bacterial cells as it has to the study of tissue cells of plants and animals. We can now compare what we know or can infer from the light microscope and biochemical fractionation studies with the high resolution of fine architecture obtainable in the electron microscope. It is unfortunate that at the present stage of development the species of bacteria that have been studied belong almost entirely to the Eubacteriales, and relatively few species of these, with only a few tentative forays into other orders and very limited studies so far of the photosynthetic bacteria and the blue-green algae. Thus the picture is necessarily incomplete. The state of our understanding was reviewed up to the year 1956 by Bradfield[28] and much of his discussion remains valid today, although a few new approaches and some new varieties of structure have been added.

The most striking feature of the cytoplasm of the true bacteria that have been studied up to now is the *lack* of membrane systems. If the cells have been taken from a culture in the logarithmic phase of growth it is also very striking that the over-all density of the cytoplasm is extremely high and of a very different order to that observed for most plant and animal cells. Cytoplasm consists of a vast conglomeration of closely packed granules that are not draped on any obvious scaffolding (Plate III, Figs. 13 and 14; Plate XI, Fig. 50; Plate XII, Fig. 52). There is an almost even packing between the cytoplasmic membrane and those central areas of the cell where we know the chromatin bodies to be situated, and which will be described in a later section. In most cases the cytoplasmic membrane is hard to demonstrate but, in the healthy cell, the cell wall seems to be intimately and closely associated with the surface of the cytoplasm and sometimes a thin differentiated layer can be detected immediately within the cell wall. Where the cytoplasmic membrane can be distinguished clearly there is no elaboration of this surface to form intrusions or lamellae that penetrate into the substance of the cytoplasm, which is such a characteristic feature of a high proportion of the cells of higher organisms. It is also true that no obvious nuclear membrane appears to separate the area that contains the chromatin bodies and the cytoplasm. This is a matter of considerable importance because nuclear membranes are associated with the nuclear structures of the fungi and the protozoa, but they have not been demonstrated unequivocally in either the bacteria or the blue-green algae. Further, there is increasing evidence that where nuclear membranes occur they are really a part of the membrane systems of the cytoplasm[62] and as such must be considered to be cytoplasmic structures. Bacteria, with one exception so far, do not show even the rudiments of a reticulum system. From this one might infer, as an argument supporting observation, that they should not have a nuclear membrane.

The granular cytoplasm has been a matter of remark or has been obvious in the micrographs published since the pioneer papers of Chapman and Hillier[63] and of Birch-Andersen, Maaløe and Sjöstrand.[64] The size of these granules, at least in minimum diameter, approximates the 100 to 200 A. described by Bradfield. It is tempting to relate these to the Palade granules or ergastoplasmic granules that are rich in ribonucleic acid and populate the cytoplasm of higher cells. But it is difficult indeed to show that the bacterial cytoplasmic granules are rich in ribonucleic acid. It is true that the density of bacterial cytoplasm to electrons is reduced by hydrolysis of the Feulgen type[65] and also is reduced by treatment with ribonuclease. However, identification of the precise site is more difficult.

Studies of the submicroscopic particles that are released from the protoplasm of cells exposed to disruption by an ultrasonic generator may give some indication of the nature of the particles and an indirect approach to the identification of granules seen in electron micrographs. The whole body of experiments involving centrifugal fractionation of bacterial cytoplasm, reviewed recently by Alexander,[66] seem to show that most fractions are very heterogeneous. However, there are a few indications that particulate fractions can be found which contain ribonucleic acid and protein and which are released by sonic treatment in amounts proportional to the rate of cell disruption. According to one observation[57] the major component of the ribonuclear protein has a sedimentation constant of 42 S and is a particle of the order of diameter of 300 A. Perhaps the 40 S particles found

---

PLATE III. FIG. 13. Cross section of *Bacillus cereus*, including the region of a chromatin body, denoted by irregular spaces in the cytoplasm. The cytoplasm, both in the periphery and near the chromatin strands, consists of closely packed granules and rodlets. The cells were fixed from a culture medium containing 0.5% NaCl; methacrylate embedding. (Reprinted[125] with the permission of the *Canadian Journal of Biochemistry and Physiology*.) (×86,000)

FIG. 14. Micrograph of *Bacillus cereus* to show cytoplasmic fine structure. Cells were grown in a medium deficient in Na and K, fixed with 1% OsO_4 in distilled water and embedded in methacrylate. The sections do not show the usual chromatin "vacuoles" and it is impossible to identify the chromatin region. (From unpublished experiments in collaboration with A. Birch-Andersen.) (×82,000)

THE INTERNAL STRUCTURE OF THE CELL 55

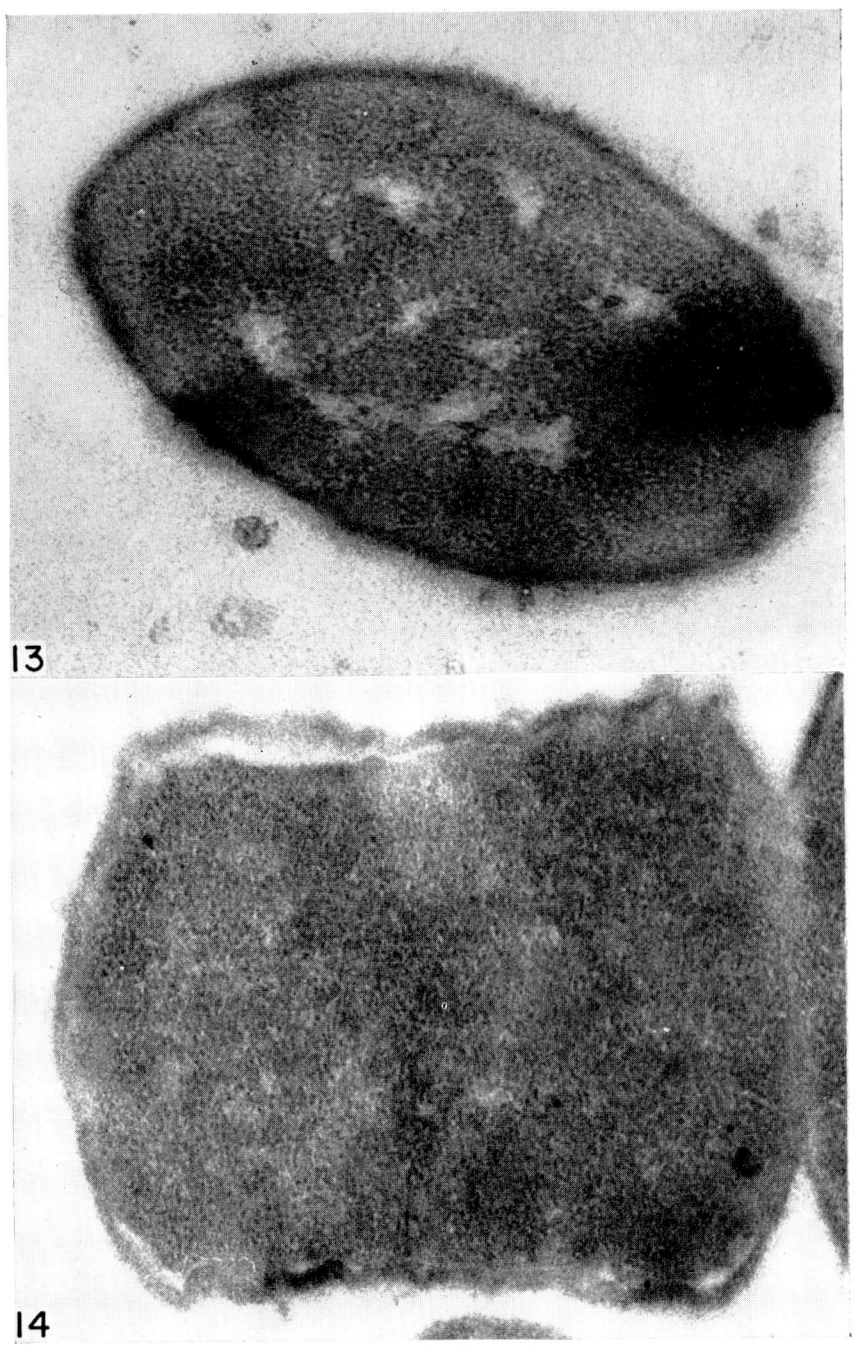

PLATE III.

by Schachman *et al.*[67] in quite a variety of bacteria are also ribonucleoprotein particles. These particles with their definite composition (approximately equal parts of ribonucleic acid and protein, based on dry weight) and size may be identified with granules that are such a constant finding in the cytoplasm of bacteria. The crude particle fractions and the soluble fractions each showed different varieties of enzymic activity. However, the ribonucleoprotein particle fraction, when separated and purified by zone electrophoresis, was found to be practically free of the enzymes that were surveyed. In general, experiments involving centrifugal fractionation of bacterial cytoplasm show that most fractions are very heterogeneous and what we can see of the fine structure in sections does not make this result too surprising. Undoubtedly it is a bit early to hope for an accurate mapping of both the structure and chemical nature of bacterial cytoplasm.

What is seen in the electron micrographs represents the electron-scattering or the electron-capturing ability of atoms making up complex organic molecules and these capabilities may be considered as roughly proportional to atomic weight. Although trace amounts of atoms of high atomic weight are present in the substance of the cell, they are of insufficient concentration to have any particular effect in producing density in the micrograph; those atoms that are most likely to produce the image that we see are sulfur and phosphorus. We know that pure polysaccharide accumulations, for instance, glycogen or starch, have a density even less than the methacrylate in which the cells are embedded. It seems likely that much of what we see of the sections represents the large quantity of ribonucleic acid known to be present in the cytoplasm of bacteria. High-resolution pictures of cytoplasm (Plate III, Figs. 13 and 14) indicate that the granule is not entirely homogeneous. The very best micrographs seem to show that the dense portion is a ring with a less dense center. Furthermore, close inspection of such micrographs would indicate that not all of these granules could be considered spherical particles and that some of them at least seem more like rodlets probably of short length.

Recognition of a structure depends on an adequate density difference between the structure and its surroundings. If by some procedure we can remove the highly scattering regions and leave the less scattering material it is then probable that the material of lesser scattering ability will appear positively as a more highly scattering and definite structure. We have made attempts to find out more about the architecture of cytoplasm by floating thin sections on enzyme solutions and on suitable diluent controls (Plate XII, Figs. 51–55). Specific conclusions are not yet possible. However, it is certain that papain, ribonuclease, and deoxyribonuclease all lead to a dramatic reduction in the density or electron scattering of the cytoplasm of several sorts of cells. This may be an indication of the impurity of the nucleases. But it was further noted that flotation of sections on water, saline, and various buffers led to a partial extraction of cytoplasm and the revelation of a somewhat different architecture. This may indicate a means of gaining a slightly different knowledge of the scaffolding on which the granules and other structures of the cytoplasm are situated. So what is then evident as a different order of structure may actually represent something that was in fact there before but was not apparent because its relative scattering was of a low order. This allows of an approach that may have some promise for the future and the principles involved should be remembered when interpreting electron micrographs of sections.

The granular cytoplasmic structure, which is so dense in the logarithmic phase, becomes less so toward the end of the growth cycle and in the phase of decline. It is true that at this stage the cytoplasm may be diluted by the accumulation and increase of reserve substances in the form of inclusions or vacuoles, but in between these structures the fine granules are still visible, although spaced apart and not so numerous. This accords with the change in staining properties of bacteria in the various phases of growth and in the often observed decline in the concentration of ribonucleic acid extractable from the cell cytoplasm. Bradfield observes that the granules in older cells are distinctly larger, approaching 300 A., but no one seems to have made more than a cursory study[68] of the changes of fine structure during the growth cycle.

The cytoplasmic granules are very similar in various kinds of cells among the Eubacteriales. However, Brieger and Glauert[69] think that the cytoplasm of the avian tubercle bacillus contains large fairly dense granules of the order of size 300 A., giving a pattern that differs from that of other bacteria in their experience.

Nearly all the observations on thin sections of bacteria have been from material fixed with osmium tetroxide, dehydrated with alcohol, and embedded in methacrylates, and the possibility exists that to some extent the architecture of the cytoplasm might be an artifact of embedding or fixation. However, it may be noted that granules of the same order of size and distribution are to be found in sections of bacteria from embeddings in not only methacrylate but also epoxy resins[70] and polyester resins.[71]

It would also be of the greatest interest to be able to follow the genesis and fate of these granules throughout the life cycle of a single organism. This does not seem to have been done by intention. A situation exists in which it would be possible to do a useful study and some indications of possibilities already appear in the literature. For instance, the fine structure of the spore cytoplasm that is the central body of the resting spore is of comparatively low density and is remarkably homogeneous and free from recognizable granules. When spores are followed in germination, granules suddenly appear within remarkably few minutes in the spore cytoplasm.[72] These increase rapidly in number up to the first vegetative cell, when the cytoplasm looks not greatly different from that of cells later in the growth cycle. With adequate techniques and preferably with concomitant biochemical fractionation studies it should be possible to arrive at some useful information. This is the period, as shown by Fitz-James,[73] of the most rapid ribonucleic acid synthesis.

Although these cytoplasmic granules are a remarkably constant finding in a variety of bacteria and are fairly constant in form, they are on occasions hard to see. This may be due to concomitant scattering from diffuse materials distributed through the cytoplasm leading to obscuration or a lack of resolution within the cytoplasm. We have observed this on occasions in sections of *Sarcina* and a few other organisms. Interestingly enough, in some of these cases flotation of the sections on saline or on some other diluents have led to some change which has revealed the aforementioned fine structure not visible before the treatments. Such technical difficulties as these are common in electron microscopy.

4. Membranous Systems and Organelles

Experience with plant and animal cells has led the microscopist to expect adumbrations of the cytoplasmic membrane that penetrate the cytoplasm—complex membranous organelles, such as the mitochondria and chloroplasts, special membrane systems, such as the Golgi apparatus and the nuclear membrane. When these systems are modified or are elaborated within special cells they become matters for intensive research; when these systems are absent a far-reaching comparative study is required. The hope is that the study of such systems and their modification will allow some understanding of their nature and their genesis, as well as contribute to taxonomy.

Until recently it would have been fair to state that membranous organelles were conspicuous by their absence. However, recent work indicates that membranous systems are not completely lacking in bacteria; this is based on three sets of observations: (1) on photosynthetic bacteria, (2) on isolated observations of cytoplasmic structure in heterotrophic bacteria, and (3) observations on *Thiovulum majus*.

Photosynthetic bacteria contain pigments that are essential to this function. Analogy with plant cells suggests that the photosynthetic pigments may possibly be organized into discrete subcellular units such as the chloroplast, which is now known to have a particular and peculiar construction. The first indication that photosynthetic pigments in a bacterium were contained in organized units was given by the fractionation of cytoplasmic particles derived from *Rhodospirillum rubrum*.[67, 74] All of the photosynthetic pigments were contained in a particle fraction with a sedimentation constant of approximately 190 S and a size approximating 1100 A. as determined by electron microscopy of dried preparations. These pigment-bearing particles were called chromatophores. The presence of such units in the cytoplasm of *R. rubrum* have been most convincingly confirmed by the observations of Vatter and Wolfe.[75] They found that the cytoplasm was packed with discrete units varying in size between 500 and 1000 A., which were less dense than the intervening cytoplasm and were surrounded by a membrane. The surprising feature of their distribution is that they appear to pack the entire substance of the cell when it is exposed to light during growth (Plate II, Fig. 10). In dark-grown *R. rubrum* these structures are almost completely absent. The cytoplasm consists of a very finely granular groundwork of particle sizes varying from 80 to 200 A. and containing, as was the case with the light-grown cells, granules of polymetaphosphate and vacuoles that could be identified with the lipid-staining areas of the intact cells.

Vatter and Wolfe also studied sections of *Rhodopseudomonas spheroides*, which exhibited chromatophores of similar construction but slightly smaller size. The green sulfur bacterium *Chlorobium limicola* seem to lack the well-defined vesicular chromatophores found in the Athiorhodeaceae. A prominent feature of the cytoplasm in this case was the presence of many opaque granules 150–250 A. in diameter, which might be the chromatophores of the species. A strain of the purple sulfur bacterium, *Chromatium*, showed small (200–400 A.) chromatophores with morphological features similar to those of *R. rubrum*.

There is no doubt from the micrographs published (see Plate II, Fig. 10) that these photosynthetic bacteria do indeed contain rather special structures that may be identified with the chromatophores. They are synthesized within the cytoplasm but there is no real information on this synthesis, whether it is diffuse or a function of restricted sites such as the cytoplasmic membrane. There is good agreement of the observations on fractions from disrupted cells and the observations on thin sections; both the estimates of size and the estimates of number are of the same order. There is no basis for agreement between the observations quoted above and those of Niklowitz and Drews[76] who described rather gross lamellar structures in the cytoplasm of *Rhodospirillum rubrum*.

Isolated observations of membranous cytoplasmic structures have been made in bacteria from several groups, the majority being in the Actinomycetales and, particularly, the genera *Mycobacterium* and *Streptomyces*. The only publications so far are those of Shinohara,[77] who described a membrane-enclosed structure in the cytoplasm containing "some fine tube-like structures, each of which consisted of two parallel, dense lines separated by a clear space." Although inspection of this paper alone would not be entirely convincing of the existence of such a structure, examination of the original and other prints indicates that it is probably a real cytoplasmic structure. This is confirmed by the experiences of other laboratories with similar but unpublished observations on various mycobacteria and streptomyces as well as in *Bacillus subtilis*. It would appear that these membranous structures consist of concentric rings of light and dense fine layers (Plate II, Figs. 11 and 12) but, despite their observations, they have not as yet hazarded any interpretation. However, Shinohara considers that the structures have a close resemblance to the mitochondria of animal cells and indeed describes them as mitochondria. The difficulty is that these structures do not appear with absolute regularity, according to comparison of the experiences in various laboratories, and their size and form are such that, if they were a regular feature of bacterial cells in general, they would have been found much more frequently than has been the experience up to now. There is little doubt that these are cytoplasmic features and are not to be confused with the chromatin areas.

Thiovulum majus, the remarkable sulfur bacterium of enormous size, has provided in some ways the most interesting evidence of a cytoplasmic system of membranes, particularly stimulating to those interested in comparative cytology. This organism, which can measure as much as 18 microns in diameter, has been studied by Fauré-Fremiet and Rouiller.[78] In this special case the cytoplasm is quite dilute and the cytoplasmic membrane stands out clearly in the sections (Plate IV, Fig. 15). The cytoplasmic membrane is not a smooth, enveloping membrane as is the case with all other bacteria that have been studied in detail so far. Small penetrating invaginations are formed, which anastomose in the superficial region about 0.25–0.5 micron in depth (Plate IV, Figs. 17 and 19). However there are deeper membranous intrusions which penetrate and ramify within the cytoplasm

---

PLATE IV. These micrographs were kindly provided by Professor E. Fauré-Fremiet[78] and Figs. 16 and 18 are reproduced with the permission of the publishers of *Experimental Cell Research*.

Fig. 15. Micrograph of a section of *Thiovulum majus*. The relatively dilute cytoplasm (compare with Plate II, Fig. 7) contains large spaces corresponding to the sulfur inclusions (removed in processing) characteristic of the organism. These spaces are bounded by a fine interfacial membrane. There is a superficial elaboration of the cytoplasmic membrane into folds and occasional intrusive invaginations forming intracytoplasmic double membranes. (×13,600)

Fig. 16. A section across the "organite antapical" showing the mass of parallel fibers and the disposition of double membranes in its vicinity. (×41,000)

Fig. 17. A view of the complex cytoplasmic surface. (×41,000)

Figs. 18 and 19. Intracytoplasmic double membranes showing the local concentration of granules which seem to be rich in RNA. (×83,300 and 28,000, respectively)

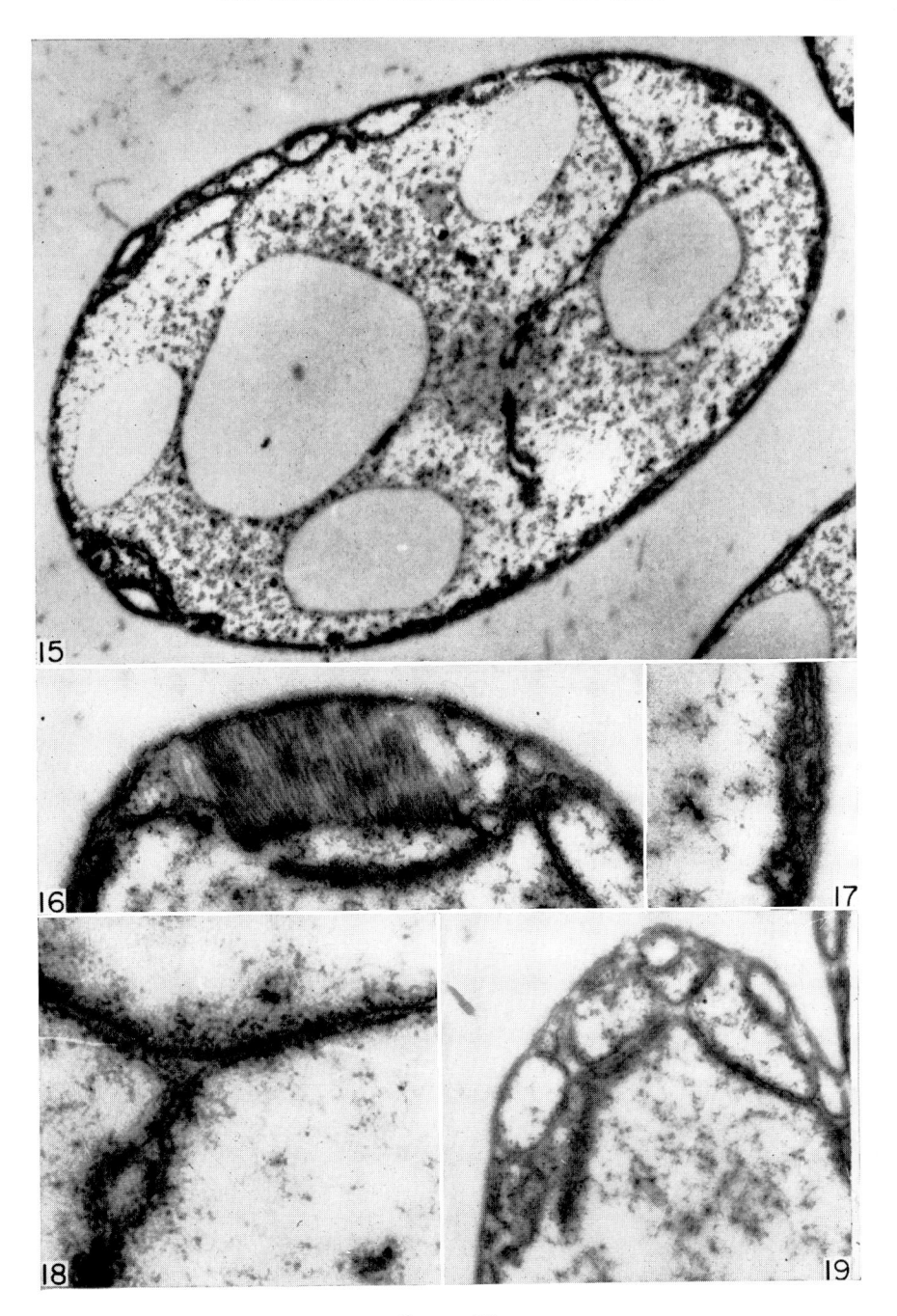

PLATE IV.

and these seem to be closely apposed layers of cytoplasmic membrane, showing as two dense lines separated by a low density space (Plate IV, Figs. 15 and 16). There do not seem to be complex outpouchings or vesicle formations from these lamellae, but the surfaces presenting to the cytoplasm either are sprinkled with fine cytoplasmic granules or have clots of attached granules, giving an irregular appearance to that surface (Plate IV, Figs. 18 and 19). In stained preparations the whole cytoplasm seems to be compartmented by basophilic planes, giving the cell an alveolar look which is heightened by the refractile sulfur inclusions. Without doubt the granule-studded membranes contribute to this appearance.

There is one other great peculiarity in the cytoplasmic structure of this organism and that is the "organite antapical," which has a strong affinity for acid fuchsin and forms a plaque about 1 micron $\times$ 0.3 microns at one pole of the cell. It is easily spotted in the light microscope as a consistent polar organelle. Thin sections show a bundle of short parallel fibrils arranged perpendicularly to the surface of the cell (Plate IV, Fig. 16). Each fibril is about 100 A. in diameter. They terminate peripherally in contact with the internal face of the cytoplasmic membrane, which seems more complicated in this region. They terminate internally about 0.3 microns from the surface on a fine basement membrane which seems to enclose the whole structure.

Thiovulum shows, then, two extraordinary features. The one is an elaboration of the cytoplasmic membrane of the surface to form a cell membrane system with some of the characters of an "endoplasmic reticulum." The other is a cytoplasmic organelle of unknown function and affinity, although it has been suggested that it may have some function in the peculiar motility of this organism as a sort of pulsatile organ.

Thus it would appear that some bacteria have membranous or fibrous systems within their cytoplasm. There is obviously much more to be learned of the extent of these systems and of their possible variations among the bacteria. Further search for more illuminating examples and careful analysis of their features and characteristics will be of inestimable value to comparative cytology.

5. Some Difficulties and Perspectives

A general concept of the organization of bacterial cytoplasm has been brought almost within our grasp by the recent and stimulating approaches afforded by electron microscopy. However, there still exists a considerable gulf between the actual and potential accomplishments of light microscopy and electron microscopy. At this time the difficulty is that one or the other approach tends to be used exclusively; it would be most desirable if the investigation of cytologically rewarding objects could be in fact combined approaches. Within the limits of our present technical methods there remains much to be done, even on the simplest plane, about the correlation of cytoplasmic components with components identifiable in micrographs of sections.

The choice of suitable test objects is probably the most important factor determining rewarding returns in cytological studies. Comparatively few bacteria are of a size and conformation suitable to the separation and identification of closely set components. Some compromises will probably be worthwhile, such as the selection of particular species or strains that are rich in one particular component or have it situated in a very characteristic position allowing identification or location in all types of preparations. An example of this approach may be afforded by a peculiar tetrad-forming coccus that has been studied in this laboratory whose cells, in some clones, have a remarkable number of polysaccharide inclusions that are recognizable by the periodate-Schiff reaction. These are arranged peripherally in the cytoplasm, which is remarkably free from other identifiable inclusions, such as lipid and metaphosphate. Micrographs of sections are remarkably disappointing but they do show in the corresponding regions of cytoplasm circular areas of extremely low density. If the sections are floated for a few minutes on a 1 % solution of lanthanum nitrate these low density areas show a remarkable uptake and change to an area of high density. Exposed in a similar fashion to a solution of phosphotungstic acid the center of these areas is not stained but the periphery shows a considerable uptake so that it is ringed by a dense line. One may wonder then, without the possibilities of a formal conclusion as yet, whether this represents a particular type of organization for inclusions of this nature with a polysaccharide center surrounded by a layer of protein.

This emphasizes one of the drawbacks of electron microscopy in cytological studies and that is the lack of any really specific cytochemical approaches to determining the nature of the structures under study. A number of laboratories have commented on the usefulness of various metal salts and other treatments after fixation for the increasing of contrast in electron microscopy,[79] but little attempt has been made to find out what the specificities of these reactions may be. The general possibilities of such approaches are only just being realized. At any rate there is some sort of pseudospecificity involved in these reactions, since some portions of structure are "stained" whereas other portions are not, and this is reproducible and regular (see Plate VII, Figs. 39 and 40). Undoubtedly we may expect the development of more useful techniques as well as some understanding of those already in use.

There may be some use, as has already been mentioned, for an approach involving the specific removal of cellular components by a biochemical extraction or by enzyme action. This has of course been of the greatest importance in the use of light microscopy and staining but the possibilities have hardly been explored in the case of electron microscopy. It may be possible to demonstrate an organized fine structure in cytoplasm by the removal of components of the all-pervasive granular cytoplasmic structure. Only a segment of the vast variety of bacteria has been studied up till now and almost infinite possibilities are to be expected in the major groups that remain to be surveyed.

A variety of methods of fixation are available for the making of preparations for light microscopy; these are fairly easily controlled by comparison of the shape and form of both the entire cell and of inclusions within the cell with what can be discerned in living cell by direct microscopy and phase microscopy. However, no such variety of approaches has yet been possible for electron miscroscopy, for which approach osmium tetroxide has not yet been surpassed as a general fixative. The only criterion available for determining the suitability of fixatives for this purpose is the apparent degree of retention not only of shape and form but also of integrity of the cellular structure. Unfortunately this is a very subjective approach and full of pitfalls for the unwary, because the processes of dehydration and embedding that must follow can also cause shrinkage or damage to structure that may be confused with fixation damage or vice versa. If these damages are not discernible in the light microscope there can be little hope of determining what is going on except by most laborious and difficult factorial experiments. Embedding damage, particularly the so-called "explosions," is easily recognizable in its worst form but minor orders of local damage are extremely hard to spot. Fortunately there are several laboratories that have been experimenting rather thoroughly with alternative embedding procedures as well as with modifications or variations of the dehydration phase of preparation, and promising alternatives are being explored. As experience with these new media is accumulated we will have a greater variety of material for comparative study and a consideration of the possible range of artifact, whether on a gross or ultramicroscopic level.

III. Chromatin Bodies

A. General Characteristics

The demonstration of a chromatin moiety within the bacterial cell, usually an easy procedure nowadays, was difficult and elusive in the early days of bacteriology. Yet, in spite of and possibly because of the difficulties, the chromatin bodies of bacteria have been the subject of more intensive studies and varieties of interpretation than any other structure of the cell. In accordance with biological tenets and experience it was to be expected that the bacterial cell should contain a definite structure that could be identified as a nucleus.

The relatively small size and the apparent homogeneity of the living cells was discouraging. A more serious hazard was the general experience that standard techniques of fixation and staining, perfectly satisfactory in the study of plant and animal cells, failed to distinguish nucleus from cytoplasm at all regularly in the bacteria. Furthermore, when chromatin was demonstrable, the range of patterns described by one author was usually completely different from those of another using an identical organism. Understanding awaited the realization that the chromatin conformation varied with the stage of the growth cycle, was susceptible to changes in the physiological state of the organism and changes in its environment, and that the accumulation of cytoplasmic inclusions could crowd and deform the chromatin.

There is no doubt that the chromatin bodies give some of the histochemical reactions that are associated with the chromosomes of higher organisms and that the chromatin bodies divide in step with and a little ahead of the division of the cell, so that portions are distributed between the daughter cells. What is now accepted was slow in development because of the difficulties in demonstrating a nuclear substance using the accepted techniques. It is not now surprising that the opinions of respected observers ranged all the way from no nucleus to a diffuse nucleus encompassing virtually the whole cell. So an essential phase of the history of bacterial cytology has been the development of methods that regularly demonstrate the chromatin bodies and preserve them in a form consistent with what can be observed in the living cell with phase microscopy. The milestones in this development of effective means of observation, changing directions, and shades of interpretation have been chronicled along the way by many outstanding contributors.[1, 4-7, 10, 18, 34, 80-83]

The majority of those who have studied the chromatin of bacteria have been content to consider that their efforts are part of the *demonstration* of a bacterial nucleus; they were not in general markedly concerned with the nature and structure of this nucleus and its relation to all other kinds of nuclei. The fact is that a nucleus is a cellular organelle with distinct character and is not a relatively undifferentiated lump of chromatinic material. This was put forward particularly forcefully by Dobell[34] as well as by Delaporte.[1] It was realized that the discernible nuclear element in bacteria must represent a "primitive" or "simple" form of nucleus and modified terms came into being, such as the "éléments nucléaires" as used by Delaporte and the "nucleoids" of Piekarski. However, in many bacteria and particularly in young, actively growing forms of some of the common bacteria, the chromatin is arranged in rather neat packets and it was perhaps not unexpected that the term "chromosome" should be used for them by both Badian[84] and Robinow.[5] The implication was that the demonstrable bodies were either nuclear equivalents or the equivalents of definite parts of a nucleus.

A critical review of the approaches to the unraveling of its structure and behavior, and a consideration of the comparative cytology of the nuclei of protists has been put forward by Robinow.[83] This appreciation of the current state of knowledge and understanding of the chromatin bodies of bacteria is most valuable in that it pulls together the divergent observations of the many workers and unifies them in a general consideration of the attributes. The conclusion (and it appears to be the most reasonable for the moment) is that the chromatin bodies of bacteria are unlike the chromosome nuclei of higher organisms and can only be related in kind to the nuclei of blue-green algae and of certain protists in which the chromosomes are not obvious. These all share a remarkable plasticity of shape, uniformity of texture, and staining reaction throughout all cycles of division, have a structure that is still obscure, and divide directly. Thus it is probably undesirable for both practical and semantic reasons to speak of bacterial nuclei or chromosomes but rather to use some neutral term which allows of no particular preconceived notions or emotions. "Chromatin body" has been used with this purpose and seems to be both sufficiently descriptive and neutral to fill these requirements, although it is a literal translation of "chromosome." A really suitable alternative is not available and an ill-chosen term would only increase the confusion.

The chromosomes of higher organisms were called chromatinic because they were rapidly and deeply stained by basic dyes and by the time-honored but relatively nonspecific hematoxylin, which have been the mainstays of more than a century of cytology. It is a peculiar fact that the stainability of the chromatin body varies with the phase of the growth cycle of the chosen bacterium and that in the stage of most active growth it is rare, almost phenomenal, to be able to stain the chromatin body directly with simple stains. Guilliermond[85] was only able to distinguish the cytoplasm and the chromatin of *Bacillus* cells in cultures older than 8 hours using iron-alum hematoxylin. This sort of experience has been constantly repeated up to the present day including the remarkable lack of affinity of the chromatin bodies for the stains most commonly used in the study of chromosomes, such as acetocarmine and methylgreen.[86] The fixed but otherwise untreated cell is easily stained except for the chromatin bodies, which thus stand out as unstained areas within a stained cytoplasm (Plate V, Figs. 21 and 22).

This primary difficulty has been avoided in many different ways. The most useful and one of the most difficult techniques for the direct staining of chromatin is the application of the Romanovsky series of complex stains, particularly Giemsa. This was effectively applied by Dobell[34] and a number of others since then, in particular Badian.[84] However, the technique is capricious and dependent upon artful differentiation, which does not seem to have been easily adopted by all those who would wish to do so. Attempts have been made to find the conditions necessary to direct staining with Romanovsky dyes.[35, 87]

Difficult as it may be to demonstrate the chromatin bodies by direct staining, they are very easily demonstrated following simple treatments. These stem from the work of Piekarski,[88] who demonstrated the dividing chromatin bodies using the Feulgen reaction. He also showed that chromatin bodies can be demonstrated in the cytoplasm of *Salmonella* after treatment with N HCl at 60° C. by staining with Giemsa's stain and that cells of this organism are uniformly stained by Giemsa if the preliminary treatment in acid is omitted. This approach was first used systematically by Robinow,[5] who took advantage of the fact that the HCl-Giemsa method not only stained the chromatinic structures more deeply than the Feulgen process but showed at the same time the outlines of the bacteria as well as their internal cell boundaries. This approach was important because it allowed a regular demonstration of chromatin bodies in bacteria at any stage of the growth cycle and particularly in the stage of most active growth (early logarithmic stage), when the cytoplasm has an extraordinarily high affinity for stains. This led to the demonstration by Vendrely and Lipardy[36] that a large proportion of the cytoplasmic basophilia of bacteria was due to their content of ribonucleic acid and that the pretreatment with hot N HCl removes all or nearly all of the ribonucleic acid within a few minutes. Tulasne and Vendrely[89] showed soon afterwards that the cytoplasmic basophilia could be removed equally effectively by treatment of the fixed preparations with ribonuclease and that this also formed a suitable pretreatment for subsequent staining. Many additional preparative techniques have been described, e.g., extraction with cold perchloric acid[90] and even with dilute solutions of sodium chloride after fixation with methanol-formalin.[91] In fact, any treatment that extracts ribonucleic acid from the cytoplasm simplifies or enhances the staining and differentiation of chromatin bodies (Plate V, Figs. 22–26). The Feulgen reaction and its most useful derivatives, using reduced thionine or azure A[92] instead of the Schiff reagent, remain the mainstay for the identification of the chromatin body.

The stainability of the chromatin body after these various pretreatments, however, is not the same in all cases and again demonstrates some peculiarity of the chromatin of bacteria. After the Feulgen hydrolysis the chromatin can be stained with almost any basic dye and it is stained intensely with Giemsa's solution, but in this case it stains purplish or bluish rather than the classic cherry red of the chromatin of chromosomes or, for that matter, of the chromatin of bacteria *directly* stained with Giemsa. After ribonuclease treatment or extraction of suitably fixed cells with neutral salts, the chromatin bodies are *not* easily stained with simple basic dyes, such as thionine or methylene blue, but they are readily stained with Giemsa or other Romanowsky stains. This underlines a point that has already been made: the chromatin of bacteria, particularly of young cultures, does not have quite the staining properties of classic chromatin and on this basis alone could be considered to have some peculiarity of chemistry and of intimate structure.

Once the general position and arrangement of chromatin bodies had been established in the fixed and stained cell it was possible to look for traces of their presence in the living cell. The cocci and rod-shaped bacteria that have usually been studied are remarkably homogeneous during life but they are not entirely featureless. Centrally placed lighter areas may be observed and these can be markedly enhanced by examination with phase-contrast microscopy (Plate V, Fig. 20). It soon became obvious that these lighter areas coincide with the position of the chromatin bodies of stained preparations. In the majority of cases the resolution of internal detail was very limited. Nowadays the requirements are much more refined and demand sufficient detail for comparison with images obtained from stained cells and from sections examined in the electron microscope. Several good attempts have been made with the phase microscope and the most excellent detail has been obtained by Mason and Powelson,[93] who mounted *Escherichia coli* and *Bacillus cereus* in media of high gelatin content. The chromatin areas are visible and correspond in remarkable detail to the very best of fixed and stained preparations (Plate VIII, Fig. 42 and Plate IX, Fig. 43). Thus the

---

PLATE V. Photomicrographs of *Bacillus cereus*. All preparations grown from spores for 2½ hours at 37°C. on tryptone agar.

FIG. 20. Appearance of living cells using dark phase contrast (without gelatin mounting medium). The central cells show the light areas corresponding to chromatin bodies. These are quite voluminous and constantly change in form. The cells also show the dark line of a developing cell septum.

FIG. 21. Cells fixed through the agar block with Bouin's fluid and stained with thionine. The cytoplasm is strongly basophilic; the chromatin areas are not. The disposition of the chromatin *areas* corresponds closely to the phase picture with this fixative. Cell wall is unstained. Chromatin is not stainable after this fixation.

FIG. 22. Cells fixed with OsO_4 vapor (1½ minutes) then stained with thionine. The nonstaining chromatin shows as clefts in the cytoplasm.

FIG. 23. The same cell group as Fig. 22 after treatment with *ribonuclease* and staining with Giemsa. Close correspondence with the previous picture is notable.

FIG. 24. Cells fixed with OsO_4 vapor and stained with Giemsa after Feulgen hydrolysis. Many division patterns of the chromatin and chromatin bridges may be noted.

THE INTERNAL STRUCTURE OF THE CELL 69

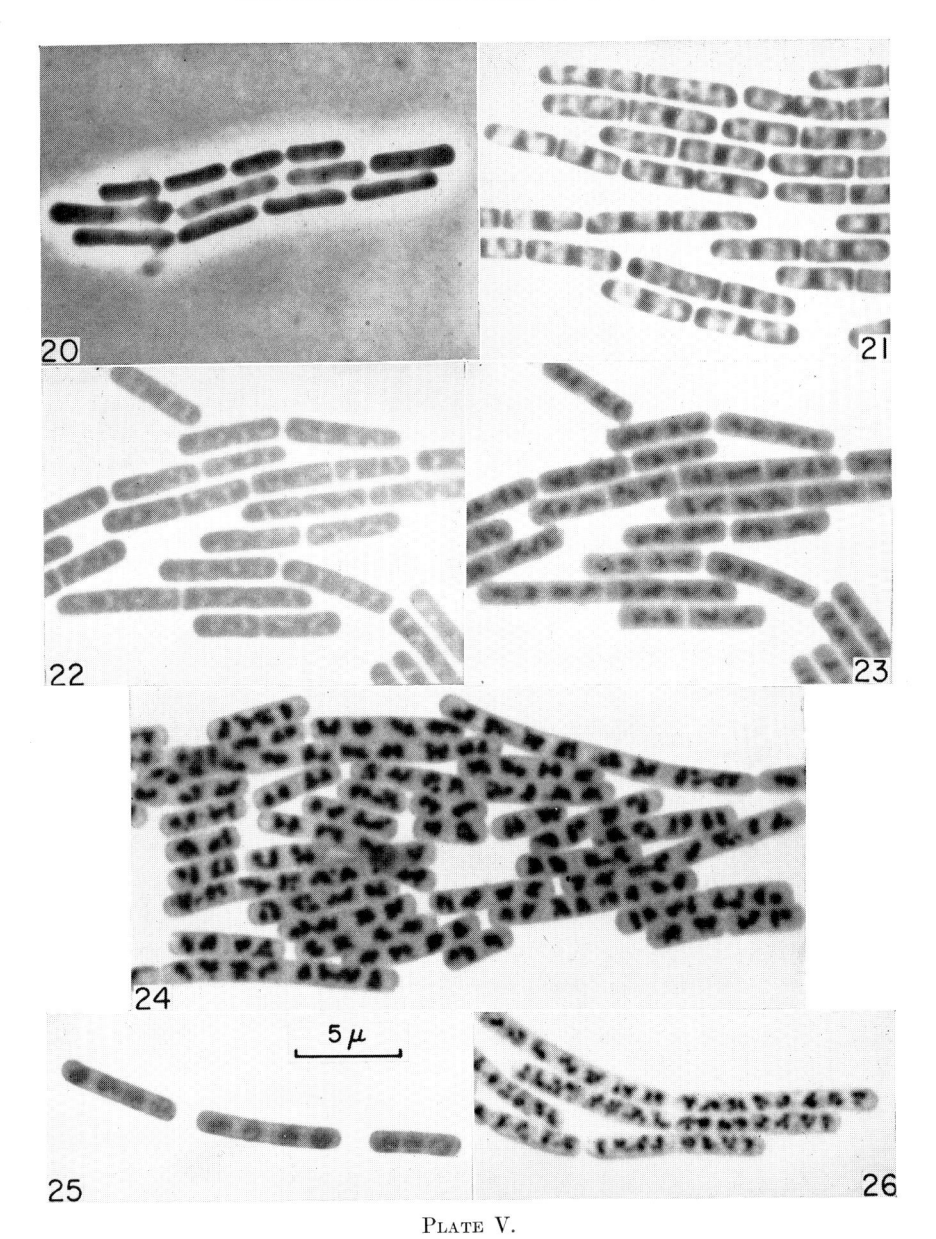

PLATE V.

PLATE V (con't.). FIG. 25. Cells fixed with methanol-formalin-picric acid, treated with N HCl at 60°C. and stained with Giemsa. This shows the appearance suggestive of a vesicular nucleus and a nuclear membrane.

FIG. 26. Cells fixed with OsO₄ vapor, treated with N HCl at 60°C. and stained by the azure-SO₂ method. A more delicate chromatin pattern than that in Fig. 24.

control and recognition of artifacts produced during the fixation and staining processes has been much improved. Here again the technique is not necessarily suitable for all stages in the growth cycle and it is generally true that phase-contrast microscopy is the most helpful in the study of young cultures. It is now the experience of many laboratories that it is possible to watch the division of chromatin bodies and to see at the same time, in suitable organisms, the relation of these divisions to the production of cell septa and the division of the cell. Cultures that are entering a phase of decline usually show cells that are more replete with reserve substances, droplets of lipid, and inclusions whose refractility blinds the observer to other differences within the cytoplasmic structure. There are limits to the utility of these techniques for observing the living cell but it may be expected that they will become more and more useful (together with ultraviolet microscopy, vital staining, and fluorescence microscopy) in providing the bridge between the living cell, the preparation for optical microscopy, and the preparation for electron microscopy.

The chromosomes of the nuclei of plant and animal cells can be effectively demonstrated by staining methods designed for the demonstration of basic protein, but these are not effective in the staining of the chromatin bodies of most bacteria. Some positive results have been obtained in the case of *Bacillus megaterium* with the ninhydrin-Schiff reagent but the chromatin bodies of *E. coli* gave negative results with the same reagent.[38] Another technique was effectively applied to *B. megaterium* allowing sequential staining of deoxyribonucleic acid (DNA) and the protein.[94] The DNA and the protein occupied the same site, but staining of the protein component required some disruption of the complex. These observations and the apparent lack of protamine or histone in bacterial nuclei[95] and the unusual properties of nucleoprotein from tubercle bacilli[96] strongly indicate the possibility that bacterial nucleoprotein may be of unusual nature. This possibility has already been voiced by both Robinow[83] and Widra.[38] However, there is remarkably little chemical evidence to give substance to this suspicion. We may hope that the techniques now available for the separation and isolation of bacterial chromatin bodies[59] will provide the opportunity for more detailed chemical and cytological studies of these structures.

One must conclude that the staining of chromatin bodies of bacteria is not directly equivalent to the staining of the chromosomes of higher cells. The common ground is that both chromosomes and chromatin bodies give a positive result with the Feulgen test and they give a bright red color with Giemsa's stain when this is applied directly, without pretreatment. However, there are major differences, because bacterial chromatin bodies do not stain at all regularly with hematoxylin, acetocarmine, or aceto-orcein, nor with the protein stains and methyl green,[86] which have been so reliable and effective in the staining of chromosomes. All the effective staining methods show one further and important distinction between chromosomal nuclei and bacterial nuclei. There are undoubtedly sequential changes in the arrangement or conformation of the chromatin in bacterial cells during division but throughout the cycle there are no changes in the texture or tinctorial qualities of the chromatin. This is a very distinct difference from the sequential changes that are so well known for the *chromosomes* of the vast majority of nuclei.

There is a general similarity in the chromatin bodies of all kinds of bacteria, and they form a single class of structure.[83] This extraordinary but useful generalization validates the necessity of discussing details of organization in terms of only a very few representative species.

B. Arrangement and Effects of Environment

The chromatin bodies are centrally placed and divide directly in the axis of division of the cell. The result is that many paired structures are seen in actively growing cells and, in rod-shaped bacteria, 2–4 or more such paired structures may be found within the confines of an individual cell (Plate V, Figs. 24 and 26; Plate VIII, Fig. 42; Plate IX, Fig. 43). However, all degrees of variation in chromatin structure have been illustrated over the years, from compact round bodies to the dense and slightly contorted axial filament, and from groups of short rodlets arranged mainly at right angles to the long axis of the cell to an almost disorderly arrangement of granules dispersed throughout the cell. The effect on the biological observer, apparently, has been one of increasing confusion.

The bacterial cytologists of today have in many cases chosen either to ignore the earlier observations or to attribute them to errors of technique. The fact is, as has been pointed out by Robinow[83] in most useful and understandable terms, that these apparent variations in the structure and form of the chromatin bodies can be reconciled and understood to a considerable extent if account is taken of the age and physiological state of the culture being observed and of the conditions applied to that culture in fixation and preparation. It is possible to reproduce without any great difficulty practically all the variations in chromatin conformations that have been described and we can now have some understanding of the factors that are involved.

Decision has to be made on the arrangement and structure of chromatin that is the most lifelike for cells in a given physiological state. This is no easy task and is certainly far from complete, but phase microscopy now provides the most useful control of methods. It is fairly certain that the methods of fixation with osmium vapor followed by hydrolysis, staining, and suitable mounting give a quite reasonable reproduction of the nuclear area as it can be defined in the very best of phase microscopy *when applied to suitable test objects*. This is not the only means of preparation and various, appropriately chosen, fixation methods have great value.[82]

Two general categories of chromatin arrangements can be recognized in the many descriptions. On the one hand, there are the neatly arranged constellations of a few interconnected granules forming bars, V, H, and butterfly shapes (Plate VI, Figs. 27, 29, and 31); on the other hand, there are the open arrangements where chromatinic granules seem to be enmeshed in a basket work of fine chromatin strands.[82] In some of these the chromatin seems to be almost peripherally arranged in the cell and the analysis of division figures is particularly difficult. The latter sort of arrangement has been observed in a variety of bacteria, such as *Azotobacter* species (Plate VI, Fig. 33) and *Sarcina ventriculi* (Plate VI, Fig. 32);[97] also among some members of the enteric group (Plate VI, Fig. 30), which usually have rather neat chromatinic arrangements. *Shigella dysenteriae* has shown quite regularly a complex pattern of chromatin granules and interconnecting bars and threads which is most difficult to analyze (Plate VI, Fig. 28). A more important example is *Escherichia coli* strain C, in which the granules of chromatin are always peripherally arranged and form an example of the open type of arrangement. Genetic recombinations can be effected between this strain and other strains of *E. coli*, notably K-12 which has a compact type of chromatin arrangement (Plate VI, Figs. 29 and 30). The effectiveness and viability of the hybrids between these two types indicate that the difference between the types of chromatin constellations is one of degree and not of kind, and the distribution of types among the recombinants would indicate that the chromatin arrangement itself is a genetic character.[98]

When active growth declines in a culture the chromatin figures usually become more compact and sharply outlined and, when there are multiple chromatin bodies within the cell, there is a tendency for them to condense, to fuse or to show chromatin "bridges." In rod-shaped bacteria it is com-

---

PLATE VI. A series of photomicrographs to show the structure and arrangement of chromatin in a variety of bacteria.

FIG. 27. *Escherichia coli* strain B.
FIG. 28. *Shigella dysenteriae* strain Sh.
FIG. 29. *Escherichia coli* strain K-12.
FIG. 30. *Escherichia coli* strain C. (Figs. 27–30, photomicrographs by Dr. J. F. Whitfield.)
FIG. 31. An undescribed tetrad-forming coccus.
FIG. 32. *Sarcina ventriculi*. (Figs. 31, 32, photomicrographs by Dr. C. F. Robinow.)
FIG. 33. *Azotobacter vinelandii* grown on blood agar, causing a great increase in size. (Photomicrograph by Dr. J. P. Truant.)

THE INTERNAL STRUCTURE OF THE CELL 73

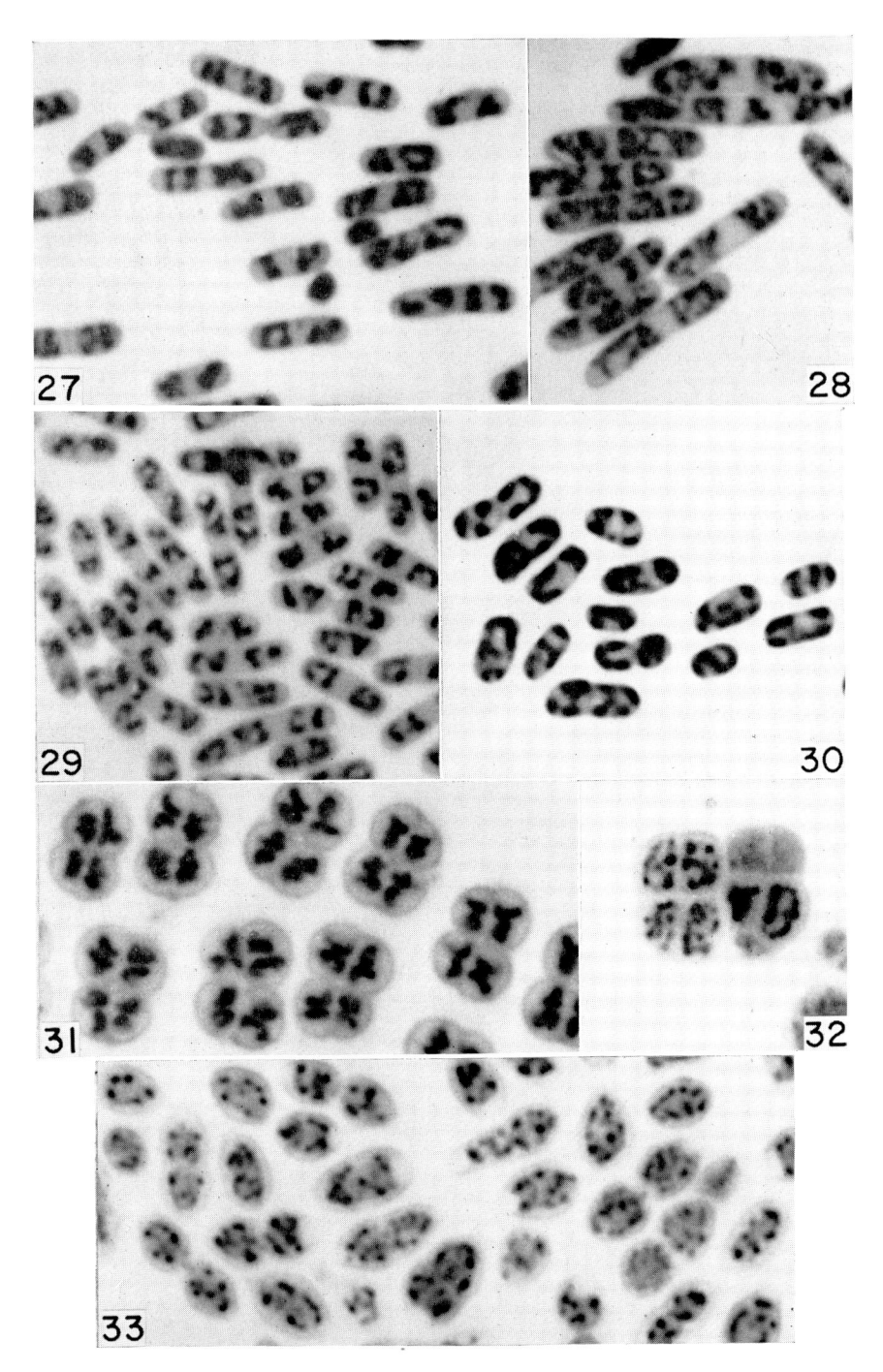

PLATE VI.

mon to find that the chromatin assumes the form of an axial filament (Plate VII, Fig. 38). It can be shown that this is a derangement due to the physiological environment and state of the cell by transferring the organisms to fresh nutrient medium. Within a few minutes the axial filaments disaggregate into two or more separate chromatin constellations of the usual kind. *Bacillus* species, which form suitable test objects for this cytological experiment, will usually show axial filament formation in regions of confluent growth on agar plates after 5 to 7 hours of incubation, and the process can be hastened by making the medium more alkaline.

Similar changes in chromatin configuration can be induced in bacterial cells by a vast variety of treatments leading to a metabolic interruption, including exposure to irradiation, cold, and metabolic poisons, such as antibiotics and dinitrophenol. The mechanism, under these various conditions as well as during actual cessation of growth in crowded cultures, would seem to be concerned with the cationic environment and the metabolic capabilities of the cell dealing with sodium and other common cations in the surrounding medium.[99] Any interference with metabolic capability disturbs the equilibrium between the inside and outside environment. The living cell, whether complete or a protoplast, responds (Plate VII, Figs. 34–36) to changes in the ionic environment.[94]

The chromatin bodies behave as an anionic gel; the state of aggregation or dispersion is determined by the availability of cations in the protoplasm and the cationic equilibrium between the interior of the cell and the environment. The formation of condensed chromatin structures or axial filaments is prevented if the environment of the cell is deficient in cations.

PLATE VII. FIG. 34. *Bacillus cereus* grown for 2½ hours on heart infusion agar containing 1% NaCl. HCl-Giemsa. Osmium vapor fixation.

FIG. 35. Companion cells to Fig. 34, fixed 5 minutes after *transfer* to a similar medium containing 3% NaCl, showing aggregated chromatin bodies.

FIG. 36. Companion cells to those in Fig. 35, fixed 30 minutes after the transfer to 3% NaCl medium, showing the recovery of normal pattern. (Figs. 34–36 are reprinted[99] with permission of the *Canadian Journal of Microbiology*.)

FIG. 37. Cells from a companion HCl-Giemsa preparation to Fig. 35. It shows the accessory chromatin structures that are revealed when the chromatin bodies are aggregated by salt treatment.

FIG. 38. *Bacillus cereus* grown on tryptone agar for 8 hours. An HCl-Giemsa preparation to show the natural development of axial filaments of chromatin.

FIG. 39. *Caryophanon latum*. Micrograph of a section showing a cell wall of low contrast and two components in the chromatin region.

FIG. 40. Companion section from same block as Fig. 37 but treated by flotation on 1% lanthanum nitrate. The cell wall shows considerable uptake as does the dense component in the chromatin region.

FIG. 41. Section of *Caryophanum latum* from a 48-hour culture showing a ringlike chromatin body incorporating a core with the structure of cytoplasm.

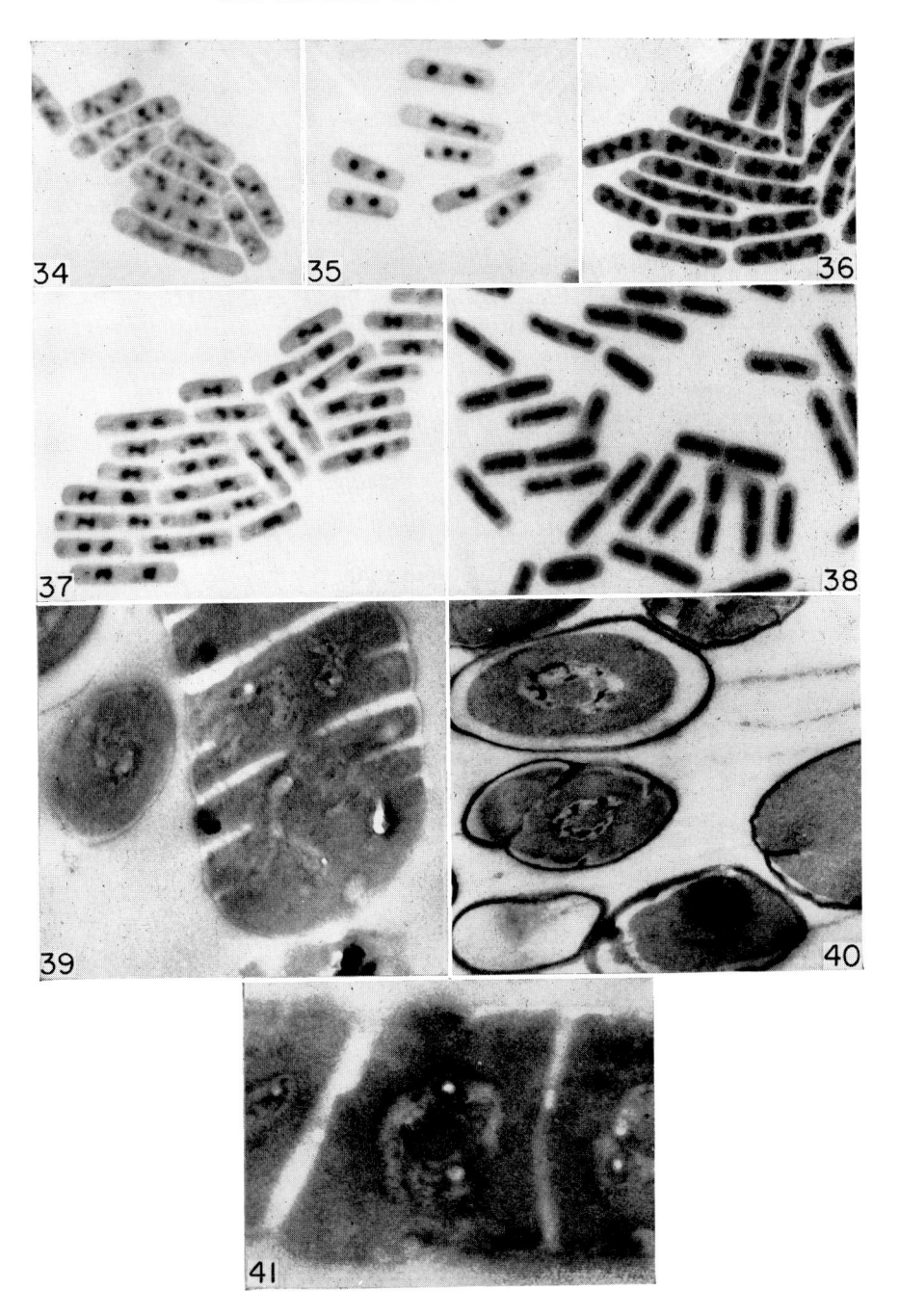

PLATE VII.

Realization of this reactivity is very helpful in understanding the variant views of chromatin arrangement. For instance, Delaporte[1, 6] has formed the conviction that the chromatin in all rod-shaped bacteria has the form of an axial filament. However, the majority of the cultures that she examined were of an age where axial filament formation can be expected and our experience is that this conformation is not the rule in actively growing cells, whether observed in the living state or in suitably fixed and stained preparations. It is further important to realize that the chromatin bodies tend to shrink or aggregate in *all* situations which impair the normal osmoregulatory mechanism *provided that* the environment is hypertonic. This is true not only of the situations mentioned but also of what happens during fixation. Dispersion occurs in an hypotonic environment.

Osmium fixation undoubtedly kills cells very rapidly, and almost immediately abolishes the semipermeability of the surface membrane. The result is that fixation in an ill-chosen cationic environment leads almost immediately to the establishment of a new cationic environment inside the cell. Some experiments have indicated to us that the chromatin bodies of *B. cereus* are still subject to aggregation when the cells are placed in an environment containing 3% NaCl despite "fixation" in osmium vapor for from 10 to 20 minutes. This indicates that the apparent structure of the chromatin bodies can be changed *during* the fixation process if the environment is not suitable. This is of the greatest importance in fixation for electron microscopy, but it must also be considered for the structures observed with the light microscope. It is common practice to put 0.5–1% NaCl in media and it would appear[99] that this is a cationic excess as far as the interior of the cell is concerned. Undoubtedly this accounts for the observation[100] that prolonged osmium fixation leads to condensation of chromatin structures and it seems likely that the very neat and compact chromatin bars and rodlets, so often described, owe much of their character to the environment during fixation.

C. Behavior during Division

There is no doubt that the Feulgen-positive chromatin bodies divide during the growth of the bacterial cell, and this was another important contribution of Piekarski's classic memoir in 1937.[88] The ensuing struggle to interpret division figures has focused the attention on (1) the need to understand the structure in order to interpret the observations, and (2) the need for fixed reference points so that the sequence of events can be followed in the order determined by the cell, rather than that dictated by the fancy of the observer. We can now see that these two needs must be satisfied before a reasonable description can be given that will be both satisfactory to the cytologist and useful to the biologist, the biochemist, and the geneticist. Yet the greatest of effort has been directed, disproportionately, to the description and involved interpretation of chromatin patterns.

Two distinct schools of interpretation have developed. One of these proposes a "true mitotic process," comparable to that of tissue cells, with a nucleus that is constructed in a similar fashion, having definite chromosomes, centromeres, spindle apparatus, centrioles, and a nuclear membrane that remains intact throughout division.[81, 101, 102] The other view is that bacterial chromatin bodies cannot be equated to chromosomal nuclei, that they divide directly by an unknown mechanism, and that they are not enclosed within a membrane.[82, 83] The detailed arguments cannot be catalogued here and these references give an adequate entry into the literature.

In essence, the concept of mitosis is based on the interpretation of individual constellations of chromatin granules and threads as chromosomal patterns that can be fitted to the sequence observed in classic mitosis. This arrangement might have had value as a working hypothesis for the design of experiments to test whether this sequence does occur in fact. However, all attempts to resolve this basic requirement have failed to be convincing. Attempts to make use of "mitotic poisons," such as colchicine and antibiotics,[103] were misleading because it was not realized that their use causes the cells to become inept at regulating the flow of cations from the environment and induces *fusion* of already divided nuclei.[99] The same problem undoubtedly plagued the attempts to synchronize cellular and chromatin division by cold shock.[104] Synchronization by some other means and in a carefully selected environment may help to tell the tale. But direct observation of division in living cells, especially with bright contrast-phase microscopy, showing dark patterns on a light background, allows the observer to follow the sequence of patterns many times over and to compare these with the stained preparations. A sequence involving a metaphase plate, spindles, and complex changes in the state of the chromatin would surely be appreciable. Yet the studies of Mason and Powelson[93] provide no such indication (Plate VIII, Fig. 42 and Plate IX, Fig. 43). The weight of evidence and conscientious observation does not support the concept of mitosis in bacteria; indeed, there is a considerable literature that questions the evidence that gave rise to it.[82, 105]

The observations on the behavior of chromatin in the living organism are more in accord with a concept of a direct form of division. This, together with the peculiar tinctorial properties already described and the remarkable constancy of staining and texture, speaks strongly for considering that these are no ordinary nuclei. It would appear that the terminology of classic cytology is not adequate to the needs in this case and can be completely misleading. It is true that, in organisms of a size and conformation

PLATE VIII, FIG. 42.

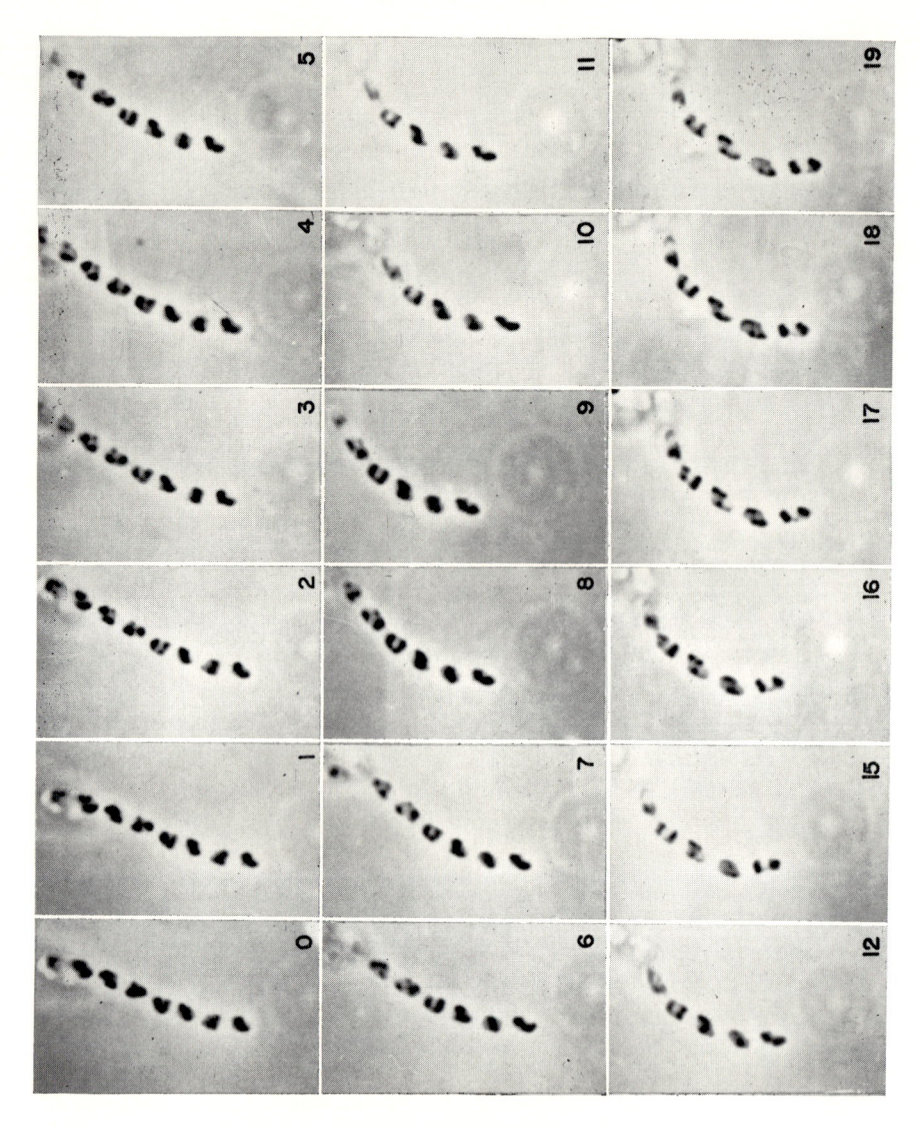

PLATE VIII, FIG. 42 and PLATE IX, FIG. 43. These figures are a series of bright phase-contrast photomicrographs of *Escherichia coli* showing the divisions of several chromatin bodies. The numbers in each frame indicate the *minutes* elapsed from the time of the initial photomicrograph. The cells were mounted and growing in brain-heart infusion broth containing 27% gelatin and were maintained at 37°C. throughout the period of observation. A Spencer bright, high-phase contrast objective (97×)

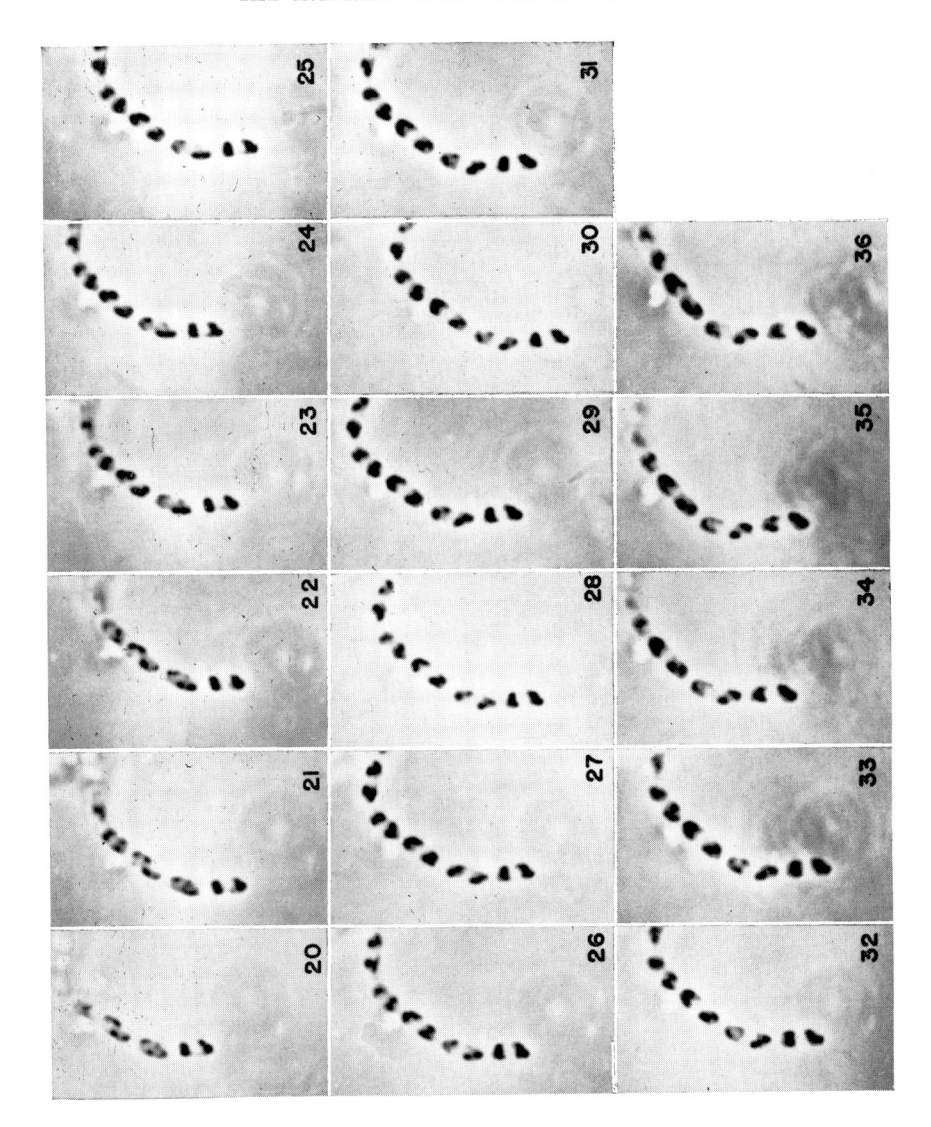

PLATE IX, FIG. 43.

was used in conjunction with a 35 mm. camera (Adox KB-14 film). The chromatin bodies are dark in this system (compare with Plate V, Fig. 20) and the flare is virtually abolished.

The figures were photographed and prepared by Dr. D. J. Mason (Department of Microbiology, The Upjohn Company, Kalamazoo, Mich.) to whom the author is most grateful. (×1980) suitable for cytological study, recurring patterns appear in the dividing chromatin bodies. This argues a certain stability and form in their structure but does not yet help us to understand how they go about their division.

Some still look upon the entire chromatin body as representing a single chromosome and the V and Y forms as part of the longitudinal splitting of such a chromosome; the "fusion nuclei" are then essentially aggregates derived from chromosomes resembling in principle the macronucleus of the ciliates.[106] This is in contrast to Robinow's view,[83] which had been hinted at by Piekarski, that the chromatin bodies themselves with their unchromosomal appearance might be likened to the macronuclei. Whatever the truth may be, and there must be many alternative possibilities to examine, the unconventional nuclei that seem to abound among the protozoa underline the *lack* of necessity for a classic process of mitosis as a prerequisite for the biological stability and activity of bacteria.[106a] Aside from the protozoa, the blue-green algae[107] and the *vegetative* structures of many of the fungi[108] have fully functional nuclei and an orderly genetics despite the most obscure mechanisms for nuclear division. Perhaps "primitive nuclei" do exist and they may teach us how the more complex mechanisms have arisen.

The chromatin bodies divide and are regularly distributed to the daughter cells, but the mechanism remains obscure!

D. Structure of Chromatin Bodies

Appropriate cytological techniques reveal the chromatin bodies as permanent organelles that divide with the cell. It is impossible to avoid thinking of them as having a particular internal structure whether the philosophy be based on cytology, biochemistry, or genetics.

The most convincing structural entity is the chromatinic or DNA-containing portion shown by the Feulgen reaction, which forms the main material in common with all kinds of nuclei. As has already been indicated, even the presence of special, sulfhydryl-rich proteins associated with chromatin in general has been hard to demonstrate. One might also suspect the possibility of an enclosing membrane to form a nucleus, an included moiety rich in RNA and forming a sort of nucleolus, and other such nuclear structures. All these have proved to be elusive or illusory, but what has been found or not found, however fragmentary the evidence, adds more substance to the thesis of the uniqueness of bacterial structure.

1. The Non-chromatinic Portion

In all except the most compact chromatin bodies (e.g., in the cocci), the chromatin appears to be basketwork—sometimes granular, sometimes threadlike—enclosing and enmeshing an area of cytoplasm. Except in the tinctorial sense, it does not seem to be sharply separated from the cytoplasm and, certainly, examination with phase microscopy does not show that the chromatin area is contained within a definite spheroidal vesicle. Nevertheless, several investigators have considered that there is slim but adequate evidence for a "nuclear membrane" and this is a necessary part[102] of the concept, already considered, of mitosis in bacteria.

The demonstration of a nuclear membrane with light microscopy is not at all convincing. The electron microscope has provided no support, as has been discussed already (Section II, C, 3). If it does exist it must be of a most unusual character both in terms of staining properties and of its preservation for electron microscopy of sections. Methods that have regularly revealed nuclear membranes in the Protozoa and in the cells of plants and animals have not been of any avail. So it must be considered that the chromatin bodies lie in the cytoplasm without any membranous enclosure. The only time at which a chromatin structure is definitely invested in a membrane is *during* spore formation (see Chapter 5).

A bounded, vesicular nucleus has been considered in the case of resting cells.[3, 109] Others[110, 111] have stated that the chromatin structures lie within a definite nuclear membrane, basing this view on occasional and suggestive appearances as well as on the apparent immiscibility of chromatin and cytoplasm. The collapse of the chromatin areas in the preparation of whole cells for electron microscopy also suggested a vesicular structure.[112]

In some optical planes the margins of dividing chromatin bodies appear to flow along an otherwise invisible boundary, and this has been taken by DeLamater[81, 102] as an indication of a nuclear membrane. Seeming support for this view was given by preparations of *B. cereus* fixed through the agar block with methanol-formalin (10:1); here the chromatin seems to be disposed within spherical, sharply bounded areas (Plate V, Fig. 25) having an hourglass form during division, whether HCl, ribonuclease, or neutral salts were used to reveal them.[91] A somewhat similar appearance was obtained when companion preparations were fixed with osmium tetroxide vapor and treated with ribonuclease for less than the usual time. In both cases the usual chromatin pattern is present but is enmeshed in a basophilic material that does not have the specific staining properties of DNA. It is this material that forms the apparently vesicular structures, and its sharp boundaries are suggestive of a nuclear membrane.[102] However, the criticism voiced by Bisset at the time and later[112a] was that the conditions of treatment may lead to "coagulation of basophilia" around the chromatin bodies, and there may indeed be some truth in this. Sections of cells fixed with methanol-formalin have demonstrated to us that there is a more than ordinary vacuolation in the chromatin area, thus creating the suspicion that we are also dealing with an artifact caused, perhaps, by extreme swelling of some constituent.

These experiments do, however, draw attention to the possibility that other components are closely associated with the Feulgen-positive material of the chromatin body. The results, probably misleading as far as a nuclear membrane is concerned, may indicate that the cytoplasm immediately surrounding and within the basketwork of chromatin is of a particular and special character. Again this is hard to support from the literature, because the individual workers have been equally individual in their methods, let alone their interpretations. Knaysi has repeatedly asserted that the chromatin units ("primary nuclei" in his terminology) have an enveloping layer of a substance chromotropic to methylene blue, stainable at a low pH, and not well revealed by the HCl-Giemsa technique. Furthermore, when chromatin bodies fuse and aggregate ("secondary nuclei") the individual chromatin elements are embedded in this substance, which he terms "nucleoplasm."[113] This is very similar in spirit to the concept[91] that the chromatin elements are embedded in a nonchromatinic matrix, only obvious in preparations fixed or handled in particular ways, and staining with the methylene blue component of Giemsa's stain. Hollande[114] may have based his often ignored interpretations on similar observations when he proposed a tripartite nuclear structure.

There is a further complication to the nonchromatinic portions of the chromatin body. When the Feulgen-positive material is viewed in the dividing state it often has the appearance of a C, U, or a complex ring, with talons or claws enclosing an area of cytoplasm. A similar appearance obtains in resting spores.[115, 116] It has been considered that the chromatin, in its matrix, is draped over the surface of this "core," which has all the staining properties of cytoplasm. This has been particularly evident, for example, in the reticular chromatin bodies of *Shigella dysenteriae*[117] and the more granular structures of *Caryophanon latum*.[118] Aggregation and fusion of chromatin bodies enhances this appearance of a central nonchromatinic portion, whether it is attained by phage infection,[117] physiological decline,[119] exposure to high salt concentrations,[99] or treatment with tetracycline drugs.[120] It would appear that the core material follows the movements of chromatin, whether natural or induced, and tends to be centrally placed in the chromatin mass. This argues for a distinct structure. The core material probably is present in all chromatin bodies but is more evident in those that are dividing because of the arrangement of the chromatin around it at this stage. It is similarly prominent when resting cells are reviving on fresh medium, as is the case in germinating spores and reviving *Caryophanon*.[82]

Another approach, fruitful even if incomplete, is the study of isolated chromatin bodies. Fitz-James has followed[121] the fate of chromatin bodies in "protoplasts," which are actively synthesizing cell materials but do not divide. The unit chromatin body is in a condensed form and shows as a ring in optical section, both in stained and living preparations. The center is non-chromatinic and Feulgen-negative. These chromatin bodies divide, even though the "protoplast" does not, and each of the new bodies is of similar appearance. Lysis of *Bacillus megaterium* "protoplasts" with lipase in a suitable environment allows separation of chromatin bodies by differential centrifugation and further chemical and cytological studies.[59] The central cores can be separated from the halo of chromatin attached to them, and they remain as discrete structures, retaining their cytological properties throughout the separation procedure. They consist mainly of ribonucleoprotein and the protein portion is distinguished only by a relative poverty of aromatic amino acids. Unfortunately, sections of this material to show the fine structure of these particular cores are not available. These workers, however, express the belief that they are structures independent of cytoplasm.

Electron micrographs of sections through the chromatin areas of a variety of bacteria show distinct inclusions which may be identified as cores. In most cases these have the structure and appearance of cytoplasm (Plate VII, Fig. 40). This, together with the basophilic staining that is abolished by ribonuclease treatment, indicates the possibilities that the cores are either merely included cytoplasm or cytoplasm of special character due to the association with the chromatin body. Where they are the most prominent, in the annular bodies in tetracycline-treated cells[122] (Plate XI, Figs. 48 and 49) and in resting *Caryophanon* (Plate VII, Fig. 41) there are no features that distinguish them from the rest of the cytoplasm. However, Chapman and Hillier[63] did find small inclusions in the axial filament shown by 7-hour cultures of *B. cereus* that were dense and different in character from the rest of the cytoplasm. Yet some sections[123, 124] show continuity at some points between cytoplasm and core. Only further study and serial sections will resolve this problem. The fact remains that cytoplasm-like material is often included as a "core" in the chromatin body; it might have special functions.

2. The Chromatinic Portion

The chromatin portion provides even more perplexing structural puzzles. It can be shown specifically with the standard Feulgen reaction and with great clarity using the azure-SO_2 modification. Even in the larger bacteria used as cytological test objects the granular and thready arrangements of chromatin are of a size and closeness of structure that strain the limit of optical resolution for decisions on detailed arrangement. The interpretations range from a continuum[125] to a finite number of chromosomes[102] or a single chromosome equivalent.[106] The arguments that the chromatin bodies do or do not resemble chromosomal nuclei have been recounted in a previous section. There is no evidence for the multiple chromosome concept other than interpretation of stained preparations. The work on isolation of chromatin bodies, cited above, indicates that the DNA portion of *B. megaterium* behaves as a continuous structure and this is an organism that DeLamater believes to have three chromosomes. It is certainly true that, in early stages of growth, the chromatin bodies often show a triad of interconnected granules.[126] Genetic studies indicate, from the operational point of view, that order exists in the hereditary determinants but they do not show clearly that there is more than a single linkage group. Unfortunately there is no coincidence of choice of genetic and cytological test objects for more than cursory examination. Yet it has been indicated[127] that there is a consistent difference in the cytological complexity of the chromatin between strains of *E. coli* that are *genetically* haploid and diploid. The chromatin bodies of diploid cells are larger and more disperse but there could be no clear interpretation in terms of chromosomes. Wide divergence of the morphological appearance of chromatin bodies is not a bar to recombination, as has been mentioned for *E. coli* strains C and K-12, so that one must suspect that a very different order of "chromosome" structure could occur in bacteria as compared to higher organisms. The field is still open because the architecture of the chromosomes of tissue cells has proved to be very difficult to analyze with the current techniques and what we believe to be the actual fine structure is more conceptual than factual.[128]

The region of the chromatin body filled with DNA is of low density and shows as a light area in the cells with dark-contrast phase optics and as a region of lesser scattering power in the electron microscope. The equivalence of the regions studied with staining, phase, and electron microscopy has been demonstrated often enough. A composite of serial sections[129] shows that the low density, often vacuolated regions build up into structures perfectly recognizable in form as the chromatin bodies of stained whole cells. They tend, however, to be rather smaller and more aggregated, which is an artifact susceptible to experiment and control. Much more difficult is the decision as to which of a number of high resolution images of the chromatin portion represents a reasonably lifelike representation.

"Accessory chromatic granules"[82] were discovered and described as "centrioles" by DeLamater and Mudd.[101] They are found quite regularly but they are often hidden by the more massive chromatin bodies. A semblance of independence is provided by the observation that they remain clear of the chromatin bodies when the latter aggregate and fuse under the influence of tetracyclines.[103] They are roundish, of a size close to the limit of resolution (Plate VII, Fig. 37) and are sometimes paired, suggesting independent division. However, they are stained by both the Feulgen and the modified azure-SO$_2$ techniques so that it is improbable that they can be considered as "centrioles" in the cytological sense. Their closeness to the chromatin body and the problems of optical resolution make for images suggestive of a spindle apparatus. Neither a centriole nor a spindle arrangement has been found, so far, in micrographs of sections. One should note that centrioles proper have a cylindrical structure resembling that of a cilium except for the lack of the central pair of fibrils;[130] bacteria do not have ciliary structures.

There are multiple hazards in the fixation and embedding of cells for sectioning and electron microscopy[70, 122] which have been a matter of concern and comment in nearly every publication. Shrinkage, and "explosion" of the cell during embedding in methacrylates, leaching and distortion of specific structures, the effect of the fixing environment on chromatin conformation, and problematic artifacts are all in the day's work and alternative or improved methods are under investigation in many laboratories. It is really too soon to give a more than temporarily useful account of the state of affairs with regard to the chromatin bodies.

Methacrylate preparations have provided images of the chromatin areas of bacteria fixed in the active growth phase and the appearance is consistent but replete with artifact. They have been termed "vacuoles"[64] because much of the region is seemingly "empty," containing a rather haphazard disposition of branching and anastomosing dense cords (Plate III, Fig. 13; Plate X, Fig. 44; Plate XII, Fig. 51). The state of this dense material varies little from study to study,[28, 123] although in somewhat older cells[63] the area corresponding to an axial filament is filled with a more delicate reticulum of dense material against a lighter background. These images suggest that, apart from the aforementioned cores, the content of the chromatin area has clotted to form the dense cords, and staining of the cells indicates that the chromatin usually aggregates under the conditions of fixation.

Most of the investigators fixed their cells with osmium tetroxide in pH 7.2–7.4 acetate-veronal buffer (following the lead of the histologist), sometimes with added sodium chloride, and grew them in media that usually contained 0.5–1.0 % of added salt. The observations of Whitfield and Murray[99] suggest that the cationic environment may play an important part during the fixation of the chromatin body. Fixation with an aqueous solution of osmium tetroxide of *Spirillum serpens* grown in a medium without added sodium chloride did not show the appearance of cords and, indeed, it is not possible to identify the chromatin area.[30] Similar experience has been obtained in this laboratory with *B. cereus*, among other organisms, grown in a broth very low in Na$^+$ and K$^+$ (Plate III, Fig. 14). By varying the salt content (using potassium chloride) of aqueous osmium tetroxide it is possible to find an end point at which cords and vacuolation just do not appear (0.4%). Then the same end point is sought at various pH levels by buffering with tris buffer to avoid the common cations; cording appears at pH 8.0 down to pH 7.5, and is not apparent below that level. The chromatin can be aggregated into masses, in cells grown in the same way, by exposing them to 3% sodium chloride for five minutes before fixation.

These experiences suggest that the DNA portion of the chromatin body is an interconnected spongework whose state of aggregation and dispersion is controlled to a considerable degree by the cationic balance during fixation and other factors affecting the net surface charge of chromatin, such

---

PLATES X and XI. A series of micrographs of sections of *E. coli* strain K-12 S, which were kindly provided by Dr. E. Kellenberger, University of Geneva, Switzerland. All except Fig. 49 were embedded in polyester resin (Vinox K3).

FIGS. 44–47. A series representing four types of preservation in the chromatin area. The varying degrees of fineness of precipitation are obvious. These are taken from the illustrations of a paper by A. Ryter and E. Kellenberger (see second footnote on page 92) and are reprinted with the permission of the *Zeitschrift für Naturforschung*. This reference should be consulted for details of fixation. (×40,000)

FIG. 44. Fixation for 1 hour at 20° with 1% OsO_4 at pH 6. This shows extreme coagulation and condensation in the chromatin area.

FIG. 45. Fixed with 1% OsO_4 under standard conditions for 16 hours but the cells were grown in a phosphate buffer-tryptone medium. The gross coagulation of the nucleoplasm is ascribed to the phosphate ions.

FIG. 46. Cells grown in tryptone, fixed under standard conditions, and then washed with distilled water before dehydration and embedding. This provides a finer nucleoplasm.

FIG. 47. Treated as in Fig. 46 but the fixative included Ca^{++} and uranyl acetate. The uranium salt has increased the contrast of the cell wall. The nucleoplasm is homogeneously filled with the finest of reticulum.

FIGS. 48 and 49. Sections showing the effect of some antibiotics in causing the aggregation of chromatin bodies into a single vesicle with a central "core" of cytoplasm. (×50,000)

FIG. 48. Cells cultivated in tryptone and then treated with chloramphenicol (25 mg./ml. for 2½ hours.) before fixation by method similar to that of Fig. 50. (Portion of Fig. 3 in Kellenberger and Ryter[71] reprinted with the permission of *Experientia*)

FIG. 49. Cells cultivated in tryptone, treated with chlortetracycline (4 μg./ml. for 2½ hours) before fixation with 1% OsO_4 at pH 7.2 (Palade buffer) for 16 hours. This shows a central mass of included cytoplasm and precipitation in the chromatin area of the order of that in Fig. 51 (Portion of Fig. 6 from Kellenberger and Ryter,[122] reprinted with the permission of the Schweizerische *Zeitschrift für Allgemeine Pathologie und Bakteriologie*.)

FIG. 50. A high-resolution micrograph of a section showing fibrous arrays in the chromatin area and the fine structure of cytoplasm. The cells were cultivated in tryptone medium without added NaCl and were pre-fixed in the presence of 2% NaCl. (This is a portion of Fig. 5 reprinted (see second footnote on page 92) with permission of the *Zeitschrift für Naturforschung*) (×100,000)

THE INTERNAL STRUCTURE OF THE CELL 87

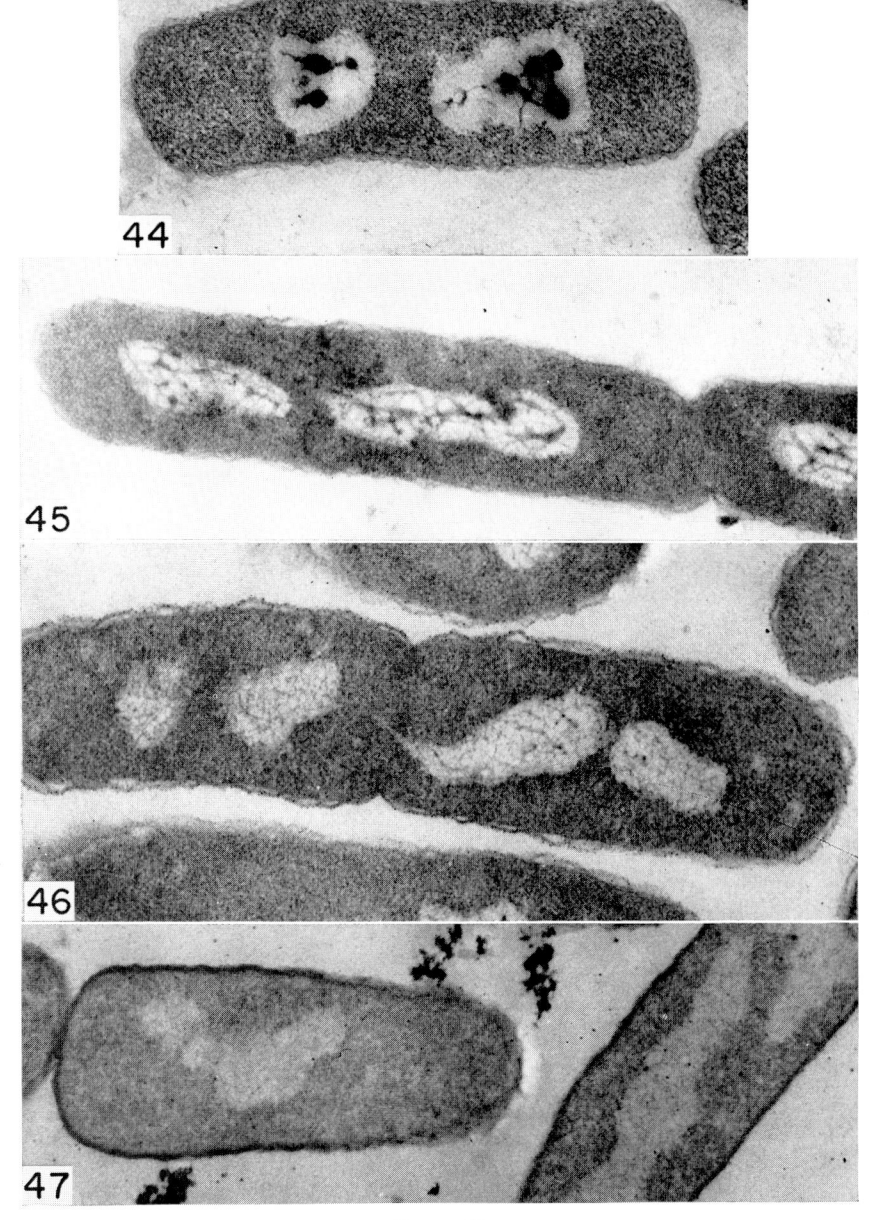

PLATE X.

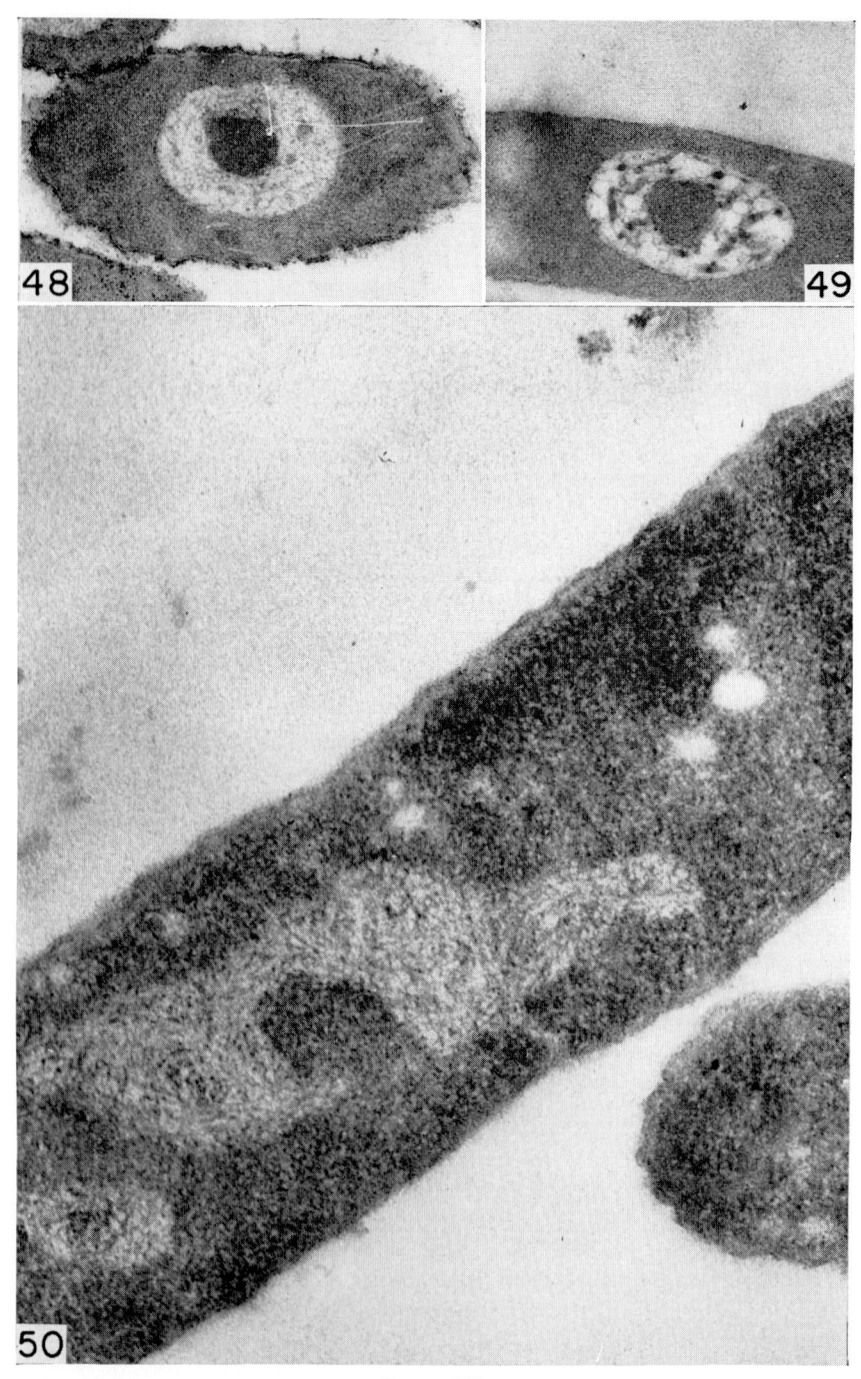

PLATE XI.

as pH. However, this is not the whole story because certain divalent cations, Ca^{++} and Mg^{++}, have important stabilizing functions for membranous structures and, perhaps, for deoxyribonucleoprotein. It may, then, be possible to balance the cationic content and the pH of the fixing environment in such a way that gross distortions of the chromatin portion are avoided. Kellenberger and his co-workers have made the greatest progress along this line. They find that by combining *fixation* for a long time (16 hours) in osmium tetroxide buffered to pH 6, in the presence of Ca^{++} and a medium rich in amino acids, with *embedding* in a polyester resin, the appearance of gross precipitation in the chromatin area is avoided.[131, 132] Treatment with lanthanum nitrate, which prevents solution of DNA, has an effect similar to calcium. By these stratagems the nuclear area is filled out (Plate X, Figs. 44–47; Plate XI, Figs. 48–50) and contains a network of fine filaments (about 50–70 A. diameter) which resembles the appearance occasionally obtained in physiologically older cells fixed by the more usual methods.

The truth must lie somewhere within these extremes. The trouble with the soft outlines and almost homogeneously reticular interiors shown by Kellenberger is that they do not permit us to visualize the patterns of chromatin concentration that we must consider to be present from our light microscope studies. Either the total outline has been changed in the process of preparation, or the internal arrangement of DNA is not visible to us or has been deranged. For that matter one must agree with Kellenberger that the vacuoles and cords, so commonly represented, are not at all satisfying or convincing. But one cannot cast out the best of methacrylate embeddings; let us see what we can learn from them.

Caryophanon, for instance, has provided us with micrographs that correspond remarkably well with the light microscope picture. The finely granular chromatin areas appear as chaplets of interconnected, small, rounded areas in the actively growing stage; at physiological rest the chromatin bodies are annular. In both one can see that there are two components: dense rings or strands within a cloud of material of lesser density (Plate VII, Figs. 39–41). Staining shows that DNA seems to fill out the entire area encompassed by the material of low density. Thus one might suspect that, if there are two components, it is not likely that the dense component is DNA.

However, Maaløe and Birch-Andersen[70] consider that the dense cords in methacrylate and the densely outlined "tubes" in epoxy-resin embeddings represent DNA. This is largely based on the dense interior of the phage particle[133] which is taken to be entirely DNA. The dark strands in the chromatin areas of *B. megaterium*[134] and the similar strands interspersed with granules in *E. coli*[124] are regarded as chromosomes and chromomeres. The strands in these cases seem to be equivalent to the dense element that we have observed in *Caryophanon* and a variety of other bacteria.

The dense portion, like the reticulum shown by Kellenberger, shows an affinity for osmium, phosphotungstic acid, and lanthanum (Plate VII, Figs. 39 and 40) whether included in the fixative or as a treatment of sections.[79] The first, at the least, is not a characteristic of DNA.[135] We have made attempts to use "crystalline" enzymes on sections to see whether we can come any closer to the nature of the components (Plate XII, Figs. 52–55). Unfortunately, differences between the effects of batches of deoxyribonuclease are indicative of probable contamination with other enzymes. Ribonuclease did not remove the high density component and always led to a smudging and swelling (judging from shadowed sections) of this element. Deoxyribonuclease usually caused the removal of both components. Taken together, these rather tenuous pieces of evidence indicate that the dense components of cords, of aggregated chromatin, and of "resting" chromatin bodies are probably rich in ribonucleic acid and in protein. The tentative interpretation of the "chromosomes" of Piekarski and the dense fine reticulum of Kellenberger is that they are various aspects of this dense component and that the DNA is of low density and distributed in the interstices of the rest of the chromatin body. One may wonder if the core isolated by Spiegelman et al.,[59] and shown to be mainly ribonucleic acid and protein, is a condensation of this dense component—as in the chromatin bodies of *B. cereus* rapidly aggregated by brine treatment—rather than the included cytoplasm observed in sections. The surrounding cloud of low density material, that they identify as DNA, would fit the concept.

3. The Total Structure

The arrangement of the chromatin in the living cell is still a puzzle. Proposals must take into account the various states, whether natural or induced by manipulation of the environment, without destroying the biological capability of the cell and the corresponding structures observed in sections of cells in those various states. There are two broad possibilities, if one

---

Plate XII. A series of sections of *Bacillus cereus* taken from the same block but treated in various ways. Cells were grown from spores (2½ hours) on tryptose medium with 0.25% NaCl, and fixed with 1% OsO_4 in veronal-acetate buffer pH 7.4. (These are taken from unpublished experiments in collaboration with B. H. Mayall.)

Fig. 51. Untreated section showing general appearance of cytoplasm. The "nuclear vacuoles" contain cords with some more dense portions.

Fig. 52. A part of a favorably thin untreated section to show: (a) the cytoplasmic granules in a matrix of lesser density; (b) the chromatin areas showing the profile of a dense component surrounded by a "cloud" of less dense material; (c) there is a light zone immediately under the cell wall. (×52,000)

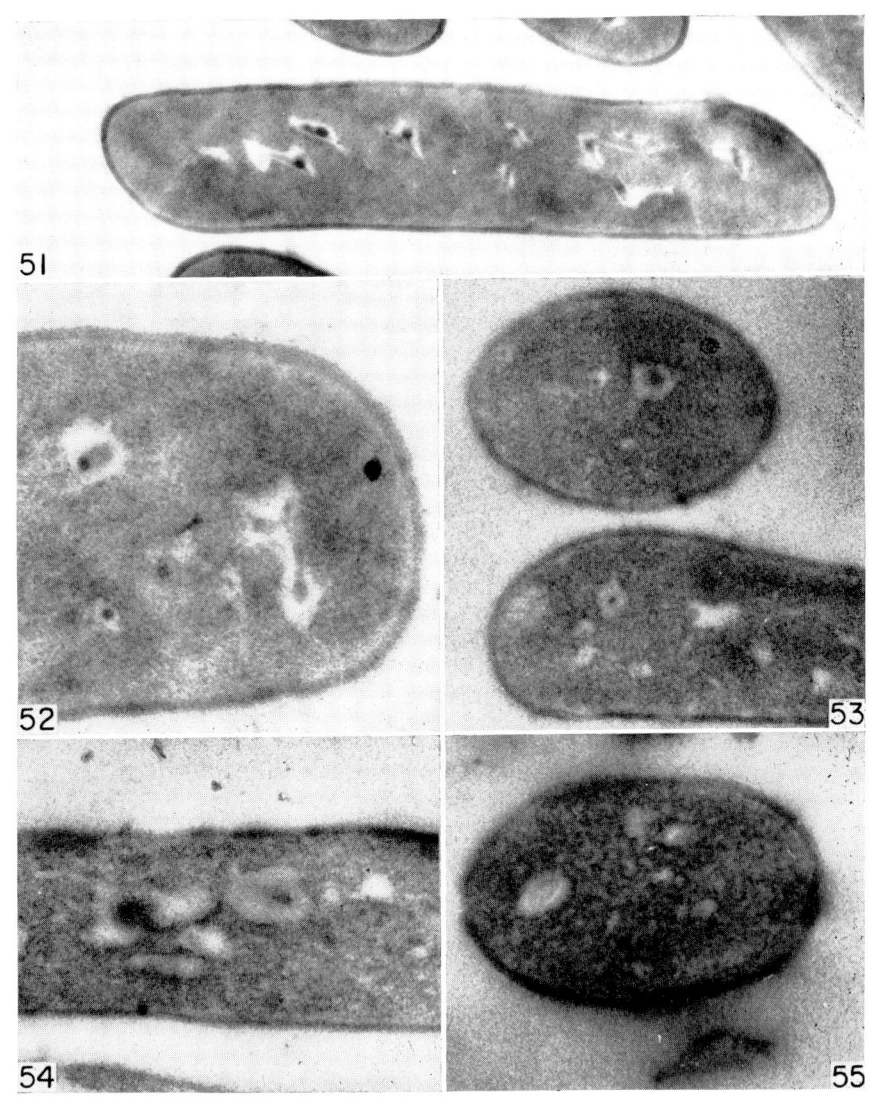

PLATE XII.

PLATE XII (con't.). FIG. 53. Section treated by floating on distilled water for 6 hours. The cytoplasmic pattern shows a more reticular appearance. Some of the chromatin areas still show two components. (×40,000)

FIG. 54. Section treated by flotation on ribonuclease solution (100 μg./ml.) for 1½ hours. The cytoplasmic pattern is altered and the dense portion in the chromatin area shows an enlarged and fuzzy outline. (×40,000)

FIG. 55. Section treated with deoxyribonuclease (100 μg./ml.) for 1½ hours. A reticular cytoplasmic structure and the faint presence of material in the "vacuole" should be noted. Specificity of action can not be assumed. (×40,000)

leaves aside the alternative that structure is so distorted that a reasonable facsimile of truth is not apparent.

1. The chromatin complex consists of a ramifying continuum; the DNA component encloses a finer continuum of RNA-protein; the whole spongework lies within and penetrates the cytoplasmic structure; its degree of aggregation or dispersion is determined by the metabolic regulation of the internal physicochemical environment at the instant of observation; there may be preferred arrangements such that the spongework forms a partial shell around an area of cytoplasm forming a "core" and the observer with the light microscope is dependent upon the optical summation of the components for the image that he gets of the total structure. There is no truly objective evidence for chromosomes.

2. The chromatin complex is held within an area of the cell characterized by low density and a complex network of dense but fine threads (RNA-protein); the DNA portion is somehow disposed in the meshes and its exact distribution is not yet apparent; the total form and the state of its elements may be modified by physicochemical factors both in life and during fixation; the chromatin area may be penetrated by peninsulae of cytoplasm, which can be cut off, by extreme aggregation or fusion of chromatin bodies, to form "cores."

It is obvious that these schemes represent the general view of the author's laboratory. I incline toward the first form because it seems to fit the variety of situations, despite the unconventional picture it presents. The second is based on the excellent work of Kellenberger and his co-workers, but is modified to accord with the arguments presented in the previous section, and has equal validity. It must be emphasized that the interpretations depend upon the exact identification of components and their distribution. The whole matter is still in the experimental stage and, indeed, to produce a semblance of order into the foregoing account it has been necessary to invoke current observation.*, †

Some will regret the abandonment of any concept involving distinct chromosomes; this is not done lightly. There seems to be no clear evidence for them and a good deal of general but telling behavioral evidence against

* The author is grateful to Professor C. F. Robinow, A. Birch-Andersen, P. C. Fitz-James, B. H. Mayall and Mrs. Gertrude Vaughan for helpful discussions of various aspects of this chapter and for the provision of illustrative material. The work of this laboratory described in this chapter has been supported and facilitated by grants from the National Research Council of Canada and the Atkinson Charitable Foundation.

† The author is grateful to Dr. E. Kellenberger for his kindness in providing papers and illustrative material in process of publication and for discussion of his experiments. See A. Ryter and E. Kellenberger, *Z. Naturforsch* **13b**, 597 (1958), and E. Kellenberger, A. Ryter, and J. Séchaud, *J. Biophys. Biochem. Cytol.* **4**, 671 (1958).

them. One should note that the technique of saline treatment that leads to aggregation and fusion of *divided* bacterial chromatin bodies is in current use by cytologists to enhance the separateness and individuality of chromosomes of higher cells for the purpose of accurate counts.

There is no evidence that a nuclear membrane encloses the chromatin body and separates it from cytoplasm. It is, so to speak, naked and unadorned within the cytoplasm. It appears to be a complex and reactive symplasm, which does not allow of the ordinary cytological terms used in describing the parts of a nucleus.

The cytoplasm that immediately surrounds the chromatin body or is enclosed as a cytoplasmic core may be modified in some particulars. It could serve a special function and form a matrix for the body. There is no absolutely regular inclusion that can be identified as a nucleolus; perhaps the dense component or reticulum within the chromatin area has this function and represents the nucleolus in its most primitive form.

The "accessory chromatinic granules" may prove to be separate entities of special significance, but little can be said of them at the moment except that they are to be found fairly regularly. It seems unlikely that they fulfill the role of centrioles. They may, perhaps, correspond to the "nucleosome" of Hollande.[114] One must always be prepared for surprises, and these could easily be the nucleus of one of them!

REFERENCES

[1] B. Delaporte, *Rev. gén. botan.* **51,** 615, 689, 748 (1939); **52,** 112 (1940); Also published as a thesis "Recherches cytologiques sur les bactéries et les cyanophycées." Imprimerie André Lesot, Paris, 1939.

[2] G. Knaysi, "Elements of Bacterial Cytology," 2nd ed. Comstock (Cornell Univ. Press), Ithaca, New York, 1951.

[3] K. A. Bisset, "The Cytology and Life-History of Bacteria." Livingstone, Edinburgh, 1950.

[4] G. Piekarski, *Ergeb. Hyg Bakteriol. Immunitätsforsch. u. Exptl. Therap.* **26,** 333 (1949).

[5] C. F. Robinow, *in* Addendum to "The Bacterial Cell" by R. J. Dubos, p. 355. Harvard Univ. Press, Cambridge, Massachusetts, 1945.

[6] B. Delaporte, *Bull. soc. botan. France* **103,** 521 (1956).

[7] A. Winkler, "Die Bakterienzelle." Fischer, Stuttgart, 1956.

[8] "Bacterial Anatomy" (E. T. C. Spooner and B. A. D. Stocker, eds.), Cambridge Univ. Press, London and New York, 1956.

[9] "Symposium—Bacterial Cytology," Fondazione Emanuele Paterno, Rome, 1953.

[10] M. A. Peshkoff, "Cytology of Bacteria," Academy of Science of the U.S.S.R., Moscow and Leningrad, 1955. (in Russian)

[11] J. F. Whitfield and R. G. E. Murray, *J. Bacteriol.* **65,** 715 (1953).

[12] E. S. Dowding, *Can. J. Microbiol.* **4,** 295 (1958).

[12a] S. Bayne-Jones and A. Petrilli, *J. Bacteriol.* **25,** 261 (1933).

[13] R. Barer K. F. A. Ross, and S. Tkaczyk, *Nature* **171,** 720 (1953).

[14] C.-G. Hedén, *Acta Path. Microbiol. Scand. Suppl.* **88**, 1 (1951).
[15] R. W. G. Wyckoff and A. L. Ter Louw, *J. Exptl. Med.* **54**, 449 (1931).
[16] J. Smiles and M. J. Dobson, *J. Roy. Microscop. Soc.* **75**, 244 (1956).
[17] K. A. Zworykin and G. B. Chapman, *J. Cellular Comp. Physiol.* **48**, 301 (1956).
[18] G. Knaysi, *Botan. Rev.* **4**, 83 (1938).
[19] E. S. Anderson, *Nature* **180**, 1355 (1957).
[20] A. Fischer, "Vorlesungen über Bakterien." Fischer, Jena, 1903.
[21] P. Mitchell and J. Moyle, in "Bacterial Anatomy" (E. T. C. Spooner and B. A. D. Stocker, eds.), p. 150. Cambridge Univ. Press, London and New York, 1956.
[22] G. Knaysi, *J. Bacteriol.* **19**, 113 (1930).
[23] C. F. Robinow and R. G. E. Murray, *Exptl. Cell Research* **4**, 390 (1953).
[24] P. Mitchell and J. Moyle, cited in reference 21.
[25] C. Weibull, in "Bacterial Anatomy" (E. T. C. Spooner and B. A. D. Stocker, eds.), p. 111. Cambridge Univ. Press, London and New York, 1956.
[26] C. Weibull, *Acta Chem. Scand.* **11**, 881 (1957).
[27] C. E. Georgi, W. E. Militzer, and T. S. Decker, *J. Bacteriol.* **70**, 716 (1955).
[28] J. R. G. Bradfield, "Bacterial Anatomy" (E. T. C. Spooner and B. A. D. Stocker, eds.), p. 296. Cambridge Univ. Press, London and New York, 1956.
[29] C. Weibull and K. G. Thorsson, in "Electron Microscopy" (F. S. Sjöstrand and J. Rhodin, eds.), p. 266. Academic Press, New York, 1957.
[30] G. B. Chapman and A. J. Kroll, *J. Bacteriol.* **73**, 63 (1957).
[31] R. G. E. Murray, *Can. J. Microbiol.* **3**, 531 (1957).
[32] E. Kellenberger and A. Ryter, *J. Biophys. Biochem. Cytol.* **4**, 323 (1958).
[33] S. R. Douglas and A. Distaso, *Zentr. Bakteriol. Parasitenk. Abt. I. Orig.* **66**, 321 (1912).
[34] C. C. Dobell, *Quart. J. Microscop. Sci.* **56**, 395 (1911).
[35] P. E. Hartman and J. I. Payne, *J. Bacteriol.* **68**, 237 (1954).
[36] R. Vendrely and J. Lipardy, *Compt. rend.* **223**, 342 (1946).
[37] H. Henry and M. Stacey, *Proc. Roy. Soc.* **B133**, 391 (1946).
[38] A. Widra, *J. Bacteriol.* **71**, 689 (1956).
[39] A. Meyer, "Die Zelle der Bakterien." Fischer, Jena, 1912.
[40] G. Knaysi, J. Hillier, and C. Fabricant, *J. Bacteriol.* **60**, 423 (1950).
[41] R. G. E. Murray and J. P. Truant, *J. Bacteriol.* **67**, 13 (1954).
[42] A. Meyer, *Botan. Ztg.* **62**, 113 (1904).
[43] A. Winkler, in "Symposium—Bacterial Cytology," p. 89. Fondazione Emanuele Paterno, Rome, 1953.
[44] M. Lemoigne, B. Delaporte, and M. Croson, *Ann. inst. Pasteur* **70**, 224 (1944).
[45] R. M. Macrae and J. F. Wilkinson, *J. Gen. Microbiol.* **19**, 210 (1958).
[46] G. B. Chapman, *J. Bacteriol.* **71**, 348 (1956).
[46a] H. Swift, *Sci. Instr. News (R.C.A.)* **3**, (1) (1958).
[47] S. Mudd, L. C. Winterscheid, E. D. DeLamater, and H. J. Henderson, *J. Bacteriol.* **62**, 459 (1951).
[48] S. Mudd, in "Symposium—Bacterial Cytology," p. 67. Fondazione Emanuele Paterno, Rome, 1953.
[49] S. Mudd, *Ann. Rev. Microbiol.* **8**, 1 (1954).
[50] A. F. Howatson and A. W. Ham, *Can. J. Biochem. and Physiol.* **35**, 549 (1957).
[51] G. Turian and E. Kellenberger, *Exptl. Cell Research* **11**, 417 (1956).
[52] A. W. Sedar and M. A. Rudzinska, *J. Biophys. Biochem. Cytol. Suppl.* **2**, 331 (1956).
[53] C. Weibull, *J. Bacteriol.* **66**, 137 (1953).
[54] S. Mudd, K. Takeya, and H. J. Henderson, *J. Bacteriol.* **72**, 767 (1956).

[55] S. Mudd, *Bacteriol. Revs.* **20**, 268 (1956).
[56] C. Weibull, *J. Bacteriol.* **66**, 688 (1953).
[57] E. H. Cota-Robles, A. G. Marr, and E. H. Nilson, *J. Bacteriol.* **75**, 243 (1958).
[58] R. Y. Stanier, in "Cellular Metabolism and Infections" (E. Racker, ed.), p. 1. Academic Press, New York, 1954.
[59] S. Spiegelman, A. I. Aronson, and P. C. Fitz-James, *J. Bacteriol.* **75**, 102 (1958).
[60] C. Shinohara, K. Fukushi, J. Suzuki, and K. Sato, *J. Electronmicroscopy (Chiba)* **6**, 47 (1958).
[61] E. Kellenberger and L. Huber, *Experientia* **9**, 289 (1953).
[62] M. L. Watson, *J. Biophys. Biochem. Cytol.* **1**, 257 (1955); See also G. E. Palade, **2**, *Suppl.* p. 85 (1956).
[63] G. B. Chapman and J. Hillier, *J. Bacteriol.* **66**, 362 (1953).
[64] A. Birch-Andersen, O. Maaløe, and F. S. Sjöstrand, *Biochim. et Biophys. Acta* **12**, 395 (1953).
[65] S. Mudd and A. G. Smith, *J. Bacteriol.* **59**, 561 (1950).
[66] M. Alexander, *Bacteriol. Revs.* **20**, 67 (1956).
[67] H. K. Schachman, A. B. Pardee, and R. Y. Stanier, *Arch. Biochem. Biophys.* **38**, 245 (1952).
[68] G. B. Chapman and A. J. Kroll, *J. Bacteriol.* **73**, 63 (1957).
[69] E. M. Brieger and A. M. Glauert, *Nature* **178**, 544 (1956).
[70] O. Maaløe and A. Birch-Andersen, in "Bacterial Anatomy" (E. T. C. Spooner and B. A. D. Stocker, eds.), p. 261. Cambridge Univ. Press, London and New York.
[71] E. Kellenberger and A. Ryter, *Experientia* **12**, 420 (1956).
[72] B. H. Mayall and C. F. Robinow, *J. Appl. Bacteriol.* **20**, 333 (1957).
[73] P. C. Fitz-James, *Can. J. Microbiol.* **1**, 525 (1955).
[74] A. B. Pardee, H. K. Schachman, and R. Y. Stanier, *Nature* **169**, 282 (1952).
[75] A. E. Vatter and R. S. Wolfe, *J. Bacteriol.* **75**, 480 (1958).
[76] W. Niklowitz and G. Drews, *Arch. Mikrobiol.* **23**, 123 (1955).
[77] C. Shinohara, K. Fukushi, and J. Suzuki, *J. Bacteriol.* **74**, 413 (1957).
[78] E. Fauré-Fremiet and C. Rouiller, *Exptl. Cell Research* **14**, 29 (1958).
[79] I. R. Gibbons and J. R. G. Bradfield, in "Electron Microscopy" (F. S. Sjöstrand and J. Rhodin, eds.), p. 108. Academic Press, New York, 1957.
[80] I. M. Lewis, *Bacteriol. Revs.* **5**, 181 (1941).
[81] E. D. DeLamater, *Intern. Rev. Cytol.* **2**, 158 (1953).
[82] C. F. Robinow, in "Bacterial Anatomy" (E. T. C. Spooner and B. A. D. Stocker, eds.), p. 181. Cambridge Univ. Press, London and New York.
[83] C. F. Robinow, *Bacteriol. Revs.* **20**, 207 (1956).
[84] J. Badian, *Arch. Mikrobiol.* **4**, 409 (1933).
[85] A. Guilliermond, *Arch. Protistenk.* **12**, 9 (1908).
[86] W. A. Cassel and W. G. Hutchinson, *Stain Technol.* **30**, 105 (1955).
[87] M. Piéchaud, *Ann. inst. Pasteur* **86**, 787 (1954).
[88] G. Piekarski, *Arch. Mikrobiol.* **8**, 428 (1937).
[89] R. Tulasne and R. Vendrely, *Nature* **160**, 225 (1947).
[90] W. A. Cassel, *J. Bacteriol.* **59**, 185 (1950).
[91] R. G. E Murray, in "Symposium—Bacterial Cytology," p. 136. Fondazione Emanuele Paterno, Rome, 1953.
[92] E. D. DeLamater, *Stain Technol.* **26**, 199 (1951).
[93] D. J. Mason and D. M. Powelson, *J. Bacteriol.* **71**, 474 (1956).
[94] P. C. Fitz-James, *J. Bacteriol.* **75**, 369 (1958).
[95] A. N. Belozersky, *Cold Spring Harbor Symposia Quant. Biol.* **12**, 1 (1947).
[96] E. Chargaff and H. F. Saidel, *J. Biol. Chem.* **177**, 417 (1949).

[97] B. Delaporte, *Rev. cytol. et biol. Végétales* **18**, 345 (1957).
[98] M. Lieb, J. J. Weigle, and E. Kellenberger, *J. Bacteriol.* **69**, 468 (1955).
[99] J. F. Whitfield and R. G. E. Murray, *Can. J. Microbiol.* **2**, 245 (1956).
[100] A. G. Smith, *J. Bacteriol.* **59**, 575 (1950).
[101] E. D. DeLamater and S. Mudd, *Exptl. Cell Research* **2**, 499 (1951).
[102] E. D. DeLamater, in "Bacterial Anatomy" (E. T. C. Spooner and B. A. D. Stocker, eds.), p. 215. Cambridge Univ. Press, London and New York, 1956.
[103] E. D. DeLamater, in "Symposium—Bacterial Cytology," p. 108. Fondazione Emanuele Paterno, Rome, 1953.
[104] M. E. Hunter-Szybalska, W. Szybalski, and E. D. DeLamater, *J. Bacteriol.* **71**, 17 (1956).
[105] K. A. Bisset, *J. Gen. Microbiol.* **8**, 50 (1954).
[106] K. A. Bisset, in "Bacterial Anatomy" (E. T. C. Sponner and B. A. D. Stocker, eds.), p. 1. Cambridge Univ. Press, London and New York, 1956.
[106a] See p. 503 in reference[101].
[107] W. A. Cassel and W. G. Hutchinson, *Exptl. Cell Research* **6**, 134 (1954).
[108] C. F. Robinow, *Can. J. Microbiol.* **3**, 771 (1957).
[109] K. A. Bisset, *Cold Spring Harbor Symposia Quant. Biol.* **16**, 373 (1951).
[110] G. Knaysi, *J. Bacteriol.* **69**, 117 (1955).
[111] E. D. DeLamater, M. E. Hunter, and S. Mudd, in "The Chemistry and Physiology of the Nucleus" (V. T. Bowen, ed.), p. 319. Academic Press, New York, 1952.
[112] S. Mudd and A. G. Smith, *J. Bacteriol.* **59**, 561 (1950).
[112a] See text figure 5 in reference[106].
[113] G. Knaysi, *J. Appl. Bacteriol.* **20**, 425 (1957).
[114] A. C. Hollande, *Compt. rend. soc. biol.* **113**, 120 (1933).
[115] P. C. Fitz-James, *J. Bacteriol.* **66**, 312 (1953).
[116] C. F. Robinow, *J. Bacteriol.* **66**, 300 (1953).
[117] J. F. Whitfield and R. G. E. Murray, *Can. J. Microbiol.* **1**, 216 (1954).
[118] M. A. Peshkoff, *J. Gen. Biol.* (*U.S.S.R.*) **1**, 598 (1940).
[119] E. G. Pringsheim and C. F. Robinow, *J. Gen. Microbiol.* **1**, 267 (1947).
[120] E. Kellenberger, in "Symposium—Bacterial Cytology," p. 45. Fondazione Emanuele Paterno, Rome, 1953.
[121] P. C. Fitz-James, *J. Biophys. Biochem. Cytol.* **4**, 257 (1958).
[122] E. Kellenberger and A. Ryter, *Schweiz. Z. Pathol. U. Bakteriol.* **18**, 1122 (1955).
[123] S. G. Tomlin and J. W. May, *Australian J. Exp. Biol. Med. Sci.* **33**, 249 (1955).
[124] G. Pontieri, *Giorn. microbiol.* **1**, 367 (1956).
[125] R. G. E. Murray, *Can. J. Biochem. and Physiol.* **35**, 565 (1957).
[126] P. C. Fitz-James, *J. Bacteriol.* **66**, 312 (1953).
[127] J. Lederberg, E. M. Lederberg, N. D. Zinder, and E. R. Lively, *Cold Spring Harbor Symposia Quant. Biol.* **16**, 413 (1951).
[128] J. H. Taylor, *Sci. American* **198**, 37 (1958).
[129] A. Birch-Andersen, *J. Gen. Microbiol.* **13**, 327 (1955).
[130] W. Bernhard and E. de Harven, *Compt. rend.* **242**, 288 (1956); See also[50].
[131] E. Kellenberger and A. Ryter, *Experientia* **12**, 420 (1956).
[132] E. Kellenberger, A. Ryter and J. Séchaud, *J. Biophys. Biochem. Cytol.* **4**, 671 (1958).
[133] O. Maaløe, A. Birch-Andersen, and F. S. Sjöstrand, *Biochim. et Biophys. Acta* **15**, 12 (1954).
[134] G. Piekarski and P. Giesbrecht, *Naturwissenshaften* **43**, 89 (1956).
[135] G. F. Bahr, *Exptl. Cell Research* **7**, 457 (1954).

CHAPTER 3

Surface Layers of the Bacterial Cell

M. R. J. SALTON

	Page
I. Introduction	97
II. Anatomy of the Bacterial Surface	98
III. Extracellular Surface Components, Slime, and Capsular Materials	99
A. General Properties and Structure of Capsules and Slime	101
B. The Chemical Constitution of Capsules, Slime, and Surface Substances	102
C. Action of Enzymes on Capsules, Slime, and Other Extracellular Surface Components	113
IV. Cell Walls	115
A. Isolation of the Bacterial Cell Wall	115
B. Physical Properties	120
C. Chemical Composition	120
D. Action of Enzymes on Bacterial Cell Walls	129
E. Interaction between Bacteriophages and Cell Walls	135
F. Immunological Reactions	137
G. Biosynthesis of the Cell Wall	138
H. Functions of the Cell Wall	142
References	144

I. Introduction

The microbial world presents to the mid-twentieth century student of microbiology a spectacular array of organisms possessing great diversity of form, structure, and biochemical potentialities. In the short space of a century we have become accustomed to the notion that microorganisms can successfully establish themselves in almost any environment presented by the earth's lithosphere. One of the most fascinating aspects of the great biochemical flexibility of bacteria is the enormous variety of substances they can utilize for growth. Biochemists accustomed to the more conservative nutritional requirements of higher plants and animals are often surprised to learn of the ability of microorganisms to grow on compounds such as thiocyanate, cresols, dinitrophenols, and even trichloroacetic acid.[1] Thus, the modern microbiologist has come to accept "diversity" as one of the central features of microbial life; the steadily growing wealth of knowledge about the nature of the bacterial surface has clearly established that this property is also shared by the superficial layers of the microbial cell.

Our interest in the bacterial surface really stems from Leeuwenhoek's discovery that bacterial cells have three main shapes: the rod, coccus, and spiral form. It must have been evident that both the rod- and helicoidal-shaped cells possessed some surface structure with a rigidity greater than that of the cytoplasm. That Leeuwenhoek anticipated such a differentiation of the surface from the internal contents of the cell seems implicit from his statement that he was unable to discern any "film" that held the cells together. Apart from gross differences in cell surface reflected by differences in cellular shape, early microbiologists soon found that some microorganisms were encased by capsular material with low affinities for dyes. Because of the demands of medical microbiology for the rapid diagnosis of pathogenic bacteria, an enormous amount of information about the immunological properties of capsular and surface antigenic substances soon accumulated. It became apparent from such studies that bacteria differed enormously in the types of immunologically specific substances arranged at the surface, although there were numerous instances of similarities as shown by serological cross reactions. However, it has been only comparatively recently that explanations of these differences and similarities in the immunological properties of the surface components have been placed on a sound chemical basis.[2]

With the development of electron microscopy and the application of biochemical and biophysical techniques to the study of microbial structure, we now have a much better appreciation of the nature of the surface components of the bacterial cell. The results of our newer knowledge of the bacterial surface have underlined its diversity from one species to another, and have also revealed some basic similarities in the nature of the "building blocks" of structure such as the cell wall.

II. Anatomy of the Bacterial Surface

Staining reactions, immunological studies, and electron microscopy have established that many bacterial cells are surrounded by layers of material that lie external to the rigid cell-wall structure. Thus, some bacteria are surrounded by a thick gel of capsular material, others possess a more diffuse layer of loose slime.[3] *Acetobacter* species are often entangled in a mesh of cellulose fibrils[4] and a number of organisms are surrounded by large numbers of thin fimbriae.[5] Many of these surface components may be dispensed with, the viability of the cells remaining unchanged. Some bacteria possess surface components that are closely associated with the walls but are not part of the rigid cell-wall structure, e.g., the M-protein antigen of Lancefield's group A streptococci.[6,7] Even by examination in the electron microscope it has not been possible to differentiate the wall from the layer of M protein which can be readily removed from the cell or cell wall by digestion with trypsin.[6,7]

The relationship of some of the above surface components to the underlying cell wall has been established in several instances. A number of the capsular polysaccharides, of pneumococci, and *Klebsiella pneumoniae* can be removed from the bacterial cell by means of specific enzymes without incurring death of the organisms.[8, 9] The M proteins of streptococci and lipids have been removed from the cell surfaces of several species without affecting the viability.[10-12] These surface layers are, therefore, independent components that are separable from the underlying cell wall without impairing the functional and structural integrity of the organisms.

Most true bacteria (members of the Eubacteriales) possess a rigid wall structure that may be differentiated from the external capsular and slime materials, and from the internal protoplasm bounded by its plasma membrane. For those bacteria devoid of capsules, slime, or other dispensable surface components, the cell wall will constitute the outermost structure. That the wall is the rigid mechanical structure responsible for the characteristic shape of the bacterial cell has been established beyond doubt by both direct electron microscopy of wall preparations and the lysis of bacteria by enzymes. Removal of the cell wall with lysozyme under conditions that stabilize the rest of the cell in the form of a bacterial protoplast[13] is accompanied by a loss of ability to form colonies on agar media. Thus, the minimum structural requirements for the surface of a bacterial cell capable of continued normal growth and binary fission are a cell wall and a protoplasmic membrane. The capsules, other surface slime, and the waxy and mucoid components may be dispensed with by either direct enzymic degradation or modification of the growth conditions without affecting the ability of the organism to multiply and survive and retain its genotypic characteristics.

Thus, between the external environment of the bacterial cell and the intracellular protoplasm and inclusions, the surface components may be anatomically classified as: (1) surface appendages—flagella, fimbriae; (2) surface adherents—capsules, slime, and extracellular substances; (3) rigid cell wall; (4) protoplasmic membrane.

The surface components (2)–(4) frequently form a series of continuous layers around the bacterial cell. The purpose of this review will be to outline some of the properties that distinguish the surface capsules and slime layers from the underlying cell walls.

III. Extracellular Surface Components, Slime, and Capsular Materials

A great variety of substances coating the outer surface of the rigid cell wall has been detected and a satisfactory definition of the anatomical status of such materials presents some difficulties. Many of these "surface" components may represent the accumulation of viscous products from the medium. Such components may have only a transient existence at the bacterial surface and others may only appear as the result of some special circumstance of the growth conditions. Deoxyribonucleic acid (DNA) has been detected on the surface of both halophilic bacteria[14, 15] and nonhalophilic organisms.[16] The detection of DNA in the surface slime of bacteria is an experimental finding that seems contrary to the generally accepted belief that cellular DNA is essentially a macromolecular component of the genetic equipment of the bacterial cell. Very careful investigations by Takahashi and Gibbons[15] showed that the "surface" DNA of the halophilic organism, *Micrococcus halodenitrificans*, was indeed of intracellular origin. However, its presence upon the surface was a consequence of the instability of the cell walls when the organism was grown on media containing less than 0.7 M sodium chloride. Thus, in some instances the presence of certain substances at the bacterial surface may represent an anatomical artifact.

The difficulties of defining the extracellular surface layers of the bacterial cell are further illustrated by studies with the cellulose-producing *Acetobacter xylinum*. This organism is surrounded by a viscous layer of amorphous cellulose which subsequently assumes a fibrillar state so that the cells become enmeshed in a network of fibrils that are no longer intimately associated with the bacterial surface.[17,18]

The effects of environmental conditions upon the production of slime and capsular substances add further to the difficulties of obtaining a satisfactory definition of all the "extramural" components of bacteria. Unlike the bacterial cell wall, most of the surface components that appear outside it are not produced under all cultural conditions. Organisms producing anatomically demonstrable capsules, slime, and serologically detectable components of the O somatic antigens can be largely deprived of these surface substances by growing under appropriate conditions. With some strains of bacteria the differences in the surface structure may be quantitative rather than qualitative. A well-defined capsule may be formed under one set of growth conditions, whereas under another set the thickness of the layer may be so reduced as to be undetectable except by immunological and chemical means. Burrows and Bacon[19] have presented evidence for this state of affairs with virulent capsulated and noncapsulated strains of *Pasteurella pestis*. Thus, in an attempt to overcome some of the difficulties of differentiating and defining the surface components of the bacterial cell, Wilkinson[20] has collectively defined the smooth antigenic substances and other related surface materials as microcapsules.

For the purpose of this review, the surface layers outside the rigid wall that will be discussed will include: (1) the cytologically demonstrable capsules and slime; (2) the immunologically detectable components that have not been resolved microscopically but, in common with capsules and slime, are outside the cell wall and may be dispensed with without affecting the integrity of the bacterial cell—the surface substances in this group constitute the microcapsules as defined by Wilkinson;[20] (3) viscous extracellular components that accumulate at the bacterial surface and fre-

SURFACE LAYERS OF THE BACTERIAL CELL 101

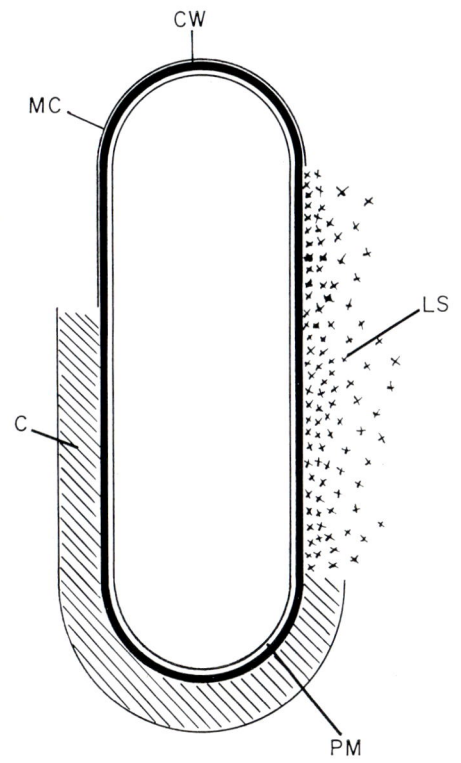

FIG. 1. A diagrammatic representation of the relationship between the capsules, slime, and microcapsular layers to the cell wall and protoplast membrane. CW = cell wall; PM = protoplasmic membrane; C = capsule; MC = microcapsule; LS = loose slime.

quently appear in the culture media—many of these substances are of unspecified anatomical significance and may be readily washed from the surface. The relationships of some of these surface layers to the bacterial wall is illustrated in Fig. 1.

A. General Properties and Structure of Capsules and Slime

Capsules are apparently formed by the accumulation of various types of polymeric substances of high viscosity around the bacterial cell wall. Most capsular and slime substances have a very low affinity for various dyes and electron microscopy has established that in general they are less electron-dense than the cell wall and cytoplasm. However, in some instances the slime may be more opaque to the electron beam than the bacterial cells, as shown in the electron micrographs of a strain of *Mycobacterium tuberculosis* investigated by Knaysi et al.[21]

Although there are a number of reliable staining techniques for the demonstration of capsules and loose slime,[3] such methods have revealed little of the fine structure of the capsule. In general many bacterial capsules have an amorphous appearance even when examined in the electron microscope. However, the investigations of Tomcsik,[22-24] Ivanovics and Horvath,[25] and Labaw and Mosley[26] have demonstrated that some bacterial capsules are by no means simple homogeneous accumulations of amorphous material. So far, the only capsule showing the presence of well-defined macromolecular components is that of the Lisbonne strain of *Escherichia coli*. Labaw and Mosley[26] have shown that within the capsular matrix of this organism there are some remarkable structures with striated fibers that are arranged roughly at right angles to the surface of the bacterial cell.

Because the refractive index of the capsule can be changed by the precipitation of proteins within the capsular substances, Tomcsik[22-24] was able to develop methods for the demonstration of capsules by phase-contrast microscopy. Tomcsik made the method a very sensitive one by using specific antibodies and in so doing he discovered a complex structure within the capsule of *Bacillus megaterium*.[23, 24] However, to reveal the bacterial capsule in the phase-contrast microscope both antibody proteins and "nonspecific" proteins (in the immunological sense) could be used, the latter proteins requiring conditions for an isoelectric precipitation.[27]

The capsule of the strain of *B. megaterium* studied by Tomcsik is a complex structure composed of two molecular species—a glutamyl polypeptide substance of the type forming the capsule of *Bacillus anthracis* together with a "polysaccharide" that is immunologically related to the substance of the underlying bacterial cell wall.[28] These capsular materials were not arranged in continuous concentric layers; the capsular polysaccharide occurred in localized patches within the polypeptide.

The strain of *B. megaterium* investigated by Ivanovics and Horvath[25] possessed a capsule that showed the presence of indentations at regular intervals along its surface.

Some of the variations in the structures of bacterial capsules mentioned above are illustrated in Fig. 2.

B. The Chemical Constitution of Capsules, Slime, and Surface Substances

1. Gram-Positive Bacteria

a. Pneumococci. As a group, the capsules of the pneumococci have been investigated in more detail than any others. A great number of different "type-specific" polysaccharides have been detected immunologically, but not all of these have been isolated and chemically characterized. The chemical basis for the immunological specificity of the various types of capsular polysaccharide stems largely from the work of Heidelberger,

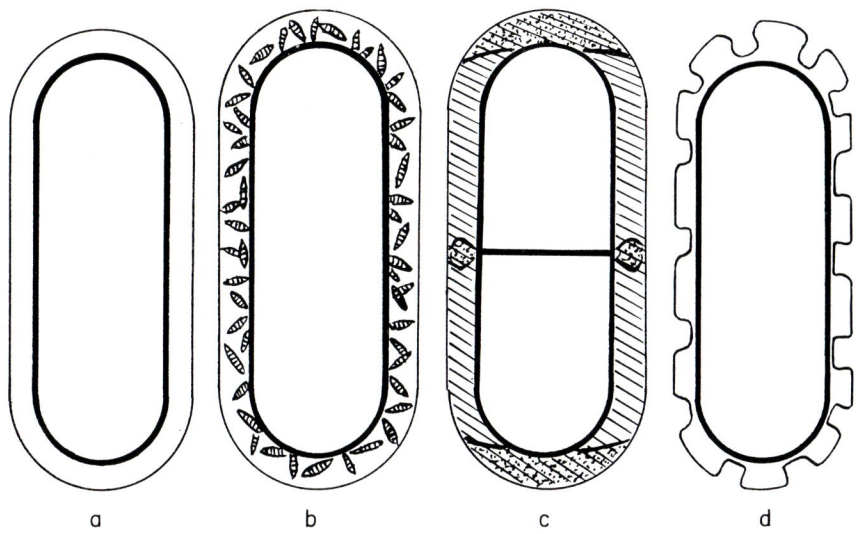

FIG. 2. An illustration of the types of capsular structure. (a) normal capsule forming a continuous layer around the cell wall; (b) capsular layer containing banded fibrils (after Labaw and Mosley[26]); (c) a complex capsule showing localized patches of polysaccharide and polypeptide (after Tomcsik[22]); (d) a discontinuous capsule of the type described by Ivanovics and Horvarth.[25]

Goebel, and Avery (see Heidelberger[2]). It appears likely from immunological studies that the capsular polysaccharide is combined with some other material on the surface of the intact bacterium. However, the isolated, highly purified polysaccharides can behave as full antigens in some species, e.g., mice and man. These polysaccharides are of high molecular weight; ultracentrifugal analyses of types I, II, and III indicate molecular weights of the order of 1,000,000 (Pasternak and Kent, quoted by Kent and Whitehouse[29]).

Some of the pneumococcal polysaccharides are polymers of sugars and are devoid of any nitrogenous constituents, whereas others contain amino sugars.[2] The products of acid hydrolysis of a number of pneumococcal polysaccharides have been investigated and some of the simple monosaccharide constituents are listed in Table I. (Data taken from Heidelberger,[2] Kent and Whitehouse,[29] and Brown.[30])

Of the capsular polysaccharides so far investigated, most is known about the molecular structure of that isolated from type III pneumococci. The type III polysaccharide is a polyaldobiuronic acid, the structural unit of which is cellobiuronic acid (glucurono-4-β-glucose). The D-glucuronic acid is linked to D-glucose in the 4-position and the cellobiuronic acid units are linked to the 3-positions of the glucuronic acid residues.[2]

Some of the extraction procedures used for the isolation of the capsular

TABLE I

Monosaccharide Constituents of Pneumococcal Capsular Polysaccharides

Type	Constituent monosaccharides
I	Galacturonic acids and amino sugars (?)
II	L-Rhamnose, D-glucose, D-glucuronic acid
III	Glucose, glucuronic acid (as cellobiuronic acid)
IV	Glucose, N-acetylamino sugar
VI	Galactose, glucose, rhamnose
VII	Galactose, glucose (?), rhamnose, amino sugar
VIII	Cellobiuronic acid (glucose, glucuronic acid), glucose, and galactose
IX	Glucose, amino sugar, uronic acid
CII	Glucose, galactose, amino sugar
XIV	Galactose, glucose, N-acetyl glucosamine
XVIII	Rhamnose, glucose

polysaccharides also yield an immunologically active material referred to as the "somatic carbohydrate or C substance." This component is apparently common to all types of pneumococci and is probably analogous to the C substance of streptococci which has now been shown to be an integral part of the bacterial cell wall.[7, 31, 32] The C substance of pneumococci was first investigated chemically by Tillet and Francis[33] and Heidelberger and Kendall;[34] more recently, Smith et al.[35] have identified glucosamine, galactose, and mannose as the sugar constituents. The acetyl content of the cellular polysaccharide studied by Smith et al.[35] indicated that all of the glucosamine was acetylated on the N position and that, in addition, some O-acetyl was present. Amino acids were consistently present in the purified cellular polysaccharide substance isolated from the rough pneumococci.[35] As yet the anatomical location of this polysaccharide has not been established.

b. *Streptococci*. The viscous capsular material produced by streptococci belonging to Lancefield's groups A and C is hyaluronic acid, a polymer of N-acetylglucosamine and glucuronic acid.[36, 37] The hyaluronic acid capsular substances of groups A and C appear to be indistinguishable from one another; these bacterial hyaluronates are very similar, if not identical, to hyaluronate of animal origin.[36] In strains of streptococci producing hyaluronic acid the capsule is often only cytologically demonstrable in the early exponential phase of growth.[37, 38] Strains of streptococci producing hyaluronidase will, of course, be free of surface hyaluronic acid and the disappearance of the capsule after a brief period at the beginning of the exponential phase of growth may be related to the production of the enzyme by exponentially growing bacteria.[39] That some streptococci devoid of a capsule can possess surface material removable with hyaluronidase has been demonstrated by Maxted.[40] The enzymic digestion of this component, presumed to be hyaluronate, was accompanied by an increased susceptibility to infection with bacteriophages.[40]

There are a number of protein constituents of antigenic importance upon the surface of certain streptococci.[36] None of these surface proteins assumes the dimensions of capsules and it is even difficult to detect any difference in the appearance of isolated cell walls of group A streptococci after the removal of the M-protein substance.[41, 42] Both M and T proteins of streptococci have been extracted from bacteria, partially purified, and their protein nature confirmed.[36]

Hobson and Macpherson[43] studied the capsular polysaccharide isolated from an amylolytic rumen streptococcus; they found that the main constituents were galactose, rhamnose, and a uronic acid, probably galacturonic acid. The polysaccharide was found to be homogeneous in the ultracentrifuge and it possessed a molecular weight of 90,000.[44]

Other streptococci, notably those belonging to group B, together with *Streptococcus viridans* and *Streptococcus salivarius*, elaborate polysaccharides as type-specific substances and it is probable that they are comparable to those produced by the pneumococci. These capsular polysaccharides have not been investigated chemically. In common with the related bacteria, *Leuconostoc mesenteroides*, *Streptococcus bovis* produces an extracellular slime of dextran.[45] Levan is the slime polysaccharide produced by some strains of *S. salivarius*.

c. *Bacillus Species*. One of the most novel substances forming the capsule of a bacterium is the glutamyl peptide produced by *Bacillus anthracis*. The composition of this capsular substance was first indicated by Ivanovics and Bruckner[46] and there has been abundant confirmation since then that the peptide is a γ-linked polymer of D-glutamic acid. Other members of the genus *Bacillus* also produce γ-glutamyl peptides but with organisms such as *Bacillus subtilis* the material is not retained as a thick, capsular gel surrounding the bacterial cell.[47] *Bacillus subtilis* produces an enzyme capable of degrading both synthetic and naturally occurring γ-glutamyl peptide and this undoubtedly accounts for the absence of a capsule and the presence of the peptide in the culture medium.[47]

The composition of the capsular peptide from *Bacillus anthracis* has been compared with glutamyl polypeptide preparations from *B. subtilis*. Thorne[47] has shown that virtually all of the glutamic acid is the D-isomer for *B. anthracis*, whereas *B. subtilis* peptide contains 50–55% L-glutamic acid. The cultural conditions had a marked effect on the relative proportions of the D- and L-isomers of glutamic acid in the polypeptide synthesized by *B. subtilis*.

The manner of linkage of the glutamic acid in the capsular polypeptide has been the subject of a number of investigations. Bovarnick[48] concluded that the glutamic acid of *B. subtilis* peptide was joined by γ-linkages and Bruckner et al.[49] believed that the polypeptides of *B. anthracis* and *B. subtilis* were also predominantly γ-linked, with few, if any, α-glutamyl links. Hanby and Rydon[50] reported the presence of α-linked glutamic acid in *B. anthracis* polypeptide. Polypeptides from *B. anthracis* and *B. licheniformis* were found to be γ-linked by Waley.[51] Although unanimity on the exclusiveness of the γ-mode of linkage has not been achieved, the weight of evidence is heavily in its favor and the more recent investigations of Chibnall et al.[52, 53] are in accord with a completely γ-linked glutamyl peptide.

Some strains of *Bacillus megaterium* also produce a capsular D-glutamyl polypeptide.[25, 54] Tomcsik[23] has concluded that the capsule of the strain of *B. megaterium* used in his studies is a complex structure in which two different molecular species, the glutamyl peptide and a polysaccharide, are in close physical association. Ivanovics and Horvath[25] have studied a strain of *B. megaterium* with the D-glutamyl peptide of the capsule in a fibrillar form.

In discussing some of the properties of the glutamyl peptide capsular substances, Ivanovics and Horvath[54] point out that the absence of free carboxyl groups (as determined by microchemical tests performed upon encapsulated bacteria) could explain the low solubility of the intact capsule in water. They suggest that intramolecular linkages are formed and that "all carboxyl groups are bound as secondary acid amides." The presence of free carboxyl groups in the isolated polypeptide is explained by hydrolysis, as shown in Formula (I).

(I)

Such a process would occur under the mild hydrolytic conditions used for decapsulation and solubilization of the peptide.[54]

Although there are a number of slime and capsular polysaccharides produced by *Bacillus* species, their chemistry has not been investigated as extensively as the pneumococcal polysaccharides. A capsulated bacillus, subsequently shown to be a strain of *Bacillus circulans*, was studied by Kleczkowski and Wierzchowski[55] and the isolated polysaccharide identified as a mannan. Later investigations by Forsyth and Webley[56] established that well-capsulated cells of *B. circulans* synthesized a complex polysaccharide of glucose, mannose, and uronic acid in the approximate ratio 3:2:2. The capsular substance of the strain of *Bacillus megaterium* studied by Aubert[56a] was found to be a polysaccharide, yielding glucose and galactose as the monosaccharide units. The "polysaccharide" substance of the strain of *B. megaterium* investigated by Guex-Holzer and Tomcsik[28] is somewhat unusual in that it yielded the amino sugars, glucosamine and galactosamine, and an unidentified amino sugar as the products of acid hydrolysis. Whether this amino sugar polymer represents a degradation product of a more complex polysaccharide has not been established.[28]

Extracellular mucilagenous polysaccharides produced by strains of *Bacillus cereus*, *B. pumilis*, *B. subtilis*, and *B. megaterium* (group I species according to the classification of Smith, Gordon, and Clark[57]) include levans.[56] In addition to levan, *B. megaterium* produces a glucose-uronic acid polymer.[56] Of the Smith *et al.*[57] group II species, only *B. polymyxa* produced extracellular levan, but this species and others produce polysaccharides composed of the following sugar components: glucose, mannose, uronic acid; glucose, uronic acid; glucose, mannose, xylose, uronic acid.[56]

In addition to polysaccharides of streptococci and *Bacillus* species, the dextrans produced by *Leuconostoc mesenteroides* and a mucilaginous polysaccharide formed by *Lactobacillus bifidus*[57a] have been investigated chemically. Some of the properties of the capsular, slime, and extracellular substances produced by Gram-positive bacteria are summarized in Table II.

2. Gram-Negative Bacteria

The components outside the cell walls of Gram-negative bacteria include both capsules and slime, and in this respect these surface layers are common to both groups of bacteria differentiated by the Gram reaction. However, Gram-negative bacteria produce additional types of surface substances that have not been encountered so far within the Gram-positive groups of microorganisms. The protein-polysaccharide-lipid (PPL) complexes responsible for the characteristic immunological reactions of the bacterial surface of nonencapsulated, Gram-negative bacteria form the group of substances generally known as the somatic O antigens, surface, or smooth antigens.

TABLE II
The Nature of Capsular, Microcapsular, Slime, and Extracellular Substances of Various Gram-Positive Bacteria

Organism	Anatomical status of surface component	Class of substance and products of acid hydrolysis
Pneumococci	Capsules	Polysaccharides: sugars and amino sugars[a]
Streptococcus spp., (Groups A and C)	Capsules	Polysaccharide: hyaluronic acid—glucosamine and glucuronic acid
Group A	Microcapsular	Protein: M-protein antigen
Rumen streptococcus	Capsule	Polysaccharide: galactose, rhamnose, and uronic acid
Streptococcus salivarius	Extracellular slime	Polysaccharide: levan—fructose
Streptococcus bovis	Extracellular slime	Polysaccharide: dextran—glucose
Leuconostoc mesenteroides	Extracellular slime	Polysaccharide: dextran—glucose
Lactobacillus bifidus	Extracellular slime	Polysaccharide: galactose, glucose, fucose
Bacillus anthracis	Capsule	Polypeptide: γ-D-glutamyl peptide
Bacillus subtilis	Extracellular slime	Polypeptide: γ-DL-glutamyl peptide
Bacillus megaterium	Capsule	Polypeptide: γ-glutamyl peptide; Polysaccharide: amino sugars
Bacillus circulans	Capsule	Polysaccharide: glucose, mannose, uronic acid
Bacillus species belonging to group I[57]	Extracellular gums and slimes	Polysaccharide: levan—fructose; Polysaccharide: glucose, uronic acid
Bacillus polymyxa	Extracellular slime	Polysaccharide: glucose, mannose, xylose, uronic acid

[a] See Table I.

Because of their toxicity on injection into animals, these surface components have also been classified with the bacterial endotoxins. In addition to the PPL complexes, some Gram-negative bacteria produce components that are believed to overlayer and cover the former surface substances. These components have been called the "Vi antigens" by Felix and Pitt[58] because of their virulence-enhancing properties for enteric bacteria. There is no reason to suspect that there is anything unique about the so-called Vi antigens, for it is now evident that they simply represent another surface layer with chemical properties differing from those of the underlying PPL substances. Moreover, there is evidence to suggest that analogous substances are present on the surface of clover nodule bacteria.[59]

The anatomical classification of the PPL substances forming the smooth antigenic surface and the polymeric components described as Vi antigens is far from satisfactory. For convenience they can be described, as Wilkinson[20] has suggested, as "microcapsular." Although Ando and Nakamura[60] were able to demonstrate a capsule around the cells of virulent typhoid bacteria after treatment with ferric chloride, it is evident that the behavior of these microorganisms differs from those possessing a capsule demonstrable by the techniques outlined by Duguid.[3]

a. Acetobacter Species. Certain *Acetobacter* species are surrounded by an amorphous layer of cellulose which later crystallizes into fibrils of true cellulose. Investigations of the composition of the bacterial cellulose have shown that it is chemically and physically similar to that obtained from plants.[61]

Acetobacter capsulatum and *Acetobacter viscosum*, when grown on media containing dextrin, produced an abundance of capsular and extracellular polysaccharide of a dextran type.[62] Subsequent work showed that the polysaccharide gum formed by *A. capsulatum* was a polymer of D-glucopyranose, α-linked through the 1:6-positions.[63]

b. Haemophilus influenzae. Zamenhof et al.[64] have indicated that the immunologically active, type-specific capsular substance of *Haemophilus influenzae* is composed of a polyribophosphate compound. This polymer possesses a polyribophosphate chain as it exists in pentose nucleic acids, but the place occupied by the purines and pyrimidines in the latter is apparently replaced by a second similar chain linked to the first in 1:1 glycosidic linkages.

c. Klebsiella-Aerobacter. The nature of the extracellular polysaccharides of *Aerobacter aerogenes* and related organisms and the factors affecting their production have received considerable attention in the recent investigations of Duguid and Wilkinson[65] and Wilkinson et al.[66] The polysaccharides of these organisms may occur as loose slime or as distinct capsules. The polysaccharide isolated from *A. aerogenes* by Wilkinson et al.[67] contained D-glucose (50%), L-fucose (10%), and an unidentified uronic acid (29%) as the three main sugar constituents. The uronic acid subsequently proved to be glucuronic acid.[68] A number of polysaccharides have been isolated by Dudman and Wilkinson[69] from various members of the *Aerobacter-Klebsiella* group and uronic acids were found in all of them. Glucose, galactose, and fucose were the monosaccharide components of some of the polysaccharides. Five immunological types of *Klebsiella* were used in a study of the nature of the extracellular polysaccharides and although the composition of the polysaccharides of each type differed either quantitatively or qualitatively, all were found to be complex polyuronides.[69]

In common with the Gram-positive pneumococci, the mucoid strains of *Aerobacter-Klebsiella* organisms have been differentiated into a large number of immunologically distinct types.[70, 71] The investigations of

Wilkinson and his colleagues are providing a chemical basis for the immunological differences between the capsular and slime polysaccharides of these organisms. The diagnostic value of infrared spectrophotometry in studies of the capsular polysaccharides has recently been demonstrated by Levine et al.[72] and most of the known *Klebsiella* types have been examined and found to be polyuronides.

d. Escherichia coli. Beiser and Davis[73] have recently obtained mucoid mutants of *E. coli* that produced the same capsular material. The isolated polysaccharide contained 0.3–0.4% nitrogen, 63–65% reducing sugar (as glucose), 25–29% methylpentose, and 23–25% hexuronic acids. Fucose, galactose, and an unidentified third component were detected by paper chromatography.

e. The Microcapsular Smooth and Vi Antigenic Substances. The smooth antigenic complexes form a layer outside the rigid cell wall and the fact that bacteria do not produce such components under certain growth conditions suggests that the O antigen is in no way associated with the mechanical rigidity of the wall structure. However, the precise relationships of the PPL smooth antigenic complexes to the underlying bacterial walls have not been investigated, although it is apparent that some of the sugars characteristic of these components are present also in the isolated cell wall fractions.[74] Whether the O antigen is chemically attached or physically adsorbed to the outer wall surface is not known.

The smooth antigenic substances are released from the surface of acetone-dried bacteria by a variety of solvents, diethylene glycol and trichloroacetic acid (TCA) being the two most widely used. Tryptic digestion of heated bacteria also releases the PPL complexes.

Much of the early work on the chemical characterization of the surface antigenic components of certain Gram-negative bacteria was carried out by Morgan and his collaborators and has been reviewed by Morgan.[75] The material extracted from acetone-dried cells of *Shigella dysenteriae* with diethylene glycol was composed of protein, polysaccharide, and lipid and the complex could be dissociated into the individual components by various procedures.[75] The polysaccharide of *S. dysenteriae* contained glucosamine, galactose, and rhamnose.[75, 76] Davies et al.[77] reinvestigated the isolation procedures and the properties of electrophoretically homogeneous O somatic antigen and confirmed the protein-polysaccharide-phospholipid nature of the product. Similar complex substances have been isolated by Goebel et al.;[78] they found acetylglucosamine, glucose, and rhamnose in the specific polysaccharides of *Shigella flexneri*.

More recent extensive investigations of the smooth antigenic components and the immunologically specific polysaccharides and the lipopolysaccharide complexes have been carried out in several schools, notably those of Davies, Goebel, and Westphal. Many of the polysaccharides isolated from these surface components of Gram-negative bacteria contain amino sugars, the methylpentoses, rhamnose, and fucose, as well as hexoses. In addition to these monosaccharide constituents several new sugar components of bacterial polysaccharides have been found, the heptoses and deoxymethylpentoses.

In their studies of the pyrogenic substances of Gram-negative bacteria, Westphal et al.[79] obtained lipopolysaccharides by extracting the organisms with warm, 45% phenol solution. Lipopolysaccharides obtained from *E. coli*, *Salmonella*, and *Shigella* species contained hexosamine, galactose, glucose, and rhamnose as the more frequently occurring constituents.[80] The properties and results of quantitative and chromatographic analyses of a number of the products of Gram-negative bacteria have been summarized by Westphal and Lüderitz.[80] In addition to the known sugar components of these lipopolysaccharides, two new sugars were found. These were subsequently identified as the bisdeoxyhexoses (deoxymethylpentoses), abequose and tyvelose,[81] and they were detected in the lipopolysaccharide complexes from strains of the *Salmonella* species and the *E. coli* group investigated by Westphal and his colleagues.[80] Davies[82] subsequently found sugars of similar chromatographic behavior to tyvelose and abequose in the specific polysaccharides of a number of *Salmonella* species. A new sugar, paratose, has been found in the polysaccharide of *Salmonella paratyphi* A;[83] the evidence suggests that paratose, together with abequose and tyvelose, belongs to the naturally occurring deoxymethylpentose sugars,[83] now characterized as $3:6$-bisdeoxy-aldohexoses.[84]

The presence of an aldoheptose in a bacterial product was first reported by Jesaitis and Goebel.[85] The polysaccharide isolated from *Shigella sonnei* contained an aldeheptose in addition to glucose, galactose, and glucosamine. Since this investigation, there have been numerous reports of the detection of aldoheptoses in bacterial polysaccharide and lipopolysaccharide complexes. Weidel et al.[86] presented chromatographic evidence for the presence of an aldoheptose in the isolated walls of *E. coli* and Weidel[87] subsequently isolated and identified the component as L-gala-D-mannoheptose. Davies and his collaborators have now detected aldoheptoses in a variety of Gram-negative bacteria,[82, 88, 89] and Salton[90] has also found heptoses in the isolated walls of a number of Gram-negative organisms. The exact anatomical status of the lipopolysaccharides containing these heptoses is at present unknown, although it seems probable that the lipopolysaccharide complexes of rough bacteria may be derived from a layer of the cell-wall structure possessing a rigidity greater than the phenol-soluble, lipoprotein part of the wall.[91]

Davies[82] has used TCA extraction of acetone-dried bacteria for the isolation of the specific polysaccharides. Ethanol fractionation of the extracts yields essentially two classes of substances: (1) a protein-polysaccharide-phospholipid antigen and (2) a free, specific, undegraded polysaccharide.[82] The degraded polysaccharide haptenes were liberated by mild acetic acid hydrolysis, a procedure believed to release the polysaccharide by splitting a relatively labile link between acetylglucosamine and other sugar residues.[82] The polysaccharides from *Salmonella* spp., *Shigella* spp., and strains of *Chromobacterium violaceum* were examined; galactose, glucose, and rhamnose were the monosaccharides most frequently encountered.[82] Other sugars found in some of the polysaccharides were xylose, arabinose, and fucose.[82]

The relationship of the smooth antigen to the underlying surface has not been investigated in any detail and, as pointed out by Davies,[92] it generally has been assumed that the subsurface layer of a smooth strain corresponds to that of the rough form. However, in a study of the nature of the rough somatic antigen from *Shigella dysenteriae*, Davies[92] was unable to substantiate this widely held belief. The polysaccharide of the rough,

TABLE III

THE NATURE OF CAPSULAR, MICROCAPSULAR, SLIME, AND EXTRACELLULAR SUBSTANCES OF VARIOUS GRAM-NEGATIVE BACTERIA

Organism	Anatomical status of surface component	Class of substance and products of acid hydrolysis
Acetobacter spp.	Extracellular slime or fibrils	Polysaccharide: cellulose (amorphous and crystalline)—glucose
Acetobacter capsulatum	Capsule	Polysaccharide: dextrin—glucose
Haemophilus influenzae	Capsule	Polyribophosphate
Aerobacter aerogenes	Capsules and loose slime	Polysaccharide: glucose, fucose, glucuronic acid
Aerobacter-Klebsiella	Capsules, loose slime extracellular	Polysaccharides: complex polyuronides
Escherichia coli	Capsule	Polysaccharide: fucose, galactose and hexuronic acid
Escherichia *Salmonella* *Shigella*	Microcapsular components: (a) smooth surface antigens and rough antigens	Protein-polysaccharide-lipid complexes and lipopolysaccharides; monosaccharide constituents frequently encountered: rhamnose, fucose, abequose, tyvelose, paratose, hexoses, heptoses, and amino sugars
Salmonella	(b) Vi antigen	Amino uronic acid

avirulent strain of *S. dysenteriae* contained glucosamine, glucose, galactose, and an aldoheptose. No rough components could be detected in the smooth organisms[92] and Davies[89] has, therefore, suggested that "the S-R change reflects a change in polysaccharide constitution and not the loss of a smooth polysaccharide to reveal beneath a substance similar to that exposed on the surface of rough organisms."

The only other surface component of gram-negative bacteria relevant to the present discussion is the so-called Vi antigen. There is abundant evidence that strains of virulent *Salmonella typhosa* possess a surface antigenic component that is located outside the O antigen. This material reacts with ferric chloride[60] and the surface structure is then detectable in the electron microscope; extraction of the cells with hot water removes the Vi component with a concomitant disappearance of the surface layer.[93]

Clark and associates[94] have recently isolated the Vi antigenic component from *S. typhosa*, *E. coli*, and *Paracolobactrum ballerup* and investigated its chemical composition. The data make it probable that the Vi antigen is principally a polymer of N-acetylated aminohexuronic acids, linked glycosidically with the α-configuration through the 1 to 3 positions.

The properties of various capsular, microcapsular, smooth and Vi antigens, and slime components of Gram-negative bacteria are summarized in Table III.

C. Action of Enzymes on Capsules, Slime, and Other Extracellular Surface Components

Because of their greater capacity for more selective action, enzymes rather than chemical reagents have proved extremely useful in the removal of bacterial surface components. The successful enzymic removal of capsules without affecting the viability of the treated cells was first demonstrated for several types of pneumococci by Avery and Dubos.[8] This selective removal of the capsule followed from the isolation of a microorganism on a medium containing the type-specific polysaccharide as the sole carbon source. Thus, Dubos[95] was able to show the production of a specific, inducible enzyme in the presence of the capsular polysaccharide or its constituent cellobiuronic acid residues. The enzyme so formed degraded the type III but not the type VIII polysaccharide substance. The types of linkage attacked by the bacterial enzyme have not been investigated. Other soil microorganisms producing enzymes capable of degrading pneumococcal capsular polysaccharides include *Rhodobacillus palustris* (attacks only type VIII) and a myxococcus.[96] Autolytic enzymes produced by pneumococci themselves are also capable of breaking down the capsular substances but the polysaccharides are still in a form precipitable with specific antisera.[97]

Various surface components of streptococci have been removed enzymically. Lancefield[10, 36] demonstrated that the M and T proteins could be removed by digesting the group A streptococci with trypsin, the viability of the cells being unaffected. However, Slade[42] has shown more recently that his strain of group A *Streptococcus pyogenes* becomes more permeable and loses viability when high concentrations of trypsin are used for digestion of the M-protein component. The hyaluronic acid capsular material of streptococci can be removed with the enzyme hyaluronidase; as mentioned above, this increases the sensitivity of some streptococci to infection with bacteriophage.[40]

The action of enzymes on the capsular and extracellular γ-glutamyl peptides of *Bacillus* species has been investigated with enzymes from a variety of cells and tissues. Enzymes from red blood cells, kidneys, and other tissues break down γ-glutamyl polypeptides, liberating free glutamic acid and small peptides.[98-100] Thorne[47] has investigated the action of an enzyme obtained from *Bacillus subtilis* on microbial γ-glutamyl peptide and a number of synthetic substrates. He has shown that γ-D-glutamyl-D-glutamic acid, γ-L-glutamyl-L-glutamic acid, and α-L-glutamyl-D-glutamic acid are hydrolyzed, as well as the polypeptide of *Bacillus anthracis*.[47] Volcani and Margalith[100] have recently isolated an organism (*Flavobacterium polyglutamicum*) which hydrolyzes the γ-L-glutamyl bond in polypeptides.

Some years ago, Humphries[101] reported a capsule lysin active against *Klebsiella pneumoniae*, the lysin being present in bacteriophage lysates. Adams and Park[9] have investigated a similar capsule-degrading enzymic system from *K. pneumoniae*. In common with the enzymes active against *Streptococcus pneumoniae* capsular polysaccharides, the *K. pneumoniae* enzyme removed the capsule without killing the bacteria and also protected mice against infection with the virulent bacteria. The partially purified enzyme was effective in reducing the viscosity of solutions of the isolated capsular polysaccharide, but Adams and Park[9] concluded that the hydrolytic products were still of relatively large size. The enzyme was unable to alter the viscosity of solutions of the capsular polysaccharide of type II pneumococcus which is serologically related to type 2 *Klebsiella*.

Although trypsin has been used for the extraction of the smooth antigenic substances from heat-killed bacteria, there is little evidence that enzyme systems analogous to those discussed above have been obtained for the selective removal of the microcapsular smooth antigens from gram-negative bacteria.

Some of the properties, action and sources of the enzymes degrading the surface capsule, slime, and extracellular components of the bacterial cell are summarized in Table IV.

TABLE IV

Action of Enzymes on Capsular and Slime Substances of Bacteria

Organism	Nature of surface component	Source of enzyme
Pneumococci	Capsular polysaccharides	(a) Autolytic enzymes (b) Microbial enzymes: specific enzymes separately degrading types III and VIII polysaccharides
Streptococci	Hyaluronic acid	Hyaluronidase from animal and microbial sources
	M protein	Trypsin, proteolytic enzymes
Bacillus anthracis	Capsular polypeptide	(a) Enzymes from various animal tissues: kidney, blood
	γ-Glutamyl peptide	(b) Enzymes from Bacillus subtilis and Flavobacterium polyglutamicum
Klebsiella pneumoniae	Capsular polysaccharide	Enzyme from bacteriophage lysed culture of K. pneumoniae

IV. Cell Walls

Underlying the surface capsules, slime, and extracellular substances of most eubacteria is a rigid cell wall. Bacteria, in common with plants, possess a well-defined wall which can be cytologically differentiated from the surface slime and capsular components on its external boundary and from the protoplasmic membrane and protoplasm encased by the wall. The existence of a cell wall has been amply demonstrated in a variety of ways—by staining, plasmolysis, microdissection,[102] disintegration of bacterial cells and examination in the electron microscope,[103-105] and by isolation of the bacterial cell wall as a single morphological entity.[106] With the development of electron microscopy and suitable methods for the isolation of homogeneous preparations of microbial cell walls, our knowledge of the chemistry, biochemistry, and functions of the walls has been greatly advanced. This review will thus deal largely with the properties of the isolated walls.

A. Isolation of the Bacterial Cell Wall

Many of the early techniques used for the isolation of material believed to be the rigid cell wall have been discussed by Knaysi[107] and Salton.[74] It is evident from the results of Salton and Horne[106] that the alkali-insoluble residues of bacteria, formerly assumed to be the cell-wall structures, do not correspond to the walls isolated by mechanical disintegration. Abundant evidence has now accumulated to show that extraction of bacteria, or indeed the isolated walls, with alkali or acids will lead to a degradation of part of the wall structure.[74] Thus, dilute sodium hydroxide removes O-acetyl groups from the walls[108] and TCA extracts a number of compounds from the isolated walls.[109, 110]

The two main methods used for isolation of bacterial walls depend upon mechanical disintegration of the cells on the one hand, and on the other, autolysis and digestion of the bacterial protoplasm to give residues of the more resistant cell walls. Other methods of isolation which may possibly involve the participation of biochemical breakage, autolytic processes, and rupture of weak bonds holding parts of the wall together have been used. Salton and Horne[106] found that the walls of certain Gram-negative bacteria rupture when the cells are suddenly raised to temperatures of 70 to 100°C. Osmotic lysis of halophilic bacteria will also result in a disintegration of the bacterial wall,[15] but to what extent such processes are purely mechanical is not known. Another method of cell rupture which is potentially applicable to the isolation of walls is the decompression method of Fraser.[111]

1. Mechanical Disintegration

a. Vibration in the Presence of Small Glass Beads. Following the work of Alexander and King[112] on the loss of viability resulting from vigorous shaking with ballotini glass beads, Dawson[105] showed that the cells of an organism such as *Staphylococcus aureus* were disintegrated to give a preparation containing residual unbroken cells, empty cell walls, and small electron-dense particles of cytoplasmic origin. The Mickle[113] tissue disintegrator has been widely used as a means of providing rapid, vigorous shaking of bacterial suspensions.[105, 106, 114] In the hands of the author, many bacterial species have been disintegrated by shaking equal volumes (10 ml.) of bacterial suspension (containing 10–20 mg. dry weight bacteria/ml.) and Ballotini beads (grade 12, 0.13 mm. diameter) for 10 minutes in the Mickle disintegrator. Exceeding a suspension density of about 20 mg. dry weight per milliliter and using larger or smaller Ballotini beads will reduce the efficiency of disintegration and the subsequent cell wall fractions will contain higher proportions of intact bacteria. Cooper[115] has investigated in some detail the factors affecting the disintegration of *Staphylococcus aureus*.

Shockman *et al.*[116] have recently described a centrifuge shaker head which can be adapted for the disruption of bacteria at low temperatures. The bacterial suspension is shaken with 0.2 mm. beads at a speed of 1900 to 2000 r.p.m. in a stainless steel capsule on an International shaker head mounted in an International refrigerated centrifuge.[116] The advantage of this method enables temperatures near 0°C. to be maintained throughout the disintegration period and as much as 6 g. of dry bacteria can be disrupted in a single, short operation (about 10 minutes).

Other mechanical methods used have included the disintegration of bacteria in a ball mill[31] or in the Hughes press.[117]

b. Sonic and Ultrasonic Disintegration. The use of sonic oscillators for the disintegration of bacteria has been known for some time but certain bacteria, in particular the cocci, are refractory.[118] Salton[119] obtained satisfactory disintegration for cell-wall isolation of a number of rod-shaped bacteria in the Raytheon sonic oscillator. Micrococci were not disrupted after treatment for 1 hour under comparable conditions that gave a satisfactory disintegration of *Escherichia coli* and *Bacillus megaterium* in 10 minutes.

It is evident from the results of Slade and Vetter,[120] working with *Streptococcus pyogenes*, and Marr and Cota-Robles,[121] studying *Azotobacter vinelandii*, that sonic and ultrasonic disintegration leads to a fragmentation and solubilization of the bacterial wall structure. Marr and Cota-Robles[121] concluded that the primary action of ultrasound on the cells of *A. vinelandii* is the disintegration of 40% of the "cell envelope" into submicroscopic particles of varying dimensions. Salton[122] found that the walls of *Rhodospirillum rubrum*, isolated by disintegrating the cells for 8 minutes in the Raytheon sonic oscillator, could be completely disintegrated when the suspensions of the isolated walls were placed back in the Raytheon for 30 minutes.

Bosco[123] has shown that the walls of *Bacillus mycoides* resist exposure to ultrasonic oscillation (400 kc./sec.) and that the walls can be isolated by rupturing the cells by exposure for 4 to 5 minutes. In an earlier study Gale and Folkes[124] has shown that controlled exposure of *Staphylococcus aureus* to ultrasound gave disrupted cell preparations in which a cytoplasmic residue was surrounded by the bacterial wall.

2. Isolation of the Walls from Mechanically Disintegrated Bacteria

After mechanical disintegration of bacteria, the beads or abrasive agents are removed by simple filtration or sedimentation. This step is of course obviated by sonic or ultrasonic disintegration. The subsequent steps in the isolation of the walls are similar for both mechanically and sonically disrupted bacteria; most of the methods used depend upon differential centrifugation. The bacterial walls have to be separated from unbroken bacteria and coarse debris, granules of intracellular origin, small particles such as those derived from the protoplasmic membrane,[125, 126] flagella fragments, and the "soluble" cytoplasmic constituents. Thus, the separation of walls is generally achieved by centrifuging at different gravitational fields to give a separation of the walls from the heavier intact cells and lighter particulate and soluble fractions.[74, 114]

Instead of using differential centrifugation, Albertsson[127] has recently shown that particle fractionation in liquid two-phase systems can be used in the isolation of the walls from disintegrated microorganisms. The cell walls of a species of *Aerobacter* separated in the salt-rich phase of a phosphate-polyethylene glycol system and the rest of the cellular contents were found at the interface.[127]

The cell-wall fractions obtained by an initial separation after differential centrifugation are, of course, contaminated with adherent soluble and denatured protein, nucleic acids, and cellular particles. The subsequent steps in the isolation procedure involve a reduction of the amount of contaminating material by washing the walls with distilled water, molar solutions of sodium chloride, buffers, and by digesting with proteolytic enzymes and ribonuclease.[31, 74, 114] The walls of gram-positive bacteria are generally rather easier to free of contaminating cytoplasmic material than are the structures isolated from Gram-negative organisms.[74] As pointed out by Barrington and Kozloff,[128] even after prolonged washing of the walls of *Escherichia coli*, small amounts of soluble material continue to appear in the washings.

The final test for the morphological homogeneity of the cell-wall prepara-

Fig. 3. The isolated cell wall of *Bacillus megaterium* as seen in the electron microscope (Salton and Williams[136]).

tions obviously must reside with the results of examination in the electron microscope, for only the latter will detect fine structures, such as fimbriae, flagella, granules, and particles of intracellular origin. Walls isolated by mechanical disintegration followed by differential centrifugation and washing generally give satisfactory preparations devoid of electron-dense cytoplasmic materials, as illustrated in Fig. 3.

3. Isolation of Bacterial Walls by Autolysis and Digestion of Bacterial Protoplasm

Weidel[129] isolated the walls of *E. coli* by allowing the bacteria to autolyze in the presence of toluene and removing the bulk of the bacterial protoplasm by digestion with proteolytic enzymes. The cell-wall material so obtained behaved in a homogeneous manner on examination in the ultracentrifuge.

Norris[130] has shown that the lytic principle associated with cultures of *Bacillus cereus* resulted in the digestion of the cell contents of a number of *Bacillus* spp. and gave very clean cell-wall residues of an organism such as *Bacillus sphaericus*. Some autolytic systems may therefore have potential value for the preparation of cell-wall residues.

The mechanical methods of disintegrating bacteria for the isolation of cell walls seem to offer several advantages over the autolytic methods. The latter methods may give preparations with particulate material which has resisted enzymic digestion, mechanically trapped inside the undamaged wall. With mechanical rupture of the bacterial cell the intracellular granules and components can be removed. Mechanical disintegration under the conditions described by Shockman *et al.*[116] offers two great advantages of temperature control and excellent capacity (6 g. bacteria at a time) for large-scale preparations. Disadvantages of mechanical disruption at ambient temperatures include the contamination of the preparations with denatured protein and effects of unspecified temperatures on autolytic enzymes that may have some cell-wall degrading capacity.

For biochemical work, the methods involving disintegration and isolation of walls by the procedures described above may be too time consuming and may also impose a severe limitation on the number of samples that can be investigated. To overcome this Hancock and Park[131] have devised a method enabling a rapid sampling and good recovery of the cell-wall material. This method involved the extraction of cells with 5% TCA at 100°C. followed by trypsin digestion and washing; under these conditions contamination with cytoplasmic material did not exceed about 5%. However, this method would not be applicable to the isolation of walls for quantitative chemical studies, as the ribitol phosphate polymers recently found by Baddiley *et al.*[110] are extractable in TCA.

B. Physical Properties

The bacterial cell wall accounts for about 20% of the dry weight of the cell.[74, 114] The contribution by the wall to the weight of the cell will vary with the age of the culture and, as shown by Shockman et al.[132] with *Streptococcus faecalis*, it may vary from 27% during the exponential phase to 38% during the stationary phase.

The isolated walls vary in thickness from about 100 to 200 A.[74, 114] Measurements performed on ultrathin sections are in general agreement with those obtained by measuring the thickness on air-dried preparations of the isolated walls. Thus, Birch-Andersen and Maaløe[133] found that the wall of *Escherichia coli* is 100 A. thick. Recent sections of intact *E. coli* taken by Kellenberger and Ryter[134] have shown that the wall of *E. coli* consists of a double structure, the two electron-dense concentric layers being separated by a space of 60 to 80 A.

Although the walls of many Gram-positive bacteria have a homogeneous appearance, as seen in the electron microscope, fine structures have been detected in some isolated walls, especially in those of Gram-negative bacteria. Houwink[135] demonstrated the presence of macromolecular components in the walls of a large *Spirillum* sp.; in *Spirillum serpens* the macromolecular spheres possessed diameters of 120 to 140 A. and were arranged in a hexagonal pattern.[135] Houwink suggested that the wall was composed of three layers, the spherical macromolecules forming the inner layer.

Similar spherical macromolecules of about 100 A. diameter were found in the walls of *Rhodospirillum rubrum* and they too were packed in a hexagonal fashion.[136] A different type of microstructure has been observed by Labaw and Mosley.[137] They found a rectangular arrangement of spherical macromolecules (diameters of 115 A.) in the wall of a Gram-positive organism. Houwink[138] has also observed a fine structure in the walls of *Halobacterium halobium*.

The isolation of spherical, macromolecular phage receptors from the walls of *E. coli*,[139] together with the detection of macromolecules in the walls of spirilla and related species, suggest that such subunits may be more common structural elements of the cell walls of Gram-negative bacteria.

C. Chemical Composition

Until suitable methods were developed for the isolation of bacterial cell walls in a satisfactory state of homogeneity for chemical analysis, the question of their composition was open to much speculation. Many of the early studies were based on qualitative tests of dubious specificity with bacterial wall residues of doubtful anatomical identity. However, over the space of the past six years or so, a great deal of qualitative information about the chemical constitution of bacterial walls has accumulated[74, 114, 140] and more is now known about the composition of the walls of bacteria than any other group of microorganisms.

One of the first most striking features to emerge from a comparative study of the composition of the walls of a number of Gram-positive and Gram-negative bacteria was the difference in amino acid composition.[32, 41] The walls of certain Gram-positive bacteria contained as few as three or four major amino acid constituents.[41] Aromatic and sulfur-containing amino acids were absent from the walls of Gram-positive bacteria, whereas those isolated from Gram-negative organisms possessed a variety of amino acids comparable to those found in many proteins.[47, 74] Another outstanding feature was the difference in the lipid contents; the walls of Gram-negative bacteria contained appreciable amounts of lipid (up to 20%), in contrast to very small amounts in the walls isolated from Gram-positive bacteria.[41, 74] In addition to differences in lipid and amino acid constituents, many Gram-positive bacteria possess higher amino sugar content, but this is by no means invariably so. Table V summarizes some comparative data on the lipid, carbohydrate, and amino sugar content of bacterial walls from both Gram-positive and Gram-negative bacteria.

Neither Gram-positive nor Gram-negative bacteria appear to possess

TABLE V

A Comparison of the Lipid, Polysaccharide, and Amino Sugar Content of the Walls of Several Gram-Positive and Gram-Negative Bacteria[a]

Organism	Dry weight of the isolated cell walls, (%)		
	Lipid	Polysaccharide (reducing values)	Amino sugar (as glucosamine)
Gram-positive			
Bacillus subtilis	2.6	34	8.5
Micrococcus lysodeikticus	<2	45	16–18
Sarcina lutea	<2	46	16
Streptococcus faecalis	2	61	22
Streptococcus pyogenes	0	55–62	18–22
Gram-negative			
Escherichia coli	22	16	3.0
Rhodospirillum rubrum	22	23	2
Pseudomonas aeruginosa	11	16–17	2.1–2.7
Salmonella pullorum	19	46	4.8
Vibrio metchnikovi	11	12	1.9
Vibrio costicolus	11.8	1.6	—

[a] Data collected from references 31, 32, 74, and 180.

nucleic acid in the isolated walls as judged by the ultraviolet absorption spectra.[74, 106] However, Barkulis and Jones[141] have recently shown that small amounts of material with an absorption maximum at 260 mμ can be extracted from the walls. It is unlikely that the nucleic acid is part of the polymeric substance of the rigid wall structure.

With the discovery of diaminopimelic acid (DAP) by Work[142] and its detection in the cell walls of various bacteria, the isolation of muramic acid by Strange and Dark[143] and its localization in the walls of all bacterial species so far investigated,[74, 114, 140, 144, 145] and the detection of D-amino acids,[145-148] a more complete picture of the chemistry of bacterial walls has emerged. It is now evident that the entire wall of many Gram-positive bacteria is composed of a polymer of three or four principal amino acids, acetyl glucosamine, and acetyl muramic acid; in addition sugars, galactosamine, and ribitol phosphate[110] may also be present. For convenience this peptide-amino sugar complex has been referred to as the mucocomplex substance[74, 114, 149] to distinguish it from (1) mucopolysaccharides[29] possessing predominantly carbohydrate properties, and (2) mucoproteins[29] whose characteristics are essentially protein in nature, in marked contrast to the peptide complexes of the bacterial wall. There is now good evidence that the mucocomplex substance characteristic of the whole cell-wall structure of Gram-positive bacteria is also found in the walls of gram-negative organisms, but in the latter structures it may account for a much smaller proportion of the cell-wall substance.[91, 140, 149-151] Thus, as emphasized by Work,[140] the walls of both Gram-positive and Gram-negative bacteria possess a common basal cell-wall structure, in which D-alanine, D-glutamic acid, muramic acid, glucosamine, and frequently DAP appear as the monomeric building blocks.

The term "mucocomplex" is not an entirely satisfactory word for describing the peptide-amino sugar-sugar polymers of the rigid wall; the recent discovery of the ribitol phosphate polymers,[110] now called teichoic acids by Armstrong et al.,[152] make the usage of "mucocomplex" even less satisfactory. For the present, the term "mucocomplex" may be used in a general sense to distinguish the "muramic peptides" from other classes of substances, such as proteins, lipids, and polysaccharides and their complexes, these latter occurring in the walls of most Gram-negative bacteria. When more is known about the preponderance of peptide or polysaccharide in the wall mucocomplexes, then more descriptive terms, such as "muramic peptides" or "muramic polysaccharides" may be useful. With the recent discovery of the teichoic acids[152] and O-acetyl groups in the bacterial wall,[108] the list of unusual features of the cell-wall mucocomplex is steadily growing. As the proportions of such components in the wall may vary widely from one species to another, the task of defining satisfactorily the cell-wall polymer in a simple chemical term may become more and more difficult.

1. Amino Acid Composition

Although a great deal is known about the qualitative amino acid constitution of bacterial cell walls, quantitative data are available for only several organisms. The quantitative information available for several Gram-positive bacteria is summarized in Table VI, with two independent assays of the amino acid composition of *Escherichia coli* walls. These results are in general accord with the qualitative data obtained from paper chromatographic studies and the data also serve to emphasize the predominance of a few amino acids in the cell walls of Gram-positive bacteria.

Alanine and glutamic acid have been found in the walls of all bacteria so far investigated; in Gram-positive bacteria, alanine quantitatively predominates over the other amino acid constituents. The molar ratios of the principal amino acids in the cell walls of a number of Gram-positive bacteria[153] are presented in Table VII. The data obviously give a more accurate idea of the relative abundance of cell-wall amino acids than that conveyed by the intensities of chromatogram "spots."[74, 114, 144] Apart from alanine and glutamic acid, the amino acids next in abundance in the walls of different Gram-positive species are lysine, glycine, and DAP.[74, 114, 140,]

TABLE VI

The Amino Acid Composition of the Isolated Cell Walls of Several Bacteria

Amino acid	*Lactobacillus casei*[147] (g./100 g.)	*Streptococcus faecalis*[147] g./100 g.	*Escherichia coli* Molecular proportions[225] (%)	*Escherichia coli* Grams per 100 grams[a]
Alanine	8.4	12.0	14.3	5.6
Arginine	—	—	—	3.8
Aspartic acid	3.1	0.8	14.6	7.1
Diaminopimelic acid	6.7	—	—	—
Glutamic acid	7.2	5.4	9.1	6.9
Glycine	1.0	0.2	10.9	3.1
Histidine	—	—	—	0.9
Isoleucine	1.4	0.4	—	3.7
Leucine	1.4	0.4	14.2	5.3
Lysine	2.3	4.5	4.5	4.0
Methionine	—	—	—	0.7
Phenylalanine	—	—	5.2	3.0
Proline	—	—	1.9	1.5
Serine	0.6	0.2	5.4	3.7
Threonine	1.1	0.2	5.5	3.8
Tyrosine	—	—	6.2	3.3
Valine	1.4	0.24	5.7	3.4

[a] Salton, unpublished data.

TABLE VII
The Molecular Proportions of the Principal Amino Acids in the Cell Walls of Several Gram-Positive Bacteria

Cell walls from	Alanine	Glutamic acid	Glycine	Lysine	Diaminopimelic acid
Bacillus cereus	3	1	0	0	1
Bacillus stearothermophilus	4	2	0	0	1
Lactobacillus arabinosus	3	1	0	0	1
Micrococcus roseus	5	1	0	1	0
Micrococcus urea	2	1	1	1	0
Sarcina lutea	2	1.5	1	1	0
Sporosarcina ureae	2	1.5	1	1	0

[144, 154, 155] Lysine and DAP do not occur together as major amino acids in the walls of the Gram-positive bacteria so far examined and they appear to be strict alternative amino acid constituents. Whether this relationship holds true for the mucocomplex part of the walls of Gram-negative bacteria has not been determined. Aspartic acid, serine, and threonine occur less frequently in the walls of Gram-positive organisms.[154-156] The extensive investigations of Cummins and Harris[144, 154-156] have shown that the principal amino acids are characteristic for a given species and this feature provides a very useful guide in the taxonomy of Gram-positive bacteria.

Smaller quantities and traces of other amino acids, such as valine, leucine, and/or isoleucine are frequently encountered in hydrolyzates of the cell walls of Gram-positive bacteria.[74, 132, 147, 157, 158] Whether these amino acids are part of the cell-wall peptide or whether they originate from residual contaminating protein has not been experimentally resolved. However, whatever their origin might be, it is evident that they account for a small fraction of the total amino acid content of the walls of Gram-positive organisms, such as *Streptococcus faecalis* and *Lactobacillus casei*.[147] Because of the greater variety of amino acids found in the walls of Gram-negative bacteria it is even more difficult to assess the significance of the minor amino acid constituents.

Several other ninhydrin-positive substances of unknown constitution have been detected in cell wall hydrolyzates.[41, 144] Cummins and Harris[144] accounted for a ninhydrin-reacting substance in hydrolyzates of lactobacilli as an acid-resistant peptide of lysine and aspartic acid. A similar peptide has been found in the polypeptide antibiotic, bacitracin,[159] and is probably identical to that occurring in the walls of *Bacillus sphaericus*.[159a, 160]

At least from qualitative information, it seems clear that the amino acid composition of walls from Gram-positive bacteria is a stable feature except under the guidance of mutation. The careful investigations of Shockman et al.[132] have established that walls isolated at different stages in the growth cycle of *Streptococcus faecalis* show no major qualitative changes in composition. However, Graziosi and Tecce[161] claim that the walls of young cells of *Bacillus megaterium* are composed of lipoprotein with a complex amino acid composition and that the composition changes to a "typical" cell-wall peptide when the cells become older. This finding is not in accord with the general belief that the qualitative composition of the wall is a stable property of a particular bacterial species and it is possible that the results of Graziosi and Tecce may have arisen from the method of isolation of wall material used by these authors.

a. Amino Acid Isomers. It has been known for some time that D-amino acids, other than those associated with the glutamyl polypeptide, occur in a number of microorganisms.[162, 163] The significance of this finding was not realized until Snell et al.[146] demonstrated the presence of D-alanine in the cell walls of *Lactobacillus casei* and *Streptococcus faecalis*. The insoluble residues of *L. arabinosus* corresponding to the cell-wall fractions also contained D-alanine.[146] The walls of *L. casei* contained 5.0% D-alanine and 8.3% total alanine, while the D-alanine values for *S. faecalis* walls were 3.9% and 13.9% respectively.[146] Snell et al. found that D-alanine occurred in two distinct forms in the bacterial cell; in the wall, and in one or more unidentified derivates extracted from the cells by hot TCA. The latter material was especially labile to alkaline hydrolysis[146] and it now seems probable that the TCA-extractable material containing D-alanine is derived from the teichoic acid polymer of the wall which Armstrong et al.[152] have recently shown to contain ester-linked alanine.

Salton[145] has shown that D-alanine occurs in the walls of many Gram-positive bacteria; smaller amounts were chromatographically detectable in the walls of several Gram-negative species. D-Alanine is one of the products released from walls digested with the enzymes Ghuysen[164] has isolated from *Streptomyces albus*.[165]

Until recently D-glutamic acid of microbial origin has only been found in the capsular polypeptide of *Bacillus* species[46-54] and in the nucleotides isolated from penicillin-treated *Staphylococcus aureus* by Park.[106, 148, 166] The presence of the D-isomer of glutamic acid in the isolated cell walls of a number of Gram-positive bacteria has now been established[147] and it is evident that for some species all of the cell-wall glutamic acid may be the D-isomer.[145] Park and Wynngate (see Park[148]) have found D-glutamic acid in the walls of *Escherichia coli* and have made the very interesting observation that there is a fairly constant molecular ratio of muramic acid to D-glutamic acid in the walls of a number of Gram-positive bacteria and in

E. coli. This further illustrates the similarity of the mucocomplex substance in both groups of bacteria.

The only other amino acid known to occur in bacterial walls in different isomeric forms is DAP. Following investigations of Hoare and Work[167] on the distribution of the various isomers of DAP in whole microorganisms, studies with isolated walls have established the presence of the corresponding isomers in the walls.

b. N-Terminal Amino Acids. The N-terminal amino acids of bacterial cell walls have been studied by Ingram and Salton,[168] Salton,[145] and Brown.[109] Alanine has proved to be the most frequently encountered N-terminal amino acid. The only other N-terminal amino acid detected in walls of Gram-positive bacteria by reaction of walls with fluorodinitrobenzene (FDNB) has been glutamic acid in the walls of *Micrococcus varians*.[90] Reaction of the walls of Gram-negative bacteria with FDNB gave a number of N-terminal amino acids; alanine, glycine, aspartic acid, and glutamic acid appeared most frequently.[145] It is thus evident that the "protein" constituents of the walls of Gram-negative bacteria are heterogeneous.

In addition to the N-terminal amino acids, Ingram and Salton[168] found that lysine and DAP possess free amino groups. The monodinitrophenyl (DNP) derivative of DAP was detected after hydrolysis of the walls of *Bacillus megaterium* reacted with FDNB. It was estimated that about one-third of the DAP residues of the wall possessed one free amino group per molecule for reaction with FDNB.

2. Amino Sugars

Amino sugars form important constituents of many bacterial cell walls. With the discovery of muramic acid by Strange[143, 169] and its detection as an amino sugar component of bacterial walls,[114, 143-145, 148] great advances have been made in an understanding of the chemistry of the cell wall.

Quantitatively, the amino sugar contents of the walls vary widely from one organism to another.[74, 149] Within the Gram-positive group of bacteria the amino sugars may account for as little as 1% of the wall of *Staphylococcus aureus*[125] and may rise to 30% of the wall of *Bacillus cereus*.[149] The amino sugar content of the walls of gram-negative bacteria[149] range from about 2% for *Vibrio metchnikovi* to 13% for the Gram-negative rumen coccus LC_1, described by Elsden *et al.*[170] A comparison of the amino sugar content of the walls of Gram-positive and Gram-negative bacteria, presented in Table VIII, shows that in general the walls of Gram-negative bacteria are poorer in amino sugars. This, no doubt, is due to the smaller contribution by the mucocomplex material and a correspondingly greater contribution by the lipid, protein, and polysaccharide components.

TABLE VIII
Amino Sugar Content of Cell Walls of Gram-Positive and Gram-Negative Bacteria[a]

Cell walls from	Amino sugar (%) (as glucosamine, 2N-HCl hydrolyzates)
Gram-positive	
Bacillus cereus	31
Bacillus megaterium	18
Micrococcus roseus	15
Micrococcus varians	10
Sporosarcina ureae	14
Staphylococcus citreus	10
Staphylococcus aureus	approx. 1[125], 17[153]
Staphylococcus saprophyticus	14
Gram-negative	
Chlorobium thiosulphatophilum	4.2
Escherichia coli	2.2–4.0
Rhodospirillum rubrum	2
Salmonella gallinarum	3.9
Spirillum serpens	6.8
Vibrio metchnikovi	1.9
Organism LC1	13

[a] Data collected from references 74, 125, 149, 153.

All of the bacterial cell walls so far examined contain both glucosamine and muramic acid.[74, 114, 144, 145, 148] Galactosamine has also been found in the walls of some organisms but it occurs in these structures much less frequently than the other amino sugars.[114, 144] It is probable that all three amino sugars occur in the walls as the N-acetyl compounds. This is supported by the absence of any appreciable quantity of DNP-amino sugars after reaction of the walls with FDNB.[168] Small amounts of a substance corresponding to DNP-muramic acid were, however, detected after reacting lysozyme-digested walls with FDNB.[168] Although several other N-acetyl-amino sugars have been recently detected in products of microbial origin (N-acetylfucosamine[171] and N-acetylneuraminic acid[172]) glucosamine, galactosamine, and muramic acid are the only amino sugars so far known to be in the walls.

The important role that muramic acid plays in the cell-wall mucocomplex became apparent when Strange[169] and later Kent[173] proposed the structure of muramic acid as: 3-O-α-carboxyethyl hexosamine. Thus, muramic acid possesses a carboxyl group available for peptide bond formation and a reducing group for glycosidic bonding to amino sugars and sugars. The evidence available from the structure of muramic acid,[169, 173] the nucleotide complex from penicillin-treated *Staphylococcus aureus*,[148, 166, 174] and the nature of the products of lysozyme digestion[175] and partial acid hydrolysis of bacterial walls[176] establishes the role of muramic acid as the link between peptide and amino sugar and polysaccharide residues in the cell wall. The functions of muramic acid are therefore similar to those performed by N-acetylneuraminic acid in linking carbohydrate residues to proteins in mucoproteins of animal origin.[177]

The release of a substance corresponding to a disaccharide of N-acetylglucosamine and N-acetylmuramic acid by lysozyme action on bacterial cell walls[175] suggests that a disaccharide may constitute a common structural unit in certain bacterial cell walls.

3. Sugars

A variety of sugars have been detected in bacterial cell walls and in several groups of Gram-positive bacteria the cell-wall sugars may be characteristic for members of a given genus.[144, 154, 155] Thus, rhamnose is the distinctive cell-wall sugar for members of the genus *Streptococcus*[144] and arabinose is the typical sugar of the walls of *Mycobacteria*, *Nocardia* sp., and *Corynebacteria*.[144, 155, 157] The cell walls of some bacteria are devoid of sugars, e.g., *Staphylococcus aureus*, *Sporosarcina ureae*.[149, 154] Glucose and galactose are present in the walls of Gram-positive bacteria from a variety of taxonomic groups.[144, 149, 154, 155, 157, 158]

So far, only the configuration of the arabinose and rhamnose from walls of *Corynebacterium diphtheriae* and *Streptococcus pyogenes*, respectively, have been determined. Holdsworth[178] found that the cell-wall oligosaccharide from *C. diphtheriae* contained D-arabinose. Barkulis and Jones[141] found that the rhamnose in the wall of *S. pyogenes* was the L-isomer.

The only cell-wall polysaccharide isolated and characterized is that of *C. diphtheriae*.[178] The purified oligosaccharide had a molecular weight of 1,000 and contained D-galactose, D-mannose, and D-arabinose in the molecular proportions of 2:1:3.[178]

The sugars detected in the isolated walls of Gram-negative bacteria correspond very closely to those found in the smooth antigens of the parent organisms. As pointed out by Salton,[74] it is thus certain that these components are retained on the surface of the wall during its isolation. So far as the author is aware, no studies have as yet been undertaken to differentiate between the sugars of the microcapsular smooth antigens and the sugars of the rigid cell-wall structure of the Gram-negative bacteria. There is, however, good evidence from the excellent work of Weidel[87] that the heptose, manno-D-galaheptose, occurs in the rigid, cell-wall structure of *E. coli* B. Heptoses have also been detected in the isolated walls of *Spirillum serpens* and *Proteus vulgaris*.[90]

Unknown sugar components have been observed in the walls of *Micrococcus varians* and *M. citreus*.[154]

4. LIPIDS

Apart from the relatively high lipid contents in the walls of Gram-negative bacteria,[74] little is known about the nature of cell-wall lipid. It is now generally agreed that the walls of Gram-positive bacteria contain very little, if any, lipid.[74] Whatever lipid may be extractable could easily be derived from contaminating lipoprotein particles originating from the protoplast membrane which in some organisms is now known to be largely composed of lipoprotein.[179] It seems likely that some of the ether extractable materials obtained after hydrolysis of the walls of Gram-positive bacteria such as *Bacillus subtilis* could be degradation products of the ribitol phosphate polymer detected by Baddiley et al.[110]

The walls of many gram-negative bacteria contain between 10–20% lipid.[74] Smithies et al.[180] also found that the walls of the halophilic bacteria, (*Vibrio costicolus*, *Pseudomonas salinaria*, and *Micrococcus halodenitrificans*) were rich in lipids (11–19.9%). The walls of *V. costicolus* and *P. salinaria* were very fragile, but the ease of preparing the walls of *M. halodenitrificans* indicated that there was nothing inherently weak in walls of lipoprotein nature.[180]

Grossbard and Preston[181] have examined the X-ray diffraction patterns of cell walls of several bacteria and they have concluded that, at least in the case of *Escherichia coli*, crystalline lipids are located in the wall and the hydrocarbon chains lie normal to the cell surface. The diffraction lines of the lipid rings in *E. coli* indicate that the crystalline regions are at least 200 A. wide. It would seem reasonable to conclude that the lipid must extend across much of the width of the double-structured wall recently observed by Kellenberger and Ryter.[134]

5. CONSTITUENTS OF ENDOSPORE WALLS

The spore walls or spore "coats," as they are frequently called, have been analyzed by Strange and Dark.[182] In contrast to the walls of vegetative cells, those derived from spores of *Bacillus cereus*, *B. subtilis*, and *B. megaterium* contain a variety of amino acids in addition to the characteristic components of the mucocomplex substance of the vegetative wall. It is evident from this work that the spore coat is composed largely of a structural protein rather than the peptide-amino sugar complex.

D. ACTION OF ENZYMES ON BACTERIAL CELL WALLS

Bacteriolytic enzymes active against a variety of intact bacterial cells have been widely found in nature; in recent years a number of these lytic enzymes have been shown to have a direct action on the cell wall. Animal tissues and secretions contain a bacteriolytic enzyme which Fleming[183] called "lysozyme." In addition to noting the presence of lysozyme in animal tissues, Fleming detected the enzyme in extracts of plants and suggested that the organism *Micrococcus lysodeikticus* also contained a bacteriolytic system akin to lysozyme.[184] There are, then, many powerful bacteriolytic enzymes that may be of animal, plant, or microbial origin.

The principal enzymes known to attack the bacterial cell wall with partial or complete dissolution include:

(1) Lysozyme of animal tissues and secretions, plants, and microorganisms.[149, 185]

(2) Lysins and enzymes associated with lytic effects of blood platelets and immune bacteriolysis; plakin investigated by Amano and co-workers[186] has rather similar properties to lysozyme.

(3) Enzymes, such as actinolysopeptidase[164, 187] and the streptolytic enzymes[188] from *Streptomyces* species.

(4) Enzymes from sporulating, resting, and germinating spores of *Bacillus* species;[189, 190] Richmond[191] has also found a bacterial lysozyme produced during the exponential growth of *Bacillus subtilis*.

(5) "Autolytic" enzymes from a variety of bacteria; some of these systems may result in a minimum of damage to the wall, e.g., the enzyme described by Norris[130] and the autolytic system of *Staphylococcus aureus*, capable of splitting the wall into two hemispheres and liberating the bacterial protoplast.[192]

(6) Cell-wall degrading enzymes associated with bacteriophages,[128, 129, 150] phage infection, and host lysis.[193] Such a list of the origin of cell-wall degrading enzymes is by no means comprehensive, but it serves to illustrate some of the main sources found in the past. There is little doubt that many additional groups of enzymes will be added and it seems likely that the endolysins associated with the defective lysogenic bacteria[194] may also contribute a further category of enzymes.

1. Lysozyme

In the past the term "lysozyme" has been generally applied to enzymes capable of lysing certain bacteria, without referring to the types of linkages attacked. However, Salton[185] has pointed out that the lysis of sensitive organisms such as *Micrococcus lysodeikticus* or *Sarcina lutea* is not a sufficiently distinctive property to warrant the classification of an enzyme as a "lysozyme." If the characteristics of egg-white lysozyme are accepted as a basis for classifying lysozymes then in addition to digestion of the walls, evidence must be sought as to the action of the enzyme on the acetylamino sugars of the bacterial mucocomplex substrate.

Egg-white lysozyme has been investigated in greater detail than any of the other enzymes described as lysozyme. It has been known for some time that its action on microbial substrates involves the liberation of reducing groups and acetylamino sugars.[195-197] Salton[175] has shown that the simplest low-molecular weight product liberated by lysozyme is a substance with the properties of a disaccharide of N-acetylglucosamine and N-acetylmuramic acid. Digestion of the cell-wall substrates is not accompanied by the release of any free amino acids, although a small increase in the number of alanine residues could be detected by reaction of the lysozyme-digested walls with FDNB.[168] However, the increase or uncovering of alanine residues with free amino groups is small in comparison with the liberation of reducing or N-acetylamino sugar-reacting groups and there is little doubt that lysozyme belongs to the N-acetylglucosaminidase group of enzymes. There is no direct experimental evidence as to whether the cell-wall amino sugars are α-or β-linked, but the recent evidence that lysozyme can degrade highly purified chitin[198] strongly supports the β-type of linkage. Meyer et al.[196] had concluded in an earlier study that lysozyme did not attack α-glycosidic bonds. Brumfitt et al.[199] have put forward the hypothesis that lysozyme splits a 1–4 link between N-acetylmuramic acid and N-acetylglucosamine. Salton and Ghuysen[199a] have obtained direct experimental proof that it is carbon atom 1 of N-acetyl muramic acid that becomes available for reduction with $NaBH_4$ after the action of lysozyme.

As the bacterial walls contain peptides linked through acetylmuramic acid to other amino sugars (either N-acetylglucosamine or N-acetylmuramic acid) and sugars, it is not surprising that the soluble products formed by the action of lysozyme are a complex mixture. With cell walls such as those from *Micrococcus lysodeikticus, Sarcina lutea,* or *Bacillus megaterium,* lysozyme completely digests them and the nondialyzable fractions are composed of materials having molecular weights of the order of 10,000 to 20,000.[175] The nondialyzable components contain a number of electrophoretically separable products.

Attempts to predict the sensitivity of bacterial cell walls to digestion with lysozyme from a study of the chemical composition of the walls has not been successful.[149] No correlation was found between the amino sugar content and sensitivity to lysozyme. A lysozyme-resistant strain of *M. lysodeikticus* has recently been isolated by Brumfitt et al.[199] and the discovery that it contains O-acetyl groups which on removal renders it sensitive to lysozyme has thrown new light on at least one of the possible factors responsible for resistance to egg-white lysozyme.

Although the mucocomplex part of the walls of Gram-negative bacteria represents a fraction of the wall in some species, it is now clear that lysozyme action is accompanied by a release of this material from the lipoprotein part of the wall structure.[149] Most of the cell-wall DAP and muramic acid is released into soluble, nondialyzable, and dialyzable fractions after treatment of the walls of several Gram-negative bacteria with lysozyme.[149] Investigations by Hartsell and his colleagues[200-202] have shown that there is a "spectrum" of sensitivities to lysozyme and it seems probable that this may be correlated with the relative proportions of the mucocomplex and lipoprotein-polysaccharide fractions of the walls of Gram-negative organisms.

Richmond[191] has shown that the enzyme produced by growing cultures of *Bacillus subtilis* has a very similar spectrum of activities to that of egg-white lysozyme when tested on various isolated cell walls of Gram-positive bacteria. The nondialyzable products from cell-wall substrates bear a striking resemblance to those formed by egg-white lysozyme and Richmond has classified the enzyme as a lysozyme.

2. Cell-Wall Degrading Enzymes from *Streptomyces* Species

The bacteriolytic enzymes produced by *Streptomyces* species have been investigated in detail by Welsch,[187, 203] and Ghuysen[164] has more recently isolated from these microorganisms several enzymes in a highly purified state. Although these bacteriolytic enzymes were originally isolated as staphylolytic systems, it is evident that they have a broad spectrum of activity against many Gram-positive bacteria[164, 165] and they appear to differ from the streptolytic enzyme isolated by McCarty.[188]

With the purified enzymes isolated by Ghuysen,[164] it was shown that the walls of staphylococci, *Micrococcus lysodeikticus*, *Sarcina* species, *Bacillus* species, and *Clostridium welchii* (*C. perfringens*) could be digested under the appropriate conditions of ionic strength.[165] Digestion of the walls was accompanied by a liberation of amino acids; alanine and glycine were the principal amino acids appearing as the free substances in the dialyzable products (glycine was, of course, not present in the walls or digests from *Bacillus* species). In addition to free amino acids some small peptides were also found in the dialyzable fractions. At least some of the alanine liberated by the action of these enzymes is the D-isomer.[165]

The work of Ghuysen[164] has established the existence of two distinct enzymes (F_1 and F_2) capable of a synergistic lytic action on intact bacteria and isolated cell walls. These enzymes are more active against the walls of *B. megaterium* than egg-white lysozyme and even the walls of the lysozyme-resistant *C. welchii* (*C. perfringens*) are attacked by this enzyme system. Since alanine is the principal amino acid of the walls of Gram-positive bacteria and is known to be present as both the D- and L-isomer, this may account for the wide spectrum of activity of the streptomyces enzymes. It is now uncertain that the liberation of amino acids is due to a "peptidase" action since recent investigations[199b] have shown that N-acetylamino sugar compounds are also released from walls of *M. lysodeikticus* digested with F_1. If these enzymes attack a specific bond between amino acids and muramic acid it is possible to imagine both types of products arising from the action of a single enzyme.

The enzymes from *Streptomyces albus* responsible for the liberation of the group-specific substances from streptococci[31, 204] differ from the F_1 enzyme system. Although the streptolytic enzyme digests the isolated cell walls of group A streptococci, McCarty[31] has shown that its action involves a liberation of acetylamino sugar substances from the mucopolysaccharide material. Thus, this enzyme resembles egg-white lysozyme in that its action liberates cell-wall acetylamino sugars, but it differs markedly from lysozyme by not attacking *Micrococcus lysodeikticus*[31] and presumably differs also from "actinolysopeptidase" in not releasing wall amino acids.

3. Enzymes from *Bacillus* Species

Lytic enzymes from members of the genus *Bacillus* have been known for a very long time. Nicolle,[205] in 1907, investigated enzymes produced by *Bacillus subtilis*. In more recent years the studies of Greenberg and Halvorson, [206] Strange and Dark,[189] Nomura,[190] Richmond,[191] and Kaufmann and Bauer[207] have focused attention on this group of enzymes.

The autolytic enzyme studied by Greenberg and Halvorson[206] was produced during the sporulation of a strain of *Bacillus cereus*. The enzyme brought about rapid lysis of young vegetative cells of various strains of *B. cereus*, but its action on isolated bacterial walls was not studied. Strange and Dark[189] have also investigated the lytic enzymes produced by sporulating cultures of *B. cereus*; their enzymes will digest the cell walls of a number of species. Dark and Strange[208] have shown that the enzyme will dissolve the walls of *B. cereus* and liberate the bacterial protoplast when digestion is allowed to take place in the presence of sucrose. The linkages attacked by the *B. cereus* enzyme are not known but the typical cell-wall amino sugar-peptide complexes are released from the walls. Strange and Dark[189] have suggested that *B. cereus* possesses two distinct enzymes; one may be concerned with release of free spores from sporangia and the other enzyme may be responsible for the lytic processes accompanying spore germination.

The autolysin liberated from *B. subtilis* when it is allowed to undergo "anaerobic lysis" has been investigated by Nomura and Hosoda.[209] Cell-wall preparations were digested by the autolysin and by lysozyme and polysaccharide substrates isolated from the walls yielded free reducing groups on digestion with the enzyme.[209] The enzymes active in the anaerobic autolysis of *B. subtilis* studied by Kaufmann and Bauer[207] appear to differ from those of Nomura and Hosoda[209] in that the former enzyme will digest the isolated walls of *M. lysodeikticus*. Nomura and Hosoda[209] were unable to demonstrate any lytic action on intact cells of a number of other lysozyme-sensitive bacteria. It remains to be seen to what extent these enzymes are similar or identical to the lysozyme liberated by growing cells of *B. subtilis* under the conditions described by Richmond.[191]

4. Autolytic Enzymes

Although a number of the above-mentioned enzymes from *Bacillus* species are associated with autolysis of the cultures, they differ from the type of autolytic system which involves only a minimum of damage to the bacterial wall. The autolytic systems described by Norris[130] and Mitchell and Moyle[192] do not bring about digestion of the bacterial wall but break the continuity of the wall, exposing the more vulnerable underlying protoplast structures of the bacterial cell. Indeed, Mitchell and Moyle[192] have used this enzyme, which is capable of "cutting" the bacterial wall into two hemispherical shells, for the release of bacterial protoplasts from the lysozyme-resistant *Staphylococcus aureus*. The manner in which these autolytic enzymes act on the bacterial wall is not known.

5. Enzymes Associated with Bacteriophages and Bacteriophage Multiplication

Lysins associated with bacteriophage infection have been known for some time, but only recently have attempts been made to correlate the lytic properties of these lysins with an action on the bacterial cell-wall structures. In discussing cell-wall degrading enzymes in relation to bacteriophage action, it is necessary to keep in mind two possible sources of the enzymes: (1) intracellular enzyme derived from the phage lysed cells; the enzyme may or may not be phage-specific, and (2) enzyme that forms an integral part of the bacteriophage particle. At the present time there is insufficient evidence to say with certainty whether enzymes present in phage lysates represent normal bacterial enzymes or whether their production is directly induced by the infecting bacteriophage particle and therefore controlled by the phage genetic material. The functions of such cell-wall degrading enzymes are quite apparent, for they would provide a means of releasing the bacteriophage from the cell or, with the phage-bound tail enzyme, the infecting particle would have an enzymic method of puncturing the wall for the entry of phage DNA.

Lytic enzymes liberated by bacteriophage infection have been reported for a number of organisms[210-213] but only two systems have been investigated for their action on the bacterial cell wall. An enzyme found during propagation of a group C streptococcal bacteriophage[214] lysed group A cells that were resistant to the phage and, moreover, released rhamnose compounds from the isolated cell walls.[215] The soluble, nondialyzable mucocomplex released by the enzyme possessed the same rhamnose: glucosamine ratio as the original cell-wall substrate and no dialyzable products could be detected.[215] In addition to this enzyme active on streptococcal walls, Murphy[193] has purified an enzyme from phage lysates of *Bacillus megaterium*. This enzyme has a direct action on the isolated walls of the host-species and releases substances containing amino sugars.[193]

It has been suspected for some time that bacteriophage infection involves the participation of an enzymic reaction to effect a penetration of the rigid bacterial wall. The investigations of Kozloff and Weidel and their colleagues have clearly established the existence of such enzymic reactions which appear to be localized in the phage tail. Weidel and Primosigh[91, 150] have recently isolated an enzyme from the bacteriophage particles. Although the isolated bacteriophage tail enzyme studied by Weidel and Primosigh[91, 150] releases only the substances from the mucocomplex layer of the wall of *Escherichia coli* B, there is reason to believe that other phage-tail enzymes may be isolated that will completely digest the entire mucocomplex walls of Gram-positive bacteria. The properties of the phage-tail enzymes will be discussed in the following section on the interaction of bacteriophages with the cell walls.

E. Interaction between Bacteriophages and Cell Walls

Various components of the bacterial surface carry the specific receptor substances for bacterial viruses. The receptor may be a capsular polysaccharide, one of the smooth O antigenic complexes of the bacterial surface, or it may be located on or in the rigid cell wall.

Weidel was the first to demonstrate the existence of a system of phage-receptor substances in the isolated walls of *Escherichia coli* B.[129] The attachment of bacteriophages to isolated cell walls of susceptible organisms has been shown for a number of bacterial species, including *Pseudomonas aeruginosa*,[216] *Staphylococcus aureus*,[217] *Bacillus megaterium* and *Micrococcus lysodeikticus*,[218] and *Salmonella typhi*.[218a] Electron microscopy of these systems has confirmed the attachment of the bacteriophage by their tails. Weidel and Kellenberger[139] have also elegantly demonstrated the tail-first attachment of T5 to the isolated macromolecular receptor spheres from *E. coli* B. The release of DNA fibers from bacteriophages mixed with walls or receptor spheres has also been observed in the electron microscope.[139, 219]

Following Weidel's report[129] that the phage attachment to the walls of *E. coli* B was followed by a gradual disintegration of the walls, Barrington and Kozloff,[220] demonstrated that the phage brought about a release of material from N^{15}-labeled walls of *E. coli*. Further investigations by Kozloff and his colleagues have also shown that *E. coli* phage brings about a release of nonsedimentable material from N^{15}-and C^{14}-labeled cell walls.[128, 221]

Kozloff and Lute[222] have found that each cell wall of *E. coli* B has about 3,000 atoms of zinc and they believe that the zinc is in the form of a Zn-protein in the wall. Removal of the zinc by extraction with TCA did not interfere with the irreversible adsorption of T2 and T4 phages to the wall. However, the amount of C^{14} nonsedimentable material released from the walls by intact phages was substantially smaller than for the unextracted walls. Preincubation of TCA-extracted walls with Zn^{++} restored the ability of intact phage to digest the walls. Bacteriophage modified by freezing and thawing were equally active in digesting the TCA-extracted and the unextracted walls. Kozloff and Lute[222] have suggested that during phage invasion the tail comes in contact with a Zn-protein in the wall, the thiolester bonds in the viral tail are thereby ruptured, and the phage-tail enzyme is thus exposed.

As to the precise nature of the substances released from the bacterial wall by the phage-tail enzyme, Weidel and Primosigh[91, 150] have shown that the principal constituents of the nonsedimentable fraction are alanine, glutamic acid, DAP, and muramic acid. Thus, Weidel and Primosigh[91, 150] have established that the phage enzyme acts on the cell-wall layer that is chemically related to the wall substance of many Gram-positive organisms. There is now abundant evidence that it is also the layer responsible for the mechanical rigidity of the walls of Gram-negative bacteria.[91, 149, 151] Earlier investigations with DNP walls led Koch and Weidel[223] to suggest that the phage enzyme is not type-specific, in contrast to the type-specific dependence of the phage on their different receptors. T2 and T4 phages attach to chemically different sites of the wall, but their enzymes apparently remove the same material from the walls.[91, 223]

The only extensive studies on the chemical nature of the bacteriophage receptors are those of Weidel and his co-workers. Weidel[87] has shown that aldoheptoses and amino sugars are common to the cell wall receptors from *E. coli* B and to the somatic antigen of *Shigella sonnei* also possessing bacteriophage receptor activity for T4.[85] Weidel and Koch[224] have also shown that the bacteriophage possesses a greater degree of specificity toward its receptor than do antibodies, for they were unable to detect any serological differences between the receptors and their analogs (antireceptors) isolated from a phage-resistant strain of *E. coli*. The amino acid compositions of the receptors and antireceptors were moreover, indistinguishable.[225]

With cell walls such as those of *Bacillus megaterium* it seems probable that the bacteriophage receptor activity resides in a part of the mosaic presented by the cell wall polymer.[218] This is supported by the observation that the receptor activity can be destroyed by enzymes, such as lysozyme and actinolysopeptidase, now known to act on different linkages in the walls of *B. megaterium*.[165, 175, 218]

F. Immunological Reactions

The development of suitable methods for the isolation of the bacterial cell wall and for the stabilization of the bacterial protoplast has greatly facilitated an anatomical approach to the immunological structure of the bacterial cell. Tomcsik and Guex-Holzer[226] were able to show that the protoplast of *B. megaterium* presented a different antigenic surface to that of the intact cell or cell wall. Vennes and Gerhardt[227] have confirmed these differences and have shown in addition that antiserum to protoplast or protoplast membranes did not react significantly with intact cells or cell walls. Intact cells adsorbed only a portion of the antibodies to cell walls.[227]

By isolating the cell walls of group A streptococci, McCarty[31] and Salton[7, 41] demonstrated that the group-specific substance formed part of the rigid cell-wall structure. Cell walls prepared by mechanical disintegration retained the M-protein substance during the isolation procedures, but it could be readily removed by digestion with trypsin, leaving behind the uncontaminated cell-wall structure.[41] The digestion of the walls of group A streptococci and the release of the group substance with the *Streptomyces* enzyme was also demonstrated.[31, 41]

McCarty[31] analyzed the group-specific polysaccharide from the walls of a number of types of group A streptococci and found that in general the rhamnose:amino sugar ratios were fairly constant. Certain strains apparently devoid of group-specific carbohydrate were investigated by McCarty and Lancefield.[228] These variants possessed chemically and serologically similar carbohydrates, but the carbohydrates from the strains exhibiting a loss of group A specificity were composed of rhamnose and glucosamine in different ratios (rhamnose/glucosamine ranging from 4.0–6.0) from those normally encountered for the group A carbohydrate (1.5–2.0).

Our knowledge of the immunochemistry of the cell-wall group-specific carbohydrates has been further extended by McCarty's[229] investigations with an enzyme from a soil organism selectively isolated on media containing the variant carbohydrate. Free N-acetylglucosamine is liberated by the enzyme. McCarty concluded that the evidence suggests that the specificity of group A carbohydrate is determined to a large extent by side chains of N-acetylglucosamine which also serve to mask underlying rhamnose-rhamnose linkages with the specificity for the variant carbohydrate.[229]

Cummins[230] investigated the immunological properties of the isolated walls of *Corynebacterium diphtheriae* and found a labile surface antigen (probably a protein) in addition to a deeper group antigen which also occurred in *Corynebacterium ovis*.

A change in the antigenic structure of the cell wall has been described in the dissociation of *Bacillus megaterium* reported by Ivanovics.[231] The alteration of the cell-wall antigen to one entirely different from that of the wall of the original organism is unusual in that it involves an S → R change without loss in the ability to produce the capsular glutamic acid polymer.[231]

No systematic study of the serological properties of the isolated walls of gram-negative bacteria has as yet been reported. Yoshida *et al.*[232] have reported that the cell walls of *Bordetella pertussis* are as effective as intact bacteria in protecting against experimental infections with that organism. Thus, the surface protective antigens are retained by the cell wall during the isolation procedures. Shafa[233] has also found that isolated cell walls of *Salmonella gallinarum* and *Vibrio metchnikovi* present a surface which is serologically identical to that of O suspensions of these organisms. Burrows[234] studied the electrophoretic properties of protein and polysaccharide components derived from the cell walls of *Vibrio cholerae*. The serological differences between O and R strains were supported by differences in the behavior of the major polysaccharide component of the wall.

Bacterial cell walls as well as the zymosan derived from yeast cell walls have been shown to play a part in immunity phenomena involving the properdin system.[235-237] Isolated walls of both gram-negative and gram-positive bacteria can adsorb properdin factors from serum and in a number of instances the properdin can be recovered from the walls.[236, 237]

G. Biosynthesis of the Cell Wall

Much of the current interest in the biosynthesis of the bacterial cell wall has been stimulated by the discovery that penicillin interferes, in some manner as yet unknown, in the formation of the rigid wall structure.[148, 174] With the knowledge of the types of compounds present in the bacterial walls it has also been possible to relate certain lytic phenomena to metabolic disturbances affecting the cell walls.[238] Thus, at the present time there is little direct biochemical information about the formation of the cell wall polymer but much circumstantial evidence about antibiotic interference with cell-wall synthesizing systems is accumulating.[148, 174] There is little doubt that the ensuing years will provide many fascinating details of the biochemical processes responsible for the synthesis of this major microbial structure as well as the mode of action of antibiotics inhibiting cell-wall biosynthesis.

1. Cell-Wall "Precursors"

With the identification of the N-acetylamino sugar in the Park nucleotides as muramic acid the possible biochemical significance of these compounds accumulating in penicillin-treated *Staphylococcus aureus* became apparent. The presence of DL-alanine, D-glutamic acid, and lysine in about the same ratios as these constituents occur in the bacterial wall therefore make it fairly certain that this uridine pyrophosphate N-acetylmuramic acid peptide is a cell-wall precursor.[174] As Park[148] has pointed out, the occurrence of such a uridine pyrophosphate glycosyl compound would provide a transglycosidation mechanism for transferring the N-acetylmuramic acid peptide to the cell-wall polymer. The nucleotide part of the complex would possess a similar biochemical function to the uridine pyrophosphate glycosyl compounds engaged in the biosynthesis of other structural polysaccharides, such as cellulose and chitin.[239-241]

Dorfman and Cifonelli[242] have demonstrated the presence of a variety of nucleotides in extracts of group A streptococci. Uridine pyrophosphate N-acetylglucosamine (UPAG), uridine pyrophosphate glucose, and uridine pyrophosphate glucuronic acid were identified; in addition a new uridine nucleotide containing an amino sugar with all the properties of muramic acid was also isolated.[243] Of these nucleotides, the UPAG and the uridine nucleotide of muramic acid could represent precursors for the streptococcal wall, although the former compound could also participate in the biosynthesis of streptococcal hyaluronate.[242, 243]

Park[244] has also found that uridine diphosphate N-acetylamino sugar compounds devoid of amino acids will accumulate in crystal violet-inhibited cells of *Staphylococcus aureus*. A variety of uridine diphosphate amino sugar compounds containing various amino acids, including alanine, lysine, glutamic acid, aspartic acid, and glycine have been isolated from penicillin-inhibited *S. aureus* by Ito et al.[245] Smaller amounts of amino acid-uridine derivatives were obtained also from untreated cells, thus suggesting that these uridine-amino sugar peptides represent normal cell-wall precursors.[245]

Following the isolation of cytidine diphosphate glycerol and cytidine diphosphate ribitol from *Lactobacillus arabinosus* by Baddiley and his colleagues,[246-248] Baddiley et al.[110] discovered that a ribitol phosphate polymer occurred in the walls of several bacteria. The probable function of the cytidine diphosphate ribitol therefore became apparent and there seems little doubt that this nucleotide also represents a cell-wall precursor. Further evidence for the participation of the cytidine diphosphate ribitol compound in the formation of the teichoic acid polymer of the wall can be deduced from the accumulation of this compound in *S. aureus* treated with chloramphenicol and crystal violet.[152]

Meadow[249] has recently described a lipomucoprotein substance that accumulates in the growth medium of the lysine-deficient mutant *Escherichia coli*, 26–26. The substance contains the normal protein amino acids as well as some of the substances found in the bacterial wall (DAP, glucosamine, glucose, and galactose). Whether this compound is a true wall precursor is not certain but, as Meadow[249] has suggested, the accumulation of such a substance may represent an abortive attempt to build the cell wall.

2. Effects of Nutritional Deficiencies and Antibiotics on Cell-Wall Synthesis

a. Amino Acid and Amino Sugar Limitation. Recent investigations have shown that limitation of the supply of certain amino acids required for the cell wall peptides can frequently terminate in lysis of the bacterial cells. Shockman *et al.*[132] have investigated the synthesis of the cell wall of *Streptococcus faecalis* under deficiency of various amino acids required for the full growth of this organism. Although depletion of the amino acid valine resulted in the cessation of exponential growth of the cultures, the cell wall continued to be synthesized for some time afterward.[132] However, limitation of lysine, which is one of the principal amino acids of the wall of this organism, led to lysis.[132]

Lysis of *Escherichia coli* mutants resulting from a deficiency of DAP was first reported by Meadow and Work.[250] DAP is apparently an essential amino acid of the rigid mucocomplex layer of the walls of this organism. Further investigations on the lysis of *E. coli* mutants requiring DAP and those possessing a requirement for DAP in addition to a partial requirement of lysine have been carried out by Meadow *et al.*[251] and Rhuland.[252] From these studies and those of Bauman and Davis[253] and McQuillen[238, 254] it is evident that DAP limitation weakens the cell-wall structure and gives rise to spherical cells in suitable stabilizing media containing sucrose. The electron micrographs of DAP-deprived and penicillin-treated *E. coli* spherical protoplasts taken by McQuillen[254] show the presence of a well-defined envelope. Subsequent investigations by Salton and Shafa[151] have shown that this envelope probably corresponds to the less rigid lipoprotein part of the wall.

Park[148] has reported similar effects on the integrity of the wall of *Lactobacillus bifidus* when the *N*-acetylglucosamine compound required as a growth factor for this organism is omitted from the medium.

b. Inhibition of Cell-Wall Synthesis by Antibiotics. There is now abundant evidence that penicillin owes its antibacterial effects, at least in part, to an interference with the formation of the rigid cell-wall structure. Although Duguid[255] suggested that penicillin specifically interfered with the formation of the outer supporting cell wall as early as 1945, experimental evidence to verify this conclusion began accumulating from the time of the isolation of the Park[166] nucleotides. Duguid's[255] investigations led him to make two further important points. These were that all bacteria (within the limits of his investigations) were susceptible in some degree and that the cellular component upon which penicillin exerted its action must be possessed in common by all bacterial species. That both Gram-positive and Gram-negative bacteria do possess a common structural component in the wall has been emphasized by Work[140] and subsequently demonstrated by the action of phage enzyme and lysozyme on cell walls.[91, 149-151] Thus, investigations of the chemistry of the bacterial cell wall and the nature of the compounds accumulating in penicillin-treated bacteria have confirmed the general thesis propounded by Duguid.[255]

The experimental evidence of the interference in cell wall synthesis by penicillin falls into three main categories: (1) Morphological changes in the cell wall; in Gram-positive bacteria the wall may become thin and eventually burst open and separate into two hemispherical fragments[256, 257] or it may lose its rigidity, allowing the bacterial protoplast to assume a spherical shape;[258] in Gram-negative bacteria a variety of morphological changes take place and on suitable media in the presence of penicillin many species give rise to uniform spherical cells still surrounded by an external limiting wall or envelope.[151, 254, 255, 259-261] (2) Accumulation of cell-wall precursors, such as the uridine diphosphate amino sugar and peptide compounds and cytidine diphosphate ribitol. (3) The loss of typical cell-wall components accompanying the formation of stable L-forms and changes in cell-wall composition during the spherical transformation in the presence of penicillin. The investigations of Sharp et al.[262] on the L-forms of group A streptococci induced by penicillin have established the absence of the group-specific polysaccharide or any variant cell-wall polysaccharide containing rhamnose. Two independent studies of the chemical composition of the normal and L-form of *Proteus vulgaris* by Kandler et al.[263] and Weibull[264] have shown a loss or reduction of the contents of hexosamines and DAP constituents known to be localized in the bacterial walls. Kandler et al.[265] have, however, found that the membrane fraction from the unstable L-form of *P. vulgaris* has a higher DAP and muramic acid content than does the wall from the normal cell. In contrast to Weibull's[264] finding, Kandler and Zehender[266] found none of the constituents of the cell-wall mucocomplex in the stable L-forms. By following the sequence of changes in cell wall composition during the spherical transformation of *Vibrio metchnikovi* and *Salmonella gallinarum*, Salton and Shafa[151] found a progressive decline in the DAP and amino sugar contents; they suggested that penicillin action involved an inhibition of the formation or incorporation of the mucocomplex component into the wall. Thus, the decrease in the content of these cell-wall components could be due to a dilution of preexisting mucocomplex or a loss due to the action of autolytic enzymes.[151]

Even more profound changes in cell-wall composition have resulted from the selection of penicillin-resistant, Gram-negative strains arising during the growth of *Staphylococcus aureus* in aerated media in the presence of penicillin.[267] The organisms so isolated possessed DAP in the walls, as well as the amino acids glutamic acid, alanine, glycine, and glucosamine, and muramic acid present in the parent strain of *S. aureus*.[159]

Although unanimity has not been reached on the question of interference in cell wall synthesis by penicillin,[265, 268] the evidence is beginning to weigh more heavily in the affirmative.

The precise stage at which penicillin may interfere with the biosynthesis of the bacterial wall is as yet unknown. As penicillin appears to bind more selectively to the lipoprotein particles probably derived from the protoplast membrane,[269] it seems feasible that it interferes with the polymerizing mechanism and prevents the incorporation of the amino sugar-peptide precursors into the wall. It is likely that other antibiotics, such as bacitracin[159a] and oxamycin,[148] have similar effects on the wall synthesizing systems to those of penicillin.

In contrast to the inhibitory effects of penicillin, chloramphenicol allows the synthesis of the cell wall but blocks the formation of bacterial protoplasm.[131, 270] Mandelstam and Rogers[270] and Hancock and Park[131] have demonstrated the incorporation of C^{14}-labeled amino acids into the cell wall of *Staphylococcus aureus* in the presence of chloramphenicol. Under the experimental conditions used, the weight of bacterial wall material may increase from 50 to 100% while the inhibition of uptake of amino acids into the cellular protein fraction may be from 85 to 98%.[131, 270] The use of chloramphenicol therefore provides a suitable experimental system for the study of certain cell wall syntheses in the absence of protein synthesis.

H. Functions of the Cell Wall

There is little doubt that the mechanical protection afforded by the rigid bacterial wall confers upon those organisms possessing such structures a considerable survival advantage over those species devoid of a wall. It is evident from the studies with bacterial protoplasts[271, 272] that the soft structures underlying the rigid outer wall require rather special conditions for the maintenance of their integrity; without these conditions the protoplast, deprived of its protective cell wall, would have a precarious and ephemeral existence.

The bacterial wall must possess sufficient mechanical strength to withstand the high pressures that can be exerted by the intracellular solutes. Mitchell and Moyle[273] have found that the solute concentration in *S. aureus* may exert an osmotic pressure equivalent to 20 atmospheres. It is therefore understandable that any serious defect in the mechanical continuity of the wall will lead to a rupture of the weaker underlying protoplast membrane; ultimately the organisms will lyse and die. Impairment of the mechanical strength of the wall by penicillin action thus frequently culminates in lysis. Many halophilic bacteria obviously possess walls unable to withstand the pressures exerted upon them when the external environment is diluted.[15]

Although the bacterial cell wall affords such great survival value, it is by no means indispensable for the biological and biochemical continuity of the cell. The recent studies on the relationships between L-forms and bacterial protoplasts[258-266, 274] illustrate the extent to which microorganisms can tolerate a loss or decrease in cell wall structures. Thus, L-forms of group A streptococci can continue a stable existence without the parental cell-wall polysaccharide components.[262] The extent to which some of these changes are reversible is not fully known but the work of Lederberg[260] has shown that the penicillin-induced spherical forms of *Escherichia coli* can revert to the original rod-shaped bacteria. However, bacteria deprived of part or the whole of their wall by lysozyme action have not been observed to revert from the protoplast or spherical form to the parental bacteria except through a sporulation process.[275] The dispensability of the wall is again illustrated by the ability of protoplasts formed from bacteria committed to sporulation to support the synthesis and maturation of the bacterial spore.[275] The possession of the wall is not an absolute requirement for survival under these experimental conditions and, as pointed out by Salton,[74] it is conceivable that such conditions may also be found to occur in nature.

As to the biochemical functions that may be an integral part of the cell-wall structure, little can be said. The continued biosynthetic abilities of bacterial protoplasts[272] make it unlikely that the wall itself plays a major part in a number of biochemical processes. Because of the presence of ionizable groups on the wall it may play some part in the adsorption and uptake of various substances by the cell. It is unlikely, however, that this structure is entirely responsible for the selective transport of materials into the cell. On the contrary, the investigations of Weibull,[271] Mitchell and Moyle,[273] and Storck and Wachsman[276] clearly establish the protoplast membrane as the structural site of the bacterial cell's permeability barrier and transporting mechanisms. Whether this situation applies also to the Gram-negative bacteria is by no means certain. It is conceivable that the walls of these organisms may possess proteins fulfilling structural and biochemical functions. Important functional structures, such as the chromatophores of photosynthetic microorganisms[277, 278] and the enzymically active small particle fractions[279] from Gram-negative bacteria, do not appear to be derived from the cell wall.

The bacterial cell wall is obviously responsible for the shape and morphological integrity of the cell and in many bacteria it participates directly in the agglutination reactions of the bacterial surface. Its involvement in the Gram strain reaction is inescapable, for the bacterial protoplasts from which the mucocomplex wall has been removed are Gram-negative.[280] Furthermore, the quantitative variations in the Gram reaction[281, 282] may well be associated with the proportion of mucocomplex in the wall. The presence of certain groups or components in the wall will also govern its sensitivity to bacteriophages and enzymes. All of these properties are related more to the chemistry and structure of the wall rather than to the biochemical functions.

Thus, at the moment the functional status of the microbial cell wall, with the known exception of yeasts,[283, 284] seems to be essentially that of a rigid framework, supporting and protecting the more labile and biochemically active parts of the bacterial cell. Although present studies point to the wall as a biochemically inert structure, there is little doubt that in the future some unsuspected biochemical functions may be found associated with this structure.

References

[1] H. L. Jensen, *Can. J. Microbiol.* **3,** 151 (1957).

[2] M. Heidelberger, "Lectures in Immunochemistry." Academic Press, New York. 1956.

[3] J. P. Duguid, *J. Pathol. Bacteriol.* **64,** 673 (1951).

[4] M. Aschner and S. Hestrin, *Nature* **157,** 659 (1946).

[5] J. P. Duguid and R. R. Gillies, *J. Pathol. Bacteriol.* **74,** 397 (1957).

[6] R. C. Lancefield, *J. Exptl. Med.* **47,** 91 (1928).

[7] M. R. J. Salton, *Biochim. et Biophys. Acta* **9,** 334 (1952).

[8] O. T. Avery and R. J. Dubos, *J. Exptl. Med.* **54,** 73 (1931).

[9] M. H. Adams and B. H. Park, *Virology* **2,** 719 (1956).

[10] R. C. Lancefield, *J. Exptl. Med.* **78,** 465 (1943).

[11] M. T. Dyar, *J. Bacteriol.* **56,** 821 (1948).

[12] H. R. Carne, N. Wickham, and J. C. Kater, *Nature* **178,** 701 (1956).

[13] C. Weibull, *J. Bacteriol.* **66,** 688 (1953).

[14] W. R. Smithies and N. E. Gibbons, *Can. J. Microbiol.* **1,** 614 (1955).

[15] I. Takahashi and N. E. Gibbons, *Can. J. Microbiol.* **3,** 687 (1957).

[16] B. W. Catlin, *Science* **124,** 441 (1956).

[17] W. van Iterson, *in* "The Nature of the Bacterial Surface" (A. A. Miles and N. W. Pirie, eds.), p. 172. Blackwell, Oxford, 1949.

[18] K. Mühlethaler, *Biochim. et Biophys. Acta* **3,** 527 (1949).

[19] T. W. Burrows and G. A. Bacon, *Brit. J. Exptl. Pathol.* **37,** 481 (1956).

[20] J. F. Wilkinson, *Bacteriol. Revs.* **22,** 46 (1958).

[21] G. Knaysi, J. Hillier, and C. Fabricant, *J. Bacteriol.* **60,** 423 (1950).

[22] J. Tomcsik, *Experientia* **7,** 459 (1951).

[23] J. Tomcsik, *in* "Bacterial Anatomy" (E. T. C. Spooner and B. A. D. Stocker, eds.), p. 41. Cambridge Univ. Press, London and New York, 1956.

[24] J. Tomcsik, *Ann. Rev. Microbiol.* **10,** 213 (1956).

[25] G. Ivanovics and S. Horvath, *Acta Physiol. Acad. Sci. Hung.* **4,** 175 (1953).

[26] L. W. Labaw and V. M. Mosley, *J. Bacteriol.* **67,** 576 (1954).
[27] J. Tomcsik and S. Guex-Holzer, *J. Gen. Microbiol.* **10,** 97 (1954).
[28] S. Guex-Holzer and J. Tomcsik, *J. Gen. Microbiol.* **14,** 14 (1956).
[29] P. W. Kent and M. W. Whitehouse, "Biochemistry of the Aminosugars." Academic Press, New York, 1955.
[30] R. Brown, *J. Immunol.* **37,** 445 (1939).
[31] M. McCarty, *J. Exptl. Med.* **96,** 569 (1952).
[32] M. R. J. Salton, *Biochim. et Biophys. Acta* **8,** 510 (1952).
[33] W. S. Tillett and T. Francis, *J. Exptl. Med.* **52,** 561 (1930).
[34] M. Heidelberger and F. E. Kendall, *J. Exptl. Med.* **53,** 625 (1931).
[35] E. E. B. Smith, G. T. Mills, E. M. Harper, and B. Galloway, *J. Gen. Microbiol.* **17,** 437 (1957).
[36] R. C. Lancefield, *in* "Streptococcal Infections" (M. McCarty, ed.), Chapter 1, p. 3. Columbia Univ. Press, New York, 1954.
[37] E. H. Kass and C. V. Seastone, *J. Exptl. Med.* **79,** 319 (1944).
[38] P. L. Bazeley, *Australian Vet. J.* **16,** 243 (1940).
[39] H. J. Rogers, *J. Gen. Microbiol.* **10,** 209 (1954).
[40] W. R. Maxted, *Nature* **170,** 1020 (1952).
[41] M. R. J. Salton, *Biochim. et Biophys. Acta* **10,** 512 (1953).
[42] H. D. Slade, *J. Gen. Physiol.* **41,** 63 (1957).
[43] P. N. Hobson and M. J. Macpherson, *Biochem. J.* **57,** 145 (1954).
[44] C. T. Greenwood, *Biochem. J.* **57,** 151 (1954).
[45] C. F. Niven, K. L. Smiley, and J. M. Sherman, *J. Bacteriol.* **41,** 479 (1941).
[46] G. Ivanovics and V. Bruckner, *Z. Immunitätsforsch.* **90,** 304 (1937).
[47] C. B. Thorne, *in* "Bacterial Anatomy" (E. T. C. Spooner and B. A. D. Stocker, eds.), p. 68. Cambridge Univ. Press, London and New York, 1956.
[48] M. Bovarnick, *J. Biol. Chem.* **145,** 415 (1942).
[49] V. Bruckner, J. Kovacs, and G. Denes, *Nature* **172,** 508 (1953).
[50] W. E. Hanby and H. N. Rydon, *Biochem. J.* **40,** 297 (1946).
[51] S. G. Waley, *J. Chem. Soc.* p. 517 (1955).
[52] A. C. Chibnall and M. W. Rees, *Biochem. J.* **68,** 105 (1958).
[53] A. C. Chibnall, M. W. Rees, and F. M. Richards, *Biochem. J.* **68,** 129 (1958).
[54] G. Ivanovics and S. Horvath, *Acta Physiol. Acad. Sci. Hung.* **4,** 401 (1953).
[55] A. Kleczkowski and P. Wierzchowski, *Soil Sci.* **49,** 193 (1940).
[56] W. G. C. Forsyth and D. M. Webley, *Biochem. J.* **44,** 455 (1949).
[56a] J. P. Aubert, *Ann. inst. Pasteur* **80,** 644 (1951).
[57] N. R. Smith, R. E. Gordon, and F. E. Clark, Aerobic mesophilic spore forming bacteria. *U. S. Dept. Agr. Misc. Publ.* **No. 559** (1946).
[57a] R. F. Norris, M. De Sipin, F. W. Zilliken, T. S. Harvey, and P. György, *J. Bacteriol.* **67,** 159 (1954).
[58] A. Felix and R. M. Pitt, *J. Pathol. Bacteriol.* **38,** 409 (1938).
[59] J. M. Vincent, *Australian J. Sci.* **15,** 133 (1953).
[60] K. Ando and Y. Nakamura, *Japan. J. Exptl. Med.* **21,** 41 (1951).
[61] H. Mark and G. Susich, *Z. physik. Chem.* **4,** (*Leipzig*) 431 (1929).
[62] E. J. Hehre and D. M. Hamilton, *Proc. Soc. Exptl. Biol. Med.* **71,** 336 (1949).
[63] E. J. Hehre, *J. Biol. Chem.* **192,** 161 (1951).
[64] S. Zamenhof, G. Leidy, P. L. Fitzgerald, H. E. Alexander, and E. Chargaff, *J. Biol. Chem.* **203,** 695 (1953).
[65] J. P. Duguid and J. F. Wilkinson, *J. Gen. Microbiol.* **9,** 174 (1953).
[66] J. F. Wilkinson, J. P. Duguid, and P. N. Edmunds, *J. Gen. Microbiol.* **11,** 59 (1954).
[67] J. F. Wilkinson, W. F. Dudman, and G. O. Aspinall, *Biochem. J.* **59,** 446 (1955).

[68] G. O. Aspinall, R. S. P. Jamieson, and J. F. Wilkinson, *J. Chem. Soc.* p. 3484 (1956).
[69] W. F. Dudman and J. F. Wilkinson, *Biochem. J.* **62**, 289 (1956).
[70] P. R. Edwards and M. A. Fife, *J. Infectious Diseases* **91**, 92 (1952).
[71] P. N. Edmunds, *J. Infectious Diseases* **94**, 65 (1954).
[72] S. Levine, H. J. R. Stevenson, R. H. Bordner, and P. R. Edwards, *J. Infectious Diseases* **96**, 193 (1955).
[73] S. M. Beiser and B. D. Davis, *J. Bacteriol.* **74**, 303 (1957).
[74] M. R. J. Salton, *in* "Bacterial Anatomy" (E. T. C. Spooner and B. A. D. Stocker, eds.), p. 81. Cambridge Univ. Press, London and New York, 1956.
[75] W. T. J. Morgan, *in* "The Nature of the Bacterial Surface" (A. A. Miles and N. W. Pirie, eds.), p. 9. Blackwell, Oxford, 1949.
[76] S. M. Partridge, *Biochem. J.* **42**, 251 (1948).
[77] D. A. L. Davies, W. T. J. Morgan, and W. Mosimann, *Biochem. J.* **56**, 572 (1954).
[78] W. F. Goebel, F. Binkley, and E. Perlman, *J. Exptl. Med.* **81**, 315 (1945).
[79] O. Westphal, O. Lüderitz, and F. Bister, *Z. Naturforsch.* **7b**, 148 (1952).
[80] O. Westphal and O. Lüderitz, *Angew. Chem.* **66**, 407 (1954).
[81] O. Westphal, O. Lüderitz, I. Fromme, and N. Joseph, *Angew. Chem.* **65**, 555 (1953).
[82] D. A. L. Davies, *Biochem. J.* **59**, 696 (1955).
[83] D. A. L. Davies, A. M. Staub, I. Fromme, O. Lüderitz, and O. Westphal, *Nature* **181**, 822 (1958).
[84] I. Fromme, K. Himmelspach, O. Lüderitz, and O. Westphal, *Angew. Chem.* **69**, 643 (1957).
[85] M. Jesaitis and W. F. Goebel, *J. Exptl. Med.* **96**, 409 (1952).
[86] W. Weidel, G. Koch, and K. Bobasch, *Z. Naturforsch.* **9b**, 575 (1954).
[87] W. Weidel, *Z. physiol. Chem., Hoppe-Seyler's Hoppe-Seyl.* **299**, 253 (1955).
[88] A. P. Maclennan and D. A. L. Davies, *Biochem. J.* **66**, 562 (1957).
[89] D. A. L. Davies, *J. Gen. Microbiol.* **18**, 118 (1958).
[90] M. R. J. Salton, *Proc. Intern. Congr. Microbiol.*, p. 114, Stockholm, Sweden, 1958.
[91] W. Weidel and J. Primosigh, *J. Gen. Microbiol.* **18**, 513 (1958).
[92] D. A. L. Davies, *Biochim. et Biophys. Acta* **26**, 151 (1957).
[93] N. Yoshida, I. Fukuya, I. Kakutani, S. Tanaka, K. Takaishi, K. Nishino, and T. Hashimoto, *Tokushima J. Exptl. Med.* **2**, 117 (1955).
[94] W. R. Clark, J. McLaughlin, and M. E. Webster, *J. Biol. Chem.* **230**, 81 (1958).
[95] R. J. Dubos, *J. Exptl. Med.* **55**, 377 (1932).
[96] C. G. Anderson, "An Introduction to Bacteriological Chemistry," 2nd ed. Livingstone, Edinburgh, 1946.
[97] R. J. Dubos, *J. Exptl. Med.* **65**, 873 (1937).
[98] M. Green and M. A. Stahmann, *J. Biol. Chem.* **197**, 771 (1952).
[99] J. Kream, B. A. Borek, C. J. DiGrado, and M. Bovarnick, *Arch. Biochem. Biophys.* **53**, 333 (1954).
[100] B. E. Volcani and P. Margalith, *J. Bacteriol.* **74**, 646 (1957).
[101] J. C. Humphries, *J. Bacteriol.* **56**, 683 (1948).
[102] L. Wamoscher, *Z. Hyg. Infecktionskrankh.* **111**, 422 (1930).
[103] S. Mudd and D. B. Lackman, *J. Bacteriol.* **41**, 415 (1941).
[104] S. Mudd, K. Polevitsky, T. F. Anderson, and L. A. Chambers, *J. Bacteriol.* **42**, 251 (1941).
[105] I. M. Dawson, *in* "The Nature of the Bacterial Surface" (A. A. Miles and N. W. Pirie, eds.), p. 119. Blackwell, Oxford, 1949.

[106] M. R. J. Salton and R. W. Horne, *Biochim. et Biophys. Acta* **7**, 177 (1951).
[107] G. Knaysi, "Elements of Bacterial Cytology," 2nd ed. Cornell Univ. Press (Comstock), Ithaca, New York, 1951.
[108] A. Abrams, *J. Biol. Chem.* **230**, 949 (1958).
[109] A. D. Brown, *Biochim. et Biophys. Acta* **28**, 445 (1958).
[110] J. Baddiley, J. G. Buchanan, and B. Carss, *Biochim. et Biophys. Acta* **27**, 220 (1958).
[111] D. Fraser, *Nature* **167**, 33 (1951).
[112] H. K. King and H. Alexander, *J. Gen. Microbiol.* **2**, 315 (1948).
[113] H. Mickle, *J. Roy. Microscop. Soc.* **68**, 10 (1948).
[114] C. S. Cummins, *Intern. Rev. Cytol.* **5**, 25 (1956).
[115] P. D. Cooper, *J. Gen. Microbiol.* **9**, 199 (1953).
[116] G. D. Shockman, J. J. Kolb, and G. Toennies, *Biochim. et Biophys. Acta* **24**, 203 (1957).
[117] D. E. Hughes, *Brit. J. Exptl. Pathol.* **32**, 97 (1951).
[118] P. K. Stumpf, D. E. Green, and F. W. Smith, *J. Bacteriol.* **51**, 487 (1946).
[119] M. R. J. Salton, *J. Gen. Microbiol.* **9**, 512 (1953).
[120] H. D. Slade and J. K. Vetter, *J. Bacteriol.* **71**, 236 (1956).
[121] A. G. Marr and E. H. Cota-Robles, *J. Bacteriol.* **74**, 79 (1957).
[122] M. R. J. Salton, unpublished observations.
[123] G. Bosco, *J. Infectious Diseases* **99**, 270 (1956).
[124] E. F. Gale and J. P. Folkes, *Biochem. J.* **53**, 493 (1953).
[125] P. Mitchell and J. Moyle, *J. Gen. Microbiol.* **5**, 981 (1951).
[126] C. Weibull, *J. Bacteriol.* **66**, 696 (1953).
[127] P. Albertsson, *Biochim. et Biophys. Acta* **27**, 378 (1958).
[128] L. F. Barrington and L. M. Kozloff, *J. Biol. Chem.* **223**, 615 (1956).
[129] W. Weidel, *Z. Naturforsch.* **6b**, 251 (1951).
[130] J. R. Norris, *J. Gen. Microbiol.* **16**, 1 (1957).
[131] R. Hancock and J. T. Park, *Nature* **181**, 1050 (1958).
[132] G. D. Shockman, J. J. Kolb, and G. Toennies, *J. Biol. Chem.* **230**, 961 (1958).
[133] A. Birch-Andersen, O. Maaløe, and F. S. Sjöstrand, *Biochim. et Biophys. Acta* **12**, 395 (1953).
[134] E. Kellenberger and A. Ryter, *J. Biophys. Biochem. Cytol.* **4**, 323 (1958).
[135] A. L. Houwink, *Biochim. et Biophys. Acta* **10**, 360 (1953).
[136] M. R. J. Salton and R. C. Williams, *Biochim. et Biophys. Acta* **14**, 455 (1954).
[137] L. W. Labaw and V. M. Mosley, *Biochim. et Biophys. Acta* **15**, 325 (1954).
[138] A. L. Houwink, *J. Gen. Microbiol.* **15**, 146 (1956).
[139] W. Weidel and E. Kellenberger, *Biochim. et Biophys. Acta* **17**, 1 (1955).
[140] E. Work, *Nature* **179**, 841 (1957).
[141] S. S. Barkulis and M. F. Jones, *J. Bacteriol.* **74**, 207 (1957).
[142] E. Work, *Biochem. J.* **49**, 17 (1951).
[143] R. E. Strange and F. A. Dark, *Nature* **177**, 186 (1956).
[144] C. S. Cummins and H. Harris, *J. Gen. Microbiol.* **14**, 583 (1956).
[145] M. R. J. Salton, *Nature* **180**, 338 (1957).
[146] E. E. Snell, N. S. Radin, and M. Ikawa, *J. Biol. Chem.* **217**, 803 (1955).
[147] M. Ikawa and E. E. Snell, *Biochim. et Biophys. Acta* **19**, 576 (1956).
[148] J. T. Park, in "The Strategy of Chemotherapy" (S. T. Cowan and E. Rowatt eds.), p. 49. Cambridge Univ. Press, London and New York, 1958.
[149] M. R. J. Salton, *J. Gen. Microbiol.* **18**, 481 (1958).
[150] W. Weidel and J. Primosigh, *Z. Naturforsch.* **12b**, 421 (1957).

[151] M. R. J. Salton and F. Shafa, *Nature* **181,** 1321 (1958).
[152] J. J. Armstrong, J. Baddiley, J. G. Buchanan, and B. Carss, *Nature* **181,** 1692 (1958).
[153] M. R. J. Salton, *Biochim. et Biophys. Acta,* in press.
[154] C. S. Cummins and H. Harris, *Intern. Bull. Bact. Nomen. Tax.* **6,** 111 (1956).
[155] C. S. Cummins and H. Harris, *J. Gen. Microbiol.* **18,** 173 (1958).
[156] C. S. Cummins, O. M. Glendenning, and H. Harris, *Nature* **180,** 337 (1957).
[157] A. H. Romano and W. J. Nickerson, *J. Bacteriol.* **72,** 478 (1956).
[158] A. H. Romano and A. Sohler, *J. Bacteriol.* **72,** 865 (1956).
[159] E. P. Abraham, "Biochemistry of Some Peptide and Steroid Antibiotics." Wiley, New York, 1957.
[159a] E. P. Abraham, personal communication.
[160] J. F. Powell and R. E. Strange, *Biochem. J.* **65,** 700 (1957).
[161] F. Graziosi and G. Tecce, *Giorn. microbiol.* **3,** 143 (1957).
[162] C. M. Stevens, P. E. Halpern, and R. P. Gigger, *J. Biol. Chem.* **190,** 705 (1951).
[163] P. Boulanger and R. Osteux, *Biochim. et Biophys. Acta* **5,** 416 (1950).
[164] J. M. Ghuysen, *Arch. intern. physiol. et biochim.* **65,** 173 (1957).
[165] M. R. J. Salton and J. M. Ghuysen, *Biochim. et Biophys. Acta* **24,** 160 (1957).
[166] J. T. Park, *J. Biol. Chem.* **194,** 877, 885, 897 (1952).
[167] D. S. Hoare and E. Work, *Biochem. J.* **61,** 562 (1955); **65,** 441 (1957).
[168] V. M. Ingram and M. R. J. Salton, *Biochim. et Biophys. Acta* **24,** 9 (1957).
[169] R. E. Strange, *Biochem. J.* **64,** 23 P (1956).
[170] S. R. Elsden, B. E. Volcani, F. M. C. Gilchrist, and D. Lewis, *J. Bacteriol.* **72,** 681 (1956).
[171] M. J. Crumpton and D. A. L. Davies, *Biochem. J.* **64,** 22 P (1956).
[172] G. T. Barry, *J. Exptl. Med.* **107,** 507 (1958).
[173] L. H. Kent, *Biochem. J.* **67,** 5 P (1957).
[174] J. T. Park and J. L. Strominger, *Science* **125,** 99 (1957).
[175] M. R. J. Salton, *Biochim. et Biophys. Acta* **22,** 495 (1956).
[176] H. R. Perkins and H. J. Rogers, *Biochem. J.* **69,** 15 P (1958).
[177] A. Gottschalk, in "Chemistry and Biology and Mucopolysaccharides" (G. E. W. Wolstenholme and M. O'Connor, eds.), p. 287. Churchill, London, 1958.
[178] E. S. Holdsworth, *Biochim. et Biophys. Acta* **9,** 19 (1952).
[179] A. R. Gilby, A. V. Few, and K. McQuillen, *Biochim. et Biophys. Acta* **29,** 21 (1958).
[180] W. R. Smithies, N. E. Gibbons, and S. T. Bayley, *Can. J. Microbiol.* **1,** 605 (1955).
[181] E. Grossbard and R. D. Preston, *Nature* **179,** 448 (1957).
[182] R. E. Strange and F. A. Dark, *Biochem. J.* **62,** 459 (1956).
[183] A. Fleming, *Proc. Roy. Soc.* **93,** 306 (1922).
[184] A. Fleming and V. D. Allison, *Brit. J. Exptl. Pathol.* **3,** 252 (1922).
[185] M. R. J. Salton, *Bacteriol. Revs.* **21,** 82 (1957).
[186] T. Amano, K. Kato, and R. Shimizu, *Med. J. Osaka Univ.* **3,** 293 (1952).
[187] M. Welsch, *J. Gen. Microbiol.* **18,** 491 (1958).
[188] M. McCarty, *J. Exptl. Med.* **96,** 555 (1952).
[189] R. E. Strange and F. A. Dark, *J. Gen. Microbiol.* **16,** 236 (1957); **17,** 525 (1957).
[190] M. Nomura, *Nippon Nôgei-kagaku Kaishi* **29,** 674 (1955).
[191] M. H. Richmond, *J. Gen. Microbiol.* **16,** iv (1957).
[192] P. Mitchell and J. Moyle, *J. Gen. Microbiol.* **16,** 184 (1957).
[193] J. S. Murphy, *Virology* **4,** 563 (1957).
[194] F. Jacob and C. R. Fuerst, *J. Gen. Microbiol.* **18,** 518 (1958).

[195] K. Meyer, J. W. Palmer, R. Thompson, and D. Khorazo, *J. Biol. Chem.* **113**, 479 (1936).
[196] K. Meyer and E. Hahnel, *J. Biol. Chem.* **163**, 723 (1946).
[197] L. A. Epstein and E. Chain, *Brit. J. Exptl. Pathol.* **21**, 339 (1940).
[198] L. R. Berger and R. S. Weiser, *Biochim. et Biophys. Acta* **26**, 517 (1957).
[199] W. Brumfitt, A. C. Wardlaw, and J. T. Park, *Nature* **181**, 1783 (1958).
[199a] M. R. J. Salton and J. M. Ghuysen, manuscript in preparation.
[199b] J. M. Ghuysen and M. R. J. Salton, manuscript in preparation.
[200] R. G. Peterson and S. E. Hartsell, *J. Infectious Diseases* **96**, 75 (1955).
[201] E. A. Grula and S. E. Hartsell, *Can. J. Microbiol.* **3**, 13 (1957).
[202] E. A. Grula and S. E. Hartsell, *Can. J. Microbiol.* **3**, 23 (1957).
[203] M. Welsch, *Ergeb. Mikrobiol. Immunitätsforsch. exptl. Therap.* **30**, 217 (1957).
[204] W. R. Maxted, *Lancet* **ii**, 255 (1948).
[205] M. Nicolle, *Ann. inst. Pasteur* **21**, 613 (1907).
[206] R. A. Greenberg and H. O. Halvorson, *J. Bacteriol.* **69**, 45 (1955).
[207] W. Kaufmann and K. Bauer, *J. Gen. Microbiol.* **18**, xi (1958).
[208] F. A. Dark and R. E. Strange, *Nature* **180**, 759 (1957).
[209] M. Nomura and J. Hosoda, *J. Bacteriol.* **72**, 573 (1956).
[210] V. Sertic, *Compt. rend. soc. biol.* **100**, 477 (1929).
[211] D. T. Ralston, B. S. Baer, M. Lieberman, and A. P. Krueger, *Proc. Soc. Exptl. Biol. Med.* **89**, 502 (1955).
[212] J. Panijel and J. Huppert, *Ann. inst. Pasteur* **90**, 619 (1956).
[213] J. Huppert and J. Panijel, *Ann. inst. Pasteur* **90**, 711 (1956).
[214] W. R. Maxted, *J. Gen. Microbiol.* **16**, 584 (1957).
[215] W. R. Maxted and H. Gooder, *J. Gen. Microbiol.* **18**, xiii (1958).
[216] M. van den Ende, P. A. Don, W. J. Elford, C. E. Challice, I. M. Dawson, and J. E. Hotchin, *J. Hyg.* **50**, 12 (1952).
[217] J. E. Hotchin, I. M. Dawson, and W. J. Elford, *Brit. J. Exptl. Pathol.* **33**, 177 (1952).
[218] M. R. J. Salton, in "The Nature of Viruses" (G. E. W. Wolstenholme and E. C. P. Millar, eds.), p. 263. Churchill, London, 1957.
[218a] D. Kay and J. Sampson, *Brit. J. Exptl. Pathol.* **38**, 548 (1957).
[219] D. Fraser and R. C. Williams, *Virology* **2**, 289 (1956).
[220] L. F. Barrington and L. M. Kozloff, *Science* **120**, 110 (1954).
[221] D. D. Brown and L. M. Kozloff, *J. Biol. Chem.* **225**, 1 (1957).
[222] L. M. Kozloff and M. Lute, *J. Biol. Chem.* **228**, 529, 537 (1957).
[223] G. Koch and W. Weidel, *Z. Naturforsch.* **11b**, 345 (1956).
[224] W. Weidel and G. Koch, *Z. Naturforsch.* **10b**, 694 (1955).
[225] G. Koch and W. Weidel, *Z. physiol. Chem., Hoppe-Seyler's* **303**, 213 (1956).
[226] J. Tomcsik and S. Guex-Holzer, *Experientia* **12**, 1 (1954).
[227] J. W. Vennes and P. Gerhardt, *Science* **124**, 535 (1956).
[228] M. McCarty and R. C. Lancefield, *J. Exptl. Med.* **102**, 11 (1955).
[229] M. McCarty, *J. Exptl. Med.* **104**, 629 (1956).
[230] C. S. Cummins, *Brit. J. Exptl. Pathol.* **35**, 166 (1954).
[231] G. Ivanovics, *Acta Microbiol. Acad. Sci. Hung.* **3**, 135 (1955).
[232] N. Yoshida, S. Tanaka, K. Takaishi, I. Fukuya, K. Nishino, I. Kakutani, S. Inoi, K. Fukui, A. Kono, and T. Hashimoto, *Tokushima J. Exptl. Med.* **2**, 11 (1955).
[233] F. Shafa, "A Study of the Surface Structure of Some Bacteria." Ph.D. Thesis, University of Manchester, England, 1958.
[234] W. Burrows, *J. Infectious Diseases* **101**, 73 (1957).

[235] D. Rowley, *Lancet* **i**, 232 (1955).
[236] L. Pillemer and O. A. Ross, *Science* **121**, 732 (1955).
[237] L. Pillemer, M. D. Schoenberg, L. Blum, and L. Wurz, *Science* **122**, 545 (1955).
[238] K. McQuillen, *J. Gen. Microbiol.* **18**, 498 (1958).
[239] L. F. Leloir, *Proc. Intern. Congr. Biochem. 3rd Congr. Brussels, 1955* p. 154. (1956).
[240] L. Glaser, *Biochim. et Biophys. Acta* **25**, 436 (1957).
[241] L. Glaser and D. H. Brown, *Biochim. et Biophys. Acta* **23**, 449 (1957).
[242] A. Dorfman and J. A. Cifonelli, *in* "Chemistry and Biology of Mucopolysaccharides" (G. E. W. Wolstenholme and M. O'Connor, eds.), p. 64. Churchill, London, 1958.
[243] J. A. Cifonelli and A. Dorfman, *J. Biol. Chem.* **228**, 547 (1957).
[244] J. T. Park, *Federation Proc.* **13**, 271 (1954).
[245] E. Ito, N. Ishimoto, and M. Saito, *Nature* **181**, 906 (1958).
[246] J. Baddiley, J. G. Buchanan, A. P. Mathias, and A. R. Sanderson, *J. Chem. Soc.* p. 4186 (1956).
[247] J. Baddiley, J. G. Buchanan, B. Carss, and A. P. Mathias, *J. Chem. Soc.* p. 4583 (1956).
[248] J. Baddiley, J. G. Buchanan, and B. Carss, *J. Chem. Soc.* p. 1869 (1957).
[249] P. Meadow, *J. Gen. Microbiol.* **18**, iii (1958).
[250] P. Meadow and E. Work, *Biochem. J.* **64**, 11P (1956).
[251] P. Meadow, D. S. Hoare, and E. Work, *Biochem. J.* **66**, 270 (1957).
[252] L. E. Rhuland, *J. Bacteriol.* **73**, 778 (1957).
[253] N. Bauman and B. D. Davis, *Science* **126**, 170 (1957).
[254] K. McQuillen, *Biochim. et Biophys. Acta* **27**, 410 (1958).
[255] J. P. Duguid, *Edinburgh Med. J.* **53**, 401 (1945).
[256] W. J. Elford, *J. Gen. Microbiol.* **2**, 205 (1948).
[257] G. Bringmann, "Bakterienunter Hemmstoffwirkung." Transmare-Photo, Berlin, 1954.
[258] J. Tomcsik, *Bibliothec. Paediat., Suppl. Ann. Paediat.* **No. 58**, 410 (1954).
[259] H. Stempen, *J. Bacteriol.* **70**, 177 (1955).
[260] J. Lederberg, *Proc. Natl. Acad. Sci. U. S.* **42**, 574 (1956).
[261] K. Liebermeister and E. Kellenberger, *Z. Naturforsch* **11b**, 200 (1956).
[262] J. T. Sharp, W. Hijmans and L. Dienes, *J. Exptl. Med.* **105**, 153 (1957).
[263] O. Kandler, C. Zehender and A. Hund, quoted by Kandler *et al.*, ref. 265.
[264] C. Weibull, *Acta Pathol. Microbiol. Scand.* **42**, 324 (1958).
[265] O. Kandler, A. Hund, and C. Zehender, *Nature* **181**, 572 (1958).
[266] O. Kandler and C. Zehender, *Z. Naturforsch.* **12b**, 725 (1957).
[267] S. Briggs, K. Crawford, E. P. Abraham, and G. P. Gladstone, *J. Gen. Microbiol.* **16**, 614 (1957).
[268] R. E. Trucco and A. B. Pardee, *J. Biol. Chem.* **230**, 435 (1958).
[269] P. D. Cooper, *Bacteriol. Revs.* **20**, 28 (1956).
[270] J. Mandelstam and H. J. Rogers, *Nature* **181**, 956 (1958).
[271] C. Weibull, *in* "Bacterial Anatomy" (E. T. C. Spooner and B. A. D. Stocker, eds.), p. 111. Cambridge Univ. Press, London and New York, 1956.
[272] K. McQuillen, *in* "Bacterial Anatomy" (E. T. C. Spooner and B. A. D. Stocker, eds.), p. 127. Cambridge Univ. Press, London and New York, 1956.
[273] P. Mitchell and J. Moyle, *in* "Bacterial Anatomy" (E. T. C. Spooner and B. A. D. Stocker, eds.), p. 150. Cambridge Univ. Press, London and New York, 1956.
[274] J. Lederberg and J. St. Clair, *J. Bacteriol.* **75**, 143 (1958).

[275] M. R. J. Salton, *J. Gen. Microbiol.* **11,** ix (1954).
[276] R. Storck and J. T. Wachsman, *J. Bacteriol.* **73,** 784 (1957).
[277] H. K. Schachman, A. B. Pardee, and R. Y. Stanier, *Arch. Biochem. Biophys.* **38,** 245 (1952).
[278] A. E. Vatter and R. S. Wolfe, *J. Bacteriol.* **75,** 480 (1958).
[279] R. Y. Stanier, I. C. Gunsalus, and C. F. Gunsalus, *J. Bacteriol.* **66,** 543 (1953).
[280] P. Gerhardt, J. W. Vennes, and E. M. Britt, *J. Bacteriol.* **72,** 721 (1956).
[281] F. Wensinck and J. J. Boevé, *J. Gen. Microbiol.* **17,** 401 (1957).
[282] D. Shugar and J. Baranowska, *Nature* **181,** 357 (1958).
[283] A. Rothstein *Discussions Faraday Soc.* **21,** 229 (1956).
[284] E. J. Conway and P. F. Duggan, *Nature* **178,** 1043 (1956).

CHAPTER 4

Movement

CLAES WEIBULL

I. Introduction.. 153
II. Theoretical Aspects of the Movements of Bacteria....................... 154
   A. Hydrodynamic Theory... 154
   B. Thermokinetic and Electrokinetic Theories........................... 157
III. Flagellar Movement... 158
   A. Morphology and Arrangement of Bacterial Flagella.................... 158
   B. Structure and Chemistry of the Flagella............................. 165
   C. Characteristics of Flagellar Movement............................... 165
   D. The Mode of Action of the Flagella.................................. 169
IV. Movements of the Spirochetes.. 174
   A. General... 174
   B. Spirochaeta... 175
   C. Cristispira... 175
   D. Saprospira.. 176
   E. Borrelia.. 176
   F. Treponema... 176
   G. Leptospira.. 177
   H. The Mechanism of the Spirochetal Movement.......................... 178
V. Gliding Movement... 180
   A. General... 180
   B. The Beggiatoaceae... 180
   C. The Myxobacteriales... 184
   D. Fusobacteria.. 187
   E. Additional Groups Exhibiting Gliding Motion......................... 187
VI. Bacterial Movements Considered as Tactic Responses to External Stimuli 188
   A. General... 188
   B. Tactic Responses to Chemical and Physicochemical Stimuli (Chemotaxis)... 189
   C. Tactic Responses to Thermal Stimuli (Thermotaxis)................... 193
   D. Tactic Responses to Gravitational Stimuli (Geotaxis)................ 193
   E. Tactic Responses to Light (Phototaxis).............................. 194
   References.. 198

I. Introduction

Many different kinds of movements are encountered in bacteria, e.g., flexions and rotations of the cell bodies and locomotion. The greater part of this chapter will deal with the last-mentioned kind of movement.

In most motile bacteria, organelles, readily interpretable in terms of motor organs, have been demonstrated, i.e., flagella or, in the case of the spirochetes, fibrils winding tightly around the entire bacterial body. Capability of locomotion is, however, also observed in the Myxobacteriales, the Beggiatoaceae, and in some fusobacteria, in spite of the fact that no well-defined motor organs have been revealed, not even with the aid of the electron microscope.[1-3] The speed that these organisms may attain is rather moderate, not more than 5 microns per second[4, 5] whereas flagellated organisms may travel more than 50 microns per second.[6-8] The former kind of locomotion is generally referred to as gliding movement.

Bacterial motility has been extensively studied by means of direct observations and measurements, but many fundamental questions remain unsolved. This is no doubt to a great extent due to the limitations of the available experimental techniques, especially the microcinematographic methods. Another method used for attacking the motility problem has been the construction and testing of large-scale models but, unfortunately, it has sometimes not been taken into account that the hydrodynamic forces acting on microorganisms and macroscopic objects are different.[9, 10]

In the following the behavior of moving microorganisms will first be discussed from a theoretical point of view. The inferences obtained will then be related to the empirical studies that have been carried out on various kinds of bacterial movements. Morphological, biochemical, and physiological facts that have a bearing on the motility problem will also be taken into account.

II. Theoretical Aspects of the Movements of Bacteria

A. Hydrodynamic Theory

Attempts at treating the propulsion of microorganisms theoretically were made early[11, 12] but not until recently have the hydrodynamics of such bodies been considered in detail. During the last few years, however, hydrodynamical analyses of the swimming of microorganisms have been carried out by Taylor,[9, 13] Hancock,[10] and Gray and Hancock.[14] A study of undulatory movement in general has also recently appeared.[15]

In his first paper Taylor discusses the forces that are of importance for an analysis of the self-propulsion of objects of varying sizes. He points out that large bodies that propel themselves, e.g., fish, make use of the inertia in the surrounding fluid. The propulsive organ pushes the fluid backward while the resistance of the body gives the fluid a forward momentum. The forward and backward momenta exactly balance in steady swimming motion. The resistance is mainly viscous but the effect of viscosity is only important in the fluid adjacent to the body, i.e., the boundary layer.

These concepts cannot be transferred to the problems of propulsion for microscopic bodies. When such objects swim in water the stresses in the liquid due to viscosity may be a thousand times greater than those due to inertia.

A quantitative estimation of the ratio

$$\frac{\text{Stress in fluid due to inertia}}{\text{Stress in fluid due to viscosity}}$$

is given by the Reynolds number R. This entity is related to the length L and the speed V of the moving object and to the density (σ) and the viscosity (μ) of the fluid according to the formula

$$R = LV\sigma/\mu$$

For a bacterium swimming in water, V has a value of about 0.01 cm./sec. and $L \simeq 10^{-4}$ cm. Since $\sigma = 1$ and $\mu = 0.01$ poise, $R = 10^{-4}$. Thus 99.99% of the stress in the water is due to viscosity.

Taylor points out that the Reynolds number is usually defined with respect to a body swimming through a liquid without oscillating or undulating. Among the bacteria, only the slowly moving, nonflagellated organisms, mentioned in the introduction, move in this way. The rapid movements of the flagellated bacteria and of the spirochetes are accompanied by oscillations and undulations of the whole bacterial cell or its fibrillar appendages. Taylor, however, also gives a formula defining the ratio

$$\frac{\text{Stress due to inertia}}{\text{Stress due to viscosity}}$$

in a fluid where vibrating objects move. It appears that in the case of microorganisms this ratio is still smaller than the corresponding Reynolds number.

It is thus evident that, for the movements of bacteria, inertia plays only a negligible role.

The greater part of Taylor's, Hancock's, and Gray's papers[9, 10, 13, 14] deal with the problem of whether sheets or filaments will propel themselves through fluids when waves of lateral displacements are propagated along them. It is assumed that forces due to inertia may be neglected. The undulating filaments are supposed to be either free or attached to an inert head. With respect to its outer shape, the resulting complex body thus resembles many motile microorganisms.

In his first paper Taylor treats a two-dimensional problem, the motion of a fluid near a sheet down which sinusoidal waves are propagated. He finds that the sheet moves forward at a rate V equal to $2\pi^2 b^2/\lambda^2$ times U, the velocity of propagation of the waves. Here b is the amplitude of the waves and λ the wavelength; b is assumed to be small compared with λ but not infinitesimal. The formula can also be written in the form $V = U \cdot \frac{1}{2}b^2k^2$ where k is the wave number $= 2\pi/\lambda$.

A three-dimensional case in which waves of a small amplitude move down an infinite filament of circular cross section is considered in Taylor's second paper. Assuming the cross-sectional radius to be infinitesimal, an expression for the propulsion is obtained which is similar to that for the two-dimensional case just mentioned. In the same paper calculations are also made for the case where waves of lateral displacement are propagated as helices along a filament. It is shown that the body is propelled at twice the speed given by plane waves of the same amplitude. Movement by means of helical waves presupposes, however, an externally applied torque to prevent the reaction of the fluid from causing the object to rotate. This is also emphasized by Hancock[10] and Gray.[15]

Taylor has performed some model experiments in order to test his theory. The working model consisted of an aluminum tube and a tail. The tail was made of a rubber tube containing a wire spiral which could be made to rotate inside the tube by means of twisted rubber rings located in the aluminum tube. The model was tested in a tank containing glycerol, a medium giving a sufficiently low Reynolds number even for macroscopic bodies. When the wire was made to rotate, helical waves were propagated along the tail and the model started to move. The observed speed was, however, considerably lower than that calculated.

Hancock, using a different mathematical approach, extends the theory of swimming microorganisms further and gives solutions, among others, of the cases when the amplitude of the waving filament is not small and when the filament is of finite length.

In the case of plane waves and when the radius d of an infinite filament approaches zero, Hancock finds that the filament moves forward according to the formula

$$V = U \cdot \frac{1}{2} \frac{b^2 k^2}{1 + b^2 k^2}$$

It can be seen that for small b^2k^2, V becomes equal to $U \cdot \frac{1}{2}b^2k^2$, i.e., the same expression which Taylor finds for a waving sheet. On the other hand, V assumes the value $U/2$ when the product b^2k^2 becomes large.

When the radius d of the filament is finite but still small in comparison with the wavelength, the value for V is found to be smaller. Thus for $d/\lambda = 0.02$ and for large values of b^2k^2, V equals $U/5$ instead of $U/2$. It is assumed here that the filament boundary is rigid.

As far as the influence of the length of a waving system is concerned, it is found that the same speed of propulsion is obtained for filaments of finite and infinite lengths as long as the radius of the tail is infinitesimal. In the case of tails of measurable length (e.g., 1 wavelength) and thickness, the velocity of propulsion is slightly less than that of an infinite filament.

Hancock also considers the motion caused by the propagation of helical waves along a filament. He obtains the expression

$$V = U \cdot \frac{b^2 k^2}{1 + 2b^2 k^2}$$

For small values of b^2k^2, the formula becomes $V = U \cdot b^2k^2$, i.e., the speed of propulsion due to helical waves is twice that for a plane wave motion, as pointed out earlier by Taylor. As b^2k^2 becomes large, V approaches $U/2$ in both cases.

Gray and Hancock,[14] using a third method of approach, obtain results similar to those of Hancock and Taylor. A formula is, in addition, given for the propulsion of an inert head by a filament waving with plane waves. It appears that the influence of the head on the rate of propulsion is not very great. The formula is used for calculating the speed of motile sea urchin spermatozoa. Close agreement is found between calculated and observed values. Thus according to the formula, the passage of each wave over a tail should propel a spermatozoon through a distance of 5.52 microns; a large number of observations showed that the distance covered was between 5 and 6 microns.

Comparisons between the theory and results from model experiments were also made by Hancock.[10] He quotes experiments carried out by Taylor on the movement of wires of varying thickness in a viscous fluid. The agreement between calculated and observed values was found to be good.

The results of the hydrodynamic analyses and the experimental studies, quoted above, could be summarized as follows:

1. Microscopic filaments may propel themselves through viscous fluids by means of undulatory movements.

2. The undulations may take place either in one plane or in the form of helical waves. In the latter case an external torque has to be applied to the filament in order to prevent it from spinning freely. Such a spinning motion would prevent propulsion.

3. Propulsion may result even when the filament is attached as a tail to an inert head. In the case of plane waves, the effect of the head on the rate of propulsion is small. On the other hand, the presence of such a head is of fundamental importance for propulsion by helical waves, since otherwise the above-mentioned free spinning would occur.

B. Thermokinetic and Electrokinetic Theories

The hydrodynamic theory given above presupposes a mechanical deformation or movement of the surface of the motile microorganisms. Recently

Mitchell has discussed two mechanisms of locomotion based upon a different principle, viz., the streaming of the external medium over the static surface of the bacteria.[16]

The first mechanism is a thermokinetic one. It is suggested that the heat generated by the metabolism of the bacteria is transferred from the interior of the cell to the medium through the flagella. A temperature gradient down these organelles would thus result. The gradient would in turn cause a streaming of water along the flagella and a movement of the bacterium in the opposite direction. Mitchell calculates that a speed of about 10 microns per second could be produced by a reasonably efficient thermokinetic process.

The other suggested mechanism is of an electrokinetic nature. Since the surface of most bacteria suspended in physiological media is negatively charged, the water within a short distance of the cells contains an excess of positive ions. When an electric potential gradient is applied to a bacterial suspension, the cells will move toward the anode and the fluid close to the cells toward the cathode. Such a gradient could be produced by the organisms themselves if positive ions were secreted at one end of the cells and absorbed at the opposite one. Due to the electric forces, the bacterium and the adjacent medium would tend to move in opposite directions. From the known electrophoretic mobility of bacteria and from metabolic data, Mitchell concludes that only about 2 % of the energy generated by the cell metabolism would be required to give a bacterium a speed of about 10 microns per second.

III. Flagellar Movement

A. Morphology and Arrangement of Bacterial Flagella

The presence of long filamentous appendages on motile bacteria was already noted by the early investigators of bacterial movements.[17-21] Later, electron micrographic studies have confirmed these observations. The length of the flagella may thus be several times that of the bacterial body. As to their diameter, measurements on electron micrographs have given values of 186 A. (*Bacillus subtilis*),[22] 120 A. (*Proteus vulgaris*),[23] 139 A. (*Brucella bronchiseptica*),[24] and 190 A. (a motile diphteroid).[25]

These dimensions lie far below the resolving power of the light microscope. The flagella have, however, a strong tendency toward aggregation, which was first pointed out by Fischer.[20] This aggregation phenomenon, which is illustrated by Fig. 1, has since been observed by numerous investigators.[12, 26-30] That which can be seen in the light microscope with ordinary or phase-contrast illumination when unstained material is studied is thus not individual flagella but bundles of many of these organelles.

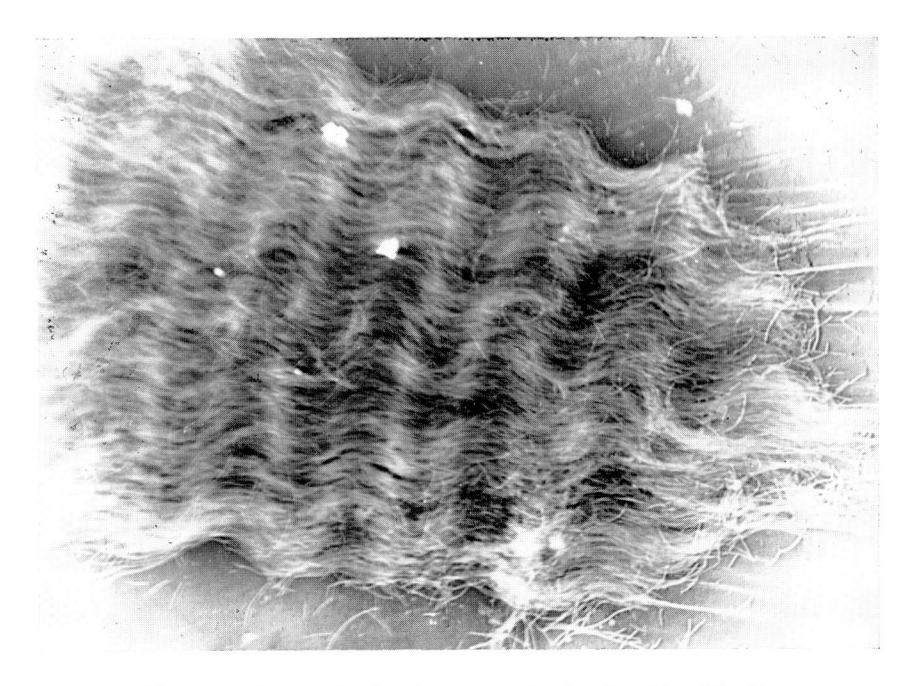

FIG. 1. Electron micrograph of a fragment of a flagellar film. The film was prepared by drying a suspension of *Proteus* flagella on a glass slide (× 11,000).[66]

Judging from the observations made by Pijper and Nunn on *Vibrio metchnikovii*[31] (a bacterium known from electron microscopic observations to have only one or two flagella) single bacterial flagella may have enough light-scattering power to make them visible in the dark field if very powerful illumination is applied. It has been pointed out, however, that *V. metchnikovii* seems to be equipped with unusually thick flagella.[28]

Staining methods have been used by numerous investigators for the study of flagellar morphology. The diameter of the flagella may be grossly enlarged by such procedures because of the precipitation of mordants and stains on the flagella. On the other hand, the staining does not necessarily very much affect a characteristic feature of the bacterial flagella, viz., their wave shape (Fig. 2). As has been shown by studies on living bacteria[12, 32-34] and on purified flagella,[29] the flagella situated on motionless or slowly moving bacteria are helical in shape. Only in exceptional cases have entirely straight flagella been demonstrated.[28, 35] Normally the pitch or wavelength of the helices is rather constant within the same species. Sometimes, however, a phenomenon which is called biplicity occurs,[36-39] i.e., two distinct wavelengths are observed, one twice the length of the other (Figs. 2 and 3).

(a) (b)

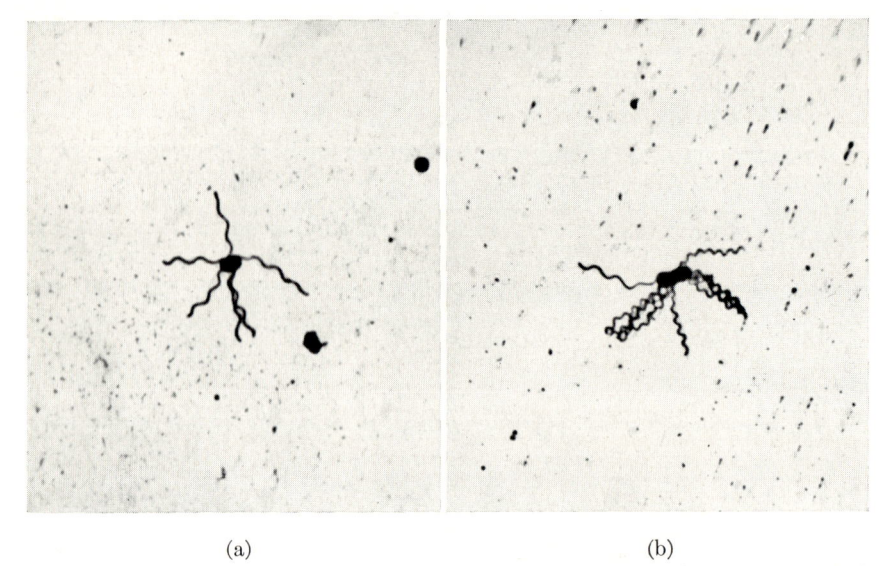

Fig. 2. Cells of *Proteus mirabilis* stained according to Leifson (×1,700).[37] (a) Cell with flagella all of which have waves of the same wavelength. (b) Cell with flagella of two different wavelengths (biplicity).

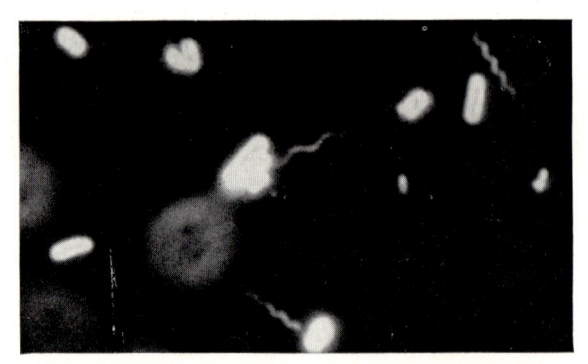

Fig. 3. Cells of *Salmonella schottmülleri*. Flagella of two kinds of wavelengths are present (biplicity). Sunlight dark field (× 2,000).[103]

The wavelength varies considerably from one bacterial species to another.[36, 37, 40]

The distribution of the points of origin of the flagella on the bacterial body varies. Two main types may be discerned: terminal (polar) and lateral (peritrichous) flagellation (Figs. 4 and 5). In the latter case the flagella seem to originate randomly from the bacterial surface (perhaps with the exception of the polar regions). The spirilla, vibrios, and pseudomonads belong to the former group, the salmonellae and bacilli to the latter.

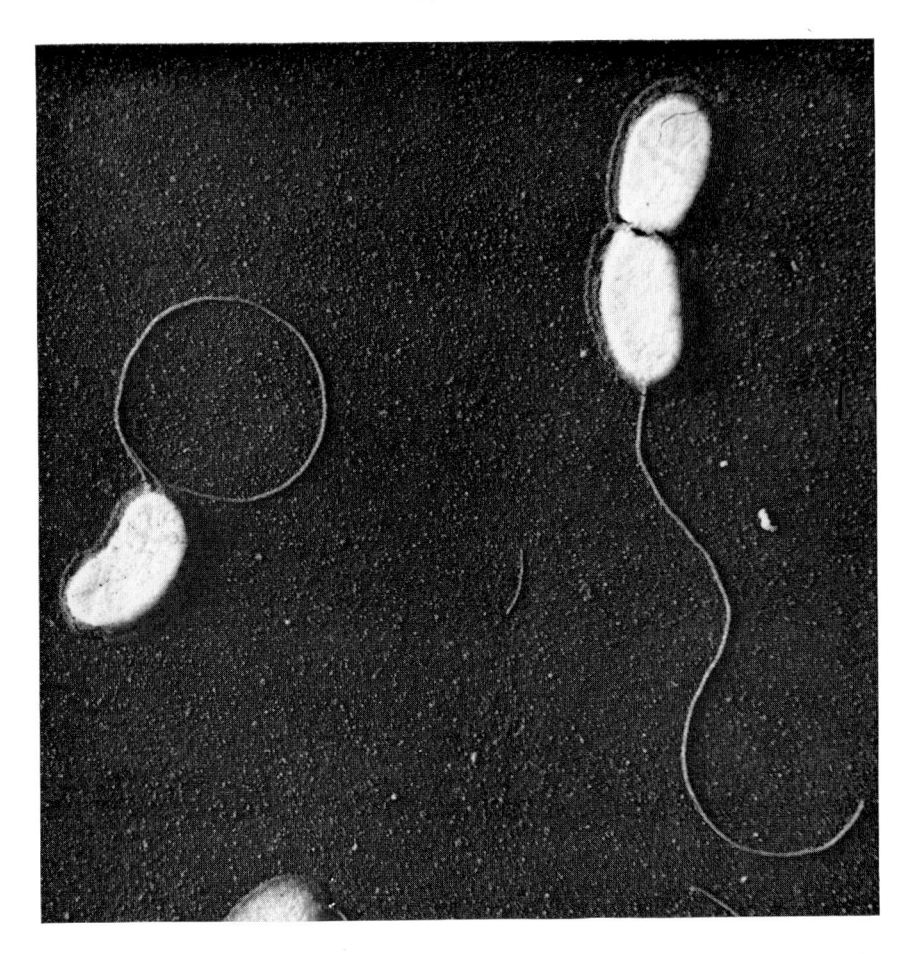

Fig. 4. Electron micrograph of a terminally flagellated bacterium, *Vibrio metchnikovii* (× 15,000). (From W. van Iterson, *Biochim. et Biophys. Acta* **1**, 527 1947.)

The origin of the bacterial flagella has been much discussed. Several electron micrographs have been published which suggest that they pierce the cell wall.[41-43] The most conclusive evidence for the protoplasmic origin of the flagella seems to be the fact that, in the case of lysozyme-sensitive species, they remain attached to the naked protoplasts after the cell wall has been digested away with this enzyme.[44, 45] This is illustrated in Fig. 6.

The flagella of higher organisms are attached to basal granules, called blepharoplasts, located within the flagella-bearing cell. Several electron micrographs have been published[28, 46-49] strongly suggesting the presence of such granules also in bacterial cells (Fig. 7). The hooklike bends that have been observed at the base of flagella detached from bacterial cells also sug-

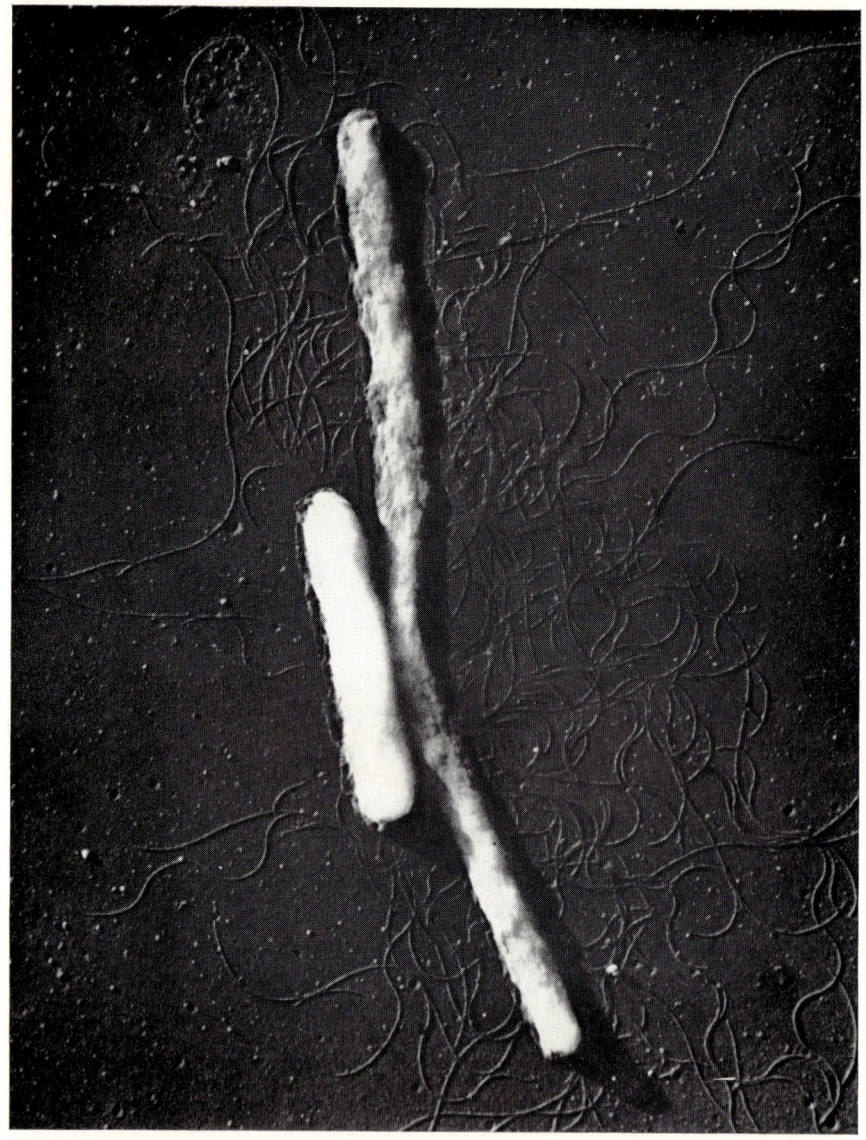

FIG. 5. Electron micrograph of a laterally flagellated bacterium, *Proteus mirabilis* (× 20,000). (From W. van Iterson, *Proc. Intern. Conf. on Electron Microscopy London*, July 1954. Royal Microscopical Society, London, England.)

gest that the flagella possess some kind of a "root."[28] On the other hand, no granules with attached flagella were found after the cell contents of a laterally flagellated bacterium, *Bacillus megaterium*, had been liberated by a dissolution of the cell wall with lysozyme.[50]

MOVEMENT 163

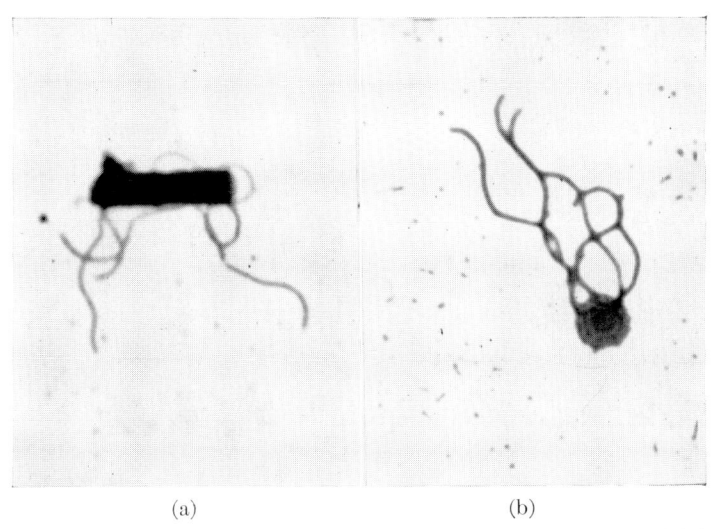

Fig. 6. (a) Cells of *Bacillus megaterium* stained according to Casares-Gil (×3000). (b) Protoplasts fixed in formaldehyde and stained in the same way[44] (×3000).

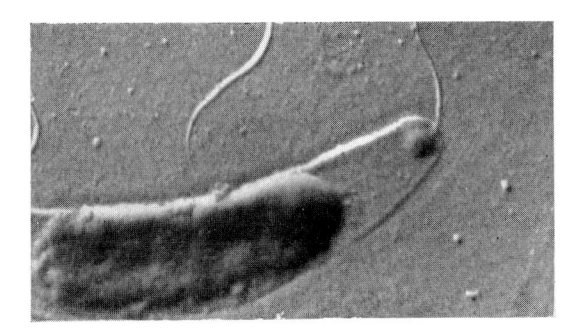

Fig. 7. Electron micrograph of a *Spirillum* sp. with a polar flagellum attached to a granule within the cell (× 12,000). (From reference[244], p. 130. With kind permission of Dr. P. E. Pease.)

It should be pointed out that, in addition to flagella, filamentous appendages of another type are sometimes observed attached to bacterial cells.[46, 51-61] These filaments, which have been named "fimbriae,"[58] are distinguished from the flagella in being shorter and thinner, more numerous, and not curved in regular waves (Fig. 8). Duguid *et al.* found fimbriae in 25 of 33 motile strains of *Escherichia coli* and in 6 of 14 nonmotile strains.[50] Thus fimbriae do not seem to be related to motility. Houwink and van Iterson have suggested that they may function as organs of attachment.[46] This view is in agreement with the observations of Duguid *et al.*, showing that the fimbriae occur in association with hemagglutinating activity.[58]

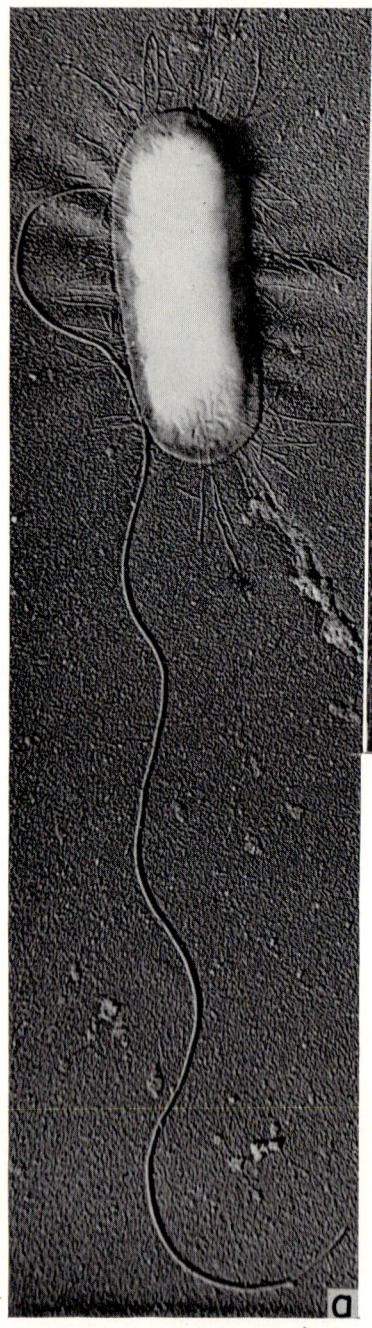

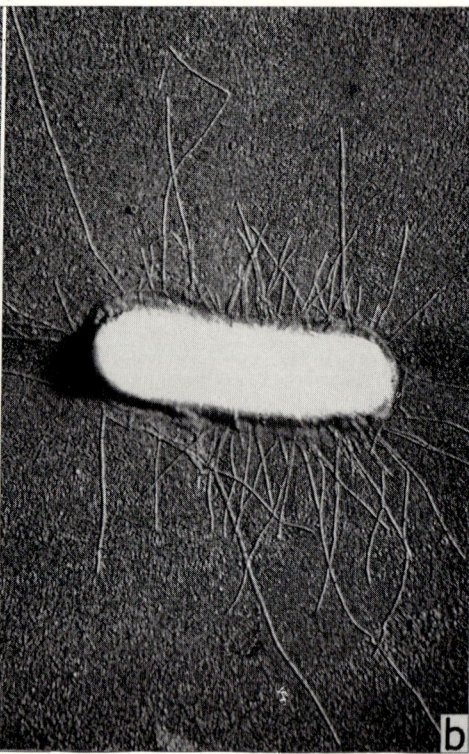

FIG. 8. Electron micrographs of a bacterium (*Escherichia coli*) equipped with flagella and fimbriae (× 20,000).[58] (a) Cell with one flagellum and numerous fimbriae. (b) A nonflagellated cell with fimbriae.

B. Structure and Chemistry of the Flagella

In most electron micrographs of bacterial flagella published so far, the filaments appear smooth and without any fine structure. In the pictures taken by Starr and Williams[25] and by Labaw and Mosley,[24, 62] however, the flagella clearly appear to be composed of two or three subfibrils twisted helically around each other. Electron micrographs suggesting a coiled structure of the flagella of *Escherichia coli* and *Bacillus brevis* have also been published.[63, 64] In the former case a fine structure consisting of a central filament and a surrounding sheath was also apparent. A similar structure was found in the flagella of *Vibrio metchnikovii*, after the cells had been partly autolyzed.[28]

X-ray investigations have revealed that the flagella of *Bacillus subtilis* and *Proteus vulgaris* belong to the keratin-myosin-epidermin-fibrinogen group of fibrous proteins.[23, 65, 66] The axial period determined from the X-ray diagrams given by bacterial flagella is most probably 410 A., or the same as that of the actomyosin complex of skeletal muscle.[66] Keratin, on the other hand, is characterized by a shorter period, viz., 198 A. The X-ray diagrams also give some tentative evidence for the presence of subfibrils in the bacterial flagella. The implications of the X-ray investigations on the motility problem have been discussed by Astbury *et al.*[66]

The chemistry and physical chemistry of the bacterial flagella have been investigated in some detail by Weibull[67-73] and Koffler and his associates.[74-77] Preparations from *Proteus vulgaris* and five other species have been studied. It appears that highly purified flagella contain 15.7–16.7 % nitrogen but less than 0.05 % phosphorus. In addition, not more than 0.0001–0.2 % carbohydrate, 0.8 % lipids, 0.0006–0.1 % nucleic acids, and 0.003–0.005 % ash were found in these specimens. It was estimated that the flagella consist of at least 99 % proteinaceous material.[74] Amino acid analyses revealed most of the acids generally found in proteins with the exception of tryptophan, histidine, proline, and hydroxyproline. Weibull was unable to detect any cysteine-cystine in the flagella of *Proteus vulgaris*,[68] whereas Koffler *et al.* reported the presence of 0.54–0.75 % of these acids in the flagella of this and three other species.[74] The flagella of at least some *Proteus*, *Salmonella*, and *Bacillus* spp. disintegrate when the acidity of the medium is brought below pH 3–4.[67-68] The fairly homogeneous decomposition products have a molecular weight of about 40,000 and are probably elongated, suggesting some kind of subfibrils in the flagella. Chemical analyses indicate a minimum molecular weight of the flagellar subunits of about 20,000.[74]

C. Characteristics of Flagellar Movement

1. Terminally Flagellated Bacteria

Among the numerous investigations published in this field,[12, 21, 78-84] those carried out by Metzner[79] seem to be the most detailed. Metzner studied

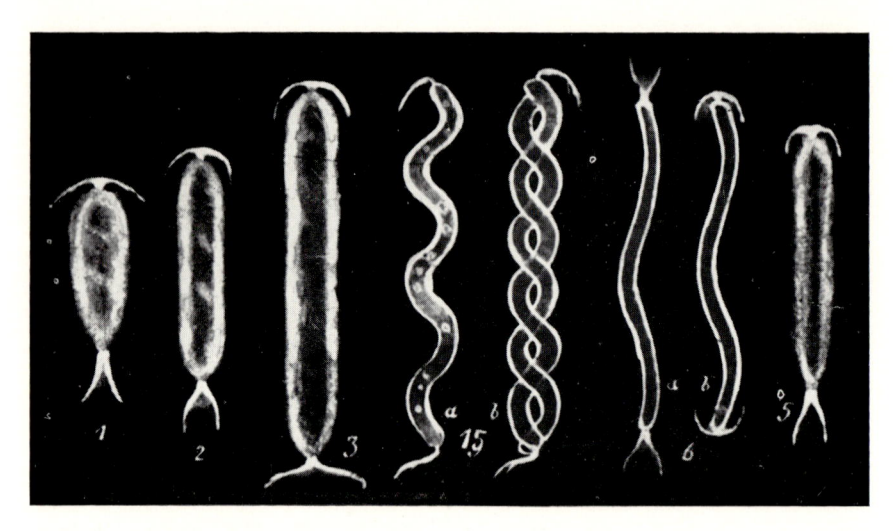

Fig. 9. Motile spirilla (1 and 2 *Spirillum undula*; 3 and 15 *S. volutans*; 5 and 6 *S. serpens*) observed in the dark field. Cells 1,2,3,5 and 6 are viewed in continuous illumination, cell 15 in flashing light. Left 13, right 26 flashes per second. Cells 6a and 6b have oppositely rotating flagella and are thus stationary.[79]

the movements of *Spirillum* and *Chromatium* spp. using dark-field microscopy. Some of his observations are illustrated in Fig. 9.

In most cases the investigated spirilla had a bundle of flagella at both poles of the cell body. The flagellar bundle at the rear end of the spirillum formed a more or less straight tail whereas the flagella at the front end were bent backward around part of the organism.

Metzner noted that the flagella at both poles of the cell body formed cones of revolution when active. During locomotion both flagella bundles rotated in a direction opposite to that of the cell body. By means of a stroboscopic method he could measure quantitatively both kinds of movement. He thus found that the flagella bundles of *Spirillum volutans* and *Spirillum undula* made about 40 revolutions per second while the body performed at the same time 12–14 r.p.s. The direction of the translatory movement of the body was changed when the rotation of both flagella bundles was reversed. Sometimes only one bundle changed the direction of its rotation. The result was that the body stopped moving and rotated only very slowly.

The currents in the medium caused by the movements of the flagella were studied by Metzner by adding collodial silver or mastic particles to the medium. Flagella forming tails were found to give rise to currents parallel to the axis of the body whereas flagella at the front end of the cell only made the water move in circles approximately at right angles to the body. The propulsive effect of the two kinds of flagella were, however, the same.

This was shown by the movement of spirilla that were flagellated at only one end. Such spirilla, which were only occasionally found, swam at approximately the same speed irrespective of whether the flagella were at the front or the rear end of the bacterium, ie., irrespective of whether the flagella formed a tail behind or an envelope around the cell.

The flagella of *Chromatium okenii* were found to make at least 40–60 r.p.s, the body, on the other hand, only 2–8 r.p.s. The bacterium is flagellated only at one pole and the flagella are longer and more like corkscrews than those in *Spirilla* spp. The direction of the rotation of the flagella could be reversed.

2. LATERALLY FLAGELLATED BACTERIA

The details of the movements of laterally flagellated bacteria are less well known than those of the terminally flagellated ones. This is probably mainly due to the smaller size of the former organisms together with their sometimes very rapid movements as viewed under the microscope. Many investigators have thus found it necessary to slow down the movements by adding viscosity-increasing substances to the medium. Such a procedure may, of course, modify the normal propulsive mechanism, especially since with the exception of methylcellulose, the added compounds tend to form precipitates around the flagella.[27, 85] On the other hand, it seems reasonable to assume that the same basic principles apply irrespective of the viscosity of the medium.

Reichert studied laterally flagellated bacteria moving slowly in viscous media.[12] It was found that the flagella were wound helically and clockwise and had very much the shape of corkscrews. They were more tightly coiled than the flagella of the spirilla. During locomotion the cell bodies rotated counterclockwise (seen in the direction of motion). The flagellar bundles had the appearance of rotating helices, i.e., the appearance characteristic of filaments down which series of helical waves travel continuously. However, the flagella always looked as though they rotated clockwise and a change in the sense of their apparent rotation was never observed. During locomotion the flagellar bundles always pointed backward. A change in the direction of movement of the bacteria was found to be a slower process than for the spirilla and required a rearrangement of the direction of the flagellar helices relative to the cell body. Similar observations were made by Uhlela.[86]

The most comprehensive studies on the movements of laterally flagellated bacteria have been carried out by Pijper with the aid of dark-field microscopy using the sun as illuminating agent. Owing to the extreme brightness of sunlight, which is hardly equaled by artificial light sources (with perhaps the exception of electronic flash), Pijper has been able to carry out his investigations without adding viscosity-increasing substances to the medium. The results of his studies have been presented as ordinary publications[27, 32, 33, 85, 87-90] and as microcinematographic recordings.[81, 91]

Pijper's observations on slowly moving bacteria agree in the main with those made by Reichert. When the bacteria move fast, on the other hand, the flagellar tails appear as straight, somewhat blurred lines on the photographic pictures and not as helices.[33, 92, 93] The disappearance of the wave structure of the flagella may be due partially to the imperfections of the existing microcinematographic technique. Apparently, however, the flagellar helices straighten, to some extent at least, when the bacteria move fast (Fig. 10). The change of shape of the flagella is reversible.[32, 92]

Pijper has paid much attention to the movements of the cell body during locomotion. These movements are described as "undulatory, gyratory". It is also pointed out that the body is able to perform semisomersaults, leaving the flagella where they were before the turn.[90]

The way in which Pijper interprets his recordings of bacterial movements will be discussed later in this chapter.

Fleming and co-workers have investigated the movements of penicillin-inhibited bacterial cells having various abnormal shapes.[94, 95] It appeared that even grossly misshapen cells were able to perform locomotory or rotatory movements (Fig. 11). Radiant heat seemed to have a stimulating effect on the motility since movements stopped if a heat filter was placed across the light beam of the microscope lamp. When the filter was removed the movements of the organisms were resumed. Then it could sometimes be observed that the flagella started to move before the cell bodies.

Kvittingen ascribes the phenomena observed by Fleming *et al.* to effects of fluid currents on the agar surface on which the bacteria were grown.[96]

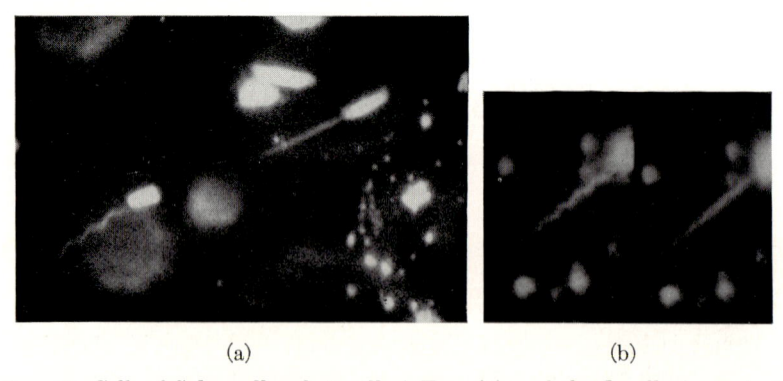

FIG. 10. Cells of *Salmonella schottmülleri*. Transition of the flagella from straight tail into helix. Sunlight dark field. (a) The bacterium was photographed twice on the same piece of film with an interval of less than a second ($\times$ 1,500). (b) Another bacterium photographed on two different frames ($\times$ 1,200).[103]

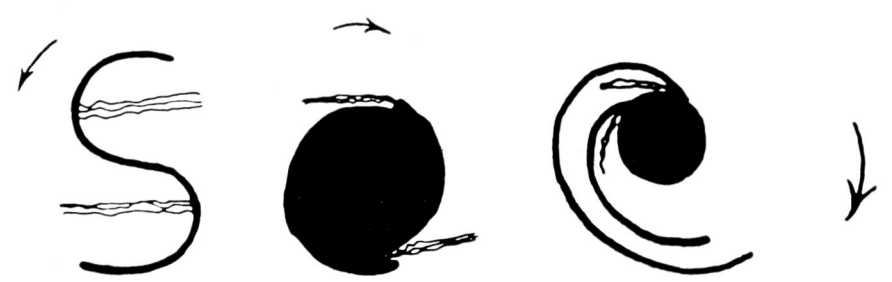

FIG. 11. Movements of penicillin-treated cells of *Proteus vulgaris*. In the drawings arrows indicate the direction of rotation; thin lines, flagella. From A. Fleming, *in* "The Nature of the Bacterial Surface" (A. A. Miles and N. W. Pirie, eds.), Chapter 9, p. 169. Blackwell, Oxford, 1949.

The method of suspending colloid particles in the medium in order to study bacterial movements has also been used in the case of laterally flagellated bacteria.[80, 97-99] It was found that particles in the vicinity of the organisms may be violently agitated even when the cell bodies did not move or moved only slowly.

D. The Mode of Action of the Flagella

1. Are the Flagella Motor Organs?

Most bacteriologists have been of the opinion that the bacterial flagella are motor organs. The fact that all rapidly moving Eubacteriales are flagellated whereas the nonmotiles generally are not has often been the main argument in favor of this view. This circumstance per se could, however, permit a reversed interpretation of the nature of the flagella, i.e., that they are the effect and not the cause of motility. Such a view was in fact early advanced by van Tieghem[100] and has later been strongly advocated by Pijper.

Pijper's main arguments for his views seem to be the following ones:[81-83, 85, 89, 101-103]

In the dark field, motile bacteria appear sometimes with a flagellar tail, sometimes without. The motility is, on the other hand, similar in both cases.

A suspension of moving bacteria showing tails may be shaken violently without affecting motility. The tails, however, disappear.

When the movements of spirilla are recorded photographically, the flagella sometimes appear immobile during locomotion, whereas the cell bodies rotate rapidly.

The following phenomena, described earlier in this chapter, are also presented by Pijper as support for his views:

The more or less helically shaped cell bodies of the swimming bacteria perform undulatory, gyratory movements. According to Pijper these movements cause the propulsion of the organisms.

Sometimes, rapidly moving bacteria perform semisomersaults. During such movements the direction of the flagellar tails remains unchanged. Pijper is of the opnion that the somersault is due to the combined effects of the momentum of the organism and a sudden reversal of the undulations of the cell body.

Objections could, however, be raised against these arguments:

As far as the visibility of the flagellar tails in the dark field is concerned, Pijper himself has pointed out that the nature and pH of the medium are important for the observing of tails.[33, 85, 89] Other factors besides the mere presence or absence of flagella may thus determine the visibility of these organelles. Generally, the light-scattering power of an object, which determines its visibility in the dark field, diminishes rapidly with the size of the object (the object is assumed to be considerably smaller than the wavelength of the light used; for spheres the light-scattering is proportional to the square of the volume, i.e., to the sixth power of the diameter). A splitting of a flagellar bundle into several thinner ones or into individual flagella may thus make it invisible in the microscope. The propulsive power, on the other hand, could remain essentially unchanged.

It is true that mechanical agitation of bacterial suspensions causes a detachment of the flagella. It has been shown, however, that not all flagella may be removed from the cell body in this way.[104] Moreover, the flagella can be rapidly resynthesized. According to Leifson they may grow at a rate of 1 micron every 2 or 3 minutes.[105] From Pijper's report it is not clear whether these circumstances have been taken into account when the experiments were carried out.[101] Stocker has repeated the experiments, the bacteria being agitated in a blender equipped with a propeller rotating at a speed of 12,000 r.p.m.[106, 107] Motility could be almost completely removed by this treatment but moving cells were again observed only a few minutes after the end of the agitation. When chloramphenicol, believed to stop specifically protein synthesis, was added, moving cells did not reappear. Control experiments showed that chloramphenicol had no influence on the motility of cells that had not been agitated. Stocker also found that the flagella could be detached from the cells by rubbing the bacteria grown on nutrient agar with a glass rod.[107] It was shown that about 1 % of the cells had flagella after this treatment and 0.6 % were motile. Control preparations showed about 93 % flagellated and about 90 % motile cells. Counts of live bacteria in control and rubbed preparations did not differ significantly.

Thus, studies carried out using proper control experiments show that nonlethal treatment which removes the flagella from bacterial cells causes loss of motility.

Pijper assumes that movements of the more or less spirally wound cell bodies cause propulsion. The thermodynamic theory quoted earlier in this chapter shows, however, that spiral waves can cause propulsion only if an externally applied torque, e.g., due to an inert head, prevents the cells from spinning freely. Since the bacterial bodies in question possess a uniform diameter it seems rather doubtful that the undulatory, gyratory movements described by Pijper could by themselves cause propulsion. In addition, the experiments performed by Fleming et al.[94, 95] and White [108] show rather clearly that deformed bacterial cells may be motile, even if their bodies are unable to perform undulatory movements. It could also be asked how the motile sarcinae and other cocci[109-111] are able, with their cell bodies, to perform any movements that might result in propulsion. Pijper's explanation in the case of the sarcinae, viz., that motility is due to the continuous tumbling of the cell packets, "brought about by cytoplasmic movements inside the cell wall, perhaps combined with inconspicuous changes in shape or relative position of the cells constituting the packet,"[112] seems rather vague.

The semisomersaults performed by swimming bacteria are explained by Pijper as a combined effect of the propulsive undulations of the cell body and its momentum. A rough calculation, based on Stokes's law and similar to that mentioned by Stocker with regard to active bacterial flagella,[106] shows, however, that the movement of the bacterium virtually comes to a standstill within about 10^{-6} seconds after propulsion has ceased. This evidently implies that momentum can play only a negligible role in connection with phenomena such as the bacterial semisomersaults, which can be directly followed under the light microscope by a human observer.

Pijper's films on moving bacteria are technically admirable. The sequences dealing with moving spirilla hardly prove, however, that the flagella are immobile while the bodies rotate rapidly. Sometimes a spirillum with one active and one inactive flagellar bundle (one bundle at each end) may have been observed. The inactive bundle is easily seen and photographed, whereas the active one moves too fast to be observed when propelling the bacterium. Another explanation would be that the flagella beat synchronously or nearly so with the changing of the film frames. Only thorough stroboscopic investigations can give us correct information concerning the speed of rapidly rotating objects. Such measurements have been performed by Metzner and Buder (see above) and they showed that the flagella of spirilla, forming cones of revolution, rotate more rapidly than, and in the opposite direction to, the cell body.

The protoplasmic origin of the flagella and their proteinaceous nature also favor the view that they are not merely passive appendages of the bacterial cell. Therefore, taking the available evidence into account and perhaps especially Stocker's recent experiments, the author concludes that the role of the bacterial flagella as motor organs must be regarded as firmly established.

2. The Mechanism of the Flagellar Propulsion

In the theoretical section it was shown that a microscopic object consisting of an inert head and a tail could be propelled by helical waves traveling down the tail. The theoretical analysis is directly applicable to the movements of laterally flagellated bacteria. The cell bodies of these organisms have various shapes (cf. salmonellae, cocci, and penicillin-treated specimens). The flagella have a pronounced helical shape and look as though they rotate around their axis like a turning corkscrew (at least when moving slowly). This appearance is characteristic for a filament down which series of helical waves are propagated. On the reasonable assumption that the base of the flagellar bundle is unable to rotate relative to the bacterial body, a wave motion of this kind will tend to cause the bacterium to rotate in a direction opposite to that of the helical waves. The viscous resistance of the medium to such rotation will, however, only permit a slow rotation of the cell. In this way a torque is applied to the flagellar helix, making propulsion possible.

The picture is somewhat different in the case of terminally flagellated bacteria, notably the spirilla. Here the helical shape of the flagella is often not very pronounced. The organelles create cones of revolution at the poles of the organisms and probably perform their role as motor organs mainly by making the body rotate. The pronounced helical shape of the cell body then makes the organism move through the medium.

Pijper explains the motility of spirilla in terms of stretching and shortening of the coils of the cell bodies with intermittent freewheeling.[91] However, the present author is unable to see how an alternating stretching and shortening of a helix could make the body move steadily and thus cover any appreciable distance. Moreover, as in the case of bacterial semisomersaults, which have been described earlier in this chapter, any motion of the spirilla due to freewheeling can go on only for extremely short periods of time after propulsion has ceased.

In bacteria equipped with many flagella these organelles beat to a large extent synchronously, several flagella waving closely side by side in each flagellar bundle. Several attempts have been made to explain this coordination. Van Iterson[28] seems to be inclined to explain the synchronous waving in terms of aggregation phenomena similar to those mentioned earlier in this chapter. Stocker discusses the problem from a hydrodynamic point of view[106] and in this connection quotes the theoretical analysis given by Taylor.[9] Stocker points out that Taylor's analysis cannot be applied directly to waving flagella since his theory deals with a simplified model of waves traveling down adjacent sheets. The analysis seems to favor the view, however, that hydrodynamic forces may tend to hold together active flagella which happen to come into contact. Stocker points out that an increase in the viscosity of the medium would, on hydrodynamic grounds, presumably enhance the aggregative tendency.

Nothing can as yet be said with certainty about the mechanism of the flagellar contractions. Reichert assumed the contractions to take place along a hypothetical "line of contraction" running helically around the originally straight flagellum. Contractions along this line would give the flagellum a helical shape. Series of waves would result if the line moved around the flagellum.[12]

Weibull put forward the hypothesis that each flagellum consists of a bundle of helically wound subfibrils.[113] Contractions were assumed to pass from one fibril to another, thus causing the phenomena described by Reichert. A system of straight subfibrils has also been proposed.[66]

As described earlier, fine structures of some bacterial flagella have been demonstrated using the electron microscope. Whether or not these structures are connected with contraction phenomena is not yet clear. X-ray investigations suggest that polypeptide chains of two different configurations, the α- and the cross β-pattern, are simultaneously present in the flagella.[66] Of these the latter one is characteristic for highly contracted or so-called supercontracted protein fibrils.

Conclusions concerning the role that may be played in the flagellar activity by the earlier mentioned basal granules must await further investigations. It has been suggested that they might serve as an energy source (cf. below) or "signal box" regulating the flagellar activity.[66]

3. SOURCE OF ENERGY FOR THE FLAGELLAR CONTRACTIONS

a. Energy Transport within the Flagella. Not very much is known about the energy supply of active flagella. It is thus an open question whether they produce the necessary energy themselves or whether it is generated in the cell body and transported into and along the flagella. Herbert has calculated that the power output required for the propulsion of bacteria corresponds to less than 0.1 % of the total metabolism of the cells.[114] He found, moreover, that if the power-to-weight ratio of the flagella is the same as that of muscle the power available should be more than sufficient for propulsion. De Robertis and Franchi claim that bacterial flagella contract in the presence of adenosine triphosphate (ATP).[115] De Robertis and Peluffo have found that ATP enhances the motility of *Proteus vulgaris*.[116] On the other hand, Barlow and Blum found no ATPase activity in flagellar preparations.[117] Moreover, these workers could not find that addition of ATP to a flagellar suspension caused any change in the shape or in the state of aggregation of the flagella.

The general enzymic properties of bacterial flagella have not been studied to any appreciable extent. On the whole, it seems doubtful that a very elaborate enzyme system could be present in the flagella because of their small diameter. Metal-containing, respiratory enzymes and flavoproteins are probably absent since purified flagellar preparations are colorless.[67] De Robertis and Peluffo found that the movements of *Proteus vulgaris* were not inhibited by cyanide, azide, and fluoride.[116] Narcotics were inhibitory only at high concentrations. On the other hand, thiol inhibitors had a very marked effect on the motility.

According to Sherris *et al.*, arginine seems to be highly specific in promoting the motility of a *Pseudomonas* sp. under anaerobic conditions.[118] When, in the absence of oxygen, arginine is given to a motionless suspension of the bacteria in question, motility is resumed but ceases when the added arginine has been converted into ornithine. Whether or not this process forms part of a reaction system which provides the flagella with the energy necessary for their activity remains to be settled.

If energy is transported from the cell body into the flagella, it could be asked how such a transport is effected. A discussion of possible mechanisms for the energy transport in polypeptide chains has recently been given by Broser and Lautsch.[119] The authors consider mainly the different ways available for a migration of electrons to occur along the chains, taking into consideration modern concepts of the structure of polypeptide molecules. A similar transport mechanism has also been discussed by Denbigh.[120]

Our present knowledge of the structure and functioning of bacterial flagella and the muscles of higher organisms hardly permits a detailed comparison between these two types of contractile material. The absence of measurable amounts of phosphorus, nucleotides, and certain amino acids in the flagella (cf. above) would indicate marked differences in the nature of these organelles and true muscles. On the other hand, the presence of an identity period of about 400 A. along the fiber axis in both flagella and muscles, as suggested by X-ray measurements, points to similarities in the organization of both kinds of fibers.

IV. Movements of the Spirochetes

A. General

The movements of the spirochetes are highly diversified. Thus, rotation and flexions of the cells are encountered both in the presence and in the absence of locomotion. Many of the movements are carried out with great rapidity. These facts, together with the minute dimensions of most spirochetes, may explain why our knowledge of the spirochetal movements is still rather unsatisfactory in spite of numerous investigations (reported in earlier reviews[121-126]).

Moreover, the morphology and the movements vary considerably from one spirochetal genus to another. Each genus will therefore be treated separately. In the concluding section, some general aspects of spirochetal movement will be considered.

B. Spirochaeta

Of this genus, the species *Spirochaeta plicatilis* seems to have been the most carefully studied as far as movements and structure are concerned. The investigators have found that the body of the organism has the shape of a true, three-dimensional spiral.[127-129] The helical waves are stable and regular. Secondary, more or less irregular, spirals or waves are superimposed on the regular ones. Light and electron microscopic investigations have not given any evidence for the existence of flagella.[130]

Locomotion is accompanied by rapid, snakelike contractions of the body. Contact with a solid surface is said to be necessary for locomotion.[127] Very often rotatory movements and vibrations of the body are noted. Rotation may take place without accompanying locomotion. The body is then wound in wide, secondary spirals.[127]

The movements of other *Spirochaeta* spp. have been described by Bach,[131] Dobell,[132] and Cantacuzène.[133] The observations agree in the main with those made by Zuelzer.[127] Thus, Bach stresses that locomotion takes place only on a solid surface.

C. Cristispira

The morphology of these large spirochetes has been studied by numerous investigators.[132, 134-141] The presence of a ridge or crista running along the entire cell body is a unique feature of this genus. The crista consists of a great number of fibrils.[136, 142] When at rest, the organism has the shape of a true, three-dimensional spiral.[143] During movement secondary, sinusoidal, or irregular waves are superimposed on the regular helical waves. According to several investigators, flagella are not present.[132, 137, 142]

Fantham has described the movements of *Cristispira balbianii* and *C. anodontae* in some detail.[138] He finds that the movements are resolvable into two components, viz., sinusoidal waves traveling down the cell body and a corkscrewlike rotation of the organism. Fantham ascribes the latter type of movement to the ridge or crista winding in spirals around the cell body. The movements were found to occur in jerks.

Dimitroff has also described the locomotory and rotatory movements of *C. balbianii* and *C. anodontae*.[141] Like Fantham, he found that secondary waves were formed along the body when it moved forward. At the same time rotation occurred together with lashing movements of the posterior third of the cell. Rotation and lashing movements were also observed when locomotion was prevented by outer forces. The investigators compared the rotating posterior of the organism with the hooked ends of rapidly moving *Leptospira* spp. (cf. Section IV, G).

D. Saprospira

Members of this genus seem to move more slowly than the cristispirae but otherwise in a similar way.[132, 139, 144]

E. Borrelia

It is generally stated that the cells of these spirochetes have a helicoid shape. Neumann,[128] Schellack,[129] and Gross,[143] however, found that the waves of the spirochetal bodies were flat. Neumann based his view on material obtained from dark-field photography.

Often the cells seem to taper into polar filaments. These filaments probably represent material remaining from completed cell divisions.[145, 146]

Several investigators, using the light or electron microscope, have reported the occurrence of a rich lateral flagellation in *Borrelia* spp.[147-149] Other workers have pointed out, however, that no such flagellation is found when osmic tetroxide fixation is applied or when repeated centrifugations and prolonged treatments with distilled water are avoided.[129, 142, 150-154] Moreover, the subterminal flagella or filaments described by Babudieri and Bocciarelli[152] could be due to an incipient disintegration of the organism.[142, 153, 154]

The movements of various *Borrelia* spp. have been repeatedly described.[33, 129, 155, 156] All three basic kinds of movements have been observed. Locomotion is accompanied by rotation, the organisms sometimes moving with such extreme rapidity that it is almost impossible to follow them in the microscopic field of sight.[155] According to Pijper[33] and others, secondary waves superimposed on the primary spiral appear during locomotion.

F. Treponema

Much of what has been said about the genus *Borrelia* is valid also for the treponemata. According to most investigators, for example, the cells of *Treponema pallidum* have the shape of helices, but it has also been stated that the waves of the organisms are flat.[128, 157] DeLamater *et al.* have reported that some individuals of a population are flat, others helicoid.[158]

Several electron micrographs have been published showing *Treponema* cells with a lateral flagellation,[146, 159-164] or a terminal one.[165-167] Both kinds of flagella could, however, well represent detached fibrils that are wound tightly around the cell bodies of living and undamaged spirochetes.[142, 153, 168, 169] (cf. Section IV, E; see also Fig. 12).

Like most other spirochetes, the treponemata perform rotatory, loco-

(a)

(b)

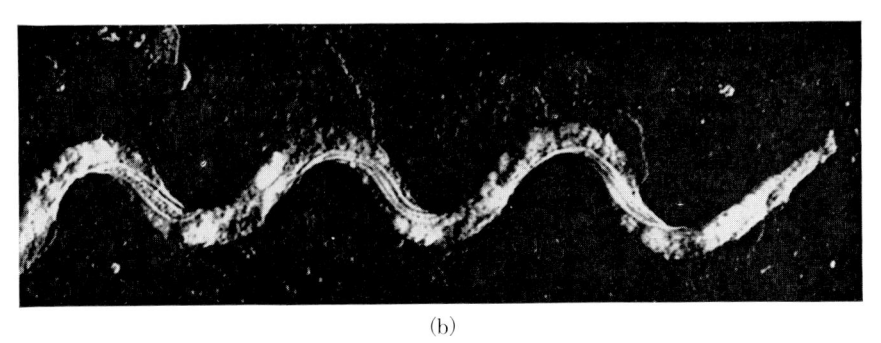

FIG. 12. Electron micrographs of *Treponema pallidum* indicating the presence of fibrils wound around the cell body. (a) Cell before trypsin treatment (× 39,000). (b) After 20 minutes trypsin digestion (× 30,000).[153]

motory, and flexing movements.[170] The cells often rotate rapidly whereas locomotion occurs less frequently.[122, 123, 171, 172] Undulatory waves have been observed traveling from one end of the organisms to the other.[121, 122, 158]

G. Leptospira

With one exception[173] all investigators state that the bodies of these organisms are wound as three-dimensional spirals.

Numerous electron micrographic studies have been carried out on various leptospirae.[142, 153, 174-178] In no case was the presence of flagella noted. On the other hand, a fiberlike structure or axistyle was found to run spirally around the spirochetal body (Fig. 13).

The movements of the leptospirae have been described by several investigators.[175, 179-183] Perhaps the most characteristic kind of movement is a rapid rotation without accompanying locomotion. Under these circumstances both ends of the organism assume the shape of a hook. Due to the rapid rotation, the moving organism thus often appears to be furnished with a buttonhole at each end. According to Zuelzer[180] and Noguchi[179] the

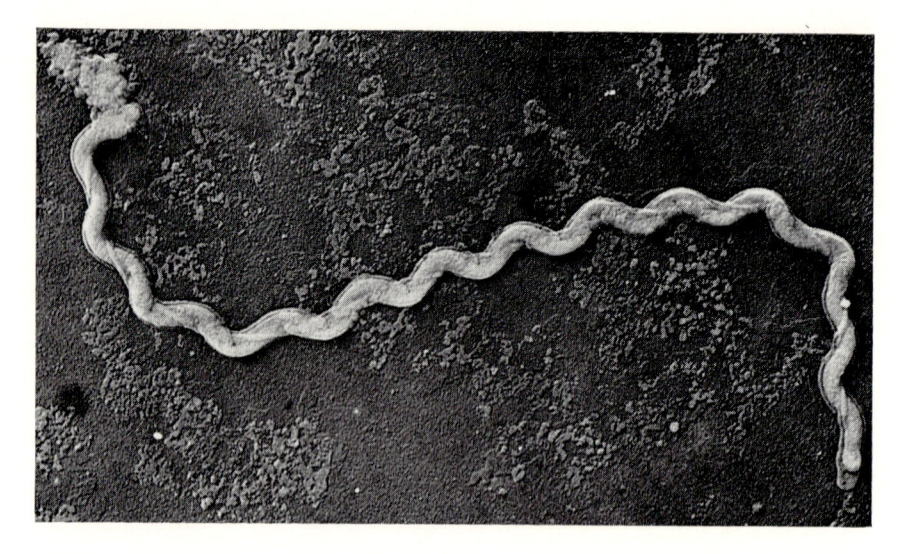

Fig. 13. Electron micrograph of *Leptospira canicola* indicating the presence of an axial filament or axistyle (× 23,000). (From R. H. A. Swain, *J. Pathol. Bacteriol.* **73**, 155, 1957.)

body is drawn out into a straight line, except for the above-mentioned hooks, when locomotion takes place in a free liquid. When, on the other hand, a leptospira penetrates a semisolid medium, the ends are seldom hook-shaped. Instead, wide, wavy undulations superimposed on the narrow primary ones are seen along the entire body.[179] Such undulations are also seen when the organisms are to some extent hindered in their movements.[182]

H. The Mechanism of the Spirochetal Movement

1. Morphological Considerations

As may be apparent from the foregoing, many fundamental questions concerning the morphology of live spirochetes are not yet very satisfactorily solved. It cannot be considered as definitely settled, for example, whether the cells of many spirochetes have a three-dimensional spiral shape or whether they should be considered as series of plane waves. Perhaps one individual may consecutively assume both shapes. As was pointed out in the theoretical section of this chapter, the hydrodynamic prerequisites for propulsion are rather different in the two cases. Only in the former one is an external torque, usually effected by an inert head, necessary for propulsion.

Another point of uncertainty has, for a long time, been the presence or absence of flagella similar to those characteristic for motile eubacteria.

According to the most recent investigations quoted above, the balance of evidence seems now to favor the latter view. Thus, another type of filamentous structure, interpretable in terms of a motor organ, seems to be present, viz., one or several fibrils wound in spirals around the entire cell body and in close contact with it.

2. LOCOMOTION

The occurrence of secondary waves superimposed on the primary body spiral during movement has been noted by several investigators and perhaps most frequently in the genera *Spirochaeta*, *Cristispira*, and *Borrelia* (see above). These waves have been described as undulatory or sinusoidal (and also as more or less irregular). This would imply that they extend preferentially in one plane. As has been pointed out by Taylor, Hancock, and Gray (cf. Section II) such waves may well cause propulsion even in the absence of an inert head connected to the waving filament. Thus the locomotion of at least some spirochetes could be fairly satisfactorily explained. The motor mechanism of the leptospirae is, however, more obscure. According to Noguchi and Zuelzer, these organisms can swim forward in a free liquid with their bodies drawn out into straight lines, except for a hook formation at one end. It could be argued that these hooks propel the organisms by means of rotatory or lashing movements. It is far from clear, however, how such movements could cause the propulsion of the filamentous body.

3. ROTATION IN COMBINATION WITH LOCOMOTION

If secondary, more or less plane, waves are assumed to cause propulsion, then rotation would be a natural consequence of locomotion, provided that the primary regular waves of the spirochetes are helicoidal.

If, on the other hand, the rotatory movement is regarded as the primary phenomenon, which seems most reasonable in the case of the leptospirae, propulsion would follow from the rotation. How the rotation, in its turn, is caused awaits, however, an explanation. As pointed out earlier, no evidence of flagella has been found in the leptospirae.

4. ROTATION IN THE ABSENCE OF LOCOMOTION

As mentioned above, this kind of movement often occurs in spirochetes, even in such cases where locomotion is not prevented by outside forces. Especially rapid rotations without accompanying locomotion are performed by the leptospirae. Theoretically, on the other hand, the rotation of these organisms would be expected to cause locomotion since almost all investigators have found the cell bodies to be helically shaped.

5. Concluding Remark

The author would like here to quote Dobell's words from 1912.[132] He stated that "the movements of the Spirochaets are still surrounded in mystery." It could be asked whether the situation has changed very much since those days.

V. Gliding Movement

A. General

This is the kind of slow locomotion that is especially characteristic of the Myxobacteriales and the Beggiatoaceae. It is not accompanied by vibrations or undulations of the bacterial body. No organelles interpretable in terms of well-defined motor organs have been observed. The locomotion seems to take place only on a solid surface, not in a liquid medium.

The above-mentioned bacteria are not the only plant microorganisms able to perform gliding movements. The phenomenon is encountered also in the Desmidiales, the Bacillariophyceae (diatoms), the Cyanophyceae (blue-green algae) and in some other algal and bacterial groups. The locomotion of the Desmidiales is generally attributed to the secretion and swelling of mucuous substances.[184-187] The most widely accepted theory for the movements of the diatoms is that advanced by Müller.[188-190] According to this worker, the motility is due to protoplasm streaming in the slits of the cell wall and generating friction with the surrounding medium. It has been suggested, however, that this theory might need some modification.[191] The movements of the blue-green algae have been explained in several ways. According to some workers, longitudinal contraction waves are propagated along the body, whereas others have attributed the movements to the secretion of slime, to osmotic forces, or to surface tension phenomena. These theories will be discussed in the following sections.

B. The Beggiatoaceae

This family is generally considered to be closely related to the Cyanophyceae. Thus the genus *Beggiatoa* has been regarded as an unpigmented counterpart to the genus *Oscillatoria* among the blue-green algae.[192-194] Consequently, in attempting to interpret the movements of the Beggiatoaceae equal consideration will be given to the results gained from investigations on the Oscillatoriaceae.

Of these two families, the genera *Beggiatoa*,[194-200] and *Oscillatoria* (referred to in earlier reviews[4, 194, 201-204]) have been the most thoroughly studied as far as motility is concerned. Besides locomotion, axial rotation and oscillation have also been observed. Some workers have reported, in addition, contractions[195, 205, 206] and contraction waves.[195, 199, 200, 207] Oscillations occur when only part of the filament is in contact with the solid substrate.[206] These movements, as observed under the microscope, may represent the optical projection of a conical circulation of the bacterial filaments distorted by the short distance between slide and coverslip.[194] Similarly, simple contraction phenomena may be due to tensions in the bacterial bodies caused by abnormal environmental conditions.[196]

The maximum translatory speed seems to be less than 5 microns per second.[4] In most *Oscillatoria* species, locomotion is accompanied by rotation, but deviations from this rule have been reported.[208] According to Schulz, a 360° rotation takes about half a minute in *Oscillatoria sancta*.[4] Correns and Schmid stated that the linear distance covered by oscillatoriae during one full turn is constant.[209, 210] Similar statements have been made with reference to *Beggiatoa*.[194, 200] It seems almost certain that locomotion takes place only when the filaments are in contact with a solid substrate or a water-air interface.[195, 196, 209, 211] Some statements have been made indicating swimming in free liquids,[201, 202] but experimental evidence supporting such a view seems to be lacking.

Several theories have been advanced to explain the movements of the Beggiatoaceae and the Oscillatoriaceae. Hansgirg[212] and Zukal[213] were of the opinion that water expelled from the cells by osmotic forces caused the locomotion. However, only under rather unnatural conditions could Hansgirg observe streaming in the fluid surrounding the filaments. Fechner reported the absence of streaming.[205] Krenner believed that osmotic forces may be of importance for the locomotion in a more indirect way.[214] He pointed out that *Oscillatoria* spp., which have highly turgid cells, move more rapidly than those having a low intracellular osmotic pressure.

Coupin[215] and Burkholder[202] thought that changes in the surface tension of the cells might cause the Oscillatoriaceae to move. Burkholder compared the movements of these organisms with those of a glass tube filled with gum camphor, sealed at one end and placed upon the surface of water. Under these conditions the glass tube glides rapidly forward. Burkholder also made the observation that dried *Oscillatoria* specimens kept in the herbarium for years exhibited characteristic gliding movements when placed between slide and coverslip in a drop of 10% glycerol.

It does not seem easy to assess the value of Burkholder's surface tension theory for the moment. It may prove useful in the future when it becomes possible to consider the details of the mechanism underlying the gliding movement.

Another explanation of the gliding movement was indicated by Engelmann.[216] From microscopic observations he concluded that the filaments of *Oscillatoria* spp. were surrounded by a thin layer of protoplasmic material. It was believed that locomotion was caused by waves of contraction in this layer.

Before Engelmann advanced his theory, Cohn described peristaltic contractions in *Beggiatoa mirabilis*.[195] He did not, however, relate these phenomena to the locomotion of the organism. Fechner described similar contractions in *Oscillatoria* spp.[205]

The contractility theory was also advocated by Schmid.[206, 207, 210] He reported[207] that sometimes light and dark bands passed alternatingly over the filaments of *Oscillatoria jenensis*. Schmid stressed, however, that the considerable amounts of slime excreted by this organism probably also play a role in the motility mechanism.

Ullrich, studying the movements of *Oscillatoria sancta*, *O. jenensis*, and *Beggiatoa mirabilis*, worked out the contractility theory in considerable detail.[199, 200] He made use of the microcinematographic method, photographing moving filaments at a rate of four exposures per second. No complete sequences are, however, reproduced in Ullrich's paper and he mentions[200] that the measured periodical displacements in various sections of a filament were often within the experimental limits of error. Most of his results were obtained in a more indirect way, viz., from stereoscopic observations made on pairs of film frames. When two frames taken with a certain time interval were viewed in the stereoscope, the observer got the impression of a three-dimensional picture. From this Ullrich concluded that the two pictures differed to some extent concerning the position of the cells in the filament and that, hence, longitudinal displacements had occurred in the organism during the time between the two exposures.

Ullrich found that the velocity of the longitudinal waves which were propagated along the filaments of *Oscillatoria sancta* was 13 microns per second. The wavelength was 25 microns = 6.5 cells and the linear speed of the organism 0.7 micron per second.

Ullrich's theories have been widely accepted but recently they have been severely criticized by Schulz.[4] Among other things Schulz shows a sequence of cinematographic pictures of *Oscillatoria sancta*. The pictures show that slight changes in the focusing of a filament may cause the septa between adjacent cells to change their appearance. The observer easily interprets the changes in terms of contraction and dilation phenomena. Schulz also gives the results of direct measurements on the dimensions of individual cells in filaments of *O. tenuis*. The measurements show that the distance between two septa changes less than 1% during 32 exposures (8 seconds).

Most workers studying the Beggiatoaceae and the Oscillatoriaceae have noted the greater or smaller amounts of slime excreted by these organisms. The presence of slime, which surrounds the filaments more or less completely,[211] has been demonstrated by suspending the organisms in India ink[210, 211, 214, 217] and in suspensions of indigo[205, 209, 218, 219] and carmine particles.[4, 209] When the particles come into contact with the surface of the sticky slime, their Brownian movement is stopped. Consequently, the slime can easily be traced in microscopic preparations. The slime has also been observed without the help of added particles, using dark-field illumination[199] and blue light filters.[220]

As already mentioned, Schmid[206, 207, 210] paid attention to the slime in connection with his studies on the motility problem. Many other investigators in this field, probably the majority of them, have considered the excretion of slime as the cause of the gliding movement.

The behavior of the slime surrounding moving and immobile filaments of *Beggiatoa* and *Oscillatoria* spp. has been studied by observing the interactions between the organisms and particles of the kind described above.[4, 205, 209, 211, 218-221] When those particles that adhere to the slime are stationary, the filaments move forward and rotate at the same time around their longitudinal axes. On the other hand, the particles start to move in spirals around the organisms when the locomotion of the filaments is stopped artificially. With respect to both the longitudinal and rotatory components of their motion, the particles close to a filament thus move in a direction opposite to that of the algal or bacterial body before it was stopped.

Sometimes particles and filaments rotate simultaneously. One section of the slime layer may then be covered with particles moving in a direction opposite to that of the particles of another section which is behind or ahead of the first one. Hence rings of accumulated particles are often formed around the organism. The rotation of the filament is as a rule opposite to that of the majority of the particles. With respect to the longitudinal axis of the organism, these particles move in a direction opposite to that of the filament.

The various kinds of movement exhibited by an *Oscillatoria* filament and by particles sticking to the surrounding slime layer are illustrated in Fig. 14.

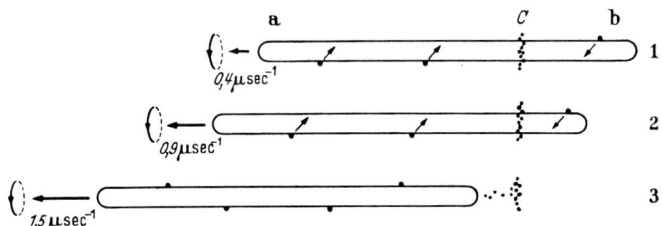

FIG. 14. Simultaneous movements of an *Oscillatoria* filament and of carmine particles adhering to the slime layer around the organism. Schematic representation. (*1*) The carmine particles of section *a* move toward end *b* of the filament. In section *b* the particles move faster than the organism and a ring of accumulated particles is thus formed at *C*. Speed of filament 0.4 microns per second. (*2*) Five minutes later than *1*. Section *a* of the filament has become larger, section *b* smaller and the filament has moved to the left. The ring of carmine particles has remained stationary. The filament moves faster than earlier (0.8–0.9 microns per second). (*3*) Five minutes later than *2*. The filament has left the ring of particles behind. The particles still located around the filament do not move any more. Speed of filament 1.5 microns per second.[4]

It would seem that the phenomena just described indicate that the filaments perform their movements inside a more or less complete envelope of slime. A filament should then move especially fast when the envelope is stationary. This conclusion is in accordance with Schulz' measurements (cf. Fig. 14) and with Correns' and Krenner's statements that the organisms have to adhere firmly to the solid substrate before movement occurs.[209, 214] It is also evident from the behavior of the adhering particles that the slime envelope starts to move when the filament is forcibly stopped.

The important role played by the slime layer in the motility process is also evident from Niklitschek's comparative investigations on various *Oscillatoria* spp.[211] He found that only those species that were completely surrounded by slime were able to perform true locomotion. With other species, excreting very little slime and only at certain parts of the body, nothing but swaying movements could be observed.

Thus movement and slime excretion seem to be closely related. It is still, however, not certain that the latter process itself directly produces a propulsive force. If this were true, one would expect the slime to be birefringent. Fechner states that birefringence can be observed[205] but later investigators have been unable to confirm his findings.[210, 211, 214] Thus, even if the excretion of slime is probably a necessary condition for the occurrence of gliding movement in the Beggiatoaceae and the Oscillatoriaceae, other processes, perhaps of the kind suggested by Mitchell[16] or Burkholder,[202] may also be of decisive importance.

C. The Myxobacteriales

The motility of these microorganisms has been less extensively studied than that of the Beggiatoaceae and Oscillatoriaceae. This may be due to their small-sized cells (width less than 1 micron) which makes detailed observations difficult.

The locomotory movements of the cells of myxobacteria are rather slow. The speed seems to average about 0.1–0.2 micron per second[222, 223] but maximum values of 2–3 microns per second have been reported.[224] Hutchinson and Clayton found that cells sometimes proceed with a sinuous motion through liquid media.[225] Stanier first reported that slight displacements of the cells can be seen in hanging drops.[223] Later, however, he stated that myxobacteria are always immobile when suspended in fluids.[224] Oxford and Norén have reported that the growth of myxobacteria in liquid media is confined to the wet glass surfaces whereas the bulk of the medium fluid remains clear.[226, 227] Since other workers have also not been able to observe cells moving in free liquids, locomotion in myxobacteria probably only occurs when the cells are in contact with a surface, either solid or liquid.

The organisms perform oscillatory movements when only part of the cell is in contact with a surface.[222, 224, 228] Flexions of the body have often been described. Stapp and Bortels believed that such movements are due to unfavorable living conditions.[229] On the other hand, Stanier has pointed out that they are especially frequent in young, healthy cells.[224] Flexing movements occur independently of locomotion.[222, 224, 230]

It is uncertain whether the myxobacteria rotate while performing locomotory movements. Baur points out that curved cells often move forward without oscillating their cell ends and he concludes that rotation probably does not occur.[222]

The cells of myxobacteria generally move in swarms consisting of a large number of similarly oriented organisms. Individually moving cells are, however, as motile as cells in swarms.[222, 224, 228, 231]

Under ordinary conditions swarming myxobacteria move evenly in all directions over a surface. Circular colonies are thus formed. When stresses are present in the gel, however, the vegetative cells move parallel to these forces. This phenomenon has been termed elasticotaxis.[232] The fruiting bodies arrange themselves at right angles to the direction of the movement. Similarly oriented growths on other gel-forming substances, such as gelatin and coagulated plasma, are shown by eubacteria,[233-235] blue-green algae,[236] and vertebrate cells cultured *in vitro*.[237] Probably preferential orientation of the macromolecular constituents of the gels in some direction causes the phenomenon.[232]

The mechanism underlying the motility of the myxobacteria must still be regarded as virtually unknown. No flagella have been found on the organisms,[222, 225, 228, 229] not even with the aid of the electron microscope.[238] Winogradsky claimed that small "haustoria" existed on the surface of the cells,[239] but this finding has not been confirmed by later workers.

The majority of the investigators studying the myxobacteria have found that a rich production of slime is a characteristic property of these organisms. Only Vahle and Meyer-Pietschmann have questioned the presence of such substances in myxobacterial cultures.[231, 240, 241] Vahle found, among other things, that no slime could be detected when myxobacteria were suspended in India ink and investigated under the microscope. The carbon particles adhered closely to the organisms and no clear zone could be found around the cells. This does not exclude, however, the possibility of a slime layer with a thickness below the resolution power of the microscope used. This was pointed out by Jahn.[242] He stressed that the slime secreted by only *one* cell cannot be detected by means of usual bacteriological methods; only the slime deposited behind a great number of swarming cells is demonstrable. According to Jahn, large slime capsules around each cell, comparable to those found in certain algal species, are not present.

Jahn believed that the slime is the direct cause of the locomotion of the myxobacteria.[242] Excretion of slime was supposed to take place through pores in the polar regions of the cells. Furthermore, the excreted substances were thought to move mainly in one direction along the elongated cells. Locomotion in the opposite direction would then follow. Kühlwein has also suggested that the movements of the cells are caused by the excreted slime.[243]

Meyer-Pietschmann has argued against the slime theory.[231, 241] Like Vahle, she finds that individual cells do not excrete considerable amounts of slime. She describes,[231] for example, how two cells may glide closely side by side but in opposite directions, apparently unhindered by any sticky substances (Fig. 15). Individual myxobacteria are found to move through colonies of eubacteria without dragging along any of the cells within the colonies.[241] Attention is also paid by Meyer-Pietschmann to the tracks that are sometimes left behind moving myxobacteria.[231] She finds that they have a lower refractive index than the surrounding agar material. Thus the tracks probably do not represent slime deposits but, instead, furrows in the agar surface. This interpretation would be in accordance with Stanier's observations on the etching of agar surfaces by *Cytophaga* spe-

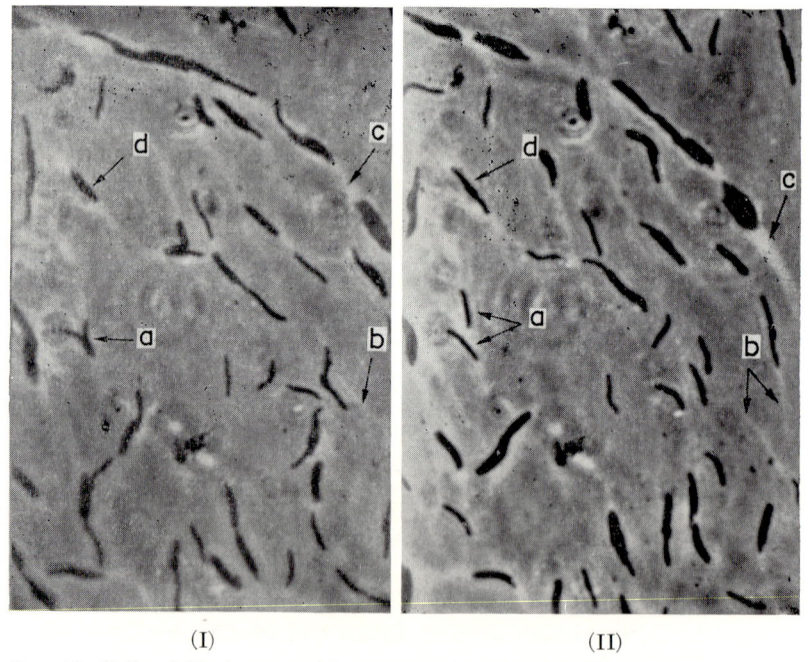

FIG. 15. Cells of *Myxococcus rubescens* moving on agar. The same field of sight at 10-minute intervals. The cells at *a* and *d* glide closely side by side for some time. Tracks of varying width (*b*, *c*) are seen behind the organisms. Dark field (× 650).[231]

cies.[224] The etching was observed even in species not capable of any enzymic decomposition of the agar.

Meyer-Pietschmann, rejecting the slime theory, suggests that the myxobacteria move by means of contractions of the flexible body. Similar views have been expressed by Bisset.[244]

It seems to the present author that the experimental evidence available for the moment does not decisively favor either of the theories so far advanced concerning the movements of the Myxobacteriales.

D. FUSOBACTERIA

Locomotory, gliding movements in this group have been observed by Prévot,[245] Macdonald et al.,[2] and Prévot et al.[3] Besides locomotion, which takes place in free liquids, rotation, flexions, and lashing around a fixed pole have been reported. The organism may attain a locomotory speed of 5 microns per second. The presence of a slime layer or slimy appendages, but not of flagella, is suggested by electron micrographs (Fig. 16) and by India ink preparations.[2, 3] Motility is inhibited by specific antisera.[2] Prévot et al. advance the hypothesis that the organisms may be propelled by wave motions in the slime appendages. On the other hand, Macdonald et al. find it hard to offer any plausible explanation of the observed movements.

E. ADDITIONAL GROUPS EXHIBITING GLIDING MOTION

The family Achromatiaceae is regarded by Pringsheim as a unicellular parallel to the Beggiatoaceae.[194] The occurrence of slow, irregular, and jerky movements has been noted in one of the genera, Achromatium, belonging

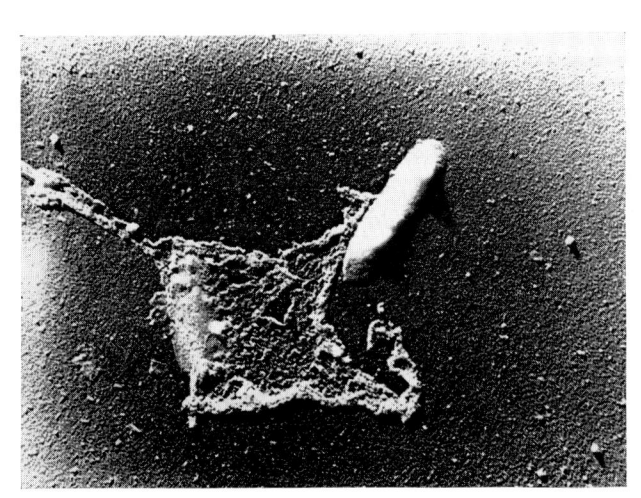

FIG. 16. Electron micrograph of a cell of *Fusocillus girans* Prévot with appendage ($\times$ 22,000).[3]

to this family.[246-248] Virieux,[248] West and Griffiths,[247] and Pringsheim have reported the existence of a thin slime layer surrounding the organisms. It does not seem quite clear whether flagella are present or not (cf. West and Griffiths[247] and Virieux[248]).

The name Vitreoscillaceae has recently been proposed for a group of filamentous forms, morphologically rather similar to the Beggiatoaceae but unable to deposit sulfur from hydrogen sulfide.[194] Because they are capable of gliding movements, these organisms may produce secondary colonies, separated from the primary one, on an agar surface.[194]

Reports on gliding movements in some members of the family Thiorhodaceae have appeared but a confirmation of the observations seems to be desirable.[194]

VI. Bacterial Movements Considered as Tactic Responses to External Stimuli

A. General

A tactic response means the movement of an organism toward or away from a source of stimulation. Movement toward the source of stimulation (the stimulus) is called positive taxis, movement away from a stimulus negative, taxis.

Higher organisms are able to respond to a source of stimulation by turning and moving directly toward or away from the stimulus. This kind of response has been named topotaxis.[249] It seems to be clearly established that bacteria are unable to respond in this way to a stimulus.[249-252] Instead, they react tactically to a stimulating agent by a so-called "shock reaction"[253] ("Schreckbewegung" in German[252]), i.e., a simple reversion of the direction of their locomotion. This kind of tactic behavior is called phobotaxis.[249]

If a bacterium reverses the direction of its movement on entering, as a consequence of random movements, a region containing a stimulating agent, it is said to exhibit a negative phobotactic response to this agent. On the other hand, a positive phobotactic response exists if a shock reaction occurs when the bacterium is about to leave a region containing a stimulating agent, but not when it enters this region. Evidently negative phobotaxis will remove a bacterial species from the neighborhood of a stimulus. Positive phobotaxis, on the other hand, may effect the trapping and accumulation of a great number of bacteria within a small region containing a stimulating agent.

It is often possible to describe tactic responses in quantitative terms. Thus one may determine the minimum concentration or intensity that a stimulating factor must possess in order to cause a response in the organism studied. This minimum concentration or intensity is called the absolute threshold. It is also possible to determine relative threshold values, i.e., the least change in an environmental factor which is needed to cause a tactic response. By determining this change at various concentrations or intensities of the factor, the validity of the so-called Weber law can be tested. This law states that the relative threshold is a constant fraction of the intensity (concentration) of the environmental factor irrespective of the absolute value of this intensity (concentration).

Various kinds of stimuli may cause phobotactic responses in bacteria. So far, most attention has been paid to the effects of chemical compounds and light in this respect.

B. Tactic Responses to Chemical and Physicochemical Stimuli (Chemotaxis)

Chemotactic responses in bacteria have generally been studied using rather simple techniques. Often a drop of a medium containing motile bacteria has been placed on a glass slide or between a slide and a coverslip and the solution of the compound to be tested has been introduced into the bacterial suspension at one point only by means of a capillary tube. The resulting sequence of events has been observed microscopically or macroscopically.[254-256] An accumulation of the bacteria around the mouth of the capillary indicates a positive chemotaxis. A negative chemotaxis, on the other hand, is revealed by a general movement of the bacteria away from the capillary tube. As has been pointed out by Clayton,[257] however, factors other than tactic responses, such as diffusion, convection, and changes in motility of the bacteria studied, may cause a congregation or dispersion of the organisms and thus have to be taken into account in order to avoid misinterpretations of the patterns of congregation observed.

When chemotaxis toward gases is studied, test tubes or capillary tubes, filled partly with a bacterial suspension and partly with gas, have been used.[118, 256] The bacteria may accumulate in a narrow band, the position of which with respect to the gas-liquid interface reflects the kind of tactic response exhibited to the gas by the suspended organisms.

Using the above-mentioned techniques, the chemotactic effects on various bacteria of a wide variety of substances have been studied. Most investigations were performed more than fifty years ago and were sometimes carried out using a mixture of species or using species whose identity is not now easily established (see works by Engelmann,[252, 258-262] Pfeffer,[255, 263, 264] Massart,[265] and Rothert[250]). Species of the following genera have, however, been investigated with respect to chemotaxis: *Aerobacter*,[266] *Bacillus*,[264-267] *Bacterium*,[256] *Beggiatoa*,[268] *Bordetella*,[266] *Chromatium*,[256, 269] *Clostridium*,[250, 256, 266] *Corynebacterium*,[266] *Escherichia*,[256, 266] *Klebsiella*,[266] *Leptospira*,[266] *Micrococcus*,[267] *Pasteurella*,[266] *Pseudomonas*,[256, 266, 267] *Pro-*

teus,[256, 266] *Rhizobium*,[256] *Rhodospirillum*,[264, 267, 270-274] *Salmonella*,[256, 264, 266, 267, 275] *Serratia*,[256, 266] *Shigella*,[266] *Spirillum*,[79, 250, 251, 255, 256, 259, 261, 262, 264, 276] *Staphylococcus*, [266] *Vibrio*.[264, 266, 267]

Many studies have been made on the chemotactic response to oxygen by bacteria. This particular kind of chemotaxis has generally been called aerotaxis. This term, although not strictly logical, will be used in the following.

The first comprehensive experiments in this field were carried out by Engelmann[258-262] using a bacterial species (or perhaps a mixture of species) derived from putrefying matter. The bacteria ("Bacterium termo" in Cohn's terminology[18]) were studied microscopically between slide and coverslip. They were highly motile immediately after the microscopic preparation had been made and gradually accumulated at the edges of the coverslip or around trapped air bubbles. After some time, however, the motility ceased but was resumed when air was permitted to come into contact with the entire suspension. This was accomplished by removing the coverslip for a moment.

Engelmann showed, in addition, that motility was enhanced by pure oxygen but was abolished when the oxygen or air was replaced by hydrogen. These facts indicated that oxygen is of decisive importance for the motility and tactic behavior of the bacteria studied. Further evidence in this respect was obtained when Engelmann showed that motionless bacteria resumed their motility in the presence of illuminated, chlorophyll-containing plant cells.[258, 260] When illumination occurred, the bacteria collected around the plant cells or, more precisely, around the chromatophores or chloroplasts of these cells. In the dark, on the other hand, no motility was observed.

Engelmann found that some bacteria, e.g., spirilla, prefer a low but definite oxygen tension to a higher one. This was evidenced by the fact that the spirilla collected at some distance from an air bubble or from a photosynthetically active plant cell. Under these conditions, other bacteria were seen to collect in the immediate vicinity of the oxygen source. Thus Engelmann's studies showed that at least some bacteria exhibit a positive tactic response to a low concentration of a substance (oxygen) but a negative one to a high concentration.

Beyerinck[256] used macroscopic techniques for the study of bacterial aerotaxis. He introduced a bacterial culture at the bottom of a test tube and filled the tube partially with distilled water (sometimes replaced by 0.1% agar-agar). After about 24 hours, a sharp band of motile organisms formed somewhere between the bottom of the tube and the surface of the liquid. The rest of the medium remained clear. When oxygen was removed from the space above the surface of the medium by a stream of hydrogen or by a germinating seed, the band moved toward the surface. Conversely, a stream of oxygen made the band descend.

Observations with various organisms showed that the position of the band relative to the gas-liquid interface varied with the species studied. Anaerobic organisms such as clostridia formed a layer of motile organisms at the bottom of the test tube, leaving the rest of the liquid clear.

Similar results were obtained by Beyerinck when he studied bacterial cultures that were placed between a slide and a circular coverslip which were partly separated from one another by a thin piece of wire. Macroscopically visible congregations of bacteria were obtained in this way. Strongly aerobic organisms accumulated at the edge of the coverslip, anaerobic ones at the center. Spirilla formed a band at a certain distance from the edge of the coverslip.

Jennings and Crosby[251] established the phobotactic character of the response of bacteria to oxygen by observing spirilla moving in the neighborhood of photosynthetically active algal cells. Rothert[250] simultaneously reached the same conclusion.

A recent comprehensive study by Baracchini and Sherris[266] has confirmed the results of the earlier investigations described above. The aerotactic behavior of 38 bacterial species were tested in two different ways, viz., by a modification of Engelmann's photosynthetic method and by following the formation of bands of motile cells in capillary tubes filled partly with gas and partly with liquid. It was found that 23 highly motile, aerobic or facultatively anaerobic species showed positive aerotaxis by both methods. Three clostridia showed negative aerotaxis when tested by Engelmann's method. In addition, 14 nonmotile organisms together with one weakly motile *Clostridium* sp. exhibited no tactic responses. A motile specimen of *Leptospira icterohaemorrhagiae* showed no tactic response to oxygen. It was thus demonstrated that, in bacteria, a close correlation exists between motility and the occurrence of tactic responses to oxygen (perhaps, however, the spirochetes should be omitted from this rule).

The relationships between bacterial aerotaxis and phototaxis, described by Clayton,[272] will be discussed in Section VI, E.

Chemotactic responses to inorganic salts by bacteria were studied by Pfeffer,[254, 255, 264] Massart,[265, 276] Miyoshi,[269] Kniep,[270] Metzner,[79] and Clayton.[274] Pfeffer found that, among salts of the alkali and alkaline earth metals, the potassium salts were the most effective ones for producing positive tactic responses in spirilla and some other bacteria. When the concentration of the salt solutions was increased, however, a negative response was often observed instead of a positive one. Pfeffer attributed this negative response at least partly to the osmotic pressure of the stimulating solution and used the term "osmotaxis" in this connection.[254] Pfeffer pointed out, however, that it is difficult to keep separate the chemotactic and osmotactic effects of a dissolved substance. Massart[265] and Rothert[250] were, perhaps to an even greater degree than Pfeffer, of the opinion that tactic responses of bacteria are caused to a great extent by osmotic effects. In support of this view, Massart published results of experiments showing that approximately isotonic solutions of urea, sugars, and salts of alkali metals produce similar tactic responses in the bacteria studied. Kniep[270] demonstrated the strong influence of environmental factors on chemotactic responses in bacteria.

Among other things, Metzner[79] made the interesting observation that a spirillum, on entering a region containing sodium chloride at a fairly high concentration, often became motionless for about half a second before reversing the direction of its locomotion. This was due to the fact that the spirillum reversed the sense of rotation of the flagella at the front end of the bacterial body a short time before changing the sense of rotation of the flagella at the rear end. Metzner assumed that the interval corresponded to the time needed for propagating a stimulating impulse from the locomotor center of the anterior flagella to that of the posterior ones.

Metzner, moreover, showed that both lead nitrate and copper sulfate are highly poisonous to spirilla, but that only the former salt produces a tactic response (a negative one) in these organisms.

The negative tactic response by bacteria to acids and alkali was first pointed out by Pfeffer.[254, 264]

A great number of organic substances have been tested with respect to their ability to produce tactic responses in bacteria: glycerol,[264, 265] ethyl alcohol,[250, 254, 274] ethyl ether,[79, 250] malate,[255, 274] tartrate,[255, 269] succinate,[274] malonate,[274] citrate,[274] ascorbate,[274] fluoroacetate,[274] chloroform,[79] urea,[255, 270] carbohydrates,[255, 264, 265, 270, 271] amino acids and amino acid derivatives,[255, 263-265, 270, 274] aromatic compounds,[79, 264, 274] alkaloids,[79, 264, 274] peptone,[255, 271] proteins,[79, 264] meat extract,[255, 263, 264, 267, 270] potato juice.[267]

The majority of these substances were found to produce a tactic response (positive or negative) in the organisms studied—glycerol being an exception to this rule. Often a substance caused a positive response when applied at a small concentration, whereas a negative reaction occurred when a more concentrated solution was tested. Substances with a high nutritional value generally produced a positive response but a strict correlation between nutritional value and tactic response could by no means be established.[264] Poisonous substances often produced a negative response but sometimes no tactic response could be observed.[79, 264] In the case of ethyl ether in low concentrations, a positive tactic response was observed.[250]

Pfeffer showed[264] that the chemotactic responses in bacteria adhere, at least to some extent, to the Weber law. Thus the investigated bacteria ("Bacterium termo"),[18] when suspended in 0.01% meat extract, were attracted by a 0.05% meat extract solution but not by weaker solutions. When the bacteria were suspended in a 1% extract solution, a 5% solution was required in order to cause a tactic response. Thus the Weber law was found to hold within the concentration range tested, the relative threshold being 500%.

The nature of the physiological processes in the bacterial cells which underlie chemotactic responses to stimuli will be discussed in connection with bacterial phototaxis (Section VI, E).

C. Tactic Responses to Thermal Stimuli (Thermotaxis)

The earliest report on thermotaxis in bacteria appears to be that of Schenk.[277] Schenk observed the movements of various bacteria in a hanging drop in which the tip of a heated wire was introduced. The organisms (bacilli, cocci) were seen to accumulate around the wire by means of what seemed to be active movements. Control experiments with India ink particles produced no accumulation.

Mast[278] was unable to observe thermotaxis in spirilla. Probably this was due to an inadequate technique since Metzner[79] later described in detail thermotactic responses in various *Spirillum* spp. Like Schenk, Metzner observed the behavior of bacteria in the neighborhood of a heated wire and he recognized the phototactic character of the bacterial responses to the heat gradient. Metzner made also the noteworthy observation that the spirilla reversed the direction of their locomotion rhythmically about twice a second when the bacterial preparation was allowed to cool uniformly. Metzner interpreted this rhythmic movement as a response to a constant stimulus (in this case the falling temperature). Similar rhythmic movements were observed by Metzner when spirilla were stimulated by chloroform and cocaine. Recently, Clayton reported their occurrence in illuminated rhodospirilla stimulated by a step-decrease in the intensity of the light source (cf. Section VI, E).[279]

D. Tactic Responses to Gravitational Stimuli (Geotaxis)

The only experiments that have been performed on bacterial geotaxis appear to be those of Massart.[276] He introduced a suspension of marine spirilla (strain A or C)[276] into a capillary tube which was open at both ends. When the tube was held horizontally, the spirilla accumulated at the two menisci in the capillary, leaving the middle part free of bacteria. These accumulations were obviously due to aerotaxis. Spirilla of the A strain accumulated, however, only at the upper end of the capillary tube if the tube was held vertically immediately after the bacteria had been introduced. If the tube was then turned upside down, the spirilla started to move upward. Some spirilla returned to the lower meniscus for some time, probably on account of aerotactic responses, but eventually all spirilla accumulated at the upper end of the capillary.

Spirilla of the C strain showed a behavior that was contrary to that of spirillum A, i.e., they concentrated at the lower end of the capillary. According to Massart, this accumulation was due to active movements of the bacteria.

Massart interpreted his findings as the effect of geotactic responses in the spirilla (negative ones in strain A, positive ones in the C strain). Geotaxis was assumed to prevail over aerotaxis. A geotactic accumulation of motile bacteria in a constant gravitational field, must, however, be due to topotactic responses, not to phobotactic ones. So far, Massart's experiments seem to be the only ones that indicate the occurrence of topotaxis in bacteria (cf. Section VI, A). A confirmation of his results seems therefore desirable before the occurrence of geotaxis (geotopotaxis) in bacteria can be considered as definitely demonstrated.

E. Tactic Responses to Light (Phototaxis)

The pioneer work on phototactic responses in bacteria was performed by Engelmann.[252, 280] Engelmann's first experiments were carried out on a purple bacterium, which he named *Bacterium photometricum*.[252] This bacterium had an ovoid form, was about 3 microns long and possessed a terminal flagellum. It was motionless when kept in the dark but, upon illumination, it began to move. A short period of "photokinetic induction" was observed between the beginning of the illumination and the first bacterial movements. When a uniformly illuminated microscopic preparation of *B. photometricum* was suddenly darkened, e.g., by lowering the microscope condensor, the bacteria exhibited a shock reaction ("Schreckbewegung"[252]), i.e., they reversed abruptly the direction of their locomotion. When a specimen of *B. photometricum* was about to leave the light region of a partly illuminated microscopic preparation, this shock reaction was also observed, i.e., the organism showed a positive phobotactic response to light. A shock reaction did not occur, however, when the illumination of the bacteria decreased slowly.

When Engelmann projected a microspectrum of gas or sunlight at the level of the bacterial suspension, the bacteria were seen to collect in bands at certain wavelengths of the spectrum. The strongest band appeared at about 8500 A., a weaker one occurring between 6100 and 5700 A. An indication of a still weaker band could be seen between 5500 and 5100 A. Measurements performed by Engelmann showed that the positions of these bands coincided with the absorption maxima given by a layer of *Bacterium photometricum* cells. The bacterial accumulations in the microspectrum were mainly formed by the phobotactic reactions of bacteria moving from one spectral region to another. Thus Engelmann was able to show that the phototactic sensitivity of *B. photometricum* was a function of the wavelength of the light that illuminated the cells. Moreover, he established the close similarity between the absorption spectrum given by the pigments of this organism and the spectral distribution of phototaxis.

Engelmann found that the phototactic behavior of *Bacterium photometricum* was characteristic also for other purple bacteria.[280] Later investigators have confirmed Engelmann's findings and have added to his results much information obtained by means of more refined experimental techniques. Thus Buder showed,[281] investigating *Rhodospirillum rubrum*, that the bacterial congregation observed by Engelmann between 5500 and 5100 A. could be more accurately described as three distinct bands with maximum densities at 5300, 4900, and 4600 A. Similarly located bands were found by Buder in the absorption spectrum given by intact *Rhodospirillum* cells. Later studies have shown that these bands are due to the carotenoid pigments of the bacterium.[253, 282] On the whole, Buder found a remarkable agreement between the action spectrum of phototaxis and the absorption spectrum of *R. rubrum*. Still later investigations, performed by Manten[253] and Clayton[283] suggest, however, that minor discrepancies may exist between the light absorption and the phototactic action spectrum of this organism. Thus it remains to be settled to what degree the predominant carotenoid pigment of *R. rubrum*, spirilloxanthin, absorbs phototactically active light.[253, 283]

Some investigations have been performed in order to establish to what extent the Weber law holds for the phototactic responses exhibited by purple bacteria. Buder made some observations concerning this problem[281] but the first comprehensive investigation appears to have been carried out by Schrammeck,[284] who studied the phototactic behavior of *Rhodospirillum rubrum*. Schrammeck found that the Weber law was followed for all values of the initial light intensity ranging from 0.05 to 250 meter candles, a change of the light intensity (the relative threshold) of about 5% being sufficient to cause a phototactic response in the bacteria. In a more recent study carried out on the same organism, Clayton[283] found that the Weber law was valid within a more limited region, viz., for light intensities of 5–500 meter candles. Clayton pointed out, however, that a potentially extensive adherence to the Weber law may be masked by factors that may vary to a considerable degree with the culture conditions and the strain of bacteria studied.

A photosynthetic metabolism proceeds in the purple bacteria when these organisms are illuminated. This kind of metabolism naturally ceases when the bacteria are transferred to a dark environment. Since the transfer of the bacteria from light to darkness also causes a phototactic response (provided that the transfer takes place rapidly), some workers have put forward the hypothesis that a close relationship exists between photosynthesis and phototaxis in the purple bacteria.[253, 283, 285, 286]

In order to test this hypothesis, Thomas[285] determined the photosynthetic action spectrum of *Rhodospirillum rubrum* and compared it with the phototactic action spectrum of the same organism which had earlier been determined by Manten.[253] Thomas found that, as well as the bacteriochlorophyll, some of the carotenoids of *R. rubrum* absorb photosynthetically active light. He found moreover that the maxima and minima of the photosynthetic and phototactic action spectra coincide. Thomas' findings were somewhat at variance with the results of a similar investigation performed earlier by French.[287] He found that the carotenoids of *R. rubrum* were inactive in photosynthesis. French's results were, however, based on less extensive experimental data than those collected by Thomas.

Manten,[253] working with *Rhodospirillum rubrum*, found that the attainment of light saturation in photosynthesis at high light intensities was connected with a decrease in the contrast sensitivity of the bacteria in phototactic experiments. Manten therefore concluded that a close connection exists between photosynthesis and phototaxis, phototactic responses being induced by a sudden decrease in the rate of photosynthesis. Similar conclusions were drawn by Thomas and Nijenhuis,[286] who studied among other things the influence of inhibitors on photosynthesis and phototaxis.

According to Clayton,[273] the ratio between the light intensities needed for the saturation of the photosynthetic and phototactic processes in *Rhodospirillum rubrum* varies with the substrate for the photosynthetic process, the saturation intensity for phototaxis being equal to or much greater than that for photosynthesis. Evidently these findings are somewhat at variance with Manten's results.[253] Clayton points out, however, that phototaxis is probably not caused by a change in the "steady state" rate of photosynthesis but is due to a transient concentration effect. This could explain why the saturation intensities for photosynthesis and phototaxis are not always equal. According to Clayton, the fact that the saturation intensity for phototaxis is influenced by the nature of the substrate for the photosynthetic process[273] favors the view that a close connection exists between phototaxis and photosynthesis.

The quantitative relations between stimulus and response in the phototaxis of *Rhodospirillum rubrum* were investigated by Clayton.[279] He found that the phototactic behavior of this organism had many properties in common with the reactions of the nerves of higher organisms. Thus, the following effects, characteristic of excitable systems, could be demonstrated in the phototaxis of *R. rubrum*: an "all-or-none" response to stimuli, accommodation, refractiveness, rhythmicity, reciprocity of strength and duration of a threshold stimulus, and summation of subliminal stimuli.

Theories concerning nerve excitation have been developed by Rashevsky[288] and Hill.[289] Clayton found a fair agreement between these theories and the quantitative characteristics of the phototaxis of *Rhodospirillum rubrum*. According to Rashevsky and Hill, one exciting and one inhibiting agent are present in a nerve. Under stimulation, these agents are formed at rates proportional to the intensity of the stimulus and are dissipated at rates proportional to their excess over the resting values. A response to a stimulus occurs when the ratio between the exciting and inhibitory agents exceeds a critical value.

Clayton has suggested[272] that the exciting and inhibitory agents postulated by the theories of Rashevsky and Hill could be represented by certain metabolites in the bacterial cells, and he advanced the hypothesis that a tactic response of the bacteria studied by him is caused by a transient disturbance which accompanies a decrease in the rate of metabolism. In this connection, Clayton referred to the close relationship existing between bacterial photosynthesis and phototaxis which have been described in the preceding paragraphs. In order to test his hypothesis further, Clayton studied the influence of phototaxis on aerotaxis and vice versa.[272] It was demonstrated that *Rhodospirillum rubrum* exhibits positive aerotaxis in the dark and in dim light, but negative aerotaxis in bright light. The transition from positive to negative aerotaxis occurs at a well-defined light intensity at which the bacteria favor an intermediate oxygen tension, accumulating in microscopic preparations in narrow bands at some distance from trapped air bubbles (cf. Engelmann[261, 262] and Beyerinck[256]). Positive aerotaxis is suppressed by moderate illumination, phototaxis by air. Negative phototaxis was never observed.

If the tactic responses of *Rhodospirillum rubrum* are associated with a decreasing rate of metabolism, it can be concluded from the experiments just described that the metabolism of this bacterium should exhibit the following properties. The rate of metabolism should decrease with the light intensity under all circumstances, but this decrease should be negligible at oxygen tensions above a critical value. In the dark and in dim light, the rate of metabolism should decrease with decreasing oxygen tension. In bright light, the rate of metabolism should, however, decrease with increasing oxygen tension. At moderate illumination, the rate of metabolism should have a maximum value at an intermediate oxygen tension.

According to experiments performed by Clayton,[272, 290] these relations between metabolism, illumination, and oxygen tension can be clearly demonstrated. Experiments with metabolic inhibitors[272, 291] also seem to support Clayton's hypothesis, if by metabolism we mean carbon assimilation. No definite relations seem to exist between tactic responses and the oxidation of substrates.

The chemotactic responses by *Rhodospirillum rubrum* to substances other than oxygen have recently been investigated by Clayton.[274] Among other things, Clayton found that *R. rubrum* exhibits a positive chemotaxis to sulfhydryl compounds and oxidizable organic substances such as malate. On the other hand, a negative chemotaxis is induced by various oxidizing substances, by solutions having an extreme pH, and by poisonous substances, especially thiol inhibitors. Nerve poisons, such as curare and lysergic acid, cause erratic movements. According to Clayton, tactic responses in bacteria are associated with a decrease in the metabolic rate or, more precisely, with a decrease in the rate of carbon assimilation, i.e., with a decrease in synthetic processes (cf. the preceding paragraph). The negative tactic responses to metabolic poisons and oxidizing compounds are then easily understood. Also the positive tactic responses to metabolic substrates and oxidizable compounds can be explained by Clayton's theory when it is kept in mind that a positive tactic response to a compound is a response caused by a decrease in the concentration of this compound.

On the basis of his comprehensive studies on photo- and chemotaxis, Clayton has formulated the following theory for the tactic responses in the purple bacteria:[274]

"Taxis in the purple bacteria is mediated through the development of an excitatory state which is transmitted to the locomotor areas, causing a coordinated motor response. The development of excitation is triggered by a decrease in the concentration of a compound whose rate of synthesis parallels the rates of anabolic activities which support growth. The synthesis or the activity of this compound requires the cooperation of sulfhydryl groups."

References

[1] F. H. Johnson and R. F. Baker, *J. Cellular Comp. Physiol.* **30,** 131 (1947).

[2] J. B. Macdonald, M. L. Knoll, and R. M. Sutton, *Proc. Soc. Exptl. Biol. Med.* **84,** 459 (1954).

[3] A. R. Prévot, J. Giuntini, and H. Thouvenot, *Ann. inst. Pasteur* **86,** 774 (1954).

[4] G. Schulz, *Arch. Mikrobiol.* **21,** 335 (1955).

[5] R. Y. Stanier, *Bacteriol. Revs.* **6,** 143 (1942).

[6] K. Ogiuti, *Japan. J. Exptl. Med.* **14,** 19 (1936).

[7] G. Sanarelli, *Ann. inst. Pasteur* **33,** 569 (1919).

[8] K. V. Thimann, "The Life of Bacteria," p. 56. Macmillan, New York, 1955.

[9] G. Taylor, *Proc. Roy. Soc.* **A209,** 447 (1951).

[10] G. J. Hancock, *Proc. Roy. Soc.* **A217,** 96 (1953).

[11] H. G. Bronn, "Klassen und Ordnungen des Thier-Reichs, wissenschaftlich dargestellt in Wort und Bild" (O. Bütschli, ed.), Vol. I, Section II, p. 846. C. F. Winter, Leipzig and Heidelberg, 1884.

[12] K. Reichert, *Zentr. Bakteriol. Parasitenk. Abt. I. Orig.* **51,** 14 (1909).

[13] G. Taylor, *Proc. Roy. Soc.* **A211,** 225 (1952).

[14] J. Gray and G. J. Hancock, *J. Exptl. Biol.* **32,** 802 (1955).

[15] J. Gray, *Quart. J. Microscop. Sci.* **94,** 531 (1953).

[16] P. Mitchell, *Proc. Roy. Phys. Soc. Edinburgh* **25,** 32 (1956).

[17] C. G. Ehrenberg, "Die Infusionsthierchen als volkommene Organismen," p. 15. Leopold Voss, Leipzig, 1838.
[18] F. Cohn, *Beitr. Biol. Pflanz.* **1,** 127 (1872).
[19] R. Koch, *Beitr. Biol. Pflanz.* **2,** 399 (1877).
[20] A. Fischer, *Jahrb. wiss. Botan.* **27,** 1 (1895).
[21] W. Migula, "System der Bakterien," p. 97. Fischer, Leipzig, 1897.
[22] S. Mudd and T. F. Anderson, *J. Immunol.* **42,** 251 (1941).
[23] W. T. Astbury and C. Weibull, *Nature* **163,** 280 (1949).
[24] L. W. Labaw and V. M. Mosley, *Biochim. et Biophys. Acta* **17,** 322 (1955).
[25] M. P. Starr and R. C. Williams, *J. Bacteriol.* **63,** 701 (1952).
[26] F. Fuhrmann, *Zentr. Bakteriol. Parasitenk. Abt. II.* **25,** 129 (1910).
[27] A. Pijper, *Zentr. Bakteriol. Parasitenk. Abt. I. Orig.* **118,** 113 (1930).
[28] W. van Iterson, *Intern. Congr., Microbiol. 6th Congr., Rome* p. 24 (1953).
[29] C. Weibull, *Arkiv Kemi* **1,** 21 (1950).
[30] C. Weibull, *Arkiv Kemi* **1,** 573 (1950).
[31] A. Pijper and A. J. Nunn, *J. Roy. Microscop. Soc.* **69,** 138 (1949).
[32] A. Pijper, *J. Pathol. Bacteriol.* **47,** 1 (1938).
[33] A. Pijper, *Symposium Soc. Gen. Microbiol.* **1,** 144 (1949).
[34] C. Weibull, *Acta Chem. Scand.* **4,** 268 (1950).
[35] E. Leifson and I. Palen, *J. Bacteriol.* **70,** 233 (1955).
[36] A. Pijper, M. L. Neser, and G. Abraham, *J. Gen. Microbiol.* **14,** 371 (1956).
[37] E. Leifson, S. R. Carhart, and M. Fulton *J. Bacteriol.* **69,** 73 (1955).
[38] F. Neumann, *Zentr. Bakteriol. Parasitenk. Abt. I. Orig.* **109,** 143 (1928).
[39] K. Pietschmann, *Arch. Mikrobiol.* **12,** 377 (1942).
[40] C. A. Peluffo, *Intern. Congr, Microbiol., 6th Congr., Rome* **1,** 370 (1953).
[41] F. H. Johnson, N. Zworykin, and G. Warren, *J. Bacteriol.* **46,** 167 (1943).
[42] M. R. J. Salton and R. W. Horne, *Biochim. et Biophys. Acta* **7,** 19 (1951).
[43] W. van Iterson, *Symposium Soc. Gen. Microbiol.* **1,** 172 (1949).
[44] C. Weibull, *J. Bacteriol.* **66,** 688 (1953).
[45] J. M. Wiame, R. Storck, and E. Vanderwinkel, *Biochim. et Biophys. Acta* **18,** 353 (1955).
[46] A. L. Houwink and W. van Iterson, *Biochim. et Biophys. Acta* **5,** 10 (1950).
[47] J. B. Grace, *J. Gen. Microbiol.* **9,** 325 (1954).
[48] P. Pease, *Exptl. Cell Research* **10,** 234 (1954).
[49] J. Tawara, *J. Bacteriol.* **73,** 89 (1957).
[50] C. Weibull, *J. Bacteriol.* **66,** 696 (1953).
[51] T. F. Anderson, *Symposium Soc. Gen. Microbiol.* **1,** 76 (1949).
[52] A. L. Houwink, *Symposium Soc. Gen. Microbiol.* **1,** 92 (1949).
[53] M. van den Ende, P. A. Don, W. J. Elford, C. E. Challice, I. M. Dawson, and J. E. Hotchin, *J. Hyg.* **50,** 12 (1952).
[54] M. L. De, A. Guha, and N. N. das Gupta, *Nature* **171,** 879 (1953).
[55] R. E. Hartman, T. D. Green, J. B. Bateman, C. A. Senseney, and G. E. Hess, *J. Appl. Phys.* **24,** 90 (1953).
[56] I. W. Smith, *Biochim. et Biophys. Acta* **15,** 20 (1954).
[57] C. C. Brinton, A. Buzzell, and M. A. Lauffer, *Biochim. et Biophys. Acta* **15,** 533 (1954).
[58] J. P. Duguid, I. W. Smith, G. Dempster, and P. N. Edwards, *J. Pathol. Bacteriol.* **70,** 335 (1955).
[59] W. Schreil, *Mikroskopie* **10,** 38 (1955).
[60] W. Schreil and F. Schleich, *Z. Hyg. Infektionskrankh.* **141,** 576 (1955).
[61] H. Braun, *Arch. Mikrobiol.* **24,** 1 (1956).

[62] L. W. Labaw and V. M. Mosley, *Biochim. et Biophys. Acta* **15**, 325 (1954).
[63] H. Braun, *Arch. Mikrobiol.* **24**, 1 (1956).
[64] E. de Robertis and C. M. Franchi, *Exptl. Cell Research* **2**, 295 (1951).
[65] C. Weibull, *Nature* **165**, 482 (1950).
[66] W. T. Astbury, E. Beighton, and C. Weibull, *Symposia Soc. Exptl. Biol.* **9**, 282 (1955).
[67] C. Weibull, *Biochim. et Biophys. Acta* **2**, 351 (1948).
[68] C. Weibull, *Biochim. et Biophys. Acta* **3**, 378 (1948).
[69] C. Weibull, *Acta Chem. Scand.* **4**, 260 (1950).
[70] C. Weibull, *Acta Chem. Scand.* **4**, 268 (1950).
[71] C. Weibull, *Discussions Faraday Soc.* **11**, 195 (1951).
[72] C. Weibull, *Acta Chem. Scand.* **5**, 529 (1951).
[73] C. Weibull, *Acta Chem. Scand.* **7**, 335 (1953).
[74] H. Koffler, T. Kobayashi, and G. E. Mallet, *Arch. Biochem. Biophys.* **64**, 509 (1956).
[75] H. Koffler and T. Kobayashi, *Arch. Biochem. Biophys.* **67**, 246 (1957).
[76] J. Adye, H. Koffler, and G. E. Mallet, *Arch. Biochem. Biophys.* **67**, 251 (1957).
[77] G. E. Mallet and H. Koffler, *Arch. Biochem. Biophys.* **67**, 256 (1957).
[78] J. Buder, *Jahrb. wiss. Botan.* **56**, 529 (1915).
[79] P. Metzner, *Jahrb. wiss. Botan.* **59**, 325 (1920).
[80] T. Y. Kingma Boltjes, *J. Pathol. Bacteriol.* **60**, 275 (1948).
[81] A. Pijper, "A Motion Picture of Bacterial Shape and Motility." Wallachs', Pretoria, 1947.
[82] A. Pijper, *Bacteriol. Proc. (Soc. Am. Bacteriologists)* p. 16 (1956).
[83] A. Pijper, *J. Bacteriol.* **57**, 111 (1949).
[84] A. Pijper, C. G. Crocker, J. P. van der Walt, and N. Savage, *J. Bacteriol.* **65**, 628 (1953).
[85] A. Pijper, *J. Bacteriol.* **53**, 257 (1947).
[86] V. Uhlela, *Biol. Zentr.* **31**, 689 (1911).
[87] A. Pijper, *Zentr. Bakteriol. Parasitenk. Abt. I. Orig.* **123**, 195 (1932).
[88] A. Pijper, *J. Biol. Phot. Assoc.* **8**, 158 (1940).
[89] A. Pijper, *J. Pathol. Bacteriol.* **58**, 325 (1946).
[90] A. Pijper, *Nature* **161**, 200 (1948).
[91] A. Pijper, *Bacteriol. Proc. (Soc. Am. Bacteriologists)* p. 16 (1956).
[92] A. Pijper, *Nature* **175**, 214 (1954).
[93] A. Pijper and G. Abraham, *J. Gen. Microbiol.* **10**, 452 (1954).
[94] A. Fleming, A. Voureka, I. R. H. Kramer, and W. H. Hughes, *J. Gen. Microbiol.* **4**, 257 (1950).
[95] A. Fleming, *J. Gen. Microbiol.* **4**, 457 (1950).
[96] J. Kvittingen, *Acta Pathol. Microbiol. Scand.* **37**, 89 (1955).
[97] J. Ørskov, *Acta Pathol, Microbiol. Scand.* **24**, 181 (1947).
[98] J. N. Rinker, C. F. Robinow, and H. Koffler, *Bacteriol. Proc. (Soc. Am. Bacteriologists)* p. 32 (1950).
[99] T. Y. Kingma Boltjes, *Antonie van Leeuwenhoek, J. Microbiol. Serol.* **14**, 251 (1948).
[100] P. van Tieghem, *Bull. soc. botan. France* **26**, 37 (1879).
[101] A. Pijper, *Science* **109**, 379 (1949).
[102] A. Pijper, *Nature* **168**, 749 (1951).
[103] A. Pijper, *Ergeb. Mikrobiol., Immunitätsforsch. u. exptl. Therap.* **30**, 37 (1957).
[104] G. E. Mallet, H. Koffler, and J. N. Rinker, *J. Bacteriol.* **61**, 703 (1951).

[105] E. Leifson, *J. Bacteriol.* **21**, 331 (1931).
[106] B. A. D. Stocker, *Symposium Soc. Gen Microbiol.* **6**, 19 (1956).
[107] B. A. D. Stocker, *J. Pathol. Bacteriol.* **73**, 314 (1957).
[108] P. B. White, *J. Gen. Microbiol.* **4**, 36 (1951).
[109] L. O. Koblmüller, *Zentr. Bakteriol. Parasitenk. Abt. I. Orig.* **133**, 310 (1935).
[110] H. Auerbach and O. Felsenfeld, *J. Bacteriol.* **56**, 587 (1948).
[111] H. Graudal, "Undersøgelser over bevaegelige streptokokker," p. 128. Christtreus Bogtrykkeri, Copenhagen, 1955.
[112] A. Pijper, C. G. Crocker, and N. Savage, *J. Bacteriol.* **69**, 151 (1955).
[113] C. Weibull, *Nature* **167**, 511 (1951).
[114] D. Herbert, *J. Gen. Microbiol.* **5**, xx (1951).
[115] E. de Robertis and C. M. Franchi, *J. Appl. Phys.* **23**, 161 (1952).
[116] E. de Robertis and C. A. Peluffo, *Proc. Soc. Exptl. Biol. Med.* **78**, 584 (1951).
[117] G. H. Barlow and J. J. Blum, *Science* **116**, 572 (1952).
[118] J. C. Sherris, N. W. Preston, and J. G. Shoesmith, *J. Gen. Microbiol.* **16**, 86 (1957).
[119] W. Broser and W. Lautsch, *Z. Naturforsch.* **11b**, 453 (1956).
[120] K. G. Denbigh, *Nature* **154**, 642 (1944).
[121] H. Noguchi, *Am. J. Syphilis* **1**, 261 (1917).
[122] E. Hofmann, *Arch. Dermatol. u. Syphilis* **144**, 306 (1923).
[123] E. Hoffmann and E. Hofmann, *in* "Handbuch der Haut- und Geschlechtskrankheiten" (J. Jadassohn, ed.), Vol. XV/1, Chapter 3, p. 15. Springer, Berlin, 1927.
[124] H. Noguchi, *in* "The Newer Knowledge of Bacteriology and Immunology" (E. O. Jordan and I. S. Falk, eds.), Chapter 36, p. 452. Univ. Chicago Press, Chicago, Illinois, 1928.
[125] G. Sobernheim and W. Loewenthal, *in* "Handbuch der pathogenen Mikroorganismen" (W. Kolle, R. Kraus, and P. Uhlenhuth, eds.), 3rd ed., Vol. VII, Chapter 1, p. 4. Fischer, Jena, and Urban & Schwarzenberg, Berlin and Vienna, 1930.
[126] M. Zuelzer, *in* "Handbuch der pathogenen Protozoen" (S. von Prowazek, ed.), Vol. III, Chapter 9, p. 1627. Barth, Leipzig, 1931.
[127] M. Zuelzer, *Arch. Protistenk.* **24**, 1 (1912).
[128] F. Neumann, *Klin. Wochschr.* **8**, 2081 (1929).
[129] C. Schellack, *Arb. kaiserl. Gesundh.* **27**, 364 (1908).
[130] M. T. Dyar, *J. Bacteriol.* **54**, 483 (1947).
[131] F. W. Bach, *Zentr. Bakteriol. Parasitenk. Abt. I. Orig.* **87**, 198 (1922).
[132] C. Dobell, *Arch. Protistenk.* **26**, 117 (1912).
[133] J. Cantacuzène, *Compt. rend. soc. biol.*, **68**, 75 (1910).
[134] A. Certes, *Bull. soc. zool. France* **16**, 95 (1891).
[135] W. S. Perrin, *Arch. Protistenk.* **7**, 131 (1906).
[136] G. Keysselitz, *Arch. Protistenk.* **10**, 127 (1907).
[137] C. Schellack, *Arb. kaiserl. Gesundh.* **30**, 379 (1909).
[138] H. B. Fantham, *Quart. J. Microscop. Sci.* **52**, 1 (1908).
[139] A. Porter, *Arch. zool. exptl. et gén.* **43**, 1 (1910).
[140] N. H. Swellengrebel, *Zentr. Bakteriol. Parasitenk. Abt. 1. Orig.* **49**, 529 (1909).
[141] W. T. Dimitroff, *J. Bacteriol.* **12**, 135 (1926).
[142] J. R. G. Bradfield and D. B. Cater, *Nature* **169**, 944 (1952).
[143] J. Gross, *Mitt. zool. Sta. Napoli* **20**, 41 (1913).
[144] J. Gross, *Mitt. zool. Sta. Napoli* **20**, 188 (1913).
[145] M. Zuelzer, *Zentr. Bakteriol. Parasitenk. Abt. I. Orig.* **85**, 155* (1921).
[146] C. Magerstedt, *Arch. Dermatol. u. Syphilis* **185**, 272 (1944).

[147] E. Zettnow, *Deut. med. Wochschr.* **32,** 376 (1906).
[148] E. G. Hampp, D. B. Scott, and R. W. G. Wyckoff. *J. Bacteriol.* **56,** 755 (1948).
[149] E. Leifson, *J. Bacteriol.* **60,** 678 (1950).
[150] R. Lofgren and M. H. Soule, *J. Bacteriol.* **50,** 679 (1945).
[151] R. Lofgren and M. H. Soule, *J. Bacteriol.* **51,** 583 (1946).
[152] B. Babudieri and D. Bocciarelli, *J. Hyg.* **46,** 438 (1948).
[153] R. H. A. Swain, *J. Pathol. Bacteriol.* **69,** 117 (1955).
[154] E. Mölbert, *Z. Hyg. Infektionskrankh.* **142,** 203 (1955).
[155] F. G. Novy and R. E. Knapp, *J. Infectious Diseases* **3,** 291 (1906).
[156] L. Norris, A. M. Pappenheimer, and T. Flournoy, *J. Infectious Diseases* **3,** 266 (1906).
[157] P. Jeantet and Y. Kermorgant, *Compt. rend. soc. biol.* **92,** 1036 (1925).
[158] E. D. DeLamater, R. H. Viggall, and M. Haanes, *J. Exptl. Med.* **92,** 239 (1950).
[159] H. E. Morton and T. F. Anderson, *Am. J. Syphilis, Gonorrhea, Venereal Diseases* **26,** 216 (1942).
[160] S. Mudd, K. Polevitzky and T. F. Anderson, *J. Bacteriol.* **46,** 15 (1943).
[161] U. J. Wile and E. B. Kearney, *J. Am. Med. Assoc.* **122,** 167 (1945).
[162] J. H. L. Watson, J. J. Angulo, F. León-Blanco, G. Varela, and C. G. Wedderburn, *J. Bacteriol.* **61,** 455 (1951).
[163] C. Levaditi, *Presse méd.* **54,** 85 (1946).
[164] A. Jakob, *Klin. Wochschr.* **25,** 882 (1947).
[165] M. Moureau and J. Giuntini, *Ann. inst. Pasteur* **90,** 728 (1956).
[166] B. Babudieri, *Rend. ist. super. sanità* **15,** 711 (1952).
[167] H. E. Morton, G. Rake, and N. R. Rose, *Am. J. Syphilis, Gonorrhea, Venereal Diseases* **35,** 503 (1951).
[168] E. Mölbert, *Z. Hyg. Infektionskrankh.* **142,** 510 (1956).
[169] W. Schmerold and B. Deubner, *Hautarzt* **5,** 511 (1954).
[170] F. Schaudinn and E. Hoffmann, *Arb. kaiserl. Gesundh.* **22,** 527 (1905).
[171] F. W. Oelze, *Münch. med. Wochschr.* **67,** 921 (1920).
[172] P. Mühlens, *Z. Hyg. Infektionskrankh.* **57,** 405 (1907).
[173] R. Gonder and J. Gross, *Arch. Protistenk.* **39,** 62 (1919).
[174] H. E. Morton and T. F. Anderson, *J. Bacteriol.* **45,** 143 (1943).
[175] B. Babudieri, *J. Hyg.* **47,** 390 (1949).
[176] E. Mölbert, *Z. Hyg. Infektionskrankh.* **141,** 82 (1955).
[177] Ph. van Thiel and W. van Iterson, *Proc. Koninkl Ned. Akad. Wetenschap.* **50,** 776 (1947).
[178] J. W. Czekalowski and G. Eaves. *J. Pathol. Bacteriol.* **69,** 129 (1955).
[179] H. Noguchi, *J. Exptl. Med.* **27,** 575 (1918).
[180] M. Zuelzer, *Arb. kaiserl. Gesundh.* **51,** 159 (1918).
[181] K. v. Angerer, *Arch. Hyg.* **91,** 201 (1922).
[182] Haendel, E. Ungermann, and Jaenisch, *Arb. kaiserl. Gesundh.* **51,** 42 (1919).
[183] E. Ungermann, *Arb. kaiserl. Gesundh.* **51,** 114 (1919).
[184] B. G. Klebs, *Biol. Zentr.* **5,** 353 (1886).
[185] B. Schroeder, *Verhandl. naturhist.-med. Ver. Heidelberg* **7,** 139 (1902).
[186] F. Oltmanns, "Morphologie und Biologie der Algen," 2nd ed., Vol. I, p. 106. Fischer, Jena, 1922.
[187] F. E. Fritsch, "The Structure and Reproduction of the Algae," Vol. I, p. 342. Cambridge Univ. Press, London and New York, 1935.
[188] O. Müller, *Ber. deut. botan. Ges.* **27,** 27 (1909).
[189] F. Oltmanns, "Morphologie und Biologie der Algen," 2nd ed., Vol. I, p. 131. Fischer, Jena, 1922.

[190] F. E. Fritsch, "The Structure and Reproduction of the Algae," Vol. I, p. 590. Cambridge Univ. Press, London and New York, 1935.
[191] P. Martens, *Cellule* **48,** 279 (1940).
[192] W. Migula, in "Die natürlichen Pflanzenfamilien" (A. Engler and K. Prantl, eds.), Vol. I, Chapter 1a, p. 41. Engelmann, Leipzig, 1900.
[193] R. S. Breed, E. G. D. Murray, and A. P. Hitchens, "Bergey's Manual of Determinative Bacteriology," 6th ed., p. 988. Williams & Wilkins, Baltimore, Maryland, 1948.
[194] E. G. Pringsheim, *Bacteriol. Revs.* **13,** 47 (1949).
[195] F. Cohn, *Arch. mikroskop. Anat.* **3,** 1 (1867).
[196] R. Kolkwitz, *Ber. deut. botan. Ges.* **15,** 460 (1897).
[197] F. Keil, *Beitr. Biol. Pflanz.* **11,** 335 (1912).
[198] W. J. Crozier and T. J. B. Stier, *J. Gen. Physiol.* **10,** 185 (1926).
[199] H. Ullrich, *Planta* **2,** 295 (1926).
[200] H. Ullrich, *Planta* **9,** 144 (1929).
[201] O. P. Phillips, *Contribs. Botan. Lab. Univ. Penn.* **2,** 237 (1904).
[202] P. R. Burkholder, *Quart. Rev. Biol.* **9,** 438 (1934).
[203] F. E. Fritsch, "Structure and Reproduction of the Algae," Vol. 2, p. 800. Cambridge Univ. Press, London and New York, 1945.
[204] G. E. Fogg. *Bacteriol. Revs.* **20,** 148 (1956).
[205] R. Fechner, *Z. Botan.* **7,** 289 (1915).
[206] G. Schmid, *Flora* **111-112,** 327 (1918).
[207] G. Schmid, *Jahr. wiss. Botan.* **62,** 328 (1923).
[208] W. J. Crozier and H. Federighi, *J. Gen. Physiol.* **7,** 137 (1925).
[209] C. Correns, *Ber. deut. botan. Ges.* **15,** 139 (1897).
[210] G. Schmid, *Jahr. wiss. Botan.* **60,** 572 (1921).
[211] A. Niklitschek, *Botan. Centr., Beih.* **A52,** 205 (1934).
[212] A Hansgirg, *Botan. Ztg.* **41,** 831 (1883).
[213] H. Zukal, *Österr. botan. Z.* **30,** 11 (1880).
[214] J. A. Krenner, *Arch. Protistenk.* **51,** 530 (1925).
[215] H. Coupin, *Compt. rend.* **176,** 1491 (1923).
[216] T. W. Engelmann, *Botan. Ztg.* **37,** 48 (1879).
[217] R. Lauterborn, "Untersuchungen über Bau, Kernteilung und Bewegung der Diatoméen," p. 130. Engelmann, Leipzig, 1896.
[218] C. T. v. Siebold, *Z. wiss. Zool.* **1,** 270 (1849).
[219] M. Schultze, *Arch. mikroskop. Anat.* **1,** 376 (1865).
[220] A. Hosoi, *Botan. Mag. (Tokyo)* **64,** 14 (1951).
[221] H. Skuja, *Nova Acta Regiae Soc. Sci. Upsaliensis* **16,** (3) (1956).
[222] E. Baur, *Arch. Protistenk.* **5,** 92 (1905).
[223] R. Y. Stanier, *J. Bacteriol.* **40,** 619 (1940).
[224] R. Y. Stanier, *Bacteriol. Revs.* **6,** 143 (1942).
[225] H. B. Hutchinson and J. Clayton, *J. Agr. Sci.* **9,** 143 (1919).
[226] A. E. Oxford, *J. Bacteriol.* **53,** 129 (1947).
[227] B. Norén, *Svensk Botan. Tidskr.* **46,** 324 (1952).
[228] L. Garnjobst, *J. Bacteriol.* **49,** 113 (1945).
[229] C. Stapp and H. Bortels, *Zentr. Bakteriol. Parasitenk. Abt. II.* **90,** 28 (1941).
[230] R. Thaxter, *Botan. Gaz.* **17,** 389 (1892).
[231] K. Meyer-Pietschmann, *Arch. Mikrobiol.* **16,** 163 (1951).
[232] R. Y. Stanier, *J. Bacteriol.* **44,** 405 (1942).
[233] H. C. Jacobsen, *Zentr. Bakteriol. Parasitenk. Abt. II.* **17,** 53 (1907).
[234] E. Sergent, *Ann. inst. Pasteur* **21,** 842 (1907).

235 H. Kufferath, *Ann. inst. Pasteur* **25**, 601 (1911).
236 R. Y. Stanier, *Nature* **159**, 682 (1947).
237 P. Weiss, *J. Exptl. Zool.* **68**, 393 (1934).
238 M. E. Loebeck and E. J. Ordal, *J. Gen. Microbiol.* **16**, 76 (1957).
239 S. Winogradsky, *Ann. inst. Pasteur* **43**, 549 (1929).
240 K. Vahle, *Zentr. Bakteriol. Parasitenk. Abt. II.* **25**, 178 (1910).
241 K. Meyer-Pietschmann, *Arch. Mikrobiol.* **24**, 297 (1956).
242 E. Jahn, "Beiträge zur botanischen Protistologie," Vol. I. pp. 16, 29. Bornträger, Leipzig, 1924.
243 H. Kühlwein, *Arch. Mikrobiol.* **19**, 365 (1953).
244 K. A. Bisset, "The Cytology and Life-History of Bacteria," 2nd ed., p. 40. Livingstone, Edinburgh, 1955.
245 A. R. Prévot, *Compt. rend. soc. biol.* **133**, 246 (1940).
246 W. Schewiakoff, *Verhandl. naturhist.-med. Ver. Heidelberg* **5**, 44 (1897).
247 G. S. West and R. M. Griffiths, *Proc. Roy. Soc.* **B81**, 398 (1909).
248 J. Virieux, *Ann. sci. nat. Botan.* **18**, 265 (1913).
249 W. Pfeffer, "Pflanzenphysiologie," 2nd ed., Vol. II, p. 755. Engelmann, Leipzig, 1904.
250 W. Rothert, *Flora* **88**, 371 (1901).
251 H. S. Jennings and J. H. Crosby, *Am. J. Physiol.* **6**, 31 (1902).
252 T. W. Engelmann, *Arch. ges. Physiol. Pflüger's* **30**, 95 (1883).
253 A. Manten, *Antonie van Leeuwenhoek. J. Mikrobiol. Serol.* **14**, 65 (1948).
254 W. Pfeffer, "Pflanzenphysiologie," 2nd. ed., Vol. 2, p. 798. Engelmann, Leipzig, 1904.
255 W. Pfeffer, *Untersuch. botan. Inst. Tübingen* **1**, 363 (1885).
256 M. W. Beyerinck, *Zentr. Bakteriol. Parasitenk.* **14**, 827 (1893).
257 R. K. Clayton, *Arch. Mikrobiol.* **27**, 311 (1957).
258 T. W. Engelmann, *Botan. Z.* **39**, 441 (1881).
259 T. W. Engelmann, *Arch. ges. Physiol. Pflüger's* **57**, 375 (1894).
260 T. W. Engelmann, *Arch. ges. Physiol. Pflüger's* **25**, 285 (1881).
261 T. W. Engelmann, *Arch. ges. Physiol. Pflüger's* **26**, 537 (1881).
262 T. W. Engelmann, *Botan. Z.* **40**, 321 (1882).
263 W. Pfeffer, *Ber. deut. botan. Ges.* **1**, 524 (1883).
264 W. Pfeffer, *Untersuch. botan. Inst. Tübingen* **2**, 582 (1888).
265 J. Massart, *Arch. biol. (Liège)* **9**, 515 (1889).
266 O. Baracchini and J. C. Sherris, *J. Pathol. Bacteriol.* **77**, 565 (1959).
267 C. H. Ali-Cohen, *Zentr. Bakteriol. Parasitenk.* **8**, 161 (1890).
268 S. Winogradsky, *Botan. Z.* **45**, 489 (1887).
269 M. Miyoshi, *J. Coll. Sci. Imp. Univ. Tokyo*, **10**, 143 (1898).
270 H. Kniep. *Jahr. wiss. Botan.* **43**, 215 (1906).
271 H. Molisch, "Die Purpurbakterien nach neuen Untersuchungen," p. 58. Fischer, Jena, 1907.
272 R. K. Clayton, *Arch. Mikrobiol.* **22**, 204 (1955).
273 R. K. Clayton, *Arch. Mikrobiol.* **19**, 125 (1953).
274 R. K. Clayton, *Arch. Mikrobiol.* **29**, 189 (1958).
275 J. Lederberg, *Genetics* **41**, 845 (1956).
276 J. Massart, *Bull. acad. roy. sci. et lettres et beaux-arts Belg.* **22**, 148 (1891).
277 S. L. Schenk, *Zentr. Bakteriol. Parasitenk.* **14**, 33 (1893).
278 S. O. Mast, *Am. J. Physiol.* **10**, 165 (1904).
279 R. K. Clayton, *Arch. Mikrobiol.* **19**, 141 (1953).

[280] T. W. Engelmann, *Botan. Z.* **46,** 661 (1888).
[281] J. Buder, *Jahrb. wiss. Botan.* **58,** 525 (1919).
[282] C. B. van Niel, *Bacteriol. Revs.* **8,** 1 (1944).
[283] R. K. Clayton, *Arch. Mikrobiol.* **19,** 107 (1953).
[284] J. Schrammeck, *Beitr. Biol. Pflanz.* **22,** 315 (1934).
[285] J. B. Thomas, *Biochim. et Biophys. Acta* **5,** 186 (1950).
[286] J. B. Thomas and L. E. Nijenhuis, *Biochim. et Biophys. Acta* **6,** 317 (1950).
[287] C. S. French, *J. Gen. Physiol.* **21,** 71 (1938).
[288] N. Rashevsky, *Protoplasma* **20,** 42 (1934).
[289] A. V. Hill, *Proc. Roy. Soc.* **A119,** 305 (1936).
[290] R. K. Clayton, *Arch. Mikrobiol.* **22,** 195 (1955).
[291] R. K. Clayton, *Arch. Mikrobiol.* **22,** 180 (1955).

CHAPTER 5

Morphology of Bacterial Spores, Their Development and Germination

C. F. ROBINOW

I. Introduction.. 207
II. Distribution of the Ability to Form Spores........................ 209
III. General Observations on the Development of Spores................ 211
IV. The Brightness of Spores... 216
V. The Interior of Spores... 216
VI. The Skin of Spores... 216
    A. Spore Wall and Cortex... 218
    B. Spore Coat and Exosporium..................................... 223
    C. Development of Spore Envelopes................................ 229
VII. The Imperviousness of Spores to Stains........................... 229
VIII. The Chromatin of the Spore...................................... 230
    A. The Behavior of the Chromatin during the Development of the Spore... 230
    B. The Arrangement of Chromatin in the Mature Spore.............. 236
IX. Germination.. 237
X. The Chromatin of Germinating Spores............................... 243
XI. Parasporal Bodies.. 243
XII. Conclusion... 245
    References... 246

I. Introduction

Several kinds of bacteria have the ability to make at the end of their period of growth a new cell or spore inside their cytoplasm (Fig. 1). The finished spore, protected by several sets of membranes, is set free by the breaking up of the mother bacterium (Fig. 2). Spores differ in important ways from their originators. They are capable of remaining alive for many years, dry and in air,[1] and they are not killed by long exposure to temperatures between 60 and 80°C. which are quickly fatal to most mesophilic bacteria. The spores of several species survive being suspended for 20 minutes and longer in boiling water. The resistance of spores to heat has received much attention from food technologists[2] and from medical men, but their hardiness and withdrawal from active life, combined with hair-trigger readiness to resume growth in a suitable environment, are of interest to all students of living cells.

The circumstances which induce bacteria to form spores and the chemical changes which they undergo during this process have been discussed at several recent symposia.[3,4] Of particular interest are the well-designed

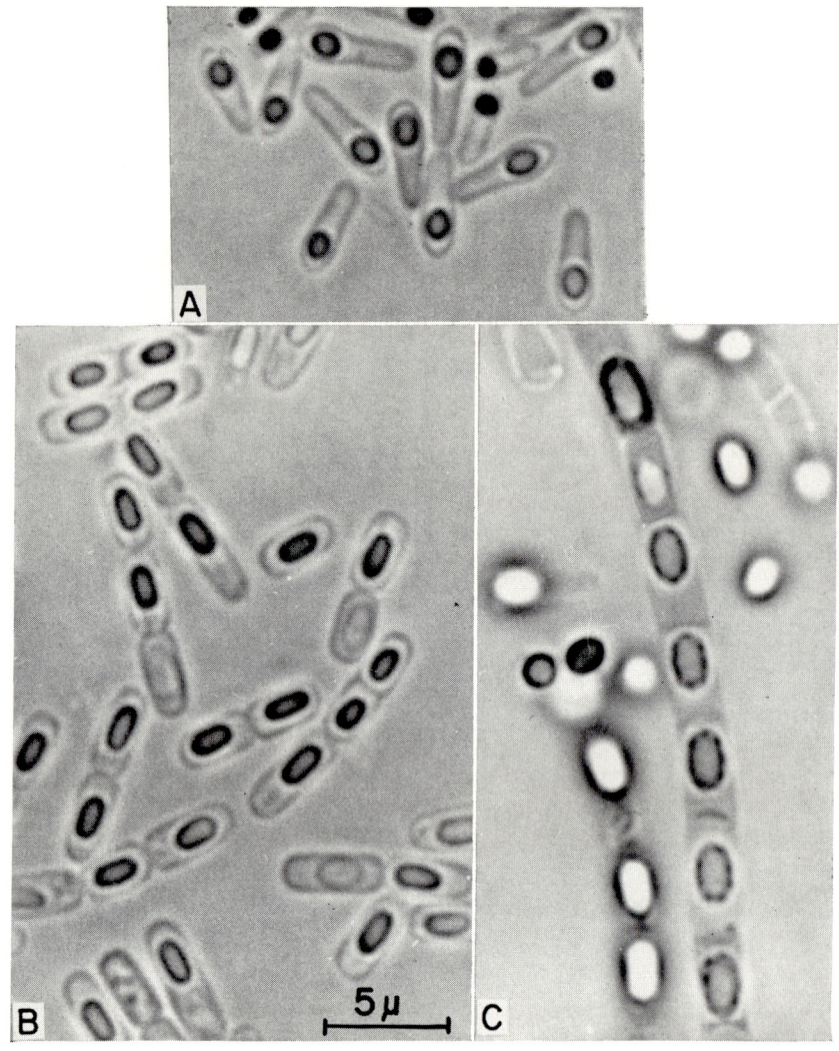

FIG. 1. Spores in *Bacillus* cells. A: unidentified bacillus isolated from soil; B: *B. cereus*; C: *B. megaterium* 350. Note the serrated contours of the spores. Details of their sculptured surface are shown in Figs. 14–16. Ordinary microscopy. A and B fixed with osmium tetroxide, photographed unstained between agar and a coverslip. C fixed with alcohol, stained lightly with 0.25% thionin.

studies of Grelet,[5-8] whose experiences have been put to good use in a combined chemical and cytological study of spore formation[9] which Dr. Elizabeth Young has recently carried out in the laboratory of Dr. P. C. Fitz-James.

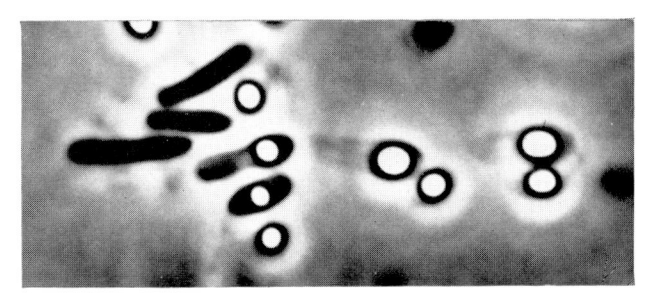

FIG. 2. Successive stages in the release of spores from dissolving vegetative forms of the bacillus from soil illustrated in Fig. 1 A. The density of the sporing rods decreases from left to right in this fortuitously but conveniently arranged sequence. Phase-contrast microscopy (× 3900).

The chemistry and physiology of spores are discussed elsewhere in this work. The present chapter is concerned with their anatomy.

II. Distribution of the Ability to Form Spores

The best-known spore formers are the rod-shaped bacteria of the two genera *Bacillus* and *Clostridium*, but certain cocci[10] and spirilla[11, 12] are also able to form spores. Shiny, oval bodies looking like large bacterial spores have repeatedly been seen extending through several cells of the trichomes of *Oscillospira guilliermondi*[12, 13] (Fig. 3). This occasional inhabitant of the cecum of guinea pigs has features reminiscent of the genus *Oscillatoria* of the blue-green algae, but is now counted among the bacteria[14] because of its lack of pigments, its peritrichous flagellation and rapid, sinuous swimming movements. Less certain is the taxonomic position of another, superficially blue-green algalike, motile spore former, *O. batrachorum*.[12]

In bacilli and clostridia only one spore is formed in one cell, but other species are known in which one cell gives rise to two or more spores. This is the rule in the spectacular, flagellated and very actively motile *Metabacterium polyspora*,[15, 16] which is common in small wild rodents in Israel[17] and is sometimes found in large numbers in the cecum of guinea pigs (Fig. 4).

Two spores, one at each end of a long cell, are also found in several kinds of filamentous bacteria which live in the proximal parts of the gut of tadpoles.[18] Examples of these curious forms, of which several new ones have recently been found by Dr. Berthe Delaporte in tadpoles in Canada, are illustrated in Fig. 5.

Persuasive evidence that heat-resistant endospores are formed by *Azotobacter* has been presented by Bisset,[19] but further descriptions, including observations of germination, must be awaited before the kinship with *Bacillus* which is now claimed for *Azotobacter* can be regarded as proven.

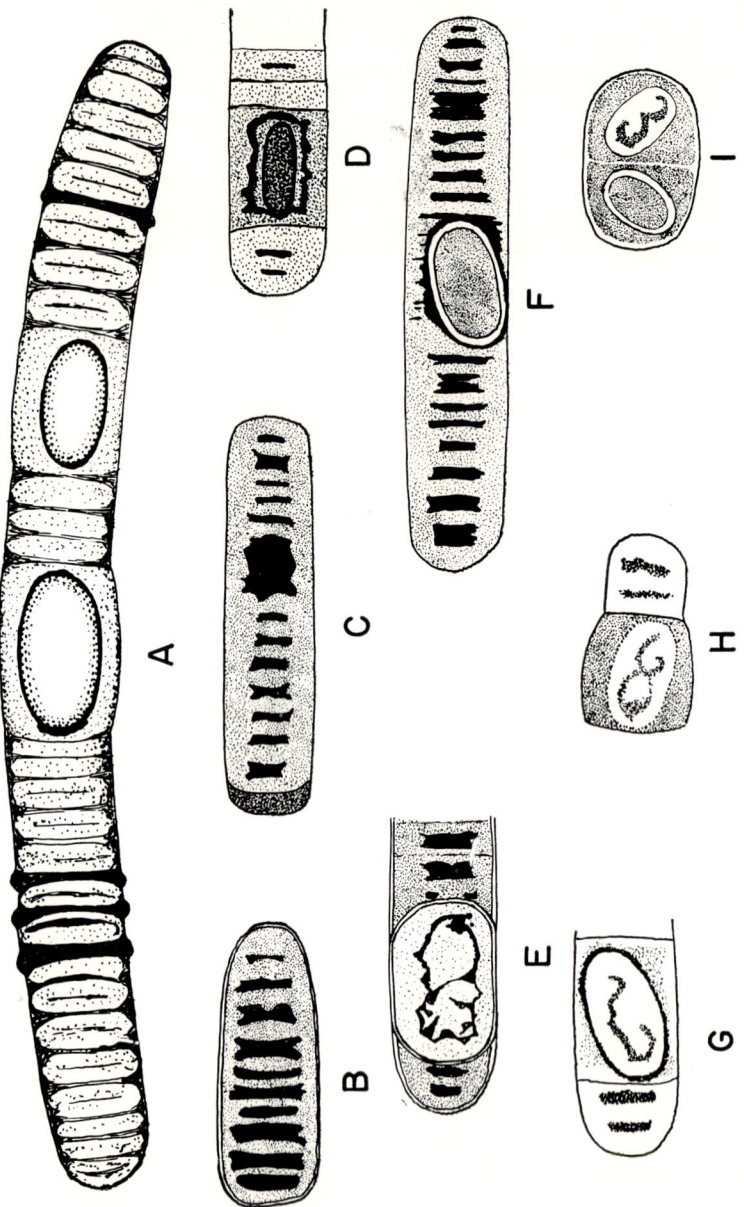

Fig. 3. *Oscillospira guilliermondi* in preparations from the caecum of the guinea pig. A: trichome with two spores; B: short vegetative trichome, iron hematoxylin; C–F: stages in the formation of a spore, stained partly with brilliant cresyl blue, partly with iron hematoxylin; G–I: developing spores with loops of chromatin. Alcohol, Feulgen. In H and I the cytoplasm also gives a positive Feulgen reaction, a curious finding which wants reinvestigating. A after Chatton and Pérard[13]; B–I after Delaporte[12]. Redrawn for this article by Dr. Berthe Delaporte.

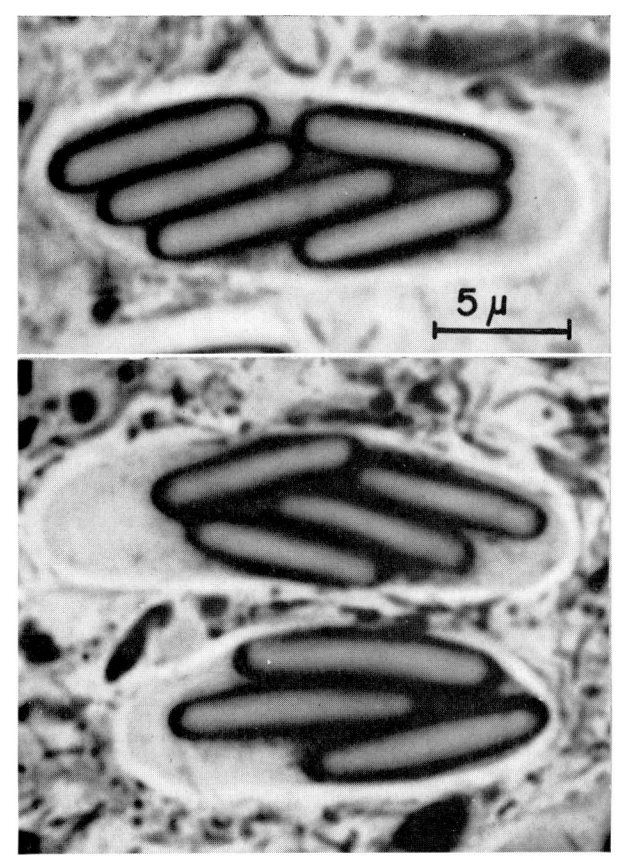

Fig. 4. Three individuals of *Metabacterium polyspora*. From the cecum of a guinea pig. Schaudinn fixation, iron hematoxylin. Compare this with Fig. 1 where the magnification is the same. Metabacteria containing as many or more spores as the three shown here are fast swimmers, despite the fact that they seem to possess little cytoplasm outside the spores. (Reprinted, with permission of the editor, from Robinow[16]).

III. General Observations on the Development of Spores

It is now well established that bacterial spores arise afresh in the cytoplasm and are not made by the packing together of preexisting visible granular inclusions, as has sometimes been held in the past. The emergence into view of the smooth, transparent "spore field" and the growth of the spore in this region is reflected in von Klinckowstroem's drawings from life[20] reproduced in Fig. 6. It has also been followed in a motion picture of living cells of *Bacillus megaterium*,[21] in photographs taken with ultraviolet light[22] and in studies employing phase-contrast microscopy.[23] At the time when they are about to form spores some bacteria contain numerous shiny drop-

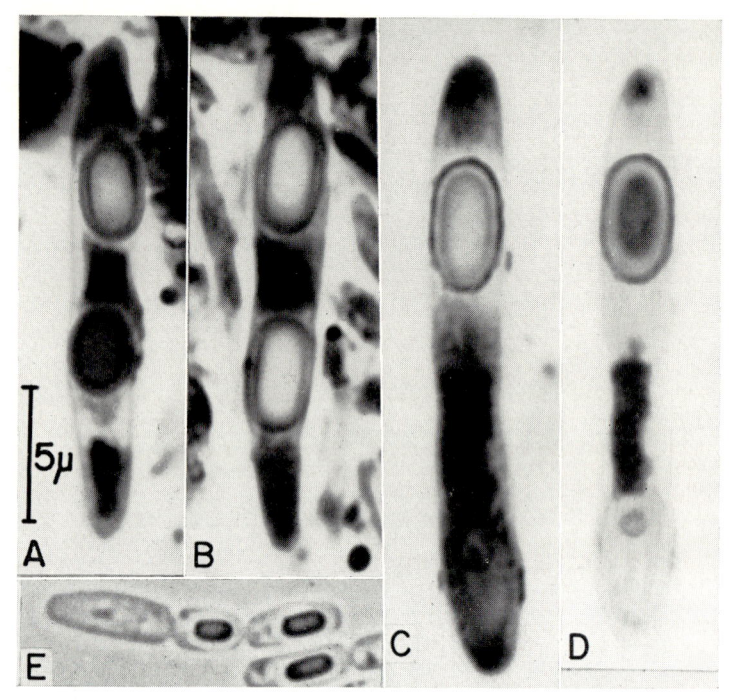

FIG. 5. Large bacilli from tadpoles of *Rana pipiens* containing two spores per individual. Schaudinn fixation, Giemsa stain. D is the same bacillus as C, as it appeared after hydrolysis with $N/1$ HCl. The light band surrounding the stained interior is probably the intine or cortex described in the text. E: Cells of *B. cereus* at the same magnification as A–D.

lets which in *B. megaterium* and *B. cereus* are known to contain lipid. The droplets, as has been well described, independently, by Klinckowstroem[20] and Bayne-Jones et al.,[21] are in continuous random motion, but do not penetrate the limits of the spore field, which suggests that the cytoplasm in this region is rather dense.

In *stained* preparations of *B. cereus* from synchronized cultures the first premonitory signs of spore formation can be detected some two hours before the spore field (the forespore of other authors) becomes visible in the living cell.[9] The interval between this latter, more immediately obvious event and the time when the ripening spore acquires its full brightness has generally been found to be four to five hours.[9, 20, 21, 23]

The making of spores seems to be a form of growth involving the synthesis of new protoplasm from raw materials in the culture medium. In some species, such as *B. cereus*, this is evident from the analysis of chemical balance sheets.[9] Synthetic activity is also suggested by the behavior of

MORPHOLOGY OF BACTERIAL SPORES 213

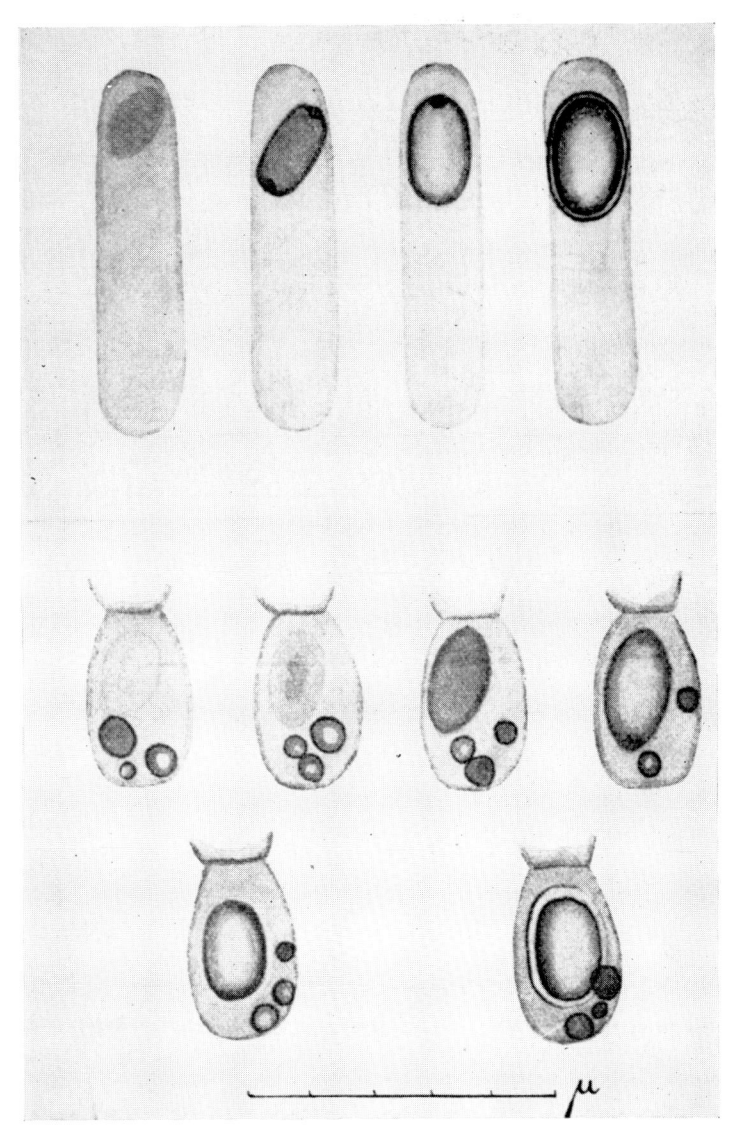

FIG. 6. Successive changes in the appearance of developing spores as seen during life in two *Bacillus* cells. The interval between first and last stages is 4–5 hours in both sequences. (Reprinted, with permission, from A. von Klinckowstroem.[20])

bacteria which become larger when they form spores. An example, illustrated in Figs. 7, 8, and 20, is provided by a *Clostridium* tentatively identified as *C. pectinovorum*. In its phase of multiplication this organism has the shape of a slender rod. Spore formation is attended by a lengthening

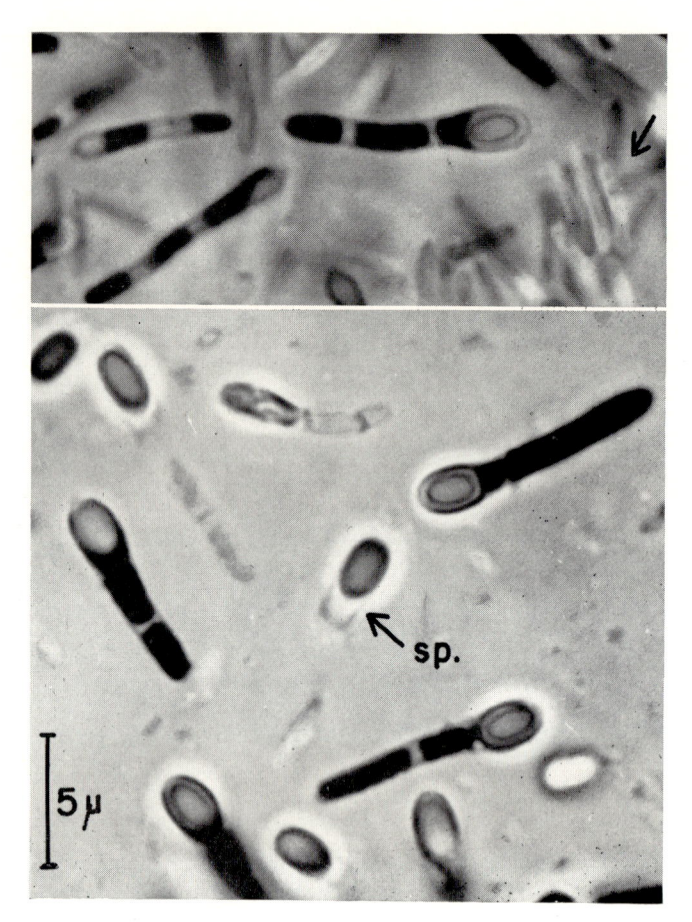

Fig. 7. *Clostridium pectinovorum*. From an anaerobic slant culture dispersed in Lugol's iodine solution. Note the difference in size between vegetative rods (arrow, right top) and sporing forms. The latter are laden with granulose which here appears homogeneously black but in reality is finely granular; sp. is the sleevelike sporangium which continues to envelop the spore after the rest of the clostridium has decayed. See also Figs, 8, 20, and 26. White bands across the clostridia indicate sites of chromatin free from granulose. Chromatin corresponding to the white bands is stained in Fig. 20.

and widening of the cell, the swelling of one end of it, the accumulation of starchlike granulose and the excretion of a voluminous capsule consisting largely of polypeptides.[24] Another view, that spore formation is a reshuffling of materials already present in the vegetative cell, is favored by Foster[25, 26] and his collaborators. These matters are further discussed elsewhere in this work.

MORPHOLOGY OF BACTERIAL SPORES 215

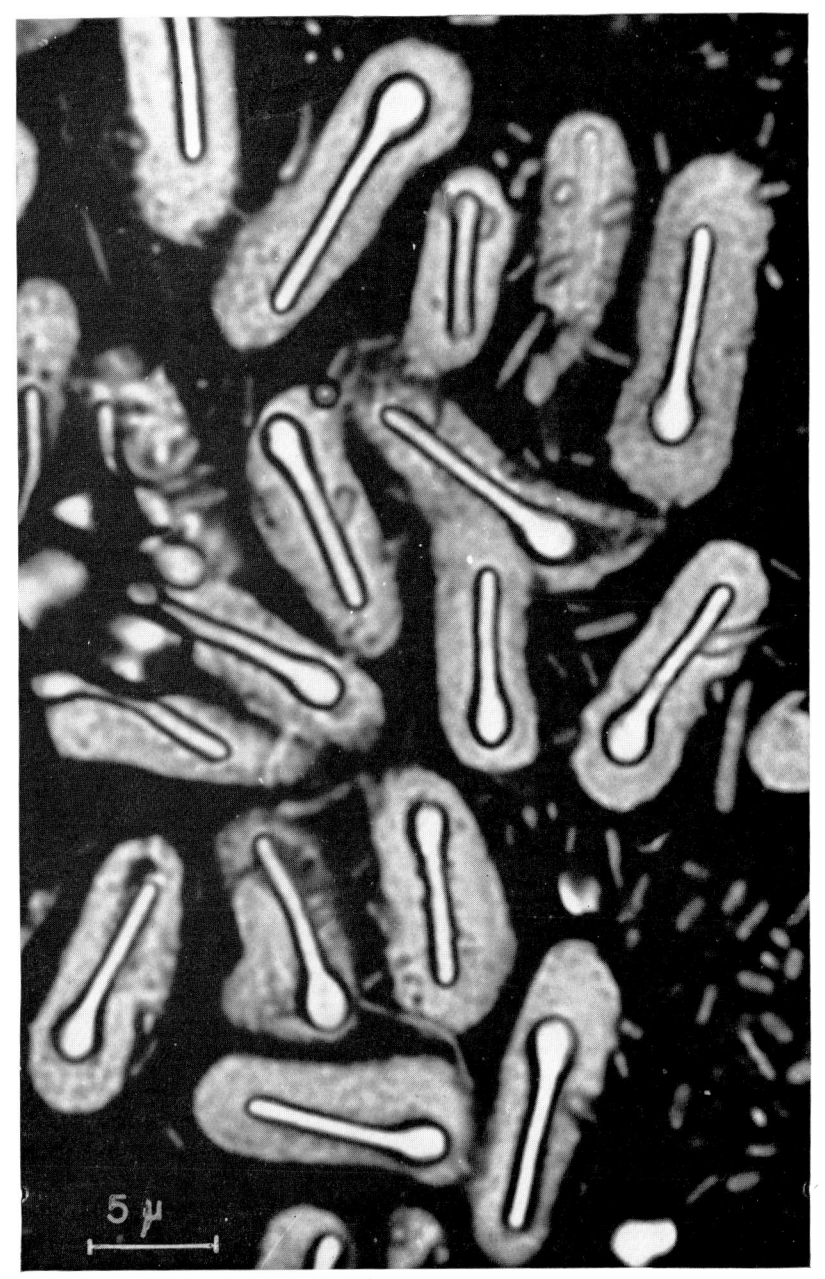

FIG. 8. *Clostridium pectinovorum.* Capsulated, sporing clostridia from a culture on grains of rice in tap water. Dispersed in nigrosin, allowed to dry, and mounted in air.

IV. The Brightness of Spores

Spores attract the eye of the microscopist because, in the manner of drops of oil suspended in water, they are edged in black when they are in focus and shine brightly when viewed in a plane slightly above it (Fig. 1 C). They are also, in some instances, colored light-yellow or brown. According to Ross and Billing[27] the refractive indices of spores are close to those of dehydrated proteins, such as casein powder or leather. Values of n between 1.512 and 1.540 have been determined for spores of *B. cereus*, *B. cereus* var. *mycoides*, and *B. megaterium*. Spores of *B. subtilis* seem equally dense because they look dark gray when they are mounted in Shillaber's synthetic immersion oil of $n = 1.5150$.[28] The quality of the medium affects the degree of refractility which spores attain on maturity. Spores that develop in the absence of calcium,[29] or have been prevented from absorbing enough of this metal from the medium are less bright and less heat-resistant than usual.[30] The dense, shiny protoplasm of the dormant cell within the spore is in all instances surrounded by several kinds of similarly or less refractile membranes, which are usually not easily distinguished in the light microscope until the spore has begun to prepare itself for germination, but are sometimes already visible during the resting stage.

V. The Interior of Spores

In electron micrographs the interior of spores, prepared by the techniques accepted today, appears diffuse and featureless except for a few shapes of low density near the surface which correspond to sites of chromatin of similar distribution, detectable with the light microscope in stained preparations of comparable material. They are represented as clearings in the stippled cytoplasm of Fig. 18 and are further discussed in Section VIII,B. Profiles of chromatin bodies are not visible in the sections of resting spores shown in Figs. 9 and 10. It is to be hoped that improvements in fixation and after-treatment will increase the visible detail in resting spores.

VI. The Skin of Spores

The older literature, thoroughly reviewed by Cook,[31] Lewis,[32] and Knaysi,[33] contains many different opinions on the number and nature of the envelopes around the spore protoplasm. The increased optical resolution now attainable with the electron microscope has raised the discussion of these matters to a higher level of accuracy. But new observations appear at such a fast rate and so rapidly refine and supersede those accepted only yesterday that the envelopes of spores can at present be discussed only in a general and provisional way without pretensions of accuracy on points of fine structure.

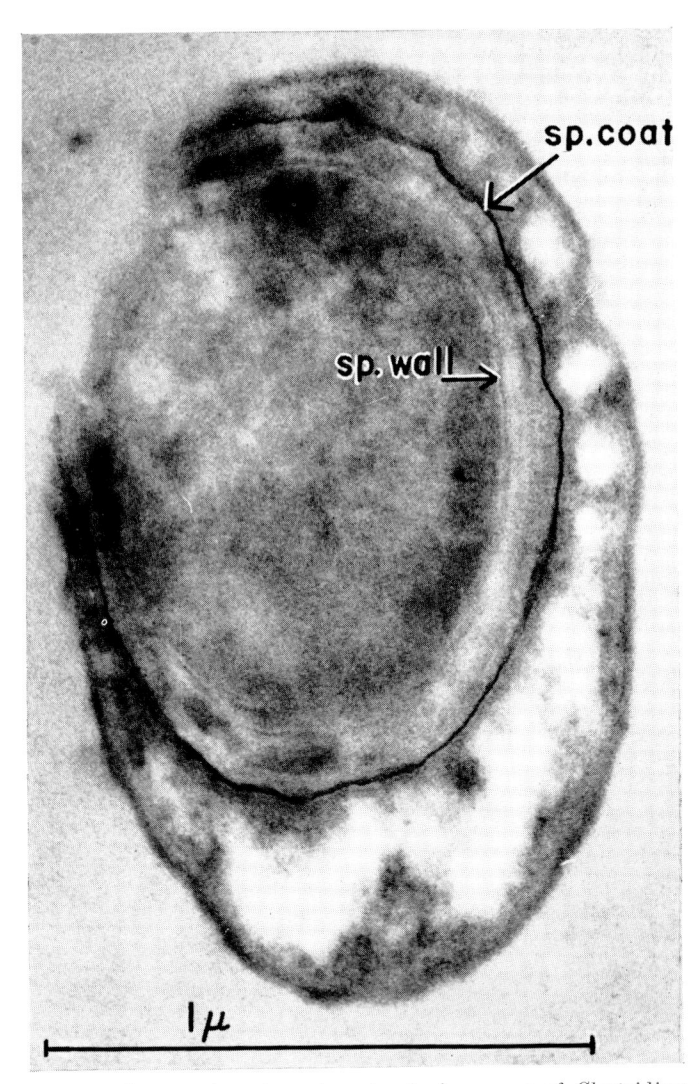

FIG. 9. Paramedian section through a developing spore of *Clostridium pectinovorum* stained with lanthanum nitrate. Between spore (sp) wall and spore coat lies the faintly laminated cortex. In mature spores the coat is more complicated than it is here. Seemingly empty, rounded spaces in the cytoplasm of the sporangium are the site of granulose which, like glycogen, has little scattering power for electrons. From an electron micrograph made by Dr. R. G. E. Murray.

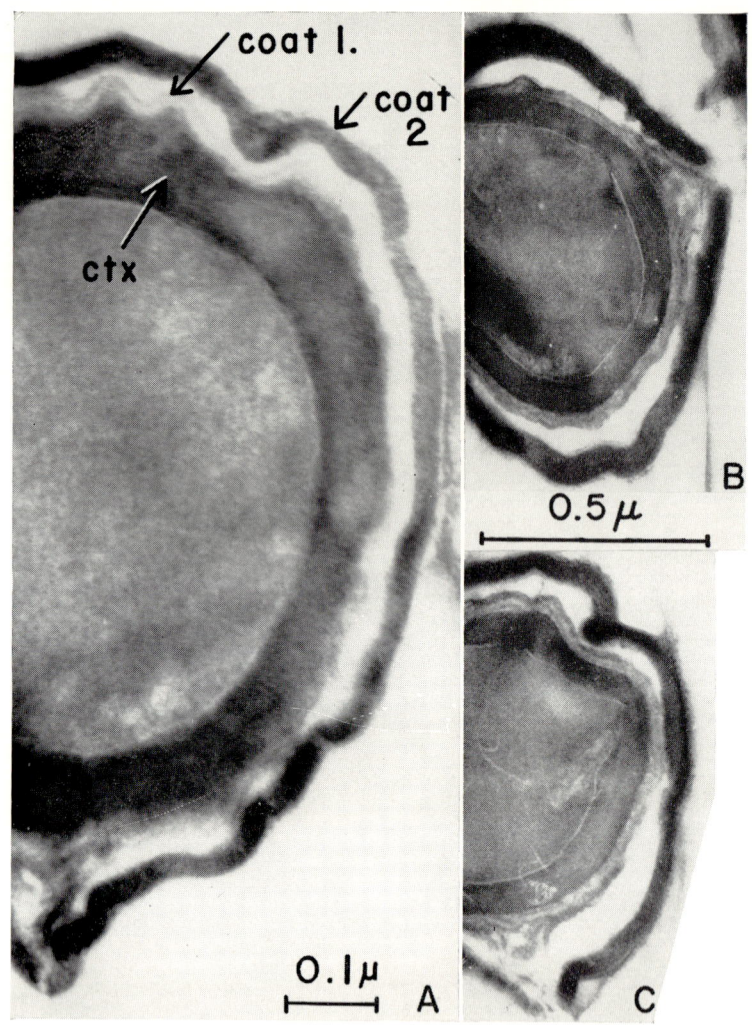

Fig. 10. Parts of sections through resting spores of *B. megaterium*. Stained with lanthanum to increase the density of the cortex (ctx). There are two separate coats. The inner one is less dense than the outer one. The wide gap between them is artificial. A: Fixed in osmium tetroxide, post-treated with $N/1$ HNO$_3$ containing $M/4$ NaF. (Reprinted, with permission of the Editor, from Mayall and Robinow,[37]); B and C: Normal spores kept for one hour in water at 60°C. before fixation with osmium tetroxide.

A. Spore Wall and Cortex

In the spores of several species the dormant cytoplasm has been found surrounded by a delicate membrane, visible in Fig. 9, which has been aptly named the "spore wall."[9, 34] It is destined to be transformed into the cell wall of the future bacterium. A spore wall is not visible, perhaps for technical reasons, in published micrographs of sections of spores of *B. polymyxa*[35, 36] and has only rarely been seen in sections of normal resting spores of *B. megaterium*,[37] although a membrane corresponding to it is regularly found in sections of acid-treated spores of this bacillus[37] (Figs. 11 and 12).

The spore cell and its wall are generally surrounded by at least two further layers, cortex and spore coat, which are in close contact with one another, but differ in organization and probably also in composition. Closest to the spore wall or, where this is lacking, close to the seemingly bare surface of the cytoplasm lies the *cortex*.[37, 38] It is visible as a faintly stratified belt of relatively low density between spore wall and spore coat in the section of Fig. 9 and as a broad band of uniform density in the *B. megaterium* spores of Fig. 10. The cortex is distinguished from the remaining spore envelopes by its low scattering power (artificially increased in the spores just referred to), its complexity, and, at least in *B. megaterium*, its behavior during germination.[37] A cortex is apparent, although not always commented on, in published electron micrographs of sections of spores of *B. polymyxa*,[35, 36] *B. licheniformis*,[39] *B. cereus* var. *alesti*,[9] and *Clostridium sporogenes*.[34] In spores of *B. megaterium* the cortex is visible during life early in germination (Fig. 24). It can generally be revealed by Feulgen hydrolysis (Fig. 11 B) or by placing resting spores for five minutes or longer in $N/1$ HNO_3 or in $N/3$ HNO_3 containing 0.1% $KMnO_4$ at room temperature.[40] Spores so treated lose their bright refractility and allow their cytoplasm and chromatin to be stained. The cortex remains unstainable and by virtue of this contrast stands out distinctly. In many instances the cortex also appears white and featureless in electron micrographs of thin sections.[38] The low density of the cortex of spores, normal or hydrolyzed, has here been accepted as natural and has been incorporated into the diagrams of Fig. 18. It is quite possible, however, that it is really unnatural and an artifact. At some time in the course of all the procedures now used to demonstrate the cortex, water is admitted to the interior of the spore. This, as Dr. Joan Powell has pointed out to the writer, must entail the immediate breakdown of the water-soluble polymers of calcium dipicolinate with which the spore is thought to be impregnated.[41] The disappearance of dipicolinic acid from acid-hydrolyzed spores has, in fact, been demonstrated by Fitz-James *et al.*[42] The loss of the calcium salts of this compound through hydrolysis of whole spores or the exposure of sections to water may well leave us with a misleading picture of the distribution of densities in the normal spore. The study of sections caught on nonaqueous liquids might prove more revealing.

The density of the cortex, or of what remains of it, in electron micrographs of thin sections can be increased by floating the sections on lan-

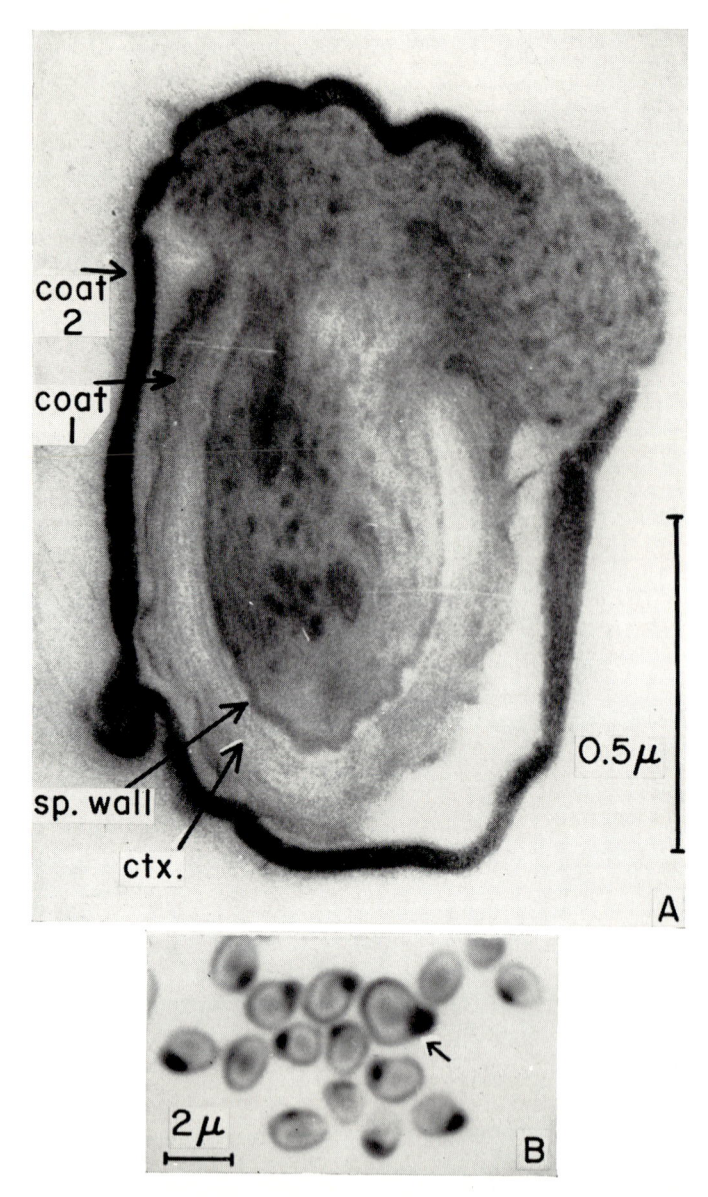

Fig. 11. Acid-treated spores of *B. megaterium*. A: Section through a resting spore that had first been fixed with osmium tetroxide and then been treated with $N/1$ HNO$_3$ at room temperature. Two spore coats, the cortex (ctx), and the spore wall can be distinguished. The plume of extruded protoplasm contains most of the spore's Feulgen-positive matter. This is apparent from B: A photo-micrograph of a Feulgen preparation of *B. megaterium* spores mounted in aceto carmine. Note that the cortex is visible in B as a narrow unstained band just inside the spore coat. (Reprinted, with permission, from Mayall and Robinow.[37])

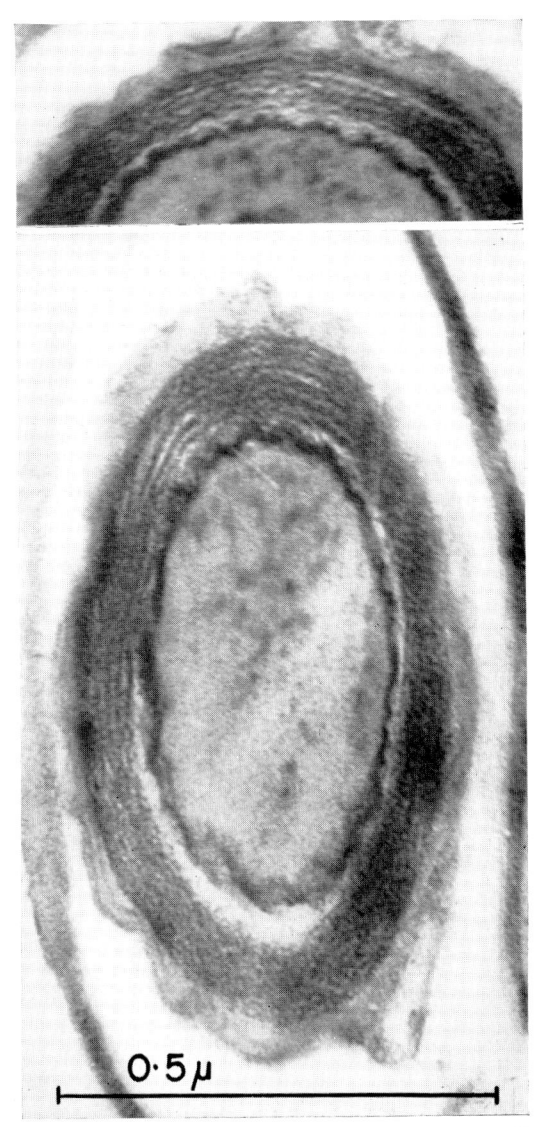

Fig. 12. Two sections of acid-treated spores of *B. megaterium*. Staining with lanthanum has revealed a concentric array of fine fibrils in the cortex. Facing the inner surface of the cortex is the dense, much folded spore wall. On the outside are two separate spore coats. (Reprinted, with permission, from Mayall and Robinow[37].)

thanum nitrate in water before placing them on carrier grids.[37] Usually, although not always, the cortex of normal spores of *B. megaterium* lacks visible organization even after it has absorbed lanthanum (Fig. 10), but intricate detail has been seen in the lanthanum-stained cortex of spores that had been treated with acid (Fig. 12). In these the lanthanophilic matter is arranged in the form of several concentric layers of what appear to be bundles of fine fibers. It is at present not known whether these fibers are unmasked or precipitated by the action of the acid. The former alternative may be true, because in developing spores of *B. cereus* the cortex appears laminated even when neither lanthanum nor acid has been used to enhance its visibility.[9] In *B. megaterium* the cortex breaks down inside the spore coat (s) early in germination and there are morphological signs that in doing so it contributes to the synthesis of the cell wall of the emerging bacillus.[37]

A distinct layer beneath the coat, separate from the spore cell and consisting of material of low density and low affinity for stains, in other words a cortex, was described long ago in spores of *B. polymyxa* by A. Meyer[43] (Fig. 13), who called it the intine. Van den Hooff and Aninga,[35] who first demonstrated the double-layered spore coat of this bacillus, regard the inner one of the two closely contiguous leaves of the coat as the equivalent of the intine. But the inner spore coat is too fine to have been visible as a separate intine in the light microscope and too dense to answer Meyer's description of it, which emphasizes its *low* density. I agree with Dondero and Holbert[36] that in their electron micrographs of thin sections of spores of *B. polymyxa* Meyer's intine is represented by a band of low scattering power just inside the two thin but dense leaves of the spore coat. A comparison of Figs. 13 and 18 A will bear out this interpretation. Dondero and Holbert give the thickness of the intine as 0.02 micron, but fail to explain how Meyer was able to distinguish from its neighboring structures a membrane one-tenth as fine as the lower limit of resolution of the light microscope. One must assume that spores of *B. polymyxa* shrink during their preparation for electron microscopy (the spores in Fig. 10 certainly have done so), and that Meyer had, perhaps, been looking at spores whose interior had lost part of its refractility as the result of having gone through an abortive process of germination. Nonviable individuals in this optically

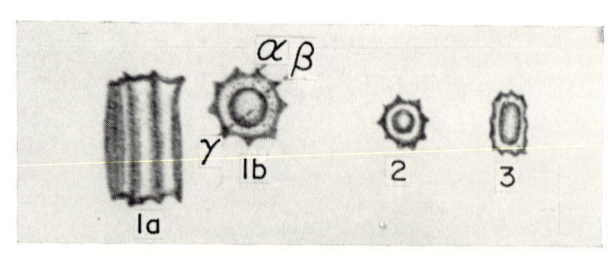

FIG. 13. Some of Meyer's drawings of the spore of "*Astasia asterospora*" now identified as *B. polymyxa*. 1b represents a transverse optical section of the spore. β marks the intine or cortex. From Meyer[43]. (Compare with Fig. 18 A.)

unusually differentiated state are often encountered among normal spores of B. *megaterium*.

B. Spore Coat and Exosporium

As far as one can now tell the spores of different species of bacteria resemble each other in their internal organization but differ in the degree of elaboration of their *outer* envelopes (Fig. 18). Thus there may be more than one spore coat, as in *B. polymyxa*[35, 36] and *B. megaterium*;[37, 38] the surface of the coat, as Bradley's technique[44] of carbon replicas has so elegantly revealed,[39] may be grooved, dimpled, or raised into ridges as in *B. polymyxa* (Figs. 18 A and 18 D) and (in a different style) in *B. megaterium* 350, illustrated in Figs. 14–16. Lastly there may be a further investment, a so-called "exosporium" beyond the spore coat(s). Good examples of this are provided by *B. cereus* and the related *B. mycoides*, *B. anthracis*, and *B. thuringiensis*, where the spores are invariably wrapped in a delicate bag which fits snugly along their sides, but protrudes considerably beyond the poles (Figs. 17 and 18 C). It used to be believed that the exosporium is the remains of the wall of the spore mother cell. Electron microscopy has now shown that the exosporium of *B. cereus* arises together with the spore in the cytoplasm of the mother bacterium.[45] This conclusion has been strengthened by observations on bacteria which form parasporal protein crystals.[9, 46] When spores and crystals are set free through the disintegration of the mother bacterium the crystals are always found outside the exosporium. If the latter were really a remnant of the old cell wall then it ought to enclose both spore and crystal in one common sack from which the crystal could presumably escape only by tearing it. But no rents have been seen in the exosporia from crystal-bearing bacilli.

The evidence for an exosporium in *B. megaterium* is conflicting. An envelope which could be so interpreted is visible, but not referred to, in some of the electron micrographs illustrating Knaysi's and Hillier's account of spore germination.[47] No loose, ill-fitting exosporium appears around any of the spores of *B. megaterium* studied in electron micrographs of thin sections by Chapman,[45] Robinow and Mayall,[37] and by Robinow,[38] unless the outer, often ragged, spore coat is regarded as one. However, an exosporium has now been demonstrated by Tomcsik and Baumann-Grace[48] in India ink mounts as well as with the aid of specific antibodies in two-thirds of more than forty strains of *B. megaterium*. Electron micrographs of the authors' material are not yet available, but a distinct, finely textured exosporium (Fig. 16) has meanwhile also been seen investing the spores of "*B. megaterium* 350," a culture obtained from the collection of Dr. John Rishbeth (Cambridge University). The spores of this variety, which has the high sensitivity to lysozyme[49] that one expects from a typical *B*.

Fig. 14. Electron micrograph of a tungsten-shadowed carbon replica of spores of *B. megaterium* 350. The pattern of ridges accounts for the serrated contours of these spores in Fig. 1 C. In several places it can be seen that the spores are wrapped in a delicate exosporium. This is shown more clearly in Fig. 16. Preparation and electron micrograph made by Dr. P. C. Fitz-James.

megaterium, differ from the spores of all other strains of this species which this writer has seen in bearing at their surface, beneath the exosporium, a honeycomb pattern of straight ridges enclosing polygonal depressions (Fig. 14). A delicate exosporium, stretching tentlike from ridge to ridge above a sculptured spore coat, has also been described by Dondero and Holbert[36] in *B. polymyxa*.

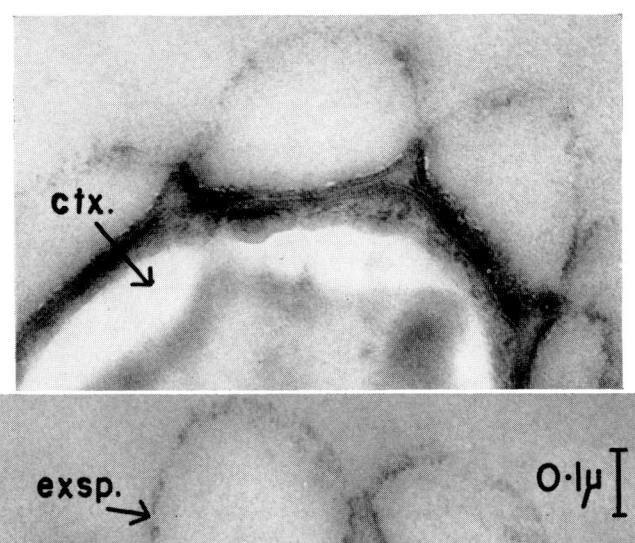

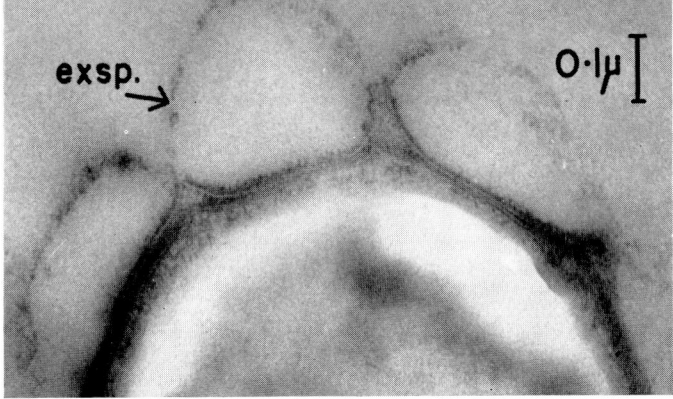

FIG. 15. Parts of equatorial sections through two spores of *B. megaterium* 350. The ridges shown in Figs. 14 and 16 are seen here as arising from the outer leaf of a double-layered spore coat. The space between adjacent crests is roofed over by the billowing exosporium (exsp) which has been lifted unnaturally high above the level of the spore coat. The wide band of low density beneath the spore coat is the cortex (ctx). Sections and electron micrographs made by Dr. R. G. E. Murray.

A spore envelope which, in contrast to the exosporia just described, is undoubtedly derived from the wall of the spore-forming cell has been found in *C. pectinovorum*. Here, as the growth of the spore gets under way, the wall of the gradually swelling head of the drumstick-shaped cell and that of the neck joining it to the shaft is reinforced by the apposition of several layers on its inner surface. The limit of the region involved is visible in stained preparations as a faint line across the neck of the head of the drumstick (Fig. 20). When the spore has been formed most of the straight shaft disintegrates, but the strengthened wall of the spore chamber remains attached to the spore in the shape of a short open sleeve (best seen at the left in Fig. 26). The spore chamber of *C. pectinovorum* with its

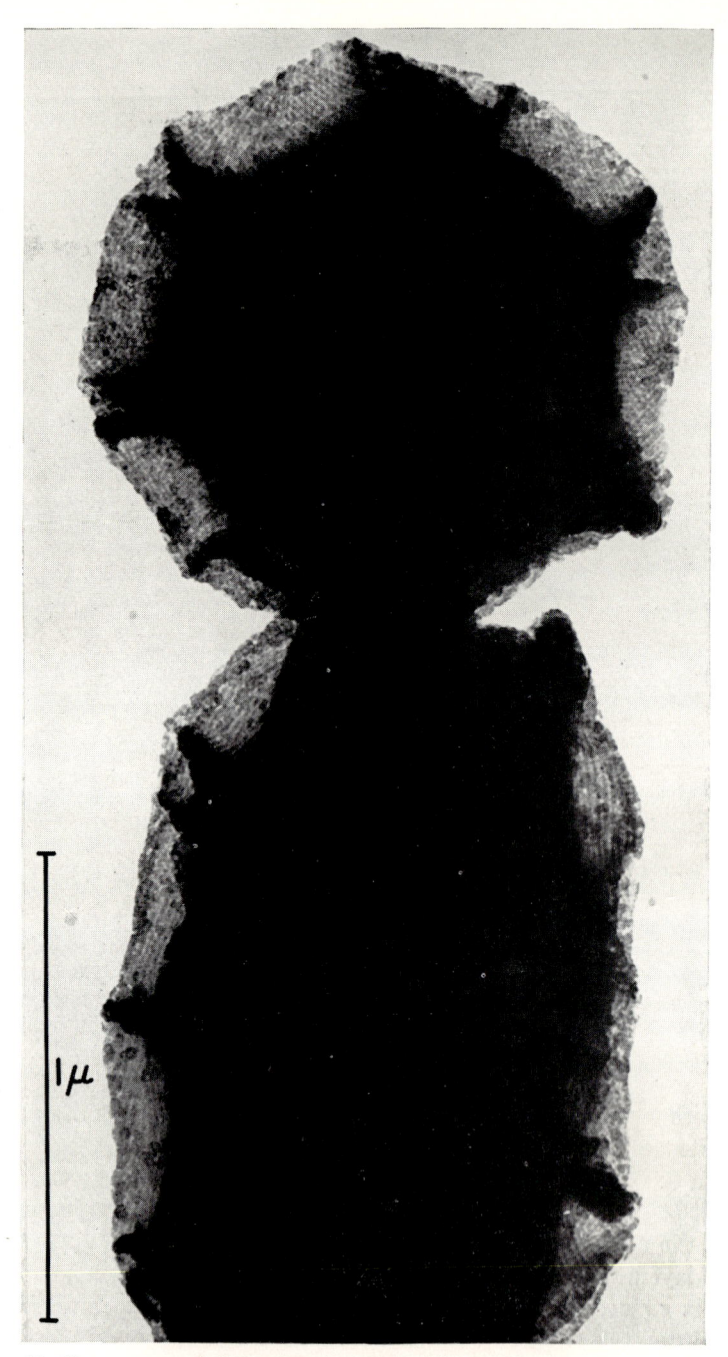

FIG. 16. Two spores of *B. megaterium* 350 with prominent spines, optical cross sections of the ridges shown in Figs. 14 and 15. Both spores are shrouded in a finely textured exosporium. Note that one spore is very much larger than the other. Such differences are characteristic of the spores of many strains of *B. megaterium* and in this one are found even between spores arising in adjacent cells in the same chain. Electron micrographs made by Mr. B. H. Mayall.

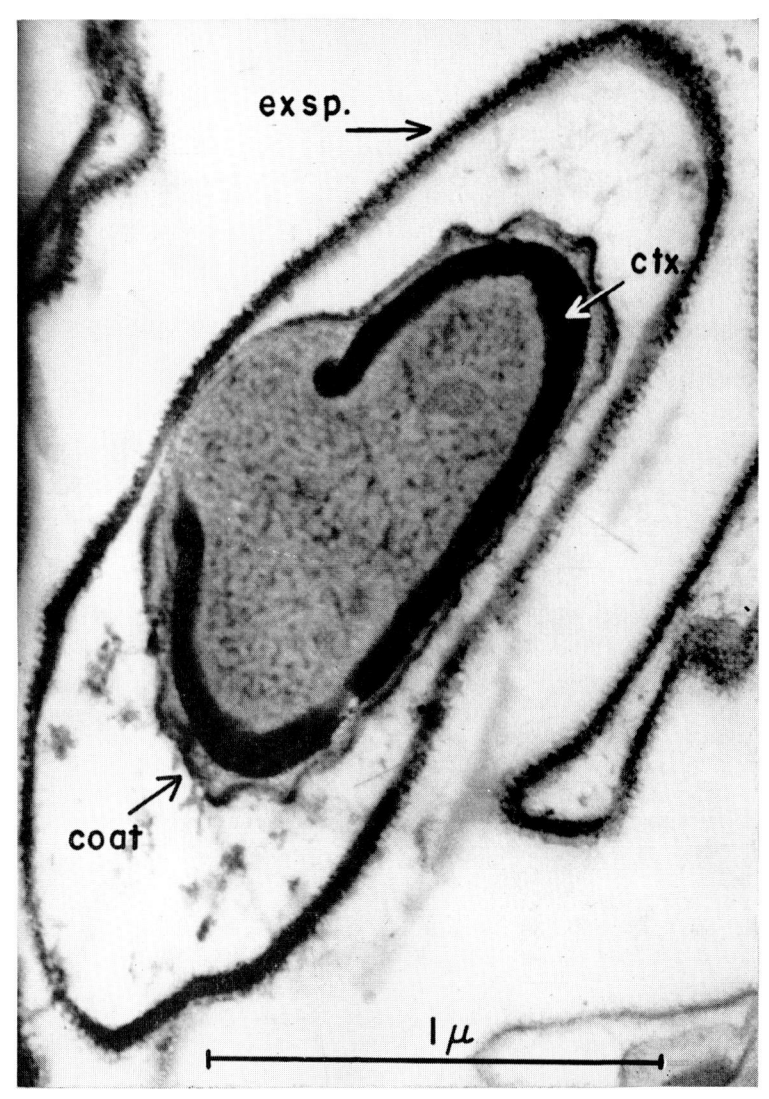

Fig. 17. Spore envelopes in *B. cereus*. Longitudinal section through a resting spore surrounded by a wide exosporium (exsp). The spore coat is thin and thrown into folds due to shrinkage. The graininess of the cytoplasm is of a degree unnatural in a resting spore and is the result of exposure to $N/1$ HNO_3, which has also caused some of the spore contents to be extruded through a gap in the cortex (ctx). The latter is so heavily stained with lanthanum that details of its structure cannot be seen. This kind of preparation is misleading as regards the normal state of the cytoplasm and chromatin of the spore but gives useful information about its envelopes.

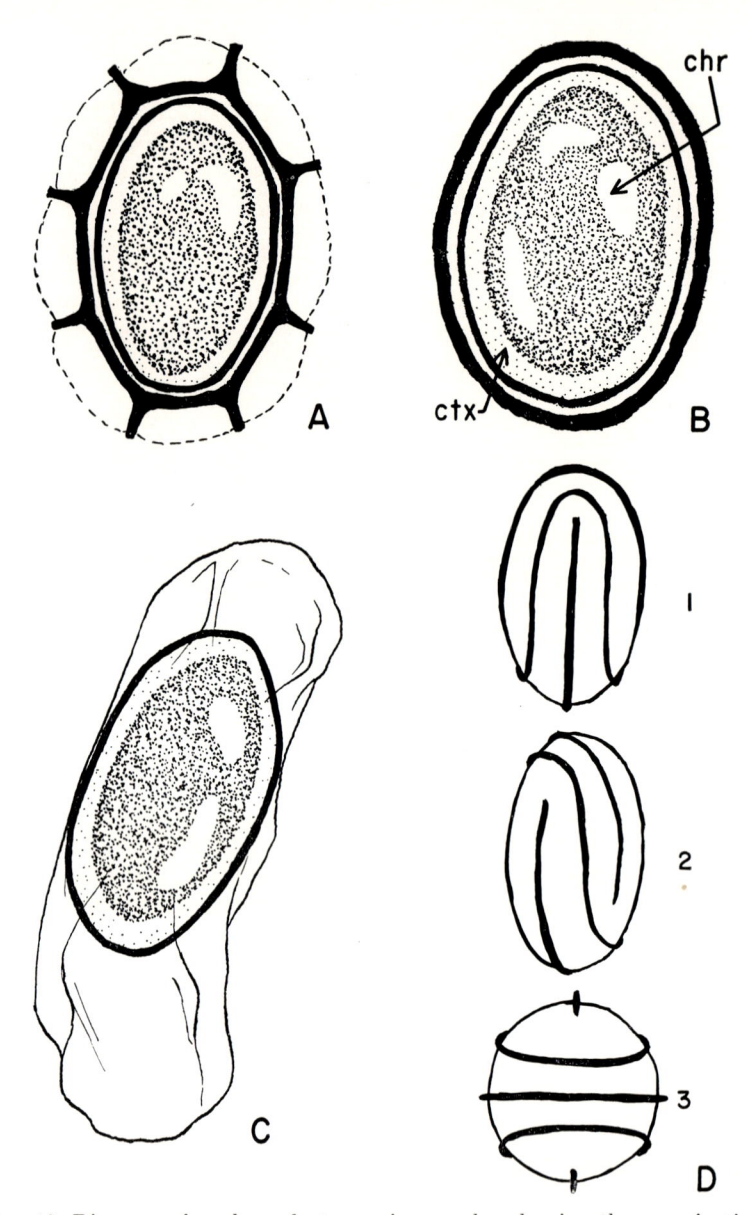

Fig. 18. Diagrams, based on electron micrographs, showing the organization of three kinds of spores differing from each other in the number and form of their outer envelopes. A: A spore of *B. polymyxa*, combined from van den Hooff and Aninga, 1956, and Dondero and Holbert, 1957. There are two spore coats and an exosporium; B: A spore of *B. megaterium*, after Mayall and Robinow.[37] There are two spore coats. The inner one is thinner and less dense than the outer one; C: A spore of *B. cereus*, after various sources. The spore has only one dense coat. It is wrapped in a wide exosporium. The interior of all three spores is essentially the same. ctx = cortex, chr = chromatin. D: Drawing of the pattern of ridges on the surface of a spore of *B. polymyxa*. (1) side view; (2) the same rotated a turn from right to left; (3) a view of a pole. (After Franklin and Bradley.[39])

reinforced wall truly deserves to be called a "sporangium" because it is an organ of the bacterium specially fashioned for the development of its spore. Limitation of the use of the term "sporangium" to instances of this kind seems more sensible than the common, somewhat pretentious practice, probably dating back to Meyer,[43] of applying it to a whole, anatomically unchanged bacterium from the moment a spore begins to develop inside it.

C. The Development of Spore Envelopes

In several laboratories the electron microscope has been employed to study the development of spores. In published descriptions of this process there is agreement on one point; the cytoplasm of the spore is at first granular and later becomes increasingly dense, diffuse and featureless.[9, 34, 45] This change in texture is reversed during germination.[37] There is disagreement about the order in which the various envelopes of the spore make their appearance. Thus, in *B. cereus* the spore wall is said to arise before spore coat and cortex become visible[9] and this was also seen in the clostridium illustrated in Fig. 9. However, in *C. sporogenes* it is apparently the spore *coat* which is laid down first, with the spore wall and cortex following later in that order. These disagreements need not be taken too seriously. Increasing standardization of the technique of electron microscopy will probably soon be reflected in increasing similarity of the results obtained by different workers.

Of great interest is the surprising observation that in certain bacteria the membrane which is going to become *the spore wall is at first continuous with and grows out of the cell wall* of the mother bacterium a short distance behind the tip of the cell.[9] Clear pictures of various stages of this process have now been obtained in sections of synchronously sporing cells of *B. cereus* and *B. mycoides* and will shortly be published by Drs. E. Young and P. C. Fitz-James. If this discovery turns out to be generally true, then spore formation will have to be regarded as a special, unequal form of cell division or internal budding in which the new cell is carved out of the cytoplasm by the activity of the existing cell wall, detaches itself from the latter after completion, and moves into the cytoplasm, there to be surrounded by further envelopes. Thus, although this chapter is supposedly dealing with the *endospore* of bacteria, we may, in truth, be concerned with something which may have to be regarded as a cross between a conidium and a chlamydospore.

VII. The Imperviousness of Spores to Stains.

It is common knowledge that stains are accepted by the *coat* of most spores, but seem unable to penetrate to the interior. The barrier to staining first becomes noticeable at the time when the cortex is laid down and may be identical with it.[9, 34] The staining properties of the cytoplasm behind the cortex are hard to assess. It has often been shown that the developing spore is strongly basophilic just before it becomes refractile and loses all affinity for stains. Is the barrier to staining superficial and does the interior of the spore retain its affinity for stains? The central portions of slices cut from resting spores are readily stainable[38] and so are intact spores minutes after they have been brought into contact with a nutrient medium. This would seem to argue in favor of the persistence of a stainable core behind the unstainable cortex. It is more probable, however, that the spore protoplasm is stainable only when it has first been allowed to take up water and that in its highly dehydrated state within the intact resting spore it has no affinity for stains at all.

VIII. The Chromatin of the Spore

A. The Behavior of the Chromatin during the Development of the Spore

Spores contain visibly organized chromatin. This material, which is Feulgen-positive, is derived from the chromatin of the mother bacillus in a way which is direct in some instances, devious in others. In certain long, slender bacteria which have several times been described in the gut of tadpoles[18] the spore acquires its chromatin in a simple manner. In these forms an unbroken filament of chromatin extends through the whole length of the cells. A spore arises near one or both ends of the bacterium and that segment of the cord of chromatin which passes through the spore area becomes the nucleus of the spore. This behavior suggests that the axial cord of chromatin contains in longitudinal sequence many replicas of the basic nuclear complement. Since no chain of spores corresponding in number to the number of putative nuclear subunits is formed in these cells one must assume that the initiative for spore formation arises in the cytoplasm and only in certain, usually distal, regions of it. These ideas gain in plausibility if the behavior of other spore-forming bacteria from tadpoles of *Rana pipiens*, recently examined in Canada by Dr. Berthe Delaporte and the writer, is also considered. Each slender cell of these long bacteria contains many regularly spaced *separate* clusters of chromatinic rodlets and granules, an arrangement which gives them a resemblance to the swarmers of *Proteus*. Here again only one spore is formed at either end of the cell and only the chromatin complex already in that area is built into it.

In the ordinary bacteria of the laboratory the origin of the spore chromatin is less certain and its behavior more complicated than it appears to be in the large bacteria in tadpoles. The most securely founded information

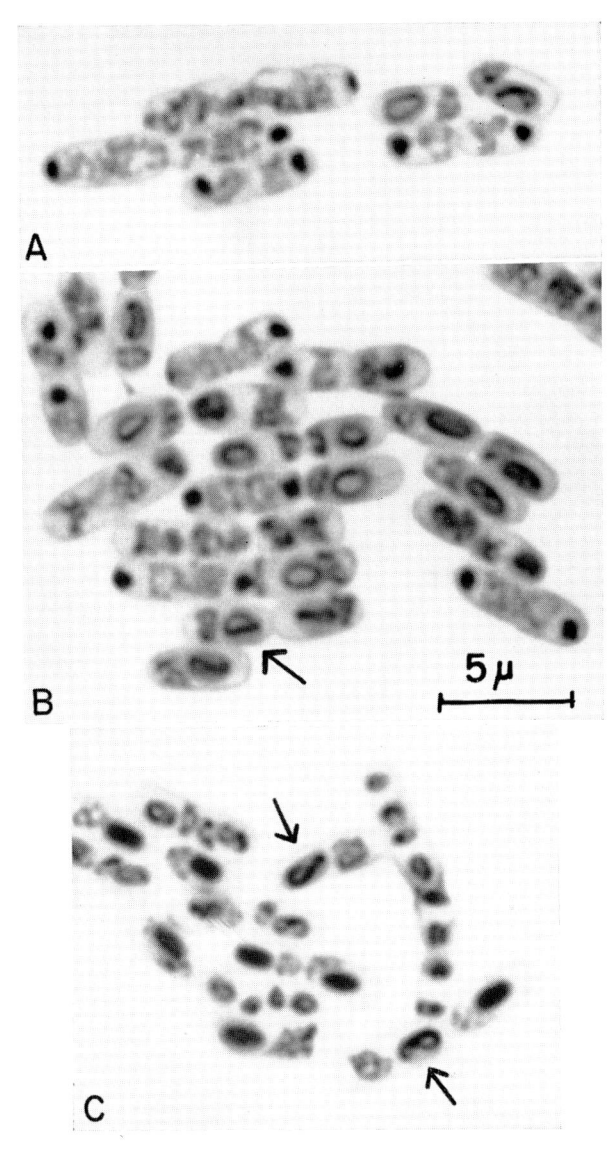

FIG. 19. Arrangement of the chromatin at different stages in the development of spores of *B. cereus*. Acetic acid/alcohol, HCl-Giemsa. Angular masses of deeply stained chromatin near the tip of the cells, numerous in A and B, open out into rings and figures of eight indicated by arrows in B and C and are finally obscured by the increasing basiphilia of the spores which has become very noticeable in many spores in C. B and C are from the same culture, the interval between them is one hour.

has been obtained from studies of spore formation in *B. cereus* and bacilli related to it. In the final stages of spore development the behavior of the chromatin seems to be the same in *B. cereus*, *B. megaterium*, and *C. pectinovorum*.

It will be convenient to start our account at the time when the spore field is first seen as a clear patch in a granular cytoplasm. The Feulgen technique and the HCl-Giemsa and HCl-Azure A/SO$_2$ stains show that at the center of that clear region there is a dense, angular grain of chromatin[9] (Fig. 19 A). As the spore elongates, the chromatin is first drawn out into a slender oval loop or open ring and is later twisted into a variety of figure-of-eight and pretzel designs (Fig. 19 C). Shortly after, one-half hour later under the conditions of Young's experiments,[9] the loop of chromatin in the spore is gradually obscured, even in hydrolyzed preparations, by the deep and even staining for which the spore is now acquiring a strong affinity (Fig. 19 C). Essentially the same sequence of events, from granule to loop to impenetrable basiphilia, has also been seen in developing spores of *B. megaterium*.[50] Helically twisted cords of chromatin are seen with beautiful distinctness in developing spores of *C. pectinovorum* (Fig. 20 B). The hydrolysis-resisting over-all basiphilia of the maturing spore has not yet been explained. It is, in my experience, not reflected in Feulgen preparations and is thus not likely to be due to materials containing an ordinary kind of deoxyribonucleic acid (DNA). Histochemical studies of this phase of spore development are needed. It is encouraging to read that the frustrating uniform staining of nearly mature hydrolyzed spores can be bypassed and their interior revealed if alcohol-fixed spores are first stained with acidified thionine and then with acidified methylene blue.[51]

The strong but fleeting affinity of the maturing spore for basic stains coincides with the rise of its refractility and may reflect chemical changes accompanying its beginning dehydration. Stained sections of mature spores reveal that the interior continues to be filled with (potentially) strongly basophilic material.[38] That the spore, emerging from the basophilic phase, becomes impervious to stains at the same time as it becomes refractile is probably due to the development of the cortex. Appearances suggesting this have been seen in electron micrographs of thin sections of spores fixed at this stage of development.[9, 34] The disposition of chromatin

---

FIG. 20. *Clostridium pectinovorum*. A: Young vegetative forms containing columns of chromatin bodies too small and too closely packed to be resolvable; B: The two panels of B are treated as one figure. They illustrate some of the shapes which the chromatin assumes in the course of spore formation. The arrow near the right top points to two cells in an early phase of spore formation. As the spore grows larger the chromatin changes first into a straight or wavy cord and later into a more or less tightly coiled helix. Ripening spores are deeply stained all through. Compare this

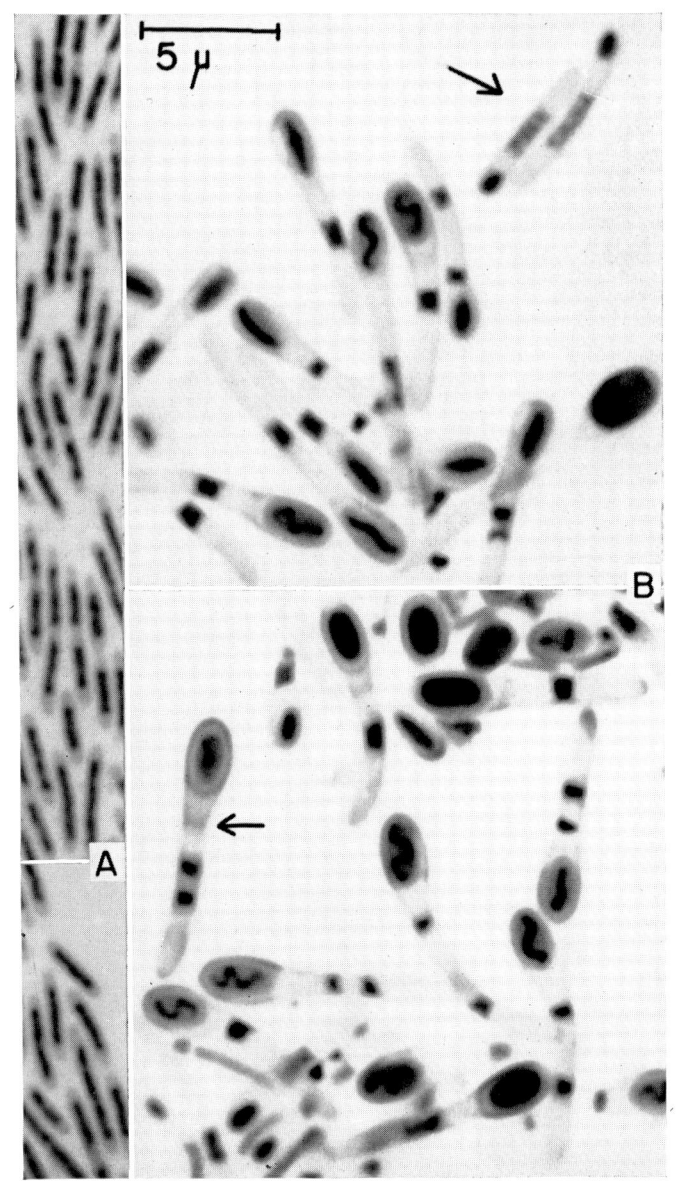

Fig. 20.

figure with Fig. 19 C and note similarities. The chromatin in the vegetative portion of the clostridia is compressed into rectangular bars by accumulations of granulose (not stained here) which now fill the cells almost completely (see Fig. 7). The arrow in the lower half of B points to the boundary where the sporangium (i.e., the spore in its sleeve of reinforced cell wall) will later break away from the more perishable straight part of the clostridium. Acetic acid/alcohol, HCl, Giemsa.

in mature, maximally refractile spores is described in Section VIII,B. It has to be studied in crushed or sectioned spores because whole spores react to hydrolysis (necessary for the Feulgen reaction) with the extrusion of their chromatin.*

The sequence of changes illustrated in Figs. 19 A–C and, diagrammatically, by the sketches to the right of the arrow in Fig. 22 has been described in detail because it is known to occur in several well-known species and is easily demonstrated in the usual more or less randomly multiplying and sporulating cultures. The sequence begins with a compact mass of chromatin in the center of the freshly emerged spore field, a point in time which marks the beginning of the spore as a structure *visibly* different from the rest of the cell. However, this easily pin-pointed event is preceded by less directly visible processes which separate the spore chromatin from that destined to remain in the vegetative part of the cell. These events will now be discussed.

The origin of the chromatin in the spore is difficult to determine in *B. megaterium* because sporulating cells of this species are usually filled with droplets of lipid which compel the chromatin bodies to arrange themselves in complicated but fortuitous patterns. *C. pectinovorum* is equally uninformative. The slender vegetative cells of this species are filled for the greater part of their length (Fig. 20 A) with a column of chromatin bodies too small and too tightly packed to be optically resolvable. When the spore is about to be formed, all that seems to happen is that a short length of the column of chromatin moves to the tip of the cell, becomes deeply stainable, and is later spun out (or grows?) into the helically twisted filament shown in Fig. 20 B. In *B. megaterium* and *C. pectinovorum* then, and we may perhaps add the large bacteria from the gut of tadpoles, there is no obvious evidence that the delegation of some of the chromatin for "spore duty" is preceded by significant rearrangements. It seems to be otherwise in *B. cereus* and its variants. Many years ago Badian described how in *B. mycoides* spore formation is preceded by the fusion of four separate chromatin bodies.[53] His observations were partly confirmed and were enlarged by Klieneberger-Nobel.[54] Evidence of nuclear fusion has been found in *C. tetani*[55] and the formation of axial cords of chromatin from formerly separate chromatin bodies has been described in great detail in *B. anthracis*[56] and, very recently, in *B. cereus*.[9] According to Young,[9] whose careful quantitative studies were made on cells from highly synchronized cultures, each cell of *B. cereus* at the end of the period of vegetative growth contains two chromatin bodies. The first intimation of impending spore formation is

* The extrusion response elicited by acid is inhibited by fluoride. In this, and in the fact that it is accompanied by the loss of dipicolinic acid and the appearance of a thick spore wall (abortive cell wall synthesis?), the response resembles germination.

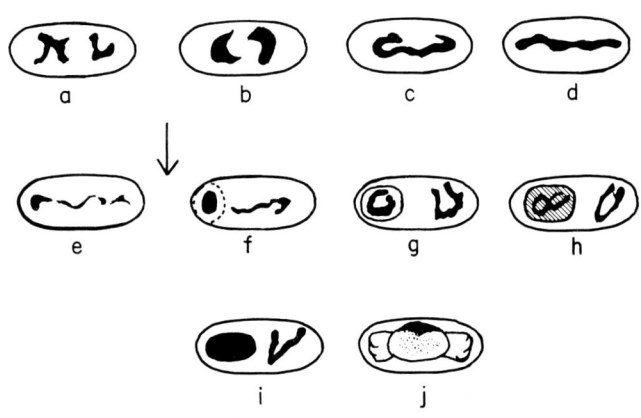

FIG. 21. *B. cereus*. Diagram illustrating the behavior of the chromatin during spore formation as seen in hydrolyzed and stained preparations. The arrow indicates the point in time where the spore field first becomes visible in the living cell. To the right of the arrow begins the development which is illustrated in Fig. 19 A–C. After Young[9]. Slightly modified. The last two figures added by the author. Cell *j* illustrates diagrammatically the mature spore with its exosporium still inside the mother bacterium. The chromatin is shown extruded to the periphery as it would have been had this spore been treated like the rest of the cells in this figure.

the contraction of these two bodies (four in *B. anthracis*, according to Flewett[56]) into more compact shapes. Later both chromatin structures unfold again and fuse to form a single wavy cord. This in turn breaks up into several unevenly staining fragments one of which, near the tip of the cell, detaches itself from the rest and becomes the angular grain of deeply stainable chromatin in the center of the forespore or "spore field" which has already been described. This sequence of events is diagrammatically illustrated by sketched bacterial forms in Fig. 21 a–i which is taken, slightly modified, from Young's thesis. Convincing photographs of fusion cylinders in bacilli approaching spore formation will be found there and also in the paper by Flewett.[56]

It is obvious from the existence of so many independent and concordant accounts that the chromatin bodies of sporing bacilli may, for a time, fuse into a single cordlike structure. There is also little doubt that this phenomenon is best studied in synchronously growing cultures of the kind which Young[9] has used so rewardingly.

The significance of the temporary formation of axial fusion cylinders is hard to assess. Two different explanations suggest themselves. First, fusion may have an important, if obscure, genetic function. That the two nuclei involved stem from the division of the same mother nucleus would not deprive their fusion from genetic usefulness if they were diploid and underwent meiosis before pairing, in the manner of the incestuous gametes of *Actinophrys*.[57] But there is at present no evidence of this happening. A second and very different explanation of the origin of fusion cylinders seems at present more plausible to the writer. The formation of axial filaments of chromatin resembling those described by Badian,[53] Klieneberger-Nobel,[54] Bisset,[55] Flewett,[56] and Young[9] can be provoked in many different ways.[58] In *Escherichia coli* the chromatin fuses into an axial cord when growing cells are transferred to a buffered salt mixture which maintains them alive but prevents further growth.[59] It is perhaps significant that Young, who grew *B. cereus* in an amino-acid yeast-extract medium diluted one in twenty with Grelet's salt mixture, noted that the bacilli cease to multiply just before the appearance of the fusion cylinders and that Flewett[56] found that the addition of 0.05% $CaCl_2$ to the culture medium boosted the incidence of fusion cords from 1 to 15% in one of his strains of *B. anthracis* and more than doubled it in another. The experiences of Whitfield and Murray[58] make it more than likely that the configuration of the chromatin would change if the balance of cations in the sporing bacilli were temporarily upset by a change in the rate of their metabolism. Comparative observations on synchronized sporing cultures growing in media of widely differing salt concentrations would probably shed light on the problem of the significance of the curious fusion cylinders.

B. The Arrangement of Chromatin in the Mature Spore

Chromatin has been stained in cracked spores of *B. megaterium*[60] and in relatively thick sections of spores of this bacillus and of spores of *B. cereus*,[38] digested with ribonuclease, and then stained with Giemsa solution. The chromatin is near the surface of the dormant cytoplasm and arranged there in the form of a few interconnected strands or beads (Fig. 22). This can probably be accepted as true because chromatin is (still) found disposed in this manner in whole mounts of germinating spores prepared by methods which are known not to cause serious artifacts.[61, 62] Areas of low density corresponding in size and distribution to the chromatin of stained preparations have been seen in electron micrographs of sections of spores of *B. polymyxa*,[35, 36] *B. cereus*,[45] and *B. megaterium*[38] (Fig. 18). But chromatin is often indistinct or invisible in electron micrographs of sections of spores, most disappointingly so in *C. pectinovorum*. The helices of chromatin which staining reveals so distinctly in developing spores of this organism (Fig. 20 B) have so far remained invisible in the electron microscope.

The interpretation of the fine structure of the chromatin extruded by spores treated with acid (Fig. 11) is still uncertain. The organization of that material is quite different from that of the chromatin in undisturbed spores.

Average values of the amount of DNA per spore tend to be remarkably constant when batches of spores of the same species grown on different media are compared.[63] In species such as *B. subtilis* and *B. cereus* where the

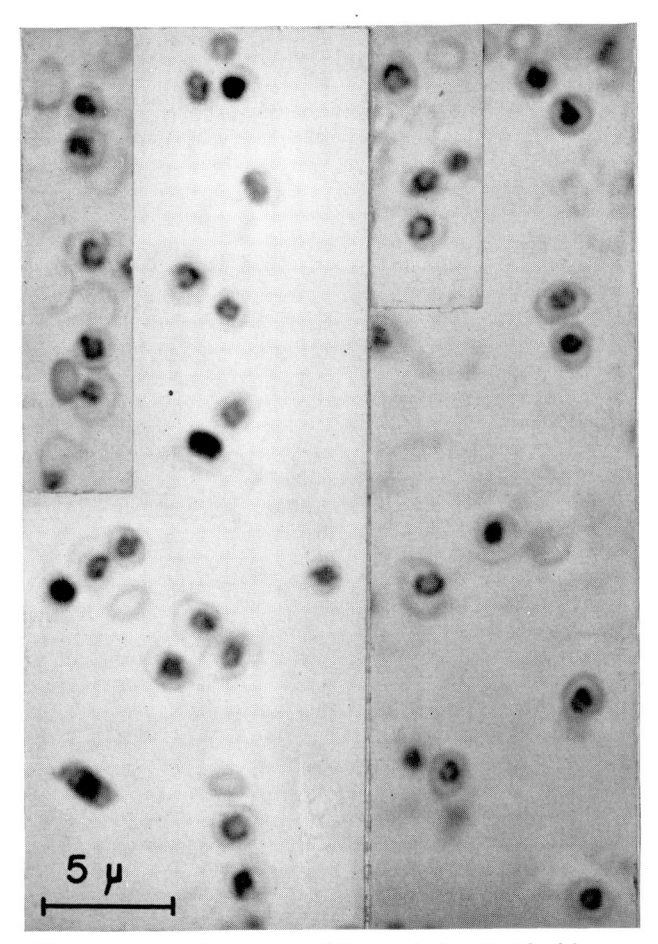

FIG. 22. Chromatin in resting spores of *B. megaterium* that had been cracked open by being shaken with ballotini beads suspended in alcohol. Ribonuclease, Giemsa. Preparations and photomicrographs by Dr. P. C. Fitz-James. (Reprinted, with permission, from Robinow[62].)

size of the spores varies within narrow limits all spores may well contain the same amount of DNA. To judge by the results of the Feulgen procedure this is probably not true of *B. megaterium* which is notorious for great variation in the size and shape of its spores (see Fig. 1 of Knaysi and Hillier (1949)[47] and Figs. 14 and 16 of the present article) and where large spores contain more chromatin than small ones (Fig. 11 B).

IX. Germination

In a suitable environment bacterial spores take up water and undergo other changes which culminate in the rending of the spore coat and the emergence of a growing bacterium. On adequate media the first visible changes, loss of refractility and increase in size, occur within a matter of minutes, but one-half to one hour may elapse before the new bacterium is released from the spore. There are differences of opinion on what should be regarded as the moment of germination. Physiologists date this event from the moment, more or less coinciding with the onset of the first visible changes, at which the spore loses two of its outstanding properties: heat resistance and dryness of its interior, and begins to secrete some of its substance into the medium. Others think that there has not been germination until the new bacterium has slipped out of the spore coat. These differences derive from the fact that germination is a lengthy and complicated process. In the course of it profound changes are initiated in the composition and visible structure of the spore, but all of these events do not occur at the same time and some are independent of each other. In fact, as Powell[64] has pointed out, germination in the physiological sense may be induced in circumstances which do not permit growth. In recent years the chemical events accompanying germination have been closely studied, notably in the laboratories of Dr. Joan Powell, Dr. P. C. Fitz-James, and Dr. H. O. Halvorson and have been discussed at several symposia.[3, 4]

What germinating spores look like has often been described in the past.[65, 66] More recently modes of germination have been carefully studied by Lamanna,[67] who has pointed out that a knowledge of their variety has uses in taxonomy. Excellent, still unsurpassed, electron micrographs of whole germinating spores of *B. megaterium* and other bacilli have been made by Knaysi and Hillier[47, 68] and details of the process of germination have been studied in electron micrographs of thin sections.[37, 69] Successive stages of the germination of spores of *B. subtilis*, *B. megaterium*, *B. cereus*, and *C. pectinovorum* are illustrated in Figs. 23–26. In three of these examples ordinary microscopy sufficed to show that the emerging vegetative cell discards its spore coat; only in *B. cereus* was it necessary to use phase-contrast microscopy to illustrate this point. The *B. subtilis* sequence requires no comment, the other series will be briefly explained.

B. megaterium. In an earlier section it has been described how in several species a stratified cortex intervenes between the dormant protoplasmic core and the spore coat. In living spores of *B. megaterium* examined early in germination the cortex can be clearly distinguished from the more refractile core as well as from the spore coat (Fig. 24 A and B). By contrast, in viable *resting* spores, examined in water, coat, cortex, and core can be distinguished from each other only with difficulty or not at all. Their clear separation early in germination and while the core is still shiny suggests that water is taken up at this time by the cortex. As ger-

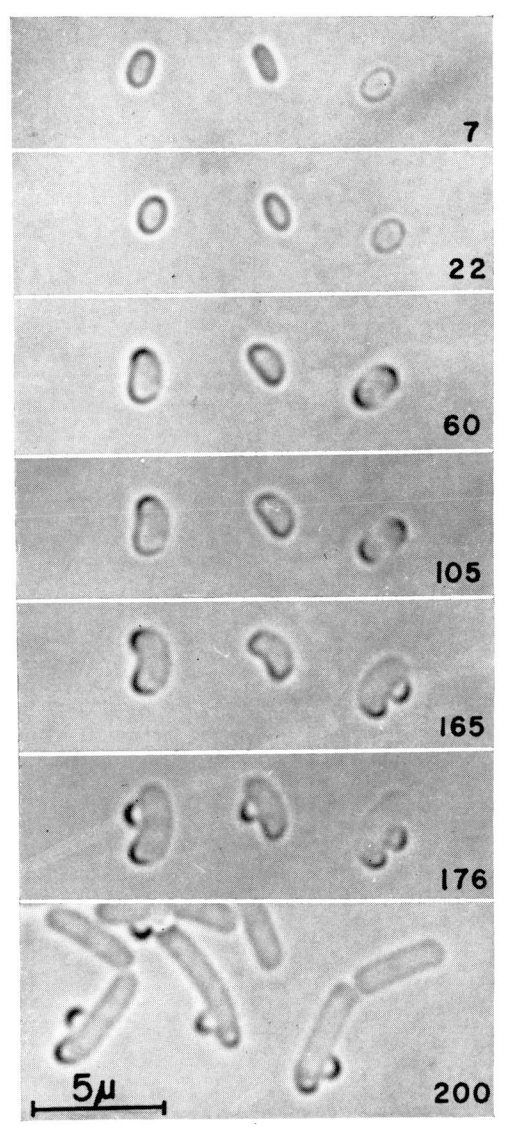

FIG. 23. Successive stages of the germination of a group of living spores of *B. subtilis*. Numbers give minutes since the start of the experiment (× 3600).

mination proceeds the cortex becomes thinner and fades from view. This corroborates the evidence of disintegration of the cortex found in electron micrographs of sections of germinating spores of *B. megaterium*. Appearances have been seen in these micrographs which are compatible with the

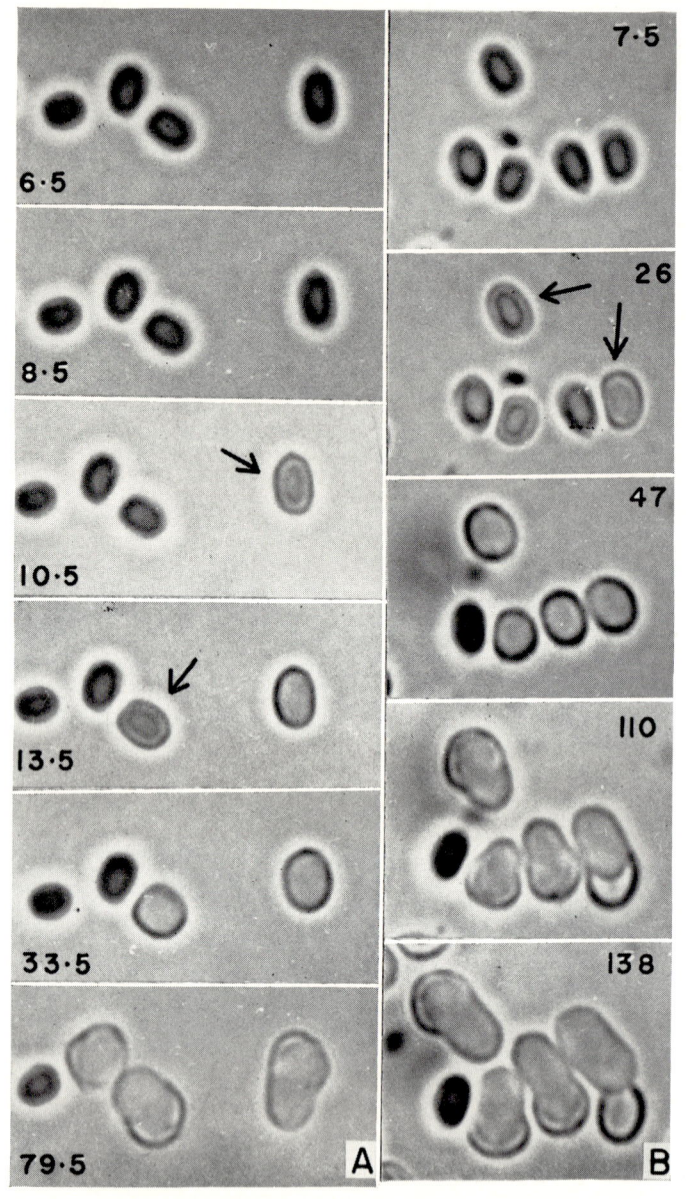

FIG. 24. Photographs illustrating successive stages in the germination of two groups of living spores of *B. megaterium*. Arrows are pointing to spores in which the cortex has become visible as a white band inside the dark contours of the spore coat. In the spore pointed out in the third picture of Column A, the cytoplasmic core is just beginning to lose its bright refractility. In the picture that follows, the cortex has become indistinct. Numbers give minutes passed since the start of the experiments ($\times$ 3600).

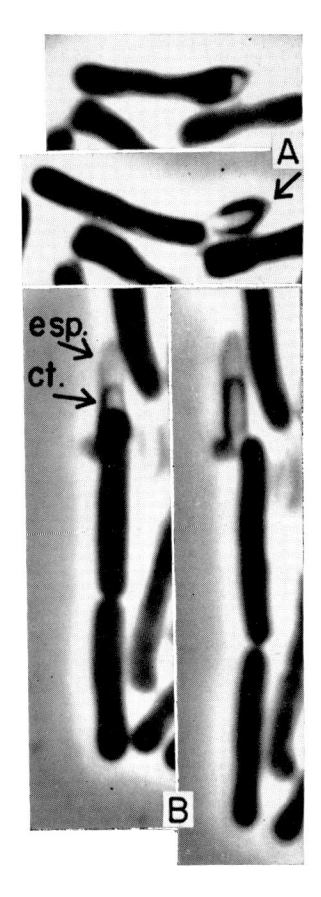

FIG. 25. *B. cereus.* Two pairs of photographs, in each of which the second one was taken five minutes after the first one. In the two B pictures it can be seen that the husk discarded by the bacillus, long after germination, consists of spore coat (ct) and exosporium (esp). Phase-contrast microscopy (× 3900).

idea that part of the cortex breaks down, inside the spore coat, and is lost from the spore and that the remainder contributes materials to the synthesis of the wall of the new vegetative cell[37] (a suggestion made by Dr. C. S. Cummins). In their turn these conclusions suggest that the cortex may be the site of some of the materials which have been shown to be excreted by the spore in large amount early in germination.[70, 71] But proof of this is lacking.

B. cereus. In this species, as is well known, the spore coat is delicate, hard to see with the ordinary microscope,[69] and liable to remain attached, together with the still more delicate exosporium, to one end of the emerging bacillus until long after the latter's first division. It is easy to understand that it should sometimes have been thought of spores which behave like those of *B. cereus* that they germinate by absorption of the spore coat into the fabric of the wall of the vegetative cell. That this ever happens seems unlikely in view of the fact that the new cell wall is derived from the spore

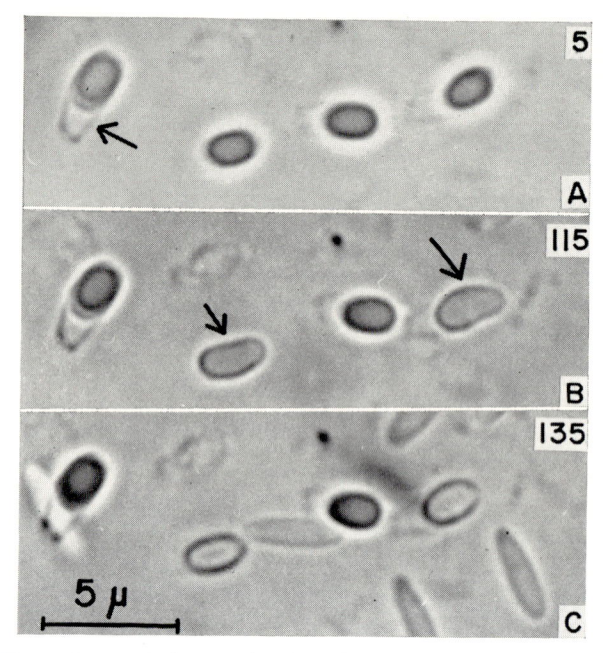

FIG. 26. Three phases of the germination of two spores of *C. pectinovorum*. In B, a cell is beginning to emerge from each of the spores indicated by arrows. In C, 20 minutes later, the spore cases are empty and there are several cigar-shaped, free, vegetative forms, the central one of which probably came from the empty spore on its left. Note the remains of the sporangium which still surround the spore near the left edge of picture A–C. The scale represents 5 microns. Numbers give minutes passed since the start of the experiment.

wall (not the spore *coat*) early in germination, at a time when the coat or coats are still largely intact. Spore coat and new cell wall are separate structures from the start and apparently differ also in composition;[72] absorption of the former by the latter is therefore highly improbable. Whether the spore coat of *B. cereus* (and of other species) ultimately disintegrates—and how—is another question and one which is at present not easily answered.

C. pectinovorum. Early stages of the germination of the spores of this species can be followed with the eye, but the moment of emergence of the new cell from the spore requires a motion picture for its adequate illustration because it happens with explosive swiftness. The liberated cell moves about very actively from the start. The violence of its extrusion is an indication of the rigidity of the spore coat and of the narrowness of the aperture through which the emerging cell is forced to make its escape.

X. The Chromatin of Germinating Spores

The behavior of the chromatin during germination has been described several times.[61, 62, 73, 74] It does not greatly differ from the behavior of the chromatin in growing vegetative cells many generations removed from germination and is in both instances equally difficult to describe with accuracy. In *B. megaterium*, *B. subtilis*, and *B. cereus* the chromatin divides early during germination and at a fast rate thereafter. In all instances the germinating cell already contains two to four separate chromatin structures when it first assumes the shape of a straight rod. The first division of the chromatin in germinating spores of *B. megaterium* is illustrated in Fig. 27.

XI. Parasporal Bodies

In several kinds of bacilli the spore is attended by a companion body which arises in the same cell and remains close to the spore when the latter is released from the mother bacterium.[75-77] Some of these bodies resemble spores, but their composition is simpler than that of protoplasm; they contain no chromatin and never grow into new bacteria. "Parasporal bodies," as the satellite structures are called, are of different kinds. Of special interest are the octrahedral crystals of protein, one per sporing cell, which Hannay recently discovered in *B. thuringiensis*.[78, 79] Rhomboid bodies accompanying the spores of this bacillus had been seen, but not recognized for what they are, by earlier workers who took them for the shriveled remains, or "Restkoerper" of the vegetative portion of the bacillus.[80, 81] Hannay showed not only that this body is a bipyramidal crystal of protein, but also that it is formed new inside the cytoplasm and grows there together with the spore.

A chain-forming bacillus, which has been maintained in culture since its isolation from cow dung eleven years ago and which in this laboratory is known as "*B. medusa*," forms parasporal bodies (Fig. 28) that are usually

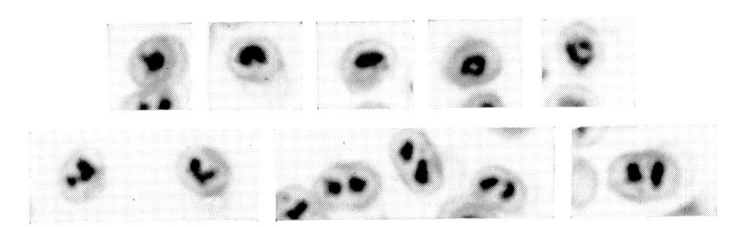

FIG. 27. Chromatin in germinating spores of *B. megaterium*. OsO_4, HCl, Azure A/SO_2. Division of the chromatin has begun in the fifth cell in the top row and has been completed in the barrel-shaped rodlet beneath it ($\times$ 3600). (Reprinted, with permission, from Robinow[62].)

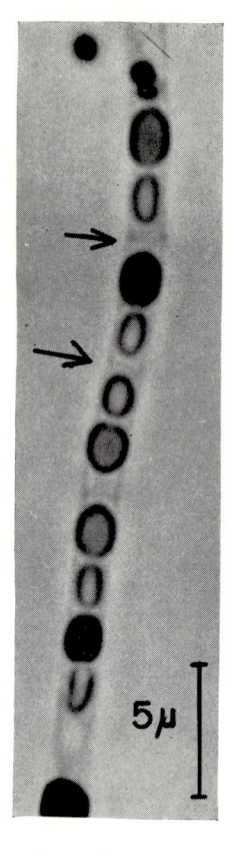

Fig. 28. Pairs of spores and parasporal bodies in cells of "*B. medusa.*" The arrows point to cell boundaries. Above the top arrow is a spore, beneath it is a darkly stained parasporal body. Air-dried film from a two-day-old culture on potato agar. Stained with 0.05% acid fuchsin in 2% acetic acid.

larger than the spore which they accompany and that differ from the protein crystals of *B. thuringiensis* by being enclosed in a distinct membrane. The Medusa bodies consist largely of protein and, like the protein crystals, are free from detectable phosphorus[82] and are unstable in alkali.

Yet another kind of companion structure confers on the spores of *B. laterosporus* a shape which may be described as that of an egg resting in a canoe appropriate to its size (Fig. 29 A and B).[83] The canoe portion of this assembly stains deeply with ordinary dyes and has a strong affinity for the Giemsa stain, even after hydrolysis with N/HCl. Figure 29 A shows unequivocally that the deeply staining parasporal canoes arise in the cytoplasm in closest contact with, if not indeed as part of, the spore. Germinating bacilli, as might be expected, escape from the spore through a tear in the ordinary, thinner half of its coat (Fig. 29 C). In electron micrographs of thin sections the canoe appears as a laminated and greatly thickened part of the spore coat.

The protein crystals of *B. thuringiensis* and of the related *B. alesti* have

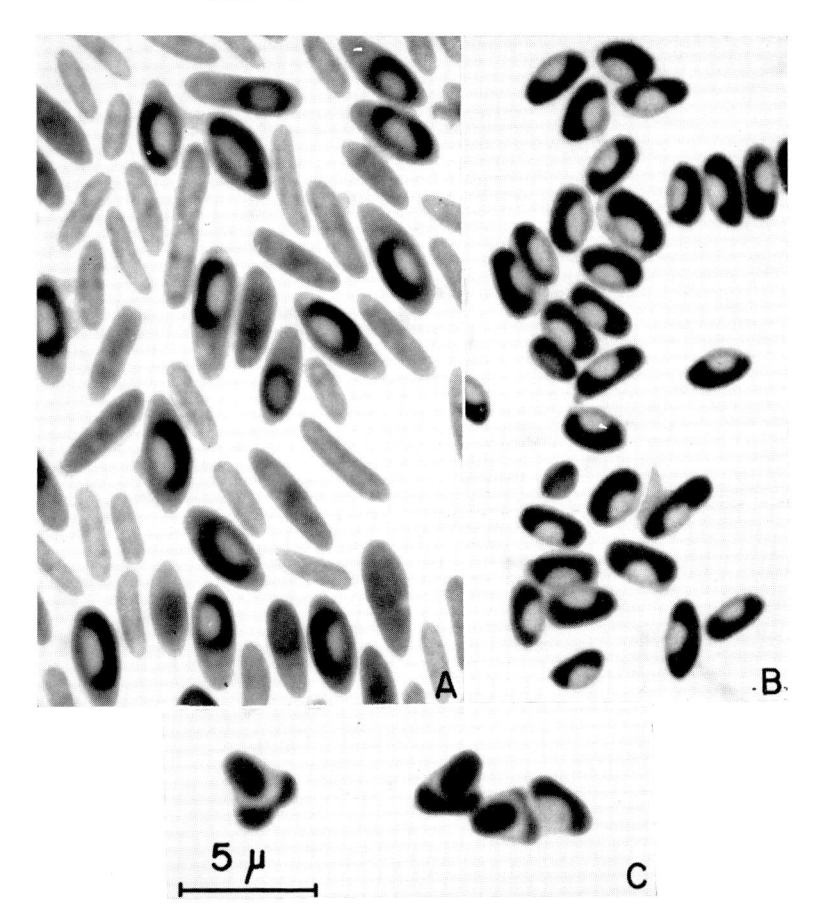

FIG. 29. *B. laterosporus*. A: Sporing bacilli; B: free mature spores; and C: germinating spores. Photograph A shows that the canoe-shaped parasporal bodies arise together with the spore in the cytoplasm of the mother bacterium. All briefly stained with 0.02% basic fuchsin. (Reprinted, with permission, from Hannay[83].)

already been the subject of detailed physicochemical studies.[84, 85] Some aspects of the chemistry of the canoes of *B. laterosporus* have recently been described by Fitz-James and Young.[86]

XII. Conclusion

The secret of the spore's great hardiness rests in the physical and chemical properties of its protoplasm. These can hardly yet be related to the parts of spores that can be distinguished by microscopy, but it is to be hoped that chemical and anatomical features of the spore will ultimately become capable of description in the same language.

References

[1] G. S. Graham-Smith, *J. Hyg.* **30**, 213 (1930).
[2] W. G. Murrell, "The Bacterial Endospore." Sidney Univ. Press, Sidney, 1955.
[3] H. O. Halvorson, ed. *Am. Inst. Biol. Sci., Publ. No.* **5**, 164 pp. (1957).
[4] S. E. Jacobs and L. F. L. Clegg, eds., Symposium on the Formation and Germination of Bacterial Spores, *J. Appl. Bacteriol.* **20**, 305 (1957).
[5] N. Grelet, *Ann. inst. Pasteur* **81**, 1 (1951).
[6] N. Grelet, *Ann. inst. Pasteur* **82**, 14 (1952).
[7] N. Grelet, *Ann. inst. Pasteur*, **82**, 310 (1952).
[8] N. Grelet, *Ann. inst. Pasteur* **83**, 71 (1952).
[9] I. E. Young, "Chemical and Morphological Changes During Sporulation in Variants of Bacillus Cereus." Ph. D. Thesis, Univ. Western Ontario, London, Canada (1958).
[10] T. Gibson, *Arch. Mikrobiol.* **6**, 73 (1935).
[11] B. Collin, *Arch. zool. exptl. et gén.* **51**, 59 (1913).
[12] B. Delaporte, Plate VI, Figs. 1–15 in *Rev. gén. botan.* **51**, 615, 689, 748 (1939).
[13] E. Chatton and C. Pérard, *Compt. rend. soc. biol.* **74**, 1159 (1913).
[14] E. G. Pringsheim, *Bacteriol. Revs.* **13**, 47 (1949).
[15] E. Chatton and C. Pérard, *Compt. rend. soc. biol.* **74**, 1232 (1913).
[16] C. F. Robinow, *Z. Tropenmed. u. Parasitol.* **8**, 225 (1957).
[17] Personal communication from Dr. Avivah Zuckerman, Department of Parasitology, Hebrew University, Jerusalem, 1957.
[18] B. Delaporte, *Bull. Soc. Bot. France* **103**, 521 (1956).
[19] K. A. Bisset, *J. Gen. Microbiol.* **13**, 442 (1955).
[20] A. von Klinckowstroem, *Arkiv Bot.* **23A**, 1 (1931).
[21] S. Bayne-Jones and A. Petrilli, *J. Bacteriol.* **25**, 261 (1933).
[22] R. W. G. Wyckoff and A. L. Ter Louw, *J. Exptl. Med.* **54**, 451 (1931).
[23] G. Knaysi, *J. Appl. Bacteriol.* **20**, 425 (1957).
[24] Personal communication from Mr. P. R. Sweeny, M.Sc., 1958.
[25] J. W. Foster and J. J. Perry, *J. Bacteriol.* **67**, 295 (1954).
[26] J. W. Foster, *Quart. Rev. Biol.* **31**, 102 (1956).
[27] K. F. A. Ross and E. Billing, *J. Gen. Microbiol.* **16**, 418 (1957).
[28] Product of R. P. Cargille Laboratories Inc., New York, New York.
[29] V. Vinter, *J. Appl. Bacteriol.* **20**, 325 (1957).
[30] N. Grelet, *Ann. inst. Pasteur.* **83**, 71 (1952).
[31] R. P. Cook, *Biol. Revs. Biol. Proc. Cambridge Phil. Soc.* **7**, 1 (1932).
[32] I. M. Lewis, *J. Bact.* **28**, 133 (1934).
[33] G. Knaysi, *Bacteriol. Revs.* **12**, 19 (1948).
[34] T. Hashimoto and H. B. Naylor, *J. Bacteriol.* **75**, 647 (1958).
[35] A. van den Hooff and S. Aninga, *Antonie van Leeuwenhoek. J. Microbiol. Serol.* **22**, 327 (1956).
[36] N. C. Dondero and P. E. Holbert, *J. Bacteriol.* **74**, 43 (1957).
[37] Figures 4 and 9 in B. H. Mayall and C. F. Robinow, *J. Appl. Bacteriol.* **20**, 333 (1957).
[38] C. F. Robinow, *J. Bacteriol.* **66**, 300 (1953).
[39] Plate 3, Figs. b, c, in J. G. Franklin and D. E. Bradley, *J. Appl. Bacteriol.* **20**, 467 (1957).
[40] C. F. Robinow, *J. Gen. Microbiol.* **5**, 439 (1951).
[41] J. F. Powell, *J. Appl. Bacteriol.* **20**, 349 (1957).

[42] P. C. Fitz-James, C. F. Robinow, and G. H. Bergold, *Biochim. et Biophys. Acta* **14,** 346 (1954).
[43] A. Meyer, *Flora* **84,** 185 (1897).
[44] D. E. Bradley, *Brit. J. Appl. Phys.* **5,** 65 (1954).
[45] G. B. Chapman, *J. Bacteriol.* **71,** 348 (1956).
[46] C. L. Hannay, *Symposium Soc. Gen. Microbiol.* **6,** 318 (1956).
[47] G. Knaysi and J. Hillier, *J. Bacteriol.* **57,** 23 (1949).
[48] J. Tomcsik and J. B. Baumann-Grace, *Abstracts Intern. Congr. Microbiol., 7th Congr. Stockholm,* **1958.**
[49] Personal communication from Dr. P. C. Fitz-James, 1958.
[50] C. F. Robinow, *Symposium, Soc. Gen. Microbiol.* **6,** 181 (1956).
[51] G. Knaysi, *J. Bacteriol.* **69,** 117 (1955).
[52] T. Hashimoto, S. H. Black, and P. Gerhardt, *Bacteriol. Proc. (Soc. Am. Bacteriol.)* abstr. G 35, p. 38 (1959).
[53] J. Badian, *Arch. Mikrobiol.* **4,** 409 (1933).
[54] E. Klieneberger-Nobel, *J. Hyg.* **44,** 99 (1945).
[55] K. A. Bisset, *J. Gen. Microbiol.* **4,** 1 (1950).
[56] T. H. Flewett, *J. Gen. Microbiol.* **2,** 325 (1948).
[57] K. Belar, *Ergeb. Fortschr. Zool.* **6,** 235 (1926).
[58] J. F. Whitfield and R. G. E. Murray, *Can. J. Microbiol.* **2,** 245 (1956).
[59] E. Kellenberger, A. Ryter, and J. Séchaud, *J. Biophys. Biochem. Cytol.* **4,** 671 (1958).
[60] P. C. Fitz-James, *J. Bacteriol.* **66,** 312 (1953).
[61] C. F. Robinow, *J. Bacteriol.* **65,** 378 (1953).
[62] C. F. Robinow, *Bacteriol. Revs.* **20,** 207 (1956).
[63] P. C. Fitz-James, *Am. Inst. Biol. Sci., Publ. No.* **5,** 85 (1957).
[64] E. O. Powell, *J. Appl. Bacteriol.* **20,** 342 (1957).
[65] A. de Bary, "Vergleichende Morphologie und Biologie der Pilze, Mycetozoen und Bakterien." Leipzig, 1884.
[66] L. Klein, *Zentr. Bakteriol. Parasitenk. Abt. I. Orig.* **6,** 377 (1889).
[67] C. Lamanna, *J. Bacteriol.* **40,** 347 (1940).
[68] G. Knaysi, R. F. Baker, and J. Hillier, *J. Bacteriol.* **53,** 525. The spores described in this paper are said to be those of *B. mycoides*, but their lack of an exosporium and their mode of germination make it likely that they are, in fact, spores of *B. subtilis*.
[69] G. B. Chapman and K. A. Zworykin, *J. Bacteriol.* **74,** 126 (1957).
[70] J. F. Powell and R. E. Strange, *Biochem. J.* **54,** 205 (1953).
[71] R. E. Strange and J. F. Powell, *Biochem. J.* **58,** 80 (1954).
[72] N. Yoshida, Y. Izumi, I. Tani, S. Tanaka, K. Takaishi, T. Hashimoto, and K. Fukui, *J. Bacteriol.* **74,** 94 (1957).
[73] B. Delaporte, *Advances in Genet.* **3,** 1 (1950).
[74] M. E. Hunter and E. D. DeLamater, *J. Bacteriol.* **63,** 23 (1952).
[75] S. R. Dutky, *J. Agric. Research* **61,** 57 (1940).
[76] E. H. Fowler and J. A. Harrison, *Bacteriol. Proc. (Soc. Am. Bacteriologists Proc.)* p. 30 (1952).
[77] A. M. Heimpel and T. A. Angus, *Can. J. Microbiol.* **4,** 531 (1958).
[78] C. L. Hannay, *Nature* **172,** 1004 (1953).
[79] C. L. Hannay, *Symposium Soc. Gen. Microbiol.* **6,** 318 (1956).
[80] E. Berliner, *Z. angew. Entomol.* **2,** 29 (1915).
[81] O. Mattes, *Sitzber. Ges. Beförd. ges. Naturw. Marburg* **62,** 381 (1927).

[82] Personal communication from Dr. P. C. Fitz-James, 1958.
[83] C. L. Hannay, *J. Biophys. Biochem. Cytol.* **3,** 1001 (1957).
[84] C. L. Hannay and P. C. Fitz-James, *Can. J. Microbiol.* **1,** 694 (1955).
[85] P. C. Fitz-James, C. Toumanoff, and I. E. Young, *Can. J. Microbiol.* **4,** 385 (1958).
[86] P. C. Fitz-James and I. E. Young, *J. Biophys. Biochem. Cytol.* **4,** 639 (1958).

Chapter 6

Bacterial Protoplasts

Kenneth McQuillen

	Page
I. Concepts and Definitions	250
A. Introduction	250
B. Action of Lysozyme on Certain Gram-Positive Species	253
C. Spherical Forms of Gram-Negative Species	255
D. Definition of the Term Protoplast	257
E. Properties of Protoplasts and Consequences of the Absence of Cell Walls	258
II. Formation of Protoplasts	261
A. The Use of Lysozyme and Related Enzymes for Preparation of Protoplasts of Certain Gram-Positive Species	261
B. Autolytic Production of Protoplasts	265
C. The Use of Lysozyme and Related Enzymes for Preparation of "Protoplasts" of Certain Gram-Negative Species	267
D. Metabolic Disturbance as a Method of "Protoplast" Formation in Gram-Negative Species	271
E. Some Factors Common to "Protoplast" Formation in Gram-Negative Species such as *Escherichia coli*	279
F. The Use of Glycine for Preparation of "Protoplasts" of Gram-Negative Species	280
G. "Protoplasts" of Yeast and Other Fungi	281
III. Morphology and Structure	282
A. Spherical Forms	282
B. Number of Protoplasts per Cell	283
C. Phase-Contrast Microscopy of Protoplasts of Gram-Positive Species	283
D. Phase-Contrast Microscopy of "Protoplasts" of Gram-Negative Species	284
E. Fixation of Protoplasts	286
F. Staining of Protoplasts	286
G. Electron Microscopy	287
IV. Physicochemical Properties of Protoplasts	288
A. Stabilizing Media	288
B. Osmotic Properties of Protoplasts	289
C. Chemical Damage to the Protoplast Membrane	292
D. Electrophoretic Studies on Protoplasts	295
V. Composition of Protoplasts	295
A. *Bacillus megaterium*	295
B. Chemical Composition of "Protoplasts" of *Escherichia coli*, L-Forms of *Proteus vulgaris*, and Pleuropneumonia-Like Organisms (PPLO)	297
C. The Protoplast Membrane	299
D. Other Components of Protoplasts	305
E. Antigenic Properties of Protoplasts	306
F. Effects of Enzymes on Protoplasts	308

	Page
VI. Physiology and Biochemistry of Protoplasts	311
A. Respiration	311
B. Incorporation of Radioactive Tracers	312
C. Synthesis of Enzymes by Protoplasts	317
D. Capabilities of Osmotically Shocked Protoplasts	319
E. Inducible Enzyme Formation in "Protoplasts" of *Escherichia coli*	322
F. Nitrogen Fixation and Oxidation by "Protoplasts" of *Azotobacter vinelandii*	323
G. Growth of Bacteriophages in Protoplasts	324
H. Bacteriophage Growth in "Protoplasts" of *Escherichia coli*	329
I. Transformation of "Protoplasts"—the Redintegration Phenomenon in *Escherichia coli*	336
J. Spore Formation in Protoplasts of *Bacillus megaterium*	338
K. Protoplasts Derived from Motile Species	338
L. Growth of Protoplasts	339
M. Reversion of Spherical Forms to Normal Morphology	343
N. Conjugation of Protoplasts	346
O. Natural Occurrence of Protoplast-Like Forms	346
References	353

I. Concepts and Definitions

A. Introduction

The component parts of most bacterial cells have dimensions in the range 100–1000 A.; since the resolving power of the light microscope is about 2000 A., it is remarkable how much information had been obtained before the development of the techniques of electron microscopy. Much of this came from the use of staining techniques but some was purely speculative, being based on analogy with the structure of larger cells. At the present time bacterial anatomy is a rapidly developing subject and is characteristically beset with terminological difficulties. Phrases are coined (or borrowed) for describing structures or functional units which are as yet incompletely identified as to location, constitution, and function. Cell wall, cell membrane, plasma membrane, cell barrier, cell envelope, osmotic barrier, cytoplasmic membrane, protoplast membrane, gymnoplast wall are all terms which are used for structures believed to surround the cytoplasm of a bacterial cell. There is not yet general acceptance (or indeed knowledge) of what is what. An attempt will, therefore, first be made to define the concept of protoplast.

It is widely agreed that a rigid or semirigid envelope maintains the shape (coccoid, rod-shaped, etc.) of the bacterial cell; this structure is usually called the *cell wall* (see Chapter 3). It can be removed by various techniques and isolated preparations show it to retain its three-dimensional form even after the cytoplasmic contents of the cell have been dispersed. There has, however, long been controversy as to whether or not the cytoplasm is bounded by a membrane (*cytoplasmic membrane, plasma membrane*) which underlies the cell wall proper. At present, a majority of writers appears to postulate the existence of such a structure. Among the arguments brought forward is the fact that many kinds of bacterial cells can be plasmolyzed when placed in hypertonic media. This occurs more readily with Gram-negative than with Gram-positive species, the cytoplasm contracting away from the cell wall. Only solutes which do not penetrate can bring about plasmolysis and the implication is that below the cell wall there exists an osmotic barrier which is semipermeable and which bounds the contracting cytoplasm during plasmolysis. Sometimes very high solute concentrations have to be used and some species of bacteria apparently cannot be plasmolyzed.

In certain media a phenomenon occurs which was called "plasmoptysis" by Fischer.[1,2] The cell wall ruptures locally and the cytoplasm is extruded to form a bubble, which either remains attached to the empty cell wall or becomes free as a spherical mass. Stähelin[3,4] has shown that such forms prepared from strain A_1 of *Bacillus anthracis* have osmotic properties suggestive of the presence of a surrounding membrane.

Staining methods were not particularly useful in the investigation of the possible presence of a cytoplasmic membrane until Robinow and Murray[5] developed a technique which differentiated cell wall and cytoplasm in certain species of Gram-positive bacteria. The method employs Victoria blue and its application shows the existence of a very thin, well-defined, intensely staining zone underlying the cell wall and at the surface of the cytoplasm.

Electron microscopy of intact bacteria shows clearly the cell wall but cannot be expected necessarily to indicate underlying membranes. In some preparations there appear to be two structures but it is uncertain whether they are independent membranes or two parts of the same cell wall. It was hoped that the technique of ultrathin sectioning of specimens would give a definitive answer, but so far it has not done so.[6] In only a few published works do authors claim that electron micrographs of thin sections demonstrate the existence of two separate envelopes in vegetative cells. Chapman and Hillier,[7] in their first report on thin sections (1953), concluded from a study of *Bacillus cereus* that "it appears that the cytoplasmic membrane is a surface phenomenon, an optical effect, or both." Tomlin and May,[8] however, reported the presence of "a very obvious plasma membrane" in a section of *Escherichia coli* and suggested that it was somewhat less than 100 A. in thickness. Murray[9] discussed the evidence for the existence of cytoplasmic membranes and pointed out that a section of *E. coli* by Birch-Andersen *et al.*[10] showed what might be such a structure, separate from the double-contoured cell wall. He also mentioned that out of hundreds of his own embeddings, the vast majority showed no sign of a cytoplasmic membrane. However, it did appear in a spirillum and was about 60 A. in thickness. Also in one out of more than a hundred embeddings of *Bacillus cereus* there was seen a very thin membrane (25–50 A.). Although never found by Murray in *E. coli*, *Caryophanon latum*, *Corynebacterium* spp., *Nocardia* spp., or *Micrococcus* spp., he believes this is due to imperfect technique rather than absence of "this physiologically important region of the cell."[9] Chapman and Kroll[11] have also reported the existence of a membrane (50–100 A.) in *Spirillum serpens*. Beautiful electron micrographs of ultrathin sections of *Escherichia coli* by Kellenberger and Ryter[181] gave a clear indication of the separate structures, the cell wall, and the cytoplasmic membrane. Similar structures were demonstrated in the photosynthetic organisms, *Rhodospirillum rubrum* and *Chlorobium limicola* by Vatter and Wolfe[182] and in *Bacillus subtilis* by Tokuyasu and Yamada.[183] The steadily increasing number of positive findings is, perhaps, more significant than the very large number of negative ones.

The relationship between osmotic function and cell structure in bacteria has been discussed by Mitchell and Moyle[12] who concluded that "both Gram-positive and Gram-negative bacteria possess an osmotic barrier impermeable to many small molecular weight solutes. The osmotically functional part of this barrier corresponds to the plasma membrane which lies close within the cell wall and is mechanically supported by it."

The over-all picture, then, of the bacterial cell is of cytoplasm bounded by a cytoplasmic membrane (functionally the main osmotic barrier), and this surrounded by the cell wall (Fig. 1). On plasmolysis the cytoplasm and its membrane contract away from the cell wall. This applies particularly to Gram-negative species, which can be readily plasmolyzed,

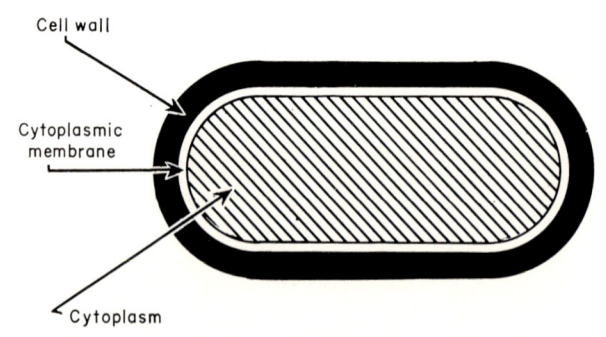

Fig. 1. Representation of a bacterial cell indicating the probable dispositions of the cell wall and the cytoplasmic membrane. The cell wall may be 10–25 mμ in thickness and the cytoplasmic membrane 5–10 mμ.

although there is more direct evidence for the existence of two separate and separable envelopes, the wall and the membrane, in certain gram-positive bacteria (see below). Hale[184] studied the plasmolysis of various Gram-positive and Gram-negative species using saturated $Ca(OH)_2$ at 90–100°C followed by staining with crystal violet. The Gram-negatives exhibited general plasmolysis except for *Azotobacter* and *Rhizobium* which behaved more like the Gram-positive genera and retracted only at the poles or else collapsed internally.

The term *protoplast* has been used since 1953 by Weibull,[13-15] Tomcsik and Baumann-Grace,[16] McQuillen,[17] and others, and the term *gymnoplast* by Stähelin,[4] and Colobert and Lenoir[18] to denote the structure remaining when the cell wall is removed from a bacterial cell. The word protoplast has also been used in other rather different senses (see Section I, C).

B. Action of Lysozyme on Certain Gram-Positive Species

Tomcsik and Guex-Holzer[19] studied the lysis of cells of a strain of *Bacillus megaterium* (*Bacillus* M) by the enzyme lysozyme. The experiments were carried out in physiological saline and demonstrated that the rods swelled, the cytoplasm became rounded within the cell wall which was gradually digested away by the enzyme, and spherical masses of cytoplasm were released into the medium. These later changed to empty membranes, which then disintegrated further. There is now no doubt but that the spherical masses were true protoplasts.

Independently, and at about the same time, Weibull[13] studied the action of lysozyme on *Bacillus megaterium* KM suspended in media containing sucrose (0.1–0.2 M) or polyethylene glycol (7.5% w./v.). Under these conditions the cell wall was completely digested and spherical protoplasts—two or three from each rod—were released into the medium, where they were stable for long periods in the absence of dilution and mechanical agitation (Fig. 2). Dilution of the medium with water resulted in lysis of the protoplasts, leaving empty membranes or "ghosts." These protoplast membranes are believed to represent the structure which, in intact bacterial cells, is called the cytoplasmic membrane and which probably functions as the osmotic barrier (Fig. 23 a, b, c; Fig. 24; Fig. 25 a, b, c).

Similar behavior, that is, the digestion of the cell wall by lysozyme and the release of spherical structures stable in media of suitable composition has been reported for *Micrococcus lysodeikticus*,[20-22] *Sarcina lutea*,[20, 21] *S. flava*,[18] and *Bacillus subtilis*.[23] However, relatively few bacterial species are amenable to such treatment and all are Gram-positive. Many workers have indicated that the site of action of lysozyme is the cell wall.[24] The isolated cell walls of *Bacillus megaterium*, *M. lysodeikticus*, and *S. lutea* are

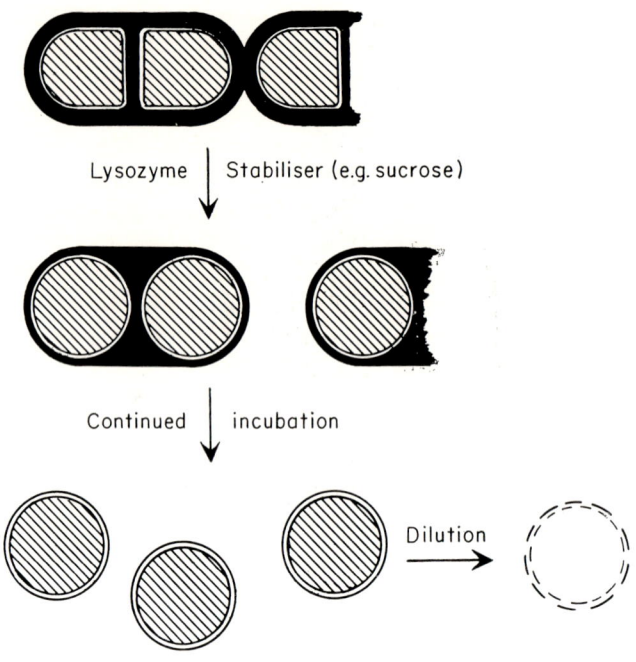

FIG. 2. Representation of formation of spherical protoplasts by lysozyme treatment of rod-shaped forms of *Bacillus megaterium*.

rendered completely soluble by lysozyme. Salton[25] and, more extensively, Cummins[26,27] have shown that the cell walls of Gram-positive bacteria are mucopolysaccharide, containing sugars, amino sugars, glutamic acid, alanine, lysine or diaminopimelic acid (DAP), with sometimes glycine, serine, or aspartic acid. Protein and lipid are virtually absent, unless a dispensable antigenic protein, such as the M protein of *Streptococcus faecalis*, is present. These proteins can be removed without impairing the viability of the cells or the structure of the wall. The peptide constituents of the mucopolysaccharide are distinguished by containing amino acids such as D-glutamic acid, D-alanine and DAP which have never been found in genuine protein.[28] A new class of compounds—teichoic acid fraction—discovered by Baddiley and his colleagues[185-187] as components of the cell walls of some Gram-positive species, consists of polymers of ribitol phosphate. Some, or all, of the ribitol units carry α-glucose residues (in *Staphylococcus aureus*) or *N*-acylglucosaminyl residues (in *Bacillus subtilis* and *Lactobacillus arabinosus*). D-Alanine, also present, is attached by an ester linkage to each sugar residue. Most or all of the phosphate of some Gram-positive cell wall preparations appears to occur in the teichoic acid fraction. This is probably negligible in amount in some strains of *Bacillus megaterium* and in *Micrococcus lysodeikticus* although substantial in some strains of *Bacillus subtilis*. This may mean that lysozyme digestion results in true protoplast release only from cells which do not contain teichoic acid in their cell wall structure. Alternatively, removal of the mucopeptide substrate of lysozyme might in some instances allow release of the teichoic acid. Lysozyme, which attacks only certain of these Gram-positive species, releases some small molecular weight fragments, the remainder of the wall being degraded to pieces about 10,000 to 20,000 in molecular weight.[24]

In the case of *Bacillus megaterium* it has been established that digestion with lysozyme in suitable media results in the removal of the entire cell wall as identified microscopically, chemically, immunochemically, and by its ability to react with bacteriophage. This is probably also true for *Micrococcus lysodeikticus* and *Sarcina lutea*. However, certain strains of *Bacillus subtilis*, although closely similar to *B. megaterium*, possess cell walls which are only partly degraded by lysozyme;[29] after exhaustive treatment of isolated walls of *B. subtilis* there still remains a structure of considerable integrity.[24] This, perhaps, explains the greater stability of "protoplasts" of *B. subtilis* as compared with protoplasts of *B. megaterium*.[17, 23]

Abrams[188] found that cell walls of *Streptococcus faecalis* 9790 contained O-acetyl groups but that the walls were susceptible to digestion by lysozyme (about 70% reduction in turbidity). Development of resistance to lysozyme by *Micrococcus lysodeikticus* has been found to be related to an increase in the O-acetyl content of the cell wall. Brumfitt *et al.*[189] showed that resistant organisms could be restored to lysozyme-sensitivity without impairing viability by displacing such O-acetyl groups by treatment with alkaline buffers (e.g. pH 11.4). Organisms inherently resistant to lysozyme such as *Staphylococcus aureus* were not rendered susceptible even after removal of these groups.

C. Spherical Forms of Gram-Negative Species

Recently there have been reports referring to the preparation of protoplasts (*sic*) of Gram-negative organisms such as *Escherichia coli* and *Salmonella typhimurium*. The criterion for protoplast formation is taken explicitly to be conversion of a rod-shaped organism to a spherical form in media of suitable composition (usually containing 5–20% sucrose) and osmotic sensitivity of these structures, coupled sometimes with inability of these forms to produce colonies when plated on the usual bacteriological media. Occasionally, the term protoplast occurs modified to protoplast-like or by quotation marks to "protoplast" (Fig. 23 d; Fig. 26; Fig. 27).

The methods used for the production of these forms include the use of lysozyme at alkaline pH values, lysozyme and chelating agents, penicillin, DAP-deprival, and bacteriophage enzymes. There is as yet no evidence that in any of these Gram-negative "protoplasts" the cell wall has been completely removed. Although less is known about the chemistry of the walls of Gram-negative than of Gram-positive bacteria, it seems that the wall, as isolated by a variety of techniques, is composed of lipoprotein and lipopolysaccharide, together with the mucopolysaccharide which is the sole component of many Gram-positive cell walls.[25, 27, 28, 30, 31] Mucopeptide components resembling those of Gram-positive cell walls have been isolated from the culture medium of *Escherichia coli* $K_{235}L + O$. O'Brien and Zilliken[190] described peptide derivatives of N-acetyl hexosamine and N-acetyl muramic acid containing lysine, alanine, glutamic acid, and aspartic acid; diaminopimelic acid, alanine, glutamic acid and aspartic acid; and alanine, diaminopimelic acid and glutamic acid. It is likely that cell walls of some Gram-negative organisms (if not all) contain such mucopeptides and that these are the substrates for lysozyme and bacteriophage enzymes.

Treatments which give rise to "protoplasts" of Gram-negative species (penicillin, lysozyme, DAP-deprival, and phage enzymes) probably all act on the mucopolysaccharide component of the wall (see Section II, E); it is probable that this component is responsible for the rigidity and form of the cell wall. Should this constituent be removed or not synthesized, the resulting structure might well be expected to be deformed into an osmotically sensitive sphere. In other words, the "protoplasts" of *Escherichia coli*, etc., may retain their cell wall, modified only insofar as the minor mucopolysaccharide component is absent. Weibull[191] has reviewed the evidence for the presence of cell wall components prepared by several means and Lederberg and St. Clair[192] reported at length on the conversion of *Escherichia coli* via a "protoplast"-like state to L-type growth. This was achieved either by use of penicillin[59] or by withholding diaminopimelic acid from a strain which required it.[31, 32, 86] There is much evidence in Lederberg and St. Clair's paper of the persistence of cell wall components in the spherical or globular forms. Microscopically "an additional envelope" is present which "may well represent some disorganized elements of the cell wall." The preparations were also "still agglutinable by homologous anti-O and anti-K serums." Receptor sites for some but not all of the T-phages appeared to be present and sexual conjugation could occur between compatible strains whether either or both of the parents were in the "protoplast"-like form. Thus "there are strong hints that much residual wall material persists." Madoff and Dienes[193] described the conversion of pneumococci to L-form growth by penicillin (but not by glycine) and discussed the differences between protoplasts and L-forms.

On the other hand, if there exists a cytoplasmic membrane distinct from the cell wall, "protoplast" formation in Gram-negative bacteria may in some instances result from removal of the entire cell wall, just as it does in lysozyme-sensitive Gram-positive organisms, to form protoplasts *stricto sensu*, which are bounded by a true protoplast membrane. Whereas there is now good evidence in certain Gram-positives for the existence of two separately isolatable envelopes, the cell wall and the protoplast membrane differing radically in composition (see Section V, C, 3), there is no proof that Gram-negative bacteria have a lipoprotein-mucopolysaccharide cell wall and a separate protoplast membrane. The evidence of phage adsorption experiments and partial degradation of cell walls for *Escherichia coli* indicates that the underlying structure in the wall is mucopolysaccharide and the overlying part is lipoprotein, penetrated in certain regions by the other components.[30] If this is so, and if there is no separate cytoplasmic membrane, it is difficult to explain plasmolysis phenomena where the cytoplasm, apparently surrounded by a semipermeable membrane, contracts away from the cell wall. It is likely that such an osmotic barrier would be lipoprotein in nature; consequently, at present it seems probable that underlying the cell wall proper in Gram-negative bacteria as well as in Gram-positives, there is a lipid membrane.

"Protoplasts" of *Escherichia coli* formed by penicillin action or deprival of DAP (see Sections II, D, 1 and 2) are surrounded by an envelope which may increase many times in area as the "protoplast" grows.[32] This may be a true protoplast membrane remaining after removal of the entire cell wall, or it may be the residual lipoprotein part of the cell wall, or both might be present. The situation has not been resolved at the time of writing (Fig. 26; Fig. 27).

D. Definition of the Terms: Protoplast and Spheroplast

For present purposes, a *protoplast* will be defined as the structure derived from a vegetative cell by removal of the entire cell wall, or alternatively, that part of the cell which lies within the cell wall and which in some species can be plasmolyzed away from it. This is in conformation with current botanical usage. In addition, the term "protoplast" with quotation marks will be used for structures in which the cell wall has been modified in such a way as to render the organism spherical, sensitive to osmotic shock, and incapable of forming colonies when plated on the usual media (compare definition given by Mahler and Fraser[33]): it is applied in cases where it is not known whether or not the cell wall is entirely absent. It should, perhaps, be added that if in Gram-negative bacteria there is no membrane underlying the cell wall, the preparation of true protoplasts from such organisms may not be possible. A recommendation that the unqualified term *protoplast* be restricted to preparations in which there is good reason to think that all cell wall components are absent was signed by thirteen workers and agreed to in principle by many others (Brenner *et al.*).[194] The need for a word to describe the globular forms in which cell wall structure has been modified (e.g. by growth in penicillin) rather than totally removed, is fairly generally recognized. The term *spheroplast* is being used in this sense, sometimes prefixed so as to indicate the method of induction, e.g. penicillin-spheroplast, DAP-less spheroplast. The comparative biochemistry of bacterial cell walls, protoplasts, and spheroplasts has recently been summarized.[195]

E. Properties of Protoplasts and Consequences of the Absence of Cell Walls

If the above definition is accepted, there are certain properties which protoplasts will exhibit and certain tests which might be applied to determine whether a preparation consists of protoplasts or whether cell wall material is present. Some of these will now be considered.

1. Microscopic Appearance (see also Section III)

Large organisms, such as *Bacillus megaterium*, can be observed under phase contrast to change from a rod-shaped form to spherical protoplasts. It is apparent that an outer layer, which normally surrounds the protoplasts, is removed during this conversion. The sequence of forms has been indicated by Tomcsik,[34] Tomcsik and Baumann-Grace,[16] Tomcsik and Guex-Holzer,[19] and McQuillen[17] (Fig. 2). Wiame *et al.*[23] show the appearance of *Bacillus subtilis* in the electron microscope at various stages of lysozyme action and the forms are similar to those of *B. megaterium*.

Small, spherical bacteria, such as *Micrococcus lysodeikticus* and *Sarcina lutea*, can be seen to undergo change but the final protoplast is similar in size and shape to the individual cells. However, whereas the cells exist as pairs or packets the protoplasts are discrete and separate. All protoplasts and "protoplasts" so far described are approximately spherical in shape and it is inferred that at least the form-determining component of the cell wall is absent.

2. Osmotic Properties (see also Section IV, B)

A general feature of protoplasts is their sensitivity to osmotic shock. Unless maintained in suitable media they are unstable. A certain concentration of a non-penetrating solute is necessary to balance the internal osmotic pressure which may range from 5 to 30 atmospheres, depending on the organism. Many bacteria are capable of surviving in media over a wide range of osmotic pressure; this is possible because in intact cells the more or less rigid cell wall takes the strain. Protoplasts, however, over a small concentration range behave as osmometers and swell or shrink as the solute concentration is altered; but below a critical concentration they lyse.

3. Chemical Composition (see also Sections V, A and B)

There are some constituents of bacterial cells which are present only in the cell wall or as intermediates in the soluble fraction of the cytoplasm. These probably include DAP, muramic acid, glucosamine, D-glutamic acid, D-alanine, and certain sugars such as rhamnose. Protoplasts *stricto sensu* should not contain any such substance except as a soluble intermediate. For example, DAP is found in the trichloroacetic acid-precipitable fraction of vegetative cells of *Bacillus megaterium* but not in the corresponding fraction of protoplasts of that organism, although such protoplasts can synthesize this compound.[31]

4. Immunochemical Properties (see also Section V, E)

Since it is now a routine matter to prepare highly purified cell wall fractions from bacteria, it is possible to use the extremely specific technique of immunochemical analysis for detection of cell wall antigens in so-called protoplast preparations. Tomcsik[16, 34] has shown how protoplasts and vegetative cells of *Bacillus megaterium* (*Bacillus* M) can be compared immunologically and it should be possible to investigate "protoplasts" of Gram-negative species, such as *Escherichia coli*, to see whether or not cell wall antigen remains.

5. Behavior of Isolated Cell Walls

When isolated cell wall preparations of an organism are completely solubilized by an enzyme system[24] there is good reason to expect that the enzyme will attack the wall of the intact, vegetative cell. This is the case with the enzyme lysozyme and the sensitive species *Bacillus megaterium*, *Micrococcus lysodeikticus*, and *Sarcina lutea*, but as yet there are few enzyme systems known which are able to degrade bacterial cell walls completely.

If, on the other hand, enzymic treatment of the isolated wall leaves a residue with a high degree of structural integrity, it is doubtful if treatment of intact cells would yield true protoplasts.

6. Staining Characteristics

Preparations of protoplasts should not give positive reactions to specific cell wall staining techniques. However, a negative reaction does not necessarily imply that a preparation consists of protoplasts since components essential for the staining reaction might be absent while other wall constituents are still present. It must be remembered also that protoplasts in general are so fragile that conventional techniques of staining on slides result in lysis.

7. Interaction with Bacteriophage

A highly specific interaction occurs between the bacteriophage tail and receptor sites in the bacterial wall. Such receptor sites do not occur in structures other than the cell wall, so that the specific adsorption reaction with phage is not given by protoplasts. This has been verified in the case of *Bacillus megaterium*. There are conflicting reports as to the interaction of *Escherichia coli* "protoplasts" with T phages (see Section VI, H, 2).

There are other agents (bacteriocines) which appear to have similar specificity of interaction with cell wall components and which might be used to test for presence of wall constituents.

8. Desirability of Having an Accepted Usage for the Term Protoplast

The above are some of the factors which have to be considered in order to determine whether or not "protoplasts" can legitimately be called protoplasts. It seems useful to use the term more strictly than has been done recently. If it is to have a reasonably precise meaning in terms of bacterial cell structure and if this is to be consistent with general usage in other fields, it is best restricted to the protoplasm surrounded by the plasma membrane or the cytoplasm enclosed by the cytoplasmic membrane or, alternatively, to that part of the cell underlying the cell wall and that part which shrinks away during plasmolysis. If the cell wall in an organism appears to act as osmotic barrier as well as form determinant, the concept of protoplast may not be relevant in that case.

There is always the possibility that because of overwhelming usage the word protoplast may come to be accepted for structures which do not satisfy the criteria set out above (as if, for instance, the "protoplasts" of *Escherichia coli* were found to possess phage receptor sites and cell wall antigens and were nevertheless admitted to the class of protoplast). If this were to happen it might, perhaps, be worth while to use another word, such as *gymnoplast*, for structures in which the cell wall is entirely absent. At present there seems a considerable measure of agreement as to the definition—the uncertainty arises because so little is known about the anatomy of Gram-negative bacterial cells.

McQuillen,[31, 32] Mitchell and Moyle,[35] Salton,[24] and Shockman *et al.*[36] have pointed out the inadvisability of using the term protoplast unless the cell wall is absent; agreement with this view has been expressed by Gerhardt, Jeynes, Kandler, Kellenberger, Klieneberger-Nobel, Rubio-Huertos, Storck, Strange and Dark, Tomcsik, Weibull, and others.

II. Formation of Protoplasts

A number of methods are now available for the preparation of protoplasts of lysozyme-sensitive Gram-positive bacteria and "protoplasts" of Gram-negative organisms. In addition, there are methods involving the use of autolytic enzymes for some Gram-positive species.

A. The Use of Lysozyme and Related Enzymes for Preparation of Protoplasts of Certain Gram-Positive Species

1. Egg-White Lysozyme

There are several different enzymes which are classed as lysozymes; of these, egg-white lysozyme is the best characterized and most easily purified. It is available commercially in crystalline form; most published work has been carried out with the material marketed by Armour Laboratories. However, it is simply prepared from egg-white.[37, 38] Raw egg-white can itself be used directly or an alternative method of preparation is as follows: add NaCl to egg-white to make 5% (w./v.), adjust to pH 9.5, seed with a crystal of lysozyme and put in the refrigerator. The enzyme crystallizes out (D. Herbert, personal communication).

2. Lysozyme and *Bacillus megaterium*

All 54 strains of *Bacillus megaterium* tested were found by Tomcsik and Baumann-Grace[16] to be lysozyme-sensitive, although to varying degrees. In contrast, none of the 35 strains of *Bacillus cereus* investigated had cell walls susceptible to digestion by lysozyme. Protoplasts can usually be formed from *B. megaterium* but different strains show different stabilities. Several studies have made use of strain KM[13, 20, 39-43] and the lysogenic strain 899(1) has been used.[39, 44] Organisms can be grown in complex or synthetic media[20, 44] and should be washed and suspended in suitable media before addition of lysozyme. The enzyme is activated by Na$^+$ and the pH optimum lies between pH 6 and 7. In order to obtain stable protoplasts it is necessary to use media containing a sufficient concentration of a solute which cannot penetrate the osmotic barrier of the cells. Weibull[13] originally used sucrose or polyethylene glycol (average molecular weight 4000). The optimal concentrations must be discovered empirically, but usually lie between 0.1 and 0.3 M for sucrose and about 7.5% (w./v.) for polyethylene glycol. Suspensions of bacteria at densities up to 50 mg. dry weight/ml. and lysozyme concentrations of 50–500 μg./ml. have been used. Other suspension media which have been used include glucose (0.3 M), raffinose, melibiose or trehalose (0.15 M), but cellobiose (0.15 M) and erythritol (0.3 M) were unsatisfactory.[15] Phosphate buffer (0.5 M, pH 7.0) has been used by McQuillen.[17, 41] The conversion to stable protoplasts occurs best with stationary phase cells[13, 15] or with cells "rejuvenated" under controlled conditions.[43] The enzymic dissolution of the cell wall will occur over a wide range of temperatures and is usually complete in less than 30 minutes if conditions are satisfactory. The process can be followed using phase contrast microscopy. Once formed, the protoplasts can be centrifuged and resuspended in suitable media (solutions containing sucrose, polyethylene glycol, NaCl, KCl, phosphate buffer, etc., at appropriate concentration). However, since these structures are extremely fragile, great care has to be taken in all manipulations. Aeration by bubbling or vigorous shaking can lead to rapid lysis.

If suspensions of *Bacillus megaterium* are heated before treatment with lysozyme, the enzymic digestion of the wall results in stubby, rod-shaped structures which are stable in distilled water instead of osmotically fragile spheres (Fig. 3; Fig. 25 e, f).

Some authors recommend the formation and washing of *Bacillus megaterium* protoplasts in complex media in order to get physiologically active preparations. For example, Landman and Spiegelman[43] used a

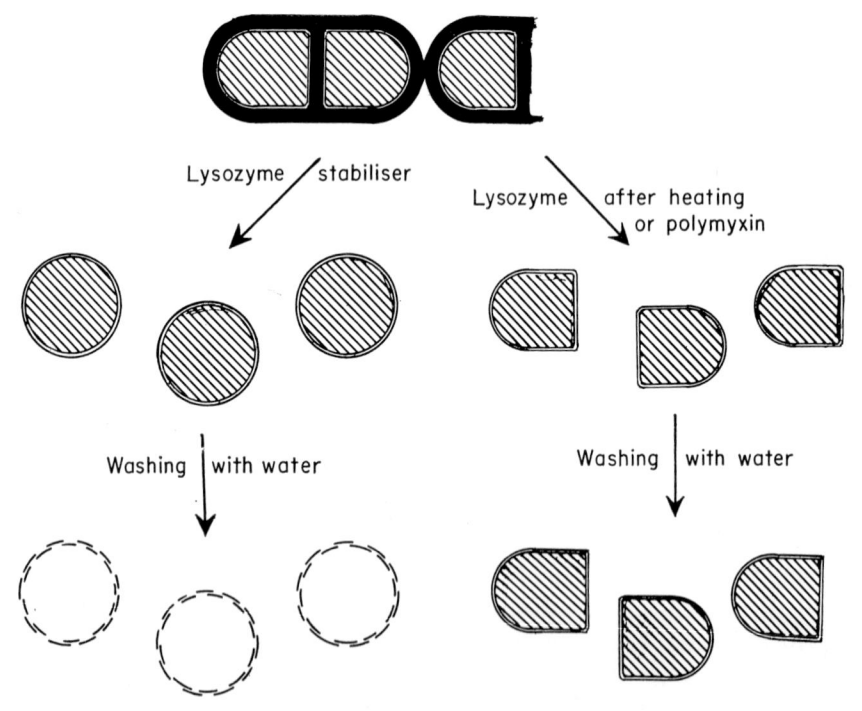

Fig. 3. Representation of effects of prefixing (by heat or polymyxin) on the subsequent release of protoplasts by treatment of *Bacillus megaterium* with lysozyme.

medium containing enzymic hydrolyzate of casein (2%), adenosine triphosphate, ATP (0.1%), hexose diphosphate, HDP (0.6%), and $MnCl_2$ (10^{-4} M), as well as Na_2HPO_4 (0.5 M, pH adjusted to 7.84). Whereas Weibull[15] advised preparing protoplasts under semianaerobic conditions, these authors say that a "continual energy source is absolutely essential and manipulations in the cold, absence of HDP, and of aeration are therefore to be avoided."

3. LYSOZYME AND *Micrococcus lysodeikticus* AND *Sarcina lutea*

Micrococcus lysodeikticus is the organism isolated by Fleming,[45] the discoverer of lysozyme, and used by him as an indicator of the lytic action of the enzyme. Lester[46] and Beljanski[47] showed that lysozyme "lysates" of *M. lysodeikticus* in sucrose (0.64 M) but not in dilute buffer, retained some ability to incorporate radioactive amino acids into their protein fraction. It is probable that these preparations contained some protoplasts as well as some totally lysed material. Shortly after the demonstration of the formation of protoplasts of *Bacillus megaterium*,[13] McQuillen[20] found that similar preparations could be made from *M. lysodeikticus* and *Sarcina lutea*. The groups and packets of cells of these organisms can be seen to separate into discrete spheres under the influence of lysozyme which is capable of totally digesting their cell walls. In media suitable for *B. megaterium* protoplasts (0.2 M sucrose), those from *M. lysodeikticus* and *S. lutea* are unstable and rapidly lyse. However, if treatment is carried out in media of higher solute concentrations, stable forms are produced which will lyse again if the suspension medium is diluted. Grula and Hartsell[48] investigated the effect of lysozyme on *M. lysodeikticus* in media containing NaCl (0.85%) but did not isolate intact protoplasts. Their work confirmed that lysozyme digests the cell wall of this organism. Mitchell and Moyle[21] and Gilby[22] have studied the influence of suspension media on the conversion to protoplasts and agree that a concentration of impermeable solute of about molar is necessary to stabilize *M. lysodeikticus* protoplasts. Similar results were obtained with *S. lutea*.[21] Suspensions of cells (10 mg. dry weight/ml.) in NaCl or sucrose (approximately 1.0 M) buffered with Na-phosphate (0.01 M, pH 6.8) were treated with lysozyme (10 µg./ml. for *M. lysodeikticus* and 40 µg./ml. for *S. lutea*).[21] Gilby[22] used sucrose (1.0 M) containing NaCl (0.05 M) and lysozyme in the proportion of 3 µg. for each milligram dry weight of *M. lysodeikticus*. The conversion to protoplasts was complete in less than 30 minutes at room temperature. Colobert[197] pointed out that unless the ionic composition of the medium is suitable, the products of digestion by lysozyme and the enzyme itself may not dissociate from the surface of *Sarcina lutea* and *Micrococcus lysodeik-*

ticus so that what he calls "angioplasts" rather than "gymnoplasts" (protoplasts) are formed.

There have been three reports of preparation of protoplasts of streptococci. The strain of *Streptococcus faecalis* used by Abrams,[188] 9790, was found to be lysozyme-sensitive, unlike some other strains. It is not known if this strain contains teichoic acid as a component of its cell wall and if so whether this is released during digestion with lysozyme. At any rate, "protoplasts" were prepared by lysozyme treatment.[203] Slade and Slamp[204] used dialyzed concentrates of *Streptomyces albus* culture filtrates to digest *Streptococcus pyogenes* and were able to release "protoplasts" which were stable in sucrose solutions (1.0 M). A further enzyme(s) in the preparation caused lysis on prolonged incubation even in sucrose solution and was thought to attack the "protoplast" membrane as does trypsin. The cell walls of actinomycetes resemble those of bacteria and, in particular, many *Streptomyces* spp. are susceptible to lysozyme. Sohler *et al.*[202] have shown that the cell walls of many species are solubilized by this enzyme unlike those of the genus *Nocardia*. It might, therefore, be possible to prepare protoplasts of *Streptomyces* spp. Gooder and Maxted[205] used enzyme preparations from *Streptomyces albus* and from bacteriophage lysates of group *C* streptococci to digest the cell walls of group *A* beta-hemolytic streptococci. The phage enzyme system was better for the release of protoplasts since the streptomyces mixture ultimately attacked the protoplast membrane. That true, naked protoplasts were formed was suggested by the osmotic fragility of the preparations and by the absence of the cell wall sugar, rhamnose, and the cell wall antigens (M and T antigens and group hapten). Moreover, appropriate bacteriophages did not adsorb nor did the preparations grow or synthesize new cell wall material. However, isolated cell walls were not completely dissolved by the enzyme treatment and it may be that teichoic acid still remained.

4. BACTERIAL LYSOZYMES

Richmond[49] described an enzyme system produced in the medium of exponentially growing cultures of *Bacillus subtilis* which has many of the properties of egg-white lysozyme. It has the same activity and specificity in attacking bacterial cell walls, produces similar fragments, and can be used for protoplast formation of susceptible species.[50] It has been considerably purified but to what degree is not yet known.

Nomura and Hosoda[51,52] described material isolated from autolysates of *Bacillus subtilis* which caused the lysis of *B. subtilis* and *B. megaterium*, but not of *B. cereus*, *Staphylococcus aureus*, *Sarcina lutea*, *Micrococcus lysodeikticus* or *Escherichia coli*. This preparation, called "autolysin," degraded isolated cell walls of *B. subtilis* but its action on living cells of this organism did not give rise to stable protoplasts. Moreover, protoplasts formed by lysozyme treatment of *B. subtilis* were caused to lyse by addition of boiled "autolysin," suggesting that the preparation contained, in addition to the heat-labile cell wall degrading system, a heat-stable component which acted subsequently to lyse the protoplasts. The electron micrographs of "protoplasts" of *B. subtilis* formed by lysozyme show structures which are far from spherical. If "autolysin" preparations do contain a lysozyme it is difficult to see why they do not attack *M. lysodeikticus* and *S. lutea*. Perhaps they are more like the autolytic systems described in the next section than like the *B. subtilis* lysozyme of Richmond.[49, 50]

5. Enzymes from Spores and Sporulating Cultures of *Bacillus* spp.

Strange and Dark[53, 54] have shown that extracts of mechanically disintegrated resting spores and partial autolyzates of sporulating cells of *Bacillus cereus* and *B. megaterium* contain enzymes capable of degrading cell walls of similar species. Among the substances in nondialyzable form released by their action on isolated cell walls were diaminopimelic acid, glutamic acid, alanine, amino sugars, and glucose. The preparations were inactivated in 15 minutes at 100° but not in 1 hour at 60°, and were optimally active at pH 7–8 in the presence of Co^{++} or Mn^{++} (10 p.p.m.) and at 58°.

Dark and Strange[55] found that cells of *Bacillus cereus* which were resistant to the action of lysozyme at concentrations up to 600 µg./ml. could be converted into protoplasts in 20 to 60 minutes when incubated in buffered sucrose (0.2 M, pH 7.0–7.5) containing Co^{++} (50 µg./ml.) and enzyme derived from sporulating cells of the same species. Although enzyme from one *Bacillus* sp. usually had activity against walls of others, the greatest activity was found with homologous preparations. It is likely that the separation of such enzymes and the exploration of their comparative action will throw light on the chemical differences between cell walls of closely related organisms, as well as extend the range of bacteria from which protoplasts can be formed. The total dissolution of walls by these preparations and the microscopic appearance of the structures resulting from digestion of intact cells make it probable that true protoplasts are formed.

B. Autolytic Production of Protoplasts

Mitchell and Moyle,[35] in trying to derive protoplasts of *Staphylococcus aureus* by lysozyme action, made the interesting observation that, under certain conditions in the absence of lysozyme, suspensions became sensitive to osmotic shock and appeared to have been converted to "protoplasts" by autodigestion of part of the cell wall. A similar phenomenon was observed with *Escherichia coli*[56] and also by Gilby[22] with *Micrococcus lysodeikticus*. The procedure recommended by Mitchell and Moyle[35] involves incubation at 25° in media containing sucrose (1.2 M) and Na-acetate (0.33 M, pH 5.8) of cells (5 mg. dry weight/ml.) harvested from rapidly growing cultures (exponential phase in rich medium). In about 2 hours the conversion of *S. aureus* cells to osmotically sensitive forms is complete, although the rate depends to a considerable extent on the growth rate at the time of harvesting. Somewhat similar conditions were used for *E. coli*, except that the incubation medium consisted of Na malonate (0.3 M, pH 7.0) in L-arabinose (0.5 molal).[56]

There are grounds for thinking that the autolytic procedure results in the formation of true protoplasts from *Staphylococcus aureus*, since a fraction can be recovered from the suspension medium, which appears to be composed of hemispherical annuli resembling cell wall preparations in appearance and equivalent to about 75% by weight of the cell wall fraction obtainable by mechanical disintegration of *S. aureus*. The authors suggest that autolysis results in the degradation of a band or ribbon of cell wall material, which allows the protoplast to emerge from the two hemispherical caps of residual cell wall. They also point out that in an examination of hundreds of these cell wall fragments the equatorial ring, if it appears, always does so at right angles to the plane of the mouth of the hemisphere.[35] Perhaps the sequence is as illustrated in Fig. 4.

In the case of *Escherichia coli* it is less certain what happens during the autolytic conversion. Osmotic fragility can develop in about an hour but at this stage the organisms are still cylindrical; it is 12 hours or longer before the change to a spherical form is complete. Since other procedures which probably involve removal of only a small component of the wall of

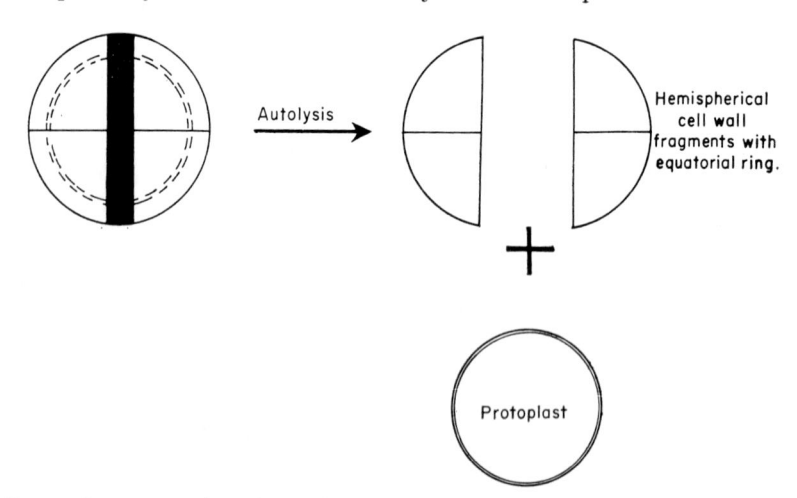

FIG. 4. Representation of possible mode of autolytic formation of protoplasts of *Staphylococcus aureus*, based on description by Mitchell and Moyle.[35]

E. coli (see Section II, C, 1) result in rapid conversion to osmotically sensitive spheres, it is premature to regard the autolytic procedure as a means of rapidly removing the entire cell wall from gram-negative bacteria.

Shockman et al.[36] have investigated the lysis of cells of the gram-positive species, *Streptococcus faecalis*, taken from the exponential phase of growth but the work is considered later in relation to amino acid deprival (see Section II, D, 3).

C. The Use of Lysozyme and Related Enzymes for Preparation of "Protoplasts" of Certain Gram-Negative Species

It has long been known that lysozyme can exert some lytic action on Gram-negative bacteria. However, this often involved applying the Nakamura technique[57]—treatment at pH 3.5 and 45°, followed by raising the pH to 9.8 to engender lysis. Recently, controlled treatment of certain Gram-negative organisms has been shown to result in transformation to osmotically sensitive, spherical forms resembling the protoplasts of lysozyme-sensitive, Gram-positive bacteria in some respects. Salton[196] has further characterized the action of lysozyme and reported on its effect on a range of Gram-positive and Gram-negative species. He emphasizes that the cell walls of the latter contain as a minor component the mucopeptide substrate which appears as a major part of the walls of lysozyme-sensitive Gram-positive species.

1. Lysozyme Treatment at pH 9 and *Escherichia coli*

Zinder and Arndt[58] found that washed cells from broth cultures of *Escherichia coli* just out of the exponential phase were converted to spheres in 5 minutes if treated with lysozyme (200 µg./ml.) in tris buffer (1.0 M, pH 9.0). At lower pH values conversion was incomplete and agglutination occurred. Between pH 8.0 and 8.8 agglutination could be prevented by adding Mg^{++} (0.1%). Other suitable media included solutions of sucrose (0.6 M) with carbonate buffer (0.1 M, pH 9.0). Each cell of *E. coli* usually yielded two spheres, "separation occurring at the point of incipient division." (Contrast this with the report that penicillin treatment produced one "protoplast" from each rod and "even cells about to divide were inhibited in further growth and division";[59] see Section II, D, 1.)

2. Lysozyme and Chelating Agents for *Escherichia coli*, *Azotobacter vinelandii*, and *Pseudomonas aeruginosa*

Almost simultaneously with the work mentioned above there appeared a report of a method of lysis of various species of Gram-negative bacteria by lysozyme which did not involve extremes of pH or temperature. Repaske[60] found that in a lysozyme/EDTA/tris system, cells of *Escherichia coli*,

Azotobacter vinelandii, and *Pseudomonas aeruginosa* were susceptible to lysozyme; Mahler and Fraser[33] went on to show that in the case of *E. coli* lysis could be prevented and the action arrested at a stage of spherical, osmotically shockable "protoplasts" if sucrose (0.5 M) were present.

Repaske[60a] has investigated in some detail the factors affecting the activity of lysozyme on the species mentioned and has found that the relative amounts of the reagents are important, as is the sequence of their addition to the bacterial suspension. For instance, increasing the concentration of one component can inhibit the reaction but with *Escherichia coli* the amounts of both EDTA and lysozyme can be increased as long as the ratio between them is kept constant. Also, in the case of *E. coli*, addition of EDTA some time before the lysozyme causes inhibition. Concentrations used were approximately as follows: tris 0.03 M; EDTA 100–500 µg./ml.; lysozyme 10–20 µg./ml.

In a later paper Fraser and Mahler[61] mention that broth-grown cells are not converted to "protoplasts" by the technique which is satisfactory for cells grown in medium 3XD.[62] This method employs sucrose (0.5 M), tris (0.03 M, pH 8.0), lysozyme (10 µg./ml.) and EDTA (200 µg./ml.). A further modification increased the lysozyme concentration to 100 µg./ml. and the EDTA to 2000 µg./ml.[61]

P. W. Wilson (unpublished observations) has prepared "protoplasts" of *Azotobacter vinelandii* using the Repaske method.[60] He found that citrate could replace EDTA or that cells could be passed through an Amberlite IRC-50 (H$^+$ form) column to make them lysozyme-sensitive. A satisfactory method was as follows: cells from a 14- to 16-hour culture were washed with distilled water and suspended in tris-buffered sucrose, to which K-citrate and lysozyme were added. Final concentrations were tris 0.033 M, pH 8.0; K-citrate 8 mg./ml.; lysozyme 13.3 µg./ml.; sucrose 0.025–0.05 M. After 10 minutes at room temperature, Mg^{++} equivalent to the citrate was added, and 5 minutes later the preparation was spun at 400 g. "Protoplasts" were resuspended in buffered media containing sucrose (0.05 M).

Aerobacter aerogenes, although closely similar in many ways to *Escherichia coli*, was not found by Gebicki and James[217] to be amenable to conversion to spheroplasts by lysozyme under a variety of conditions of pH, EDTA, tris and Mg^{++} concentrations. Growth in the presence of penicillin was, however, successful.

Colobert[198, 199] showed that cell walls of *Eberthella typhi* and *Salmonella* para-B also became susceptible to digestion by lysozyme after heat treatment, or addition of EDTA or teepol. These three kinds of pretreatment were believed to remove lipid and make accessible the lysozyme substrate in the wall. This allowed the formation of "protoplasts" of *E. typhi*.[200] Photosynthetic bacteria have been converted to "protoplasts" by Karunairatnam et al.[201] Suspensions of *Rhodospirillum rubrum* (0.4 mg. dry weight/ml.) in sucrose (10% w./v.) containing phosphate buffer (0.01 M, pH 6.8) and lysozyme (0.5 mg./ml.) and EDTA (1.6 mg./ml., pH 7.0) were incubated at 30° for 30 minutes. MgSO$_4$ was added to give 0.05 M and after a few minutes, ½ volume of sucrose solution (40% w./v.) was added. The resulting "protoplasts" had metabolic activities comparable with those of intact cells in most instances. Growth of *R. rubrum* in penicillin media was less effective in producing "protoplasts."

3. Leucozyme C and *Escherichia coli*

Amano et al.[63] fractionated leucocyte extracts into a lysozyme-like factor and three preparations, called leucozymes A, B, and C. Of these, leucozyme C was active in causing lysis of *Escherichia coli*. Lyophilized leucocytes were extracted with phosphate buffer (0.03 M, pH 7.3) and the clear extract brought to pH 5.2 by addition of HCl (0.1 N) at 0°. The precipitate was washed with citrate buffer (pH 5.2), dissolved in phosphate buffer (0.03 M, pH 7.3), and centrifuged to yield a clear supernatant which was used as leucozyme C.

The preparations were not active on *Micrococcus lysodeikticus* and, therefore, had no lysozyme activity, nor did lysozyme digest the isolated cell walls of *Escherichia coli*. However, leucozyme C degraded preparations of *E. coli* walls and this action was accelerated by further addition of lysozyme. Leucozyme C was inactivated by heating to 100° for 10 minutes.

When cells of *Escherichia coli* were treated with leucozyme C in sucrose solutions (9% w./v.) buffered with phosphate (0.06 M, pH 7.3), they were converted into spherical, osmotically sensitive forms or "protoplasts." Lysozyme under the same conditions did not result in such transformation but, if added together with leucozyme C, it increased the rate of conversion.

4. Phagocytin and *Escherichia coli*

Hirsch[64] isolated "phagocytin," a substance chemically similar to lysozyme, from rabbit leucocytes. This material kills *Escherichia coli*; Zinder and Arndt,[58] using crude phagocytin (provided by Dr. Hirsch) found that it rendered cells suspended in hypertonic media at pH 5.0 sensitive to osmotic shock. However, they were unable to demonstrate the formation of "protoplasts."

5. Bacteriophage Enzymes and *Escherichia coli*

It has been shown by Weidel and Primosigh[30, 65] that the cell wall of *Escherichia coli* consists of lipoprotein (80% by weight) and lipomucopolysaccharide (20% by weight). Phenol (90%) dissolves the lipoprotein, which is believed to be a plastic outer layer, but not the rigid mucopolysaccharide component which underlies it. The separated mucopolysaccharide plus lipopolysaccharide fraction is attacked by phage T2 enzyme to yield soluble material which includes all the alanine, glutamic acid, diaminopimelic acid (DAP), muramic acid, and smaller amounts of glycine and lysine, together with most of the glucosamine. The sedimentable residue is lipopolysaccharide and contains glucose, L-gala-D-mannoheptose, lipid, and some glucosamine. This residue also possesses all the T4 receptor site activity. The material solubilized by the phage enzyme is chemically like the cell wall of many Gram-positive species (see Section I, B and Chapter 3). It would seem, therefore, that phage T2 can release from the cell wall of the Gram-negative *E. coli* constituents of a mucopolysaccharide closely similar in composition to the whole cell wall of some Gram-positive bacteria. From a consideration of the accessibility of receptor sites for various T phages in native and more or less degraded cell walls, Weidel and Primosigh[30] concluded that the lipomucopolysaccharide forms the rigid structure of the cell wall and is overlaid by lipoprotein except in certain patches. Removal of the mucopolysaccharide ("Gram-positive layer") by phage enzymes eliminates the stiffening component and leads to "lysis from without."

McQuillen[20] noted that phage-infected cells of *Bacillus megaterium* went through a protoplast-like stage before lysis and Zinder and Arndt[58] found that phage ghosts (prepared by osmotic shock) converted *Escherichia coli* to a protoplast-like form before lysis occurred. Kellenberger (unpublished observations) has made use of this kind of system to produce spherical, osmotically sensitive "protoplasts" of *E. coli* by treatment of cells in media containing a stabilizing concentration of buffered sucrose with high multiplicities of phage T2. In view of the fact that lysozyme acts on mucopolysaccharide substrates in which muramic acid and glucosamine are probably linked to peptides of glutamic acid, alanine and lysine or DAP, it seems reasonable to include phage enzymes in this category, at least until the precise locus of action is determined. The existence of lysozyme-like enzymes in certain bacteriophages has been established beyond doubt. Murphy[210] purified more than 100-fold an enzyme associated with phage G infection of *Bacillus megaterium* KM which is capable of destroying the isolated bacterial cell wall and releasing nondialyzable reducing sugars. It was not serologically related to the phage. Panijel and Huppert[211, 212] found that osmotically shocked phage T2r preparations were able to convert *Escherichia coli* B to "protoplasts" and, similarly, phage D4 was active on *Salmonella enteritidis*. Sucrose solution (20% w./v.) was used as a stabilizing agent and EDTA was necessary in the case of *S. enteritidis*.

Carey et al.[213] also used osmotically shocked T2r+ phage to convert *Esch-*

erichia coli B to spherical "protoplasts." Bolle and Kellenberger[214] independently showed that high multiplicities of phage T2 could convert *E. coli* to spheroplasts. The enzyme in T2 responsible for these activities has been shown by Koch and Dreyer[215] to be a lysozyme. It released material containing glucosamine, muramic acid, glutamic acid, diaminopimelic acid, and alanine from isolated cell walls of *E. coli* B, caused lysis of intact cells of *E. coli* via a spheroplast form and also lysed *Micrococcus lysodeikticus*, the test substrate for lysozyme. (Panijel and Huppert[211] reported that their preparations from T2r, T4r, and T3 infection and from induced *E. coli* K 12λ were all unable to attack *Bacillus megaterium*.) It should be pointed out that egg-white lysozyme itself will not cause lysis or spheroplast formation of *E. coli* B at neutral pH values and in the absence of a chelating agent such as EDTA. The phage enzyme is not, therefore, identical in all respects.

Jacob and Fuerst[216] reported that the λ-endolysin found in lysates of induced lysogenic *Escherichia coli* K 12 (λ) could in the presence of EDTA convert *E. coli* rods into spheroplasts.

D. Metabolic Disturbance as a Method of "Protoplast" Formation in Gram-Negative Species

If it were possible to prevent the synthesis of cell wall material while allowing all other components of the cell to be formed normally, then it might be expected that protoplasts would outgrow their original walls and emerge surrounded only by the protoplast membranes. These membranes would be subject to the internal osmotic pressure of the cellular constituents (5–30 atmospheres) and in the absence of the hydrostatic pressure normally exerted by the cell wall, would have to be supported by an external osmotic pressure due to nonpenetrating solutes sufficiently great to prevent osmotic explosion of the protoplasts.

Something along these lines has been achieved by the use of penicillin and also by deprival of diaminopimelic acid.

1. Action of Penicillin on *Escherichia coli* and other Gram-Negative Species

The antibiotic penicillin causes death by lysis of growing bacteria. It has little if any action on organisms unable to grow, or, more specifically, unable to make protein,[31, 66, 67] and is, in general, more active against Gram-positive than Gram-negative species, although some of the latter are quite sensitive. It has long been suggested that penicillin might have its major effect at some site involving the cell wall or membrane.[68-72] Park[73] demonstrated the accumulation in penicillin-inhibited cells of *Staphylococcus aureus* of peptide derivatives of uridine nucleotides; recently Park and

TABLE I

Comparison of Analyses of Cell Wall and a Uridine Nucleotide Derivative Which Accumulates in Penicillin-Inhibited *Staphylococcus aureus*[a]

Component	Cell wall	Uridine nucleotide
Uracil	0	1
Ribose	0	1
N-Acetyl muramic acid	1	1
D-Glutamic acid	1	1
L-Lysine	1	1
DL-Alanine	3	3
Glucosamine	2	0
Glycine	c. 4	0

[a] Data are from Park and Strominger[74] and are expressed as approximate relative molar proportions.

Strominger[74] have shown that one of these substances has several components in common with and in the same relative proportions as has the cell wall of *S. aureus*. Table I shows the analyses of cell wall and uridine nucleotide derivative and Fig. 5 indicates a possible structure for the latter. The authors advance the thesis that penicillin may act by interfering with the normal utilization of a cell wall precursor, which accordingly accumulates in the cytoplasm, resulting in impaired maintenance or growth

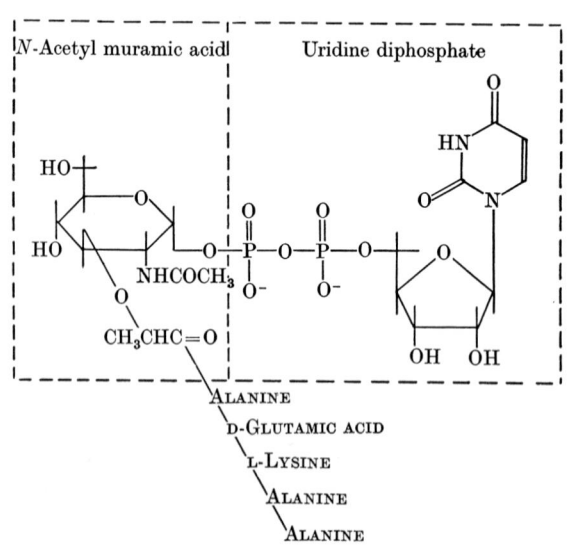

Fig. 5. Suggested structure of a uridine nucleotide which accumulates during inhibition of growth of *Staphylococcus aureus* by penicillin (Park and Strominger[74]).

of the wall. Ultimately, lysis of the cell will occur. Falcone and Graziosi[218] reported the conversion of *Staphylococcus aureus* to protoplasts by growth in the presence of penicillin although the only criterion used to detect the change was osmotic fragility. Despite the greater sensitivity to penicillin of Gram-positive species there do not seem to have been other reports of attempts to derive protoplasts of Gram-positive species by growth in penicillin.

It has been shown by Park,[219] however, that penicillin as well as bacitracin, oxamycin (cycloserine), and glycine (0.6 M) causes inhibition of synthesis of the cell wall of *Staphylococcus aureus* and accumulation of uridine 5′pyrophosphate N-acetyl amino sugar derivatives which are believed to be wall precursors. Trucco and Pardee[220] on the other hand, could not find appreciable differences in cell walls of *Escherichia coli* grown with and without penicillin. They did not, however, particularly examine the minor mucopeptide component, the formation of which is thought by some people to be inhibited by penicillin.

Liebermeister and Kellenberger,[75] Lederberg,[59] and Hahn and Ciak[76] independently showed that penicillin-inhibited growth of *Proteus vulgaris* and *Escherichia coli* could lead to spherical, osmotically sensitive "protoplasts." The possible relationship of these to L-forms was discussed by Kellenberger *et al.*[77] (see also Chapter 7). The experiments of Liebermeister and Kellenberger[75] and of Kellenberger *et al.*[77] were carried out with *P. vulgaris* growing in nutrient broth containing NaCl (0.5% w./v.) and penicillin (500–1000 I.U./ml.).

Lederberg[59] added overnight aerated cultures (3 volumes) to fresh broth (10 volumes) containing sucrose (20% w./v.), Mg-sulfate (0.2% w./v.) and penicillin (1000 I.U./ml.). After 2 to 3 hours' continued incubation at 37°, conversion to spheres was complete. The constituents could be present at lower concentrations—5%, 0.1%, and 100 I.U./ml., respectively—without much impairment of the transformation. Conversion to "protoplasts" was not nearly so good in synthetic media.

The sequence of forms is similar in *Proteus vulgaris* and *Escherichia coli* (Fig. 6). Liebermeister and Kellenberger[75] illustrate their paper with excellent light microscope and electron microscope photographs. Hahn and Ciak[76] published a series of eight phase contrast pictures of the transformation in *E. coli* and the process has also been studied by McQuillen[31, 32] (Fig. 26).

Shortly after penicillin is added, the rod-shaped cells develop an osmotically sensitive swelling, either centrally or near one end. This enlarges and can form a sphere equal in diameter to the length of the original rod. The remnants of the rods often appear protruding from the sphere like "rabbit ears" (the "Hasenform" of Liebermeister and Kellenberger[75]). Eventually, only the sphere is visible; it increases in size and often develops crescent-

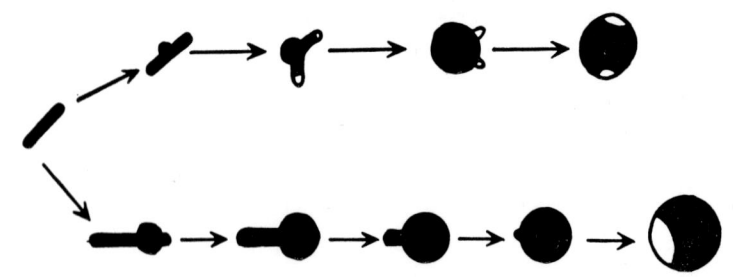

Fig. 6. Representation of sequence of forms of *Escherichia coli* and *Proteus vulgaris* growing in the presence of penicillin[31, 32, 59, 75, 76] or of *E. coli* deprived of diaminopimelic acid.[31, 32]

shaped vacuoles. It is difficult to be certain as to what happens to the residue of the cell wall which originally surrounded the rod-shaped cell. Lederberg[59] claims that it "then withers away and disappears." Hahn and Ciak[76] say "later the globes either separated from the cell walls or retained parts of them attached, giving a typical "rabbit-ear" appearance . . . eventually they released their entire contents, leaving as formed elements only circular "ghosts" that probably represent empty cytoplasmic membranes." It is obviously of considerable interest to know whether the envelope surrounding the spherical form induced by penicillin consists of the cytoplasmic membrane or whether it is cell wall material lacking only the specific mucopolysaccharide-stiffening component whose synthesis is interfered with by penicillin.[31, 32]

McQuillen[31] analyzed penicillin-induced spherical forms of *Escherichia coli* and found them to contain much less DAP—a known constituent of the mucopolysaccharide. Kandler and Zehender[78] found no DAP in stable L-forms of *Proteus vulgaris* induced by penicillin—preparations possibly derived via a "protoplast-like" stage.[79] Nor did they find a fraction containing DAP, glutamic acid, alanine, muramic acid, and glucosamine, corresponding to a component found in the walls of bacillary forms and in L-forms which were capable of reverting. However, Weibull[80] reported that the stable L-forms of *P. vulgaris*, which he analyzed quantitatively, contained both hexosamine and DAP in the trichloroacetic acid-precipitable fraction, unlike lysozyme-produced protoplasts of *Bacillus megaterium* from which DAP is entirely absent[17] (see also Sections V, A, and B). So far there are few reports of analyses, chemical or immunochemical, to determine the presence or absence of other components of the vegetative cell wall of Gram-negative species in these spherical forms.

The penicillin method for producing osmotically sensitive, spherical forms can be applied to *Salmonella typhimurium*, as well as to many strains of *Escherichia coli*, *Proteus vulgaris*, and probably other Gram-negative species. The general procedure is to incubate a broth culture containing sucrose (approximately 10% w./v.), Mg-sulfate (0.01 M) and penicillin (100–1000 μg./ml.) with adequate aeration—shaking rather than bubbling—so as to get rapid growth. When conversion to spheres is complete (about 2 hours), cultures can be centrifuged and the "protoplasts" resuspended in sucrose-containing media, with or without the addition of penicillin. *P. vulgaris* preparations can be made in nutrient broth containing NaCl (0.5%) but no sucrose.[75] *Aerobacter aerogenes* yields spherical cells in about 3 hours at 37° in a broth containing 1000 units/ml. penicillin and about 0.4 M sucrose.[217]

Landman *et al.*[221] described methods involving defined media for the almost quantitative conversion by penicillin of *Proteus mirabilis* and *Escherichia coli* to the spheroplast form and thence to the L-form. *Alcaligenes faecalis* is another Gram-negative species which can be converted to a spherical form by growth in penicillin media. Lark[222, 223] studied the development of such forms and by using synchronously dividing cultures showed that sensitivity to induction of spheroplast formation varied during the cell-division cycle, being least shortly before cell division.

Other antibiotics besides penicillin can induce formation of spheroplasts—bacitracin[224], oxamycin (cycloserine)[225], and novobiocin.[226] It is possible that all act by interfering with normal synthesis of the mucopeptide stiffening component of the cell wall. McQuillen[195] has discussed some of the attributes common to these substances. (See also Table A.)

TABLE A

COMPARATIVE EFFECTS OF ANTIBIOTICS ON PROCESSES CONNECTED WITH CELL WALL SYNTHESIS AND PROTEIN SYNTHESIS[a]

Antibiotic	Protein synthesis inhibited	Surface charge altered *E. coli* *S. aureus*	Cell wall synthesis inhibited *S. aureus*	UDP-Amino-sugar derivatives accumulate *S. aureus*	Spheroplast formation induced *E. coli*
Penicillin	−	+	+	+	+
Bacitracin	−		+	+	+
Cycloserine (oxamycin)	−		+	+	+
Glycine	−		+	+	+
Tetracycline	+	−	−	−	−
Oxytetracycline	+	−			−
Chlortetracycline	+	−	−+[b]	−+[b]	−
Chloramphenicol	+	−	−	−	−

[a] Data from various sources including references 31, 32, 59, 71, 96, 98, 192, 195, 219, 224, 225, 226, 229.

[b] Higher concentrations are necessary to give the positive response.[219]

2. Diaminopimelic Acid-Deprival and *Escherichia coli*

α-ε-Diaminopimelic acid (DAP) was discovered in 1951 by Work[81] as a constituent of *Corynebacterium diphtheriae* and *Mycobacterium tuberculosis*. Since that time it has been found in nearly all species of bacteria except Gram-positive cocci and related organisms. It has never been found in a true protein, often occurs in the cell wall of bacteria, and has never been detected in any other fraction except the soluble cytoplasmic fraction.[28] The presence of DAP in protoplasts of *Bacillus megaterium* has been excluded by McQuillen[17] and there are good grounds for thinking that it occurs uniquely as a cell wall constituent. If this is the case, and if there is no functional need for DAP in any other part of the cell, then it should be possible to grow DAP-exacting bacteria in the absence of this amino acid and derive "protoplasts," if not true protoplasts.

A mutant, 173-25 of the W strain of *Escherichia coli*, is blocked in the synthesis of DAP[82] and lyses if grown in media containing limiting amounts of DAP but adequate lysine, which, in the parent organism, is derived by decarboxylation of DAP.[83–85] Lysis is believed to be caused by continued cytoplasmic growth after DAP becomes exhausted and normal cell wall synthesis is prevented.[31, 32] However, if the medium is supplemented with sucrose (10–20% w./v.), growth in the absence of DAP (as in the presence of penicillin) results in transformation of the bacillary form into spherical bodies.[31, 32] Bauman and Davis[86] mention the production of "protoplasts" of the same mutant of *E. coli* by DAP-deprival. McQuillen[31, 32] has shown that the same strain of *E. coli* can be converted to "protoplasts" in the same complex medium either by adding penicillin or by omitting DAP. Phase-contrast pictures of these forms show the typical spherical appearance, often with one or two crescent-shaped vacuoles. Electron micrographs clearly indicate the presence of a surrounding envelope which may be a true protoplast membrane (cytoplasmic membrane) or which may be cell wall material lacking the mucopolysaccharide component of which DAP is a constituent. During development of these "protoplasts" the surface area can increase tenfold or more, so that preexisting cell wall material could not be stretched to cover the whole surface of the resulting sphere (Fig. 23 d; Fig. 27). Lederberg and St. Clair[192] have confirmed earlier findings[31, 32, 86] that growth of the DAP-requiring strain 173-25 of *Escherichia coli* in the absence of this amino acid leads to formation of spheroplasts if the medium affords suitable osmotic protection.

Davis[87] and Bauman and Davis[86] used addition of penicillin or deprival of DAP for selection of auxotrophic mutants. When a mixed population is grown on minimal medium, either containing the antibiotic or deficient in the amino acid, the wild-type cells grow and lyse, due to formation of unstable "protoplasts," whereas the auxotrophic cells, being unable to grow in the minimal medium, are able to survive.

A satisfactory method for producing "protoplasts" by DAP-deprival is as follows: mutant 173–25 of the W strain of *Escherichia coli* (which is best maintained on slopes of solid medium rather than in freeze-dried preparations, Meadow and Work, personal communication) is grown overnight with aeration at 37° in synthetic medium supplemented with lysine (100 μM) and DAP (100 μM). Various synthetic media are suitable, e.g., the glucose/NH_3/salts medium, C/G of McQuillen and Salton;[44] the glucose/citrate/NH_3/salts medium of Meadow *et al.*;[84] or the minimal medium A of Davis and Mingioli.[88] Samples of overnight cultures can be centrifuged and washed, or transferred directly to 10 volumes or more of new medium lacking DAP but containing adequate lysine. This new medium can be synthetic or, better, broth medium (casein digest or peptone, with or without yeast extract). The exact medium does not appear to be critical but it should contain sucrose (10–20 % w./v.). Continued incubation at 37° with aeration by shaking results in conversion to "protoplasts" in 2 to 4 hours.[32] If there is continued growth of rod-shaped organisms, this may be due either to reversion to DAP-independence or to carryover of unused DAP with the inoculum.

3. AMINO ACID-DEPRIVAL AND *Streptococcus faecalis*

Toennies and his colleagues[36, 89–93] studied the consequences of limiting growth of *Streptococcus faecalis* 9790 by a range of amino acids to which the organism is exacting. After increasing exponentially until the particular amino acid was exhausted from the medium, cultures continued in most cases to grow slowly for a further period.[91] For instance, if valine were limiting, postexponential increase occurred but seemed to be restricted to cell wall material—rhamnose, glucosamine, cell wall N, and cell wall lysine all approximately doubled in amount, while cytoplasmic constituents such as protein could not be synthesized owing to depletion of valine.[36, 92] It appears that if growth is limited by an amino acid which is not a constituent of the cell wall, then, after this amino acid is all used, unbalanced synthesis of wall can occur until there is about twice the normal amount of wall per cell (normal here referring to exponential phase cells). Very different behavior was encountered when growth was limited by lysine which is one of the few amino acids present in the cell wall of *S. faecalis*. In this case, gradual lysis of the cells began immediately after maximum culture density was reached[36, 89, 91] (Fig. 7).

It is not entirely clear why limitation of lysine, which is a constituent of both cell wall peptide and cytoplasmic protein, should cause this lysis.

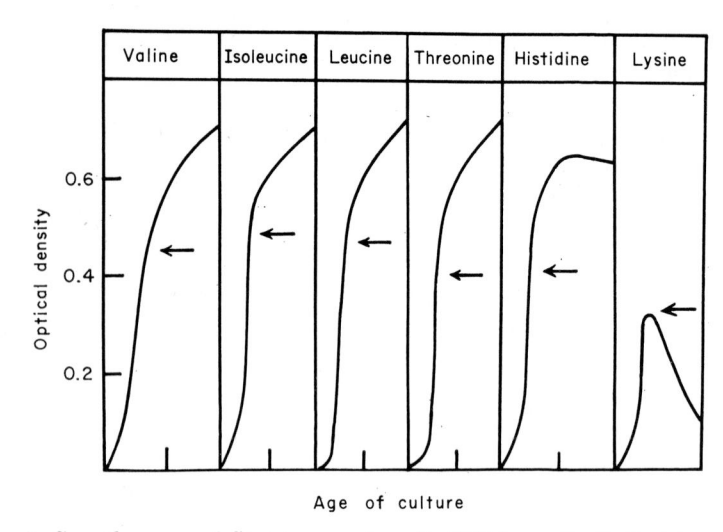

FIG. 7. Growth curves of *Streptococcus faecalis* 9790 in media limited by various amino acids. Arrows indicate where limiting amino acid is exhausted. Data of Toennies and Shockman.[91]

It is a somewhat different situation from DAP-deprival in *Escherichia coli* since DAP occurs only in the cell wall. It may be a more complex phenomenon and be related to the fact that cells taken from the exponential phase of growth of *Streptococcus faecalis* have a tendency to lyse, irrespective of the nature of the amino acid which would ultimately limit growth.

McQuillen[31] has discussed these effects and pointed out that since alanine and glutamic acid are present in the cell wall, changes in the amounts of these substances during postexponential growth would also be expected. Moreover, since D-glutamic acid and D-alanine may be present exclusively in the wall, it might be possible to limit growth of cell wall without affecting protoplast growth by using a medium containing L-alanine and L-glutamic acid but not vitamin B_6 (pyridoxal phosphate is the coenzyme of amino acid racemases). This has not yet been done but very recently Shockman *et al.*[36] reported that in media containing limited D-alanine, maximum growth was followed by lysis (as expected on this thesis) but the picture was complicated by a subsequent new wave of growth.

Holden and Holman[227] reported that the omission of vitamin B_6 from media containing adequate amounts of L-amino acids and D-alanine resulted in an altered morphology of *Lactobacillus arabinosus*—a lactic organism related to *S. faecalis* in metabolism. Most of the cells became swollen and elliptical although a few round and very large swollen forms were seen.

This may have resulted from an alteration in the chemistry or structure of the cell wall.

4. Comments on a Possible General Method for Producing Protoplasts by Metabolic Disturbance

It is apparent that some substances are probably present exclusively in bacterial cell walls. These include diaminopimelic acid; D-glutamic acid; muramic acid and other amino sugars; and certain sugars, such as rhamnose, arabinose, and manno-heptose. If strains requiring such substances, or inhibitors of the formation or utilization of them, can be found, then there is the possibility of obtaining protoplast growth with impaired cell wall synthesis. In ordinary media this will lead to lysis but in suitably protective media it may lead to protoplasts lacking cell walls or structures with some cell wall constituents missing. Park[94] has discussed the possibility of designing chemotherapeutic agents which would specifically prevent cell wall synthesis and McQuillen[31] has considered the general implications of unbalanced growth of certain cellular components as compared with others.

However, since Gram-negative bacteria contain protein and lipid in their cell walls and almost certainly also in their cytoplasmic membranes (if such structures have a separate existence, see Section I, C), it may be difficult or even impossible to devise techniques for preventing the synthesis of all of the wall without affecting the membrane. Nevertheless, there remains the possibility that the lipid components are different in the two structures. Little is yet known about the localization of the various kinds of lipid which occur in bacterial cells.

E. Some Factors Common to "Protoplast" Formation in Gram-Negative Species such as *Escherichia coli*

Four methods are available for producing spherical forms of *Escherichia coli*; it seems likely that they have some common features. The methods are: (1) lysozyme with or without chelating agents; (2) phage enzymes; (3) growth in the presence of penicillin; (4) growth in the absence of diaminopimelic acid.

The site of attack in each case is the mucopolysaccharide component (the "Gram-positive layer"[65]) which contains muramic acid, glucosamine, glutamic acid, alanine, DAP, and perhaps other substances. The enzyme lysozyme and the phage enzymes are known to act on such a substrate; obviously, deprival of DAP must lead to absence of increase in this component (or possibly production of a modified substance, perhaps containing lysine instead of DAP?); and finally, penicillin appears to prevent the incorporation of such a component and in certain other organisms leads to the accumulation in the cytoplasm of its probable precursor. Prestidge and

Pardee[67] have suggested that penicillin induces the formation in *Escherichia coli* of a specific protein (enzyme?) which attacks the cytoplasmic membrane. Their observations are of great interest and confirm some of the earlier findings of Harrington[66] (for a résumé of the latter, see McQuillen[31]). But the main conclusion—that lysis caused by penicillin is due to active destruction of the membrane rather than inability to maintain the internal osmotic pressure against a weakened wall—does not appear to be very firmly based. It is difficult to reconcile with the increases in membrane area of ten times or more which are possible in the presence of high concentrations of penicillin in sucrose-media (see Section III, D and Fig. 6). It is also assumed that the protoplast membrane of *Bacillus megaterium* has a counterpart in *E. coli*. This may or may not be the case.

At present all that can be said of "protoplasts" of Gram-negative species is that they bear a superficial resemblance to protoplasts of lysozyme-sensitive Gram-positive species but that in view of the various claims which are discussed later (reversion to bacillary forms, ability to form colonies, interaction with bacteriophage) and since the anatomy of the cell wall/cytoplasmic membrane structure is not understood, it is premature to be dogmatic about their exact nature.

F. The Use of Glycine for Preparation of "Protoplasts" of Gram-Negative Species

Glycine solutions have been used to disrupt bacterial cells[95] and are believed to cause damage to the protoplast membrane in some instances.[12] This amino acid has also been used to induce the formation of spherical forms of various Gram-negative species. Rubio-Huertos and Desjardins,[96] for example, added a small amount of glycine solution (1% w./v.) to the bottom of an agar slope seeded with *Agrobacterium* (*Phytomonas*) *tumefaciens*. This induced the formation of spherical forms which were surrounded by some kind of membrane and still possessed flagella. It was reported that they later developed into L-forms (see Chapter 7). In appearance they resembled the "gonidial stage" in *Spirillum* spp. and *Vibrio* spp.[97] and could be induced in various Gram-negative species (Rubio-Huertos, unpublished observations). Glycine was also used by Jeynes[98] to convert *Vibrio cholerae, Salmonella typhi, S. typhimurium,* and *S. paratyphi* B into "protoplasts." His procedure was to grow bacteria at 34° in rotating bottles (22 revolutions per hour) containing liquid medium supplemented with glycine (3% w./v.). It was claimed that the bacteria "lost their cell walls in a few hours" but no evidence was presented for this statement.

The addition of various D-amino acids (0.03 M) other than D-glutamic acid and D-aspartic acid which were bacteriostatic, was found by Lark and Lark[228] to cause conversion of *Alcaligenes faecalis* to "protoplast"-like structures. Many species of Gram-negative bacteria can be converted to spheroplasts by inclusion of glycine (1–5% w./v.) in their growth medium.[96, 98, 195] Welsch[229] has studied this process in *Escherichia coli* and reported that serine will also bring about such a transformation. Gerhardt[230] converted brucellae to spheroplasts under the influence of glycine.

Remarkable morphological changes of *Proteus P* 18 bacillus were described by Mandel *et al.*[231] as a result of beta radiation from incorporation of large amounts of P^{32}. Long filaments were formed and globular formations developed at one end. Later, forms resembling spheroplasts were common and these increased in size until they were 30–40 μ in diameter. Ultimately there was reversion to the normal rod form. These changes occurred in media containing horse serum (20%) and P^{32} (8 μc./100 ml.).

Growth of the halophilic bacterium, *Micrococcus halodenitrificans*, in media of various salt concentrations has been studied by Takahashi and Gibbons.[232] A minimum concentration of 0.7 M NaCl was necessary for normal morphology but at 0.6 M the cells appeared swollen and sensitive to osmotic shock. Addition of Ca^{++} or Mg^{++} to the medium allowed normal growth in lower salt concentrations. There was a change in the chemical constitution of the cell walls of the organisms grown in suboptimal salt concentrations—in particular the diaminopimelic acid content decreased to about 1/80 of the normal amount when cells were grown in 0.3 M NaCl plus Ca^{++}. Phase contrast and electron micrographs of the preparations indicate that the swollen, osmotically sensitive forms are of the nature of spheroplasts.

G. "Protoplasts" of Yeast and other Fungi

1. Autolysis and Mechanical Breakage of *Saccharomyces cerevisiae*

Nečas[99–102] found that mechanical breakage of yeast cells produced "plasma droplets," a small proportion of which contained Feulgen-positive material. Sometimes, also, in autolyzing cultures it seemed that the surface structures of yeast cells became digested and that "protoplasts" were released. In suitable media these "protoplasts" would then break up into spherical "plasma droplets," a few of which contained a nucleus. The droplets containing nuclei could enlarge in volume 1000-fold and some could then undergo transitions and eventually "bud off" structures which reverted to normal yeast cells.

2. The Use of Snail Enzymes to Prepare "Protoplasts" of *Saccharomyces carlsbergensis* and *S. cerevisiae*

The gut juice of the snail *Helix pomatia* contains enzymes which attack the yeast cell wall;[103] it has been used to prepare "protoplasts" of *Saccharomyces carlsbergensis* and *S. cerevisiae*.[104] Buffered rhamnose (0.55 M in

0.005 M citrate-phosphate, pH 5.8) was used as suspension medium; after 5 hours at 25° more than 90 % conversion had occurred of the elongated cells into spherical forms which were sensitive to osmotic shock. The "protoplasts" appeared to be extruded through holes which the enzymes made in the cell wall—possibly at the sites of bud scars. Later the entire residual cell wall appeared to dissolve. The "protoplasts," although very fragile, could be centrifuged at low speed and washed and resuspended in buffered rhamnose solution. Concentrations of rhamnose above 0.5 M caused the spheres to become crenate; at lower concentrations they swelled and burst; but in 0.5 M rhamnose and at pH 5.8 they were stable for several hours.

3. MOLD PROTOPLASTS

Horikoshi and his colleagues[206-208] have shown that the cell walls of *Aspergillus oryzae* can be digested by a combination of an enzyme system produced during sporulation of *Bacillus circulans* and a crude chitinase preparation from a *Streptomyces* sp. Such a method might release protoplasts from fungi. Emerson and Emerson[209] reported production of "protoplast"-like structures from all tested strains of *Neurospora crassa* which carried the osmotic mutant gene, *os*. Treatment was either with a commercial hemicellulase preparation or with a crude preparation of snail hepatic juice. Hyphae or hyphal fragments were digested in a medium containing sucrose (1–2 % w./v.) or rhamnose or sorbose (5–10 % w./v.). The "protoplasts" were extruded through pores in the side or end walls of the hyphal cells and were spheres about 10 μ in diameter, containing several nuclei. Most grew to 50–100 μ and then contained a hundred or more nuclei. The spherical forms lysed in distilled water to leave a delicate membrane or wall.

III. Morphology and Structure

A. SPHERICAL FORMS

The word protoplast is sometimes used for any spherical form of a habitually rod-shaped organism, but even this inadequate definition is not universally accepted (e.g., Nomura and Hosoda[51] published electron micrographs of "protoplasts" of *Bacillus subtilis* which were manifestly not spherical). It does seem, however, that all true protoplasts and most "protoplasts" are spherical in form (the exception of dumbbell forms is described in Section VI, L, 2). The reason for this is that it is the mucopolysaccharide cell wall in Gram-positive bacteria and a chemically similar component of the wall in Gram-negative organisms which imparts and maintains the characteristic nonspherical form of many bacterial cells. In the absence of this material the internal osmotic pressure (5–30 atmospheres) causes the protoplast to assume a spherical form. However, there is no *a priori* reason why a nonspherical protoplast should not exist. It would have to have a membrane capable of withstanding the high osmotic pressure without buckling or, alternatively, the cytoplasm itself would have to have sufficient internal cohesion to maintain its shape. But there is no evidence that either of these situations obtains (Figs. 23–28).

B. Number of Protoplasts per Cell

The number of protoplasts which emerge from a bacterial cell depends on how "cell" is defined. In the case of coccoid forms, one spherical protoplast is derived from each coccus; but in the case of rod-shaped bacteria one, two, and, sometimes, more spheres can emerge from a single rod. If the process of digestion of the cell wall is watched under the microscope, the individual protoplasts can be seen clearly before the removal of the wall is complete; it is apparent that cross walls or septa may divide a rod into two or more units, each of which yields a protoplast. Since staining methods indicate that each unit also contains a chromatinic body (see, e.g., Robinow[105]) it is logical to consider each unit as a cell, whether one calls the rod a multicellular organism or an association of organisms. At any rate, Robinow concludes that each cellular unit gives rise to one free protoplast, which must be considered to represent a complete cell lacking only its cell wall.

Fitz-James[233] has studied the cytology of protoplasts of *Bacillus megaterium*. He showed that the suspension media used during formation and stabilization of protoplasts caused aggregation of the chromatin. No DNA was released during the conversion but about 12% of the RNA was lost (contrast the earlier reports that neither P nor UV-absorbing substances were released during conversion of cells to protoplasts[13, 120]). A short rod containing two recently divided nuclear bodies gave a single protoplast while longer rods in which each of the two nuclear structures was itself dividing, usually yielded two protoplasts on lysozyme treatment. This occurred in spite of apparent incomplete septum formation. The chromatin which had condensed on transfer of the cells to sucrose-phosphate gradually expanded after the protoplasts were formed until its distribution after some hours resembled that in actively growing cells. Protoplasts in strongly cationic suspension media (e.g. succinate-citrate buffer) showed condensation of the nuclear bodies into a double structure with the chromatin arranged on a core. The cytology of osmium and formaldehyde fixed protoplasts was also studied.

C. Phase Contrast Microscopy of Protoplasts of Gram-Positive Species

When protoplasts are observed with the phase contrast microscope they are seen to have a diameter about equal to that of the rod or coccus from which they were derived. Sometimes, those formed from bacilli are somewhat larger in diameter, as would be expected if a cylinder whose length was more than two-thirds of its diameter were converted into a sphere of the same volume.

The contrast and internal structure visible are often greater in protoplasts than in whole cells, particularly with the larger forms; sometimes there appears a clear differentiation between a central core and a peripheral zone. It is not known whether this is related to the chromatinic body which must lie within the protoplast. Granules ranging in size from several hundred mμ down to the limit of resolution are also visible in some species and may be lipid granules, as in *Bacillus megaterium* (composed of a polymer of β-hydroxybutyric acid.[106]) (Fig. 23 a).

D. Phase Contrast Microscopy of "Protoplasts" of Gram-Negative Species

Digestion of *Escherichia coli* and other Gram-negative organisms with lysozyme at alkaline pH values or in the presence of chelating agents to yield "protoplasts" can be followed with phase contrast microscopy but not much detail can be seen because of the small size of the structures involved. Ultimately more or less all of the cells become spherical. Fraser et al.[107] who use a lysozyme/EDTA method, point out that macroscopically a useful semiquantitative criterion for the degree of "protoplasting" (*sic*) is the lack of streaming birefringence of the spherical forms as compared with the initial rod-shaped cells.

The methods involving growth of Gram-negative species in the presence of penicillin or absence of DAP give a clearer picture of what happens during conversion to "protoplasts" since there is considerable increase in size during the process. The sequence of forms has been described above (Fig. 6) and involves swelling, development of a bubble or bladder and ultimately conversion to a spherical structure. This may be 4 μ or more in diameter and since it is derived from a rod perhaps 0.75 μ × 2 μ, the surface area may have increased 10-fold and the volume 40-fold.[31, 32] The spherical forms produced from *Escherichia coli* by penicillin treatment or DAP-deprival are often vacuolated and some typical forms are illustrated in Fig. 8 (Figs. 23 d, 26 a, b, 27 c). It is not clear what causes the crescent-shaped vacuoles and whether or not a membrane bounds the dense part of the cytoplasm. Similar forms were found with penicillin-treated cultures of *Proteus vulgaris*; Liebermeister and Kellenberger[75] described and illustrated the change from rods to spheres, suggesting that the remnants of empty cell wall which were attached to the "protoplasts" in the rabbit ear stage were eventually sloughed off or otherwise destroyed.

On continued incubation, "protoplasts" of Gram-negative organisms enlarge, often lyse to leave a visible "ghost," or become progressively less

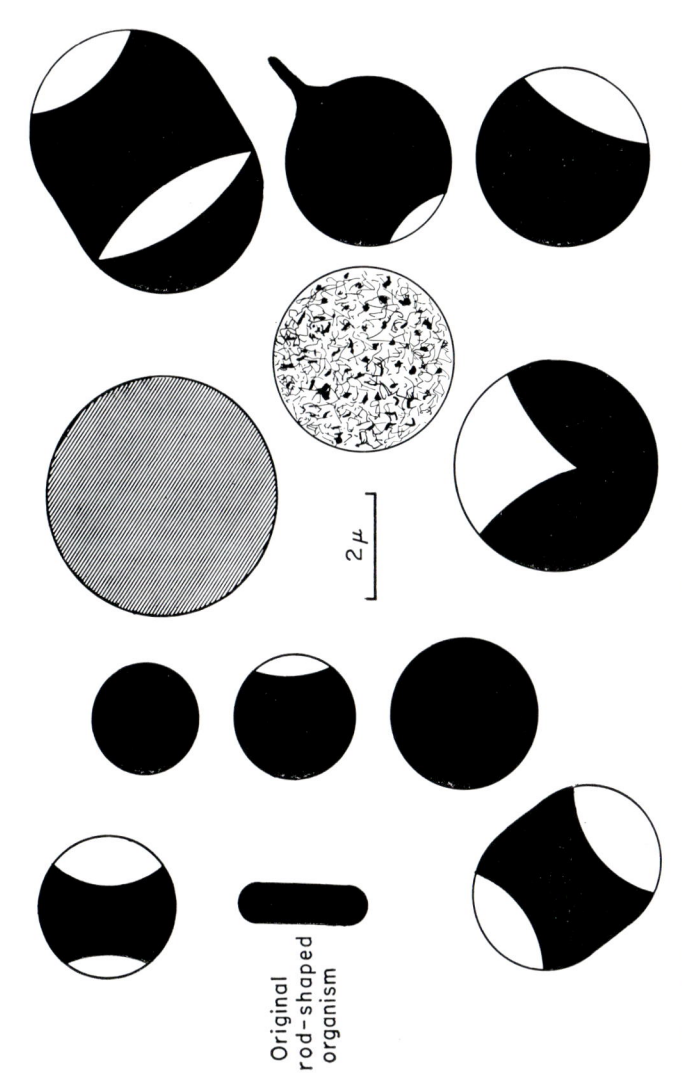

FIG. 8. Representation of typical forms of *Escherichia coli* grown in the presence of penicillin or deprived of diaminopimelic acid.

dense. Some become highly granular and may be related to the "large bodies" which precurse L-forms (see Chapter 7 for a more detailed account of L-forms).

E. Fixation of Protoplasts

Because of their great fragility, it is not possible to dry films of protoplasts without lysis occurring. For some purposes, therefore, it is necessary to fix preparations. Heating suspensions before or after enzymic treatment can be used. If vegetative cells of *Bacillus megaterium* are heated before lysozyme treatment, the resulting protoplasts are prefixed and emerge as stubby, rod-shaped forms instead of spheres (see Figs. 3, 25 e, f). The polypeptide antibiotic, polymyxin, has an effect similar to heating.[108] Formalin, osmium tetroxide, etc., can be used for fixation and in all cases specimens can subsequently be washed with distilled water.

F. Staining of Protoplasts

No general studies have been reported of the staining characteristics of any kind of bacterial protoplasts but two papers mention Gram-staining of protoplasts of *Bacillus megaterium*. Gerhardt et al.[109] dried smears of preparations of protoplasts, cell walls, and protoplast membranes without heat fixing and found them all to be Gram-negative by the Hucker modification of the reaction. So were formalin-fixed preparations. On the other hand, Amano et al.[110] reported that lysozyme-prepared protoplasts of the same strain, KM, of *B. megaterium* were Gram-positive if fixed for 5 to 10 minutes with formalin (10%) in sucrose solution (5%). Protoplasts and intact cells of *B. megaterium* and *Escherichia coli* were all treated on the same slide. Their results are shown in Table II. Gerhardt et al.[109] did show, however, that cells which had been subjected to the Gram stain and

TABLE II

Gram-Staining Reaction of Protoplasts of *Bacillus megaterium*[a,b]

Decolorization time (seconds)	*Bacillus megaterium* Cells	*Bacillus megaterium* Protoplasts	*Escherichia coli* Cells
30	++	++	+
60	++	++	+
75	++	+	−
90	++	+	−
120	++	+	−

[a] Data of Amano et al.[110]

[b] Protoplasts fixed in sucrose solution (7.5%) containing formalin (10%) for 5–10 minutes before Gram-staining on same slide as cells of *B. megaterium* and *E. coli*.

++: deep violet; +: faint violet; −: Gram-negative.

then treated with lysozyme remained Gram-positive and resisted a second decolorization, unless they were crushed under a coverslip.

Weibull[111] used Indian ink mounting to investigate the relationships between cells and protoplasts and also showed by flagellar stains that the organs of motility were still attached to the protoplasts after the cell wall was removed.[13]

G. Electron Microscopy

1. Protoplasts of Gram-Positive Species

Electron micrographs of fixed preparations (osmium tetroxide or formalin) of protoplasts of *Bacillus megaterium*,[13, 14, 17, 41, 108] *B. subtilis*,[23] and *Micrococcus lysodeikticus*[22] show spherical bodies, very electron-dense and still bearing flagella (in motile species) but lacking a cell wall such as can be seen in intact cells. Weibull and Thorsson[112] have cut thin sections of *B. megaterium* protoplasts embedded in methacrylate and found that these preparations showed the same internal structure as did sections of vegetative cells. The structures of protoplasts of *Bacillus megaterium*, penicillin-spheroplasts of *Escherichia coli* and L-forms of *Proteus vulgaris* have been investigated by Thorsson and Weibull[234] using ultrathin sections. The absence of all cell wall structure from the protoplasts was noted although the cytoplasmic membrane (c. 80 A. thick) was observed. Spheroplasts of *E. coli* showed a residual wall structure and a separate inner membrane, probably the cytoplasmic membrane. Salton and Shafa[235] found that disintegrated spheroplasts of *Vibrio metchnikovii* formed by growth in penicillin showed two concentric surface layers, the outer of which was thought to be residual cell wall and the inner to be the cytoplasmic membrane. In addition, there sometimes appeared a peripheral zone around the cytoplasm which was probably the membrane but the cell wall was entirely absent. Sections were also made of protoplast "ghosts"—the membrane plus granule fraction obtained by osmotic lysis of protoplasts. Weibull[14] and McQuillen[17, 41] show electron micrographs of such "ghosts" as they appear without sectioning (Figs. 24; 25 a, b, c, d).

2. "Protoplasts" of Gram-Negative Species

Liebermeister and Kellenberger[75] reproduce excellent electron micrographs of globular forms of *Proteus vulgaris* prepared by the agar filtration method of Kellenberger.[113, 114] It is clear that these forms are surrounded by some kind of membrane but its relationship to the cell wall is not apparent. This is also the case with "protoplasts" of *Escherichia coli* as is seen from McQuillen's[32] electron micrographs of penicillin-treated and DAP-deprived organisms. Here the fixed protoplasm is seen to be contracted away from a very delicate membrane through which pass flagella and also fibers considerably thinner than flagella (perhaps "fimbriae"[115-117]). These fibers vary in length from about 0.2 to 2 microns (Figs. 26 c, d; 27 a, b).

IV. Physicochemical Properties of Protoplasts

A. STABILIZING MEDIA

1. SUCROSE AND POLYETHYLENE GLYCOL

Most studies on protoplast formation, physiology, and biochemistry have been carried out in media of comparatively high osmotic pressure since this seems to be required for maintenance of structural integrity. However, it is not simply a matter of using a particular concentration of solute. As Weibull[15] pointed out "the stabilization phenomenon can hardly be explained in exclusively osmotic terms. Equimolecular concentrations of sucrose, sodium chloride and cellobiose give good, moderate and no stability, respectively, to the protoplasts of *Bacillus megaterium*." The solute must not be able to penetrate the osmotic barrier of the protoplast at an appreciable rate if the solution is to give protection from lysis. However, sucrose (molecular weight 342) at a concentration of 3.4 to 6.8% (w./v.) has about the same stabilizing effect as has polyethylene glycol (average molecular weight 4000) at a concentration of 7.5% (w./v.). Media containing one or the other of these two solutes have commonly been used for stabilization but the required concentration of, for example, sucrose varies from 0.1 M for *B. megaterium* to more than 1.0 M for some cocci (e.g., *Micrococcus lysodeikticus*[21, 22] and *Sarcina lutea*[21]). This can be correlated to some extent with differences in internal osmotic pressure in different species.

A wide variety of other stabilizing media has been used from time to time by different workers, but sucrose solutions are most generally used, particularly for Gram-negative "protoplasts." Much work has been done with *Escherichia coli* preparations treated with lysozyme and EDTA or with penicillin; it is usual to supplement media with 10 to 20% sucrose.

2. EFFECTS OF Mg^{++}

When penicillin (100–1000 μg./ml.) is employed, Lederberg[59] advises the addition of Mg-sulfate (0.01 M)—"the high requirement may depend partly on binding with the sodium citrate used in the compounding of the penicillin." However, there is probably much more to the Mg^{++} effect than this. Harrington[66] (see McQuillen[31]) found that 0.02 M Mg^{++} was necessary to prevent the lytic action of penicillin (5 μg./ml.) added to cultures of *Escherichia coli* B growing in synthetic media. This small concentration of antibiotic could not conceivably require such an addition of Mg^{++} to combine with citrate. The presence of added citrate in the absence of Mg^{++} enhanced the lytic effects of penicillin but this could be overcome by addition of a stoichiometric amount of Mg^{++}. It appears that the Mg ion may play some role in stabilization, perhaps acting on the cytoplasmic membrane.

Weibull[15, 118] found that protoplasts of *Bacillus megaterium* lysed in distilled water to give membranes which rapidly disintegrated into small fragments. If, however, Mg^{++} (0.01 M) were added to protoplasts, lysis resulted in much better preservation of the intact protoplast membrane. About 10% less N and 260 mμ-absorbing material was left in the supernatant after centrifuging lysates made in the presence of Mg^{++}, as compared with those made in its absence. Amano et al.[119] found that Mg^{++} (0.02 M) prevented the action of "plakin" (a substance extracted from blood platelets) which is believed to act on the cytoplasmic membrane of Gram-positive bacteria. Plakin destroys protoplasts of *B. megaterium* and, if used on intact cells, prevents their subsequent transformation to protoplasts by lysozyme.[110]

The exact point of involvement of Mg^{++} is not known but B. D. Davis (unpublished observations) found that it was not necessary for preparing "protoplasts" of *Escherichia coli* by DAP-deprival and that furthermore, such globular forms were apparently much tougher than the corresponding forms produced under the influence of penicillin.

3. SERUM ALBUMIN

Zinder and Arndt[58] found that "protoplasts" of *Escherichia coli* formed by the action of lysozyme at pH 9.0 would lyse in broth media whether or not a high concentration of sucrose, NaCl, or dextran were used to raise the osmotic pressure. However, the addition of bovine serum albumin prevented lysis; their medium (P-broth) consisted of Difco Penassay broth containing sucrose (0.6 M), serum albumin (2% w./v.), and Mg^{++} (0.1%). The use of serum albumin in media for L-forms is discussed in Chapter 7.

B. OSMOTIC PROPERTIES OF PROTOPLASTS

Stähelin,[4] Weibull,[15, 111, 118, 120] Mitchell and Moyle[12, 21, 35, 121] and Gilby[22] have investigated the osmotic properties of bacterial protoplasts and have reached similar conclusions.

1. *Bacillus megaterium*

Weibull[111] first established that in vegetative cells of *Bacillus megaterium* there exists an osmotic barrier at the surface of the protoplasm and that the cell wall is in all probability permeable to small molecular weight solutes. The osmotic barrier is likely to correspond with the cytoplasmic membrane and is impermeable to phosphate, sucrose, and diphosphopyridine nucleotide. He then showed that protoplasts prepared by digesting away the wall with lysozyme had similar permeability properties but that whereas the intact bacilli resisted change in size or shape in media of varying sucrose or polyethylene glycol concentration by virtue of their rigid walls, the spherical protoplasts shrank or swelled as the solute concentration was raised or lowered over a certain range.[120] Sucrose was studied over the range 0.125 M to 0.500 M and polyethylene glycol between 5.8% and 16% (w./v.). The protoplasts behaved as osmometers with a permeability barrier at their surface and a "water space" which was less than their total volume by an amount equivalent to the osmotically inert components of the cytoplasm. The volume, V, of the protoplasts was related to the osmotic pressure, π, over the stable range by the Boyle-Van't Hoff relation: $\pi(V - V_o) = C$ (constant) where V_o is the volume of the osmotically inert proteins, lipids, etc., of the protoplast.[120] A curious observation by Weibull[120] which has not been explained or found by workers with other species, was that the volume of a protoplast was independent of the osmotic pressure of the medium in which it was produced, but varied as described above if the osmotic pressure was altered subsequently. Stähelin[4] had earlier shown that the osmotic properties of "protoplasts" of *Bacillus anthracis* formed by plasmoptysis were qualitatively similar to those just described. The osmotic behavior of protoplasts of *Bacillus megaterium* both fixed and unfixed, was studied by Fitz-James.[233] Both formaldehyde-fixed and osmium-fixed protoplasts could be expanded and contracted by altering the nature of the suspending medium.

2. *Micrococcus lysodeikticus*

Gilby[22] has worked with lysozyme-prepared protoplasts of *Micrococcus lysodeikticus* suspended in NaCl. The optical density (O.D.) at 500 mμ (O.D.$_{500}$) was found to increase with the osmotic pressure, π, so long as the protoplasts were stable. (In sucrose the change in refractive index of the medium masked changes of O.D. due to size variation of the protoplasts.) Assuming that the volume, V, of the globular forms varies inversely with the O.D.$_{500}$, i.e., $V = K/\text{O.D.}_{500}$ where K is a constant, then the Boyle-Van't Hoff relation can be written:

$$\pi/\text{O.D.}_{.500} = \pi \cdot V_o/K + C/K$$

and $\pi/\text{O.D.}_{.500}$ when plotted against π should be linear. When this was done it was found that for *M. lysodeikticus* protoplasts there was linearity over the range 30 to 90 atmospheres.

3. Penetration of Solutes into Protoplasts of *Staphylococcus aureus*, *Micrococcus lysodeikticus*, and *Sarcina lutea*

Mitchell and Moyle[12] studied the conditions necessary for stability of protoplasts of *Staphylococcus aureus*, *Micrococcus lysodeikticus*, and *Sarcina lutea* in NaCl and sucrose solutions and concluded that protoplasts have membranes at their surface resembling the osmotic barrier of intact cells of *S. aureus*. All three species required a concentration of about 1.0 M NaCl or sucrose for stability, whereas in this concentration of glycerol the protoplasts lysed very rapidly. The authors have compiled many data on the permeability of cells and protoplasts to a variety of inorganic ions, sugars, polyhydric alcohols, etc. "There was little evidence for any structural rigidity within the 'protoplast' of *Micrococcus lysodeikticus* or *Sarcina lutea* and there is, therefore, little doubt that the cell wall must be capable of withstanding a hydrostatic pressure of 20 atmospheres or more."[12] By observing the rate of lysis (measured by reduction in light scattering at 700 mμ) of protoplasts suspended in solutions of various solutes at a concentration of 1.5 molal, the relative penetration rates were estimated. Some of the results obtained are recorded in Table III. They are all consistent with the idea that the protoplast membrane is the structure which in intact cells acts as the osmotic barrier and is located at the surface of the cytoplasm.

TABLE III

Relative Penetration Rates of Solutes into Protoplasts of *Micrococcus lysodeikticus*, *Sarcina lutea*, and *Staphylococcus aureus*[a]

M. lyso	S. lutea	Staph. aureus	Solute	Time for 50% lysis
+	+	+	Glycerol	<3 sec.
+	+	+	Erythritol	20 sec.
+	+	+	D-Ribose	5 min.
		+	D-Sorbitol	20 min.
+	+	+	L-Arabinose	30 min.
+	+		D-Sorbitol	60 min.
+	+	+	KCl, NaCl, NH$_4$Cl, KBr, Na-acetate (pH 9), K-acetate (pH 9), K$_2$SO$_4$, (KH$_2$PO$_4$ + K$_2$HPO$_4$), Na-glutamate, lysine·HCl, D-glucose, D-fructose, D-mannose, D-galactose, D-sorbose, sucrose.	Very long
+	+		MgCl$_2$, NaBr	Very long
		+	KNO$_3$	Very long
+	+		Urea[b], glycine[b]	<3 sec.

[a] Data of Mitchell and Moyle.[12, 21]

[b] It is thought that urea and glycine act directly on the protoplast membrane rather than cause lysis by rapidly penetrating through the membrane. Results based on light scattering measurements (700 mμ) of suspensions of protoplasts in 1.5 molal solutions of the various solutes.

C. Chemical Damage to the Protoplast Membrane

1. Surface Active Substances and *Bacillus megaterium*

It is generally believed that surface active agents and certain other disinfectants probably act on the cytoplasmic membrane of bacteria and abolish its semipermeable properties, causing leakage of cytoplasmic constituents into the medium. Tomcsik[122] studied the effect of pretreatment of *Bacillus megaterium* (*Bacillus* M) with bactericidal substances, including phenol, cationic and anionic detergents, etc., on the subsequent action of lysozyme. Concentrations of these agents which did not affect the viability of the cells, did not prevent transformation to spherical protoplasts by later digestion of the cell wall. On the other hand, if bacteria were killed by the disinfectant, the lysozyme treatment resulted in pale, cell wall-free rods or, sometimes, pale or empty spheres. A group of mercuric compounds which caused bacteriostasis did not affect the transformation to spherical protoplasts; Tomcsik suggested that inability to undergo transformation to spheres was a sign of death.

Cetyltrimethylammonium bromide (CTAB) and digitonin were found by McQuillen (unpublished observations) to cause rapid lysis of protoplasts of *Bacillus megaterium*.

2. Plakin

The substance "plakin", extracted from blood platelets by Amano *et al.*[110,119] and active against various Gram-positive bacteria, seems to cause damage to the cytoplasmic membrane. Although it is activated by Ca^{++} or Mg^{++} and can be inactivated by EDTA, a concentration of $0.02\ M\ Mg^{++}$ entirely prevents its effect.

3. Ionic Detergents and *Micrococcus lysodeikticus*

Gilby[22] and Gilby and Few[123,239] compared the concentrations of ionic detergents which killed 99.9% of cells of *Micrococcus lysodeikticus* with the concentrations which gave 90% of maximum reduction in O.D. (optical density) at 500 mμ (this is roughly equivalent to the concentration for 90% lysis). The sequence was the same and the relative concentrations similar for both killing and lysis. It was concluded that these agents act on the protoplast membrane. Table IV records some of the results.

Gilby[22] has also shown that uranyl nitrate inhibits lysis of protoplasts by cationic detergents. A critical concentration of $4 \times 10^{-4}\ M\ UO_2^{++}$ caused agglutination of these forms, as well as inhibition of lysis. He suggested that "anionic detergents react with the protein of the protoplast membrane and by destruction of its mechanical rigidity and cohesion can cause complete dispersion. The action of cationic detergents, on the other hand, is confined to the membrane lipid."

TABLE IV

COMPARISON OF CONCENTRATIONS OF IONIC DETERGENTS NECESSARY TO KILL CELLS AND LYSE PROTOPLASTS[a] OF *Micrococcus lysodeikticus*[b]

	Detergent	Concentration for 90% lysis of protoplasts, $M \times 10^4$	Concentration for 99.9% killing of intact cells, $M \times 10^4$
Cationic	Dodecylamine·HCl	1.8	2.1
	Dodecyltrimethyl ammonium bromide	3.2	4.0
Anionic	Sodium dodecyl sulfate	6.0	12.0
	Sodium dodecyl sulfonate	14.2	26.0

[a] Protoplasts suspended in sucrose/NaCl medium. Lytic concentration is concentration which gave 90% of maximum reduction in optical density at 500 mμ in 30 minutes.

[b] Data of Gilby and Few.[123]

4. POLYMYXIN

Newton[124] has reviewed the evidence which suggests that the site of action of the polypeptide antibiotic, polymyxin, is also the protoplast membrane. Few and Schulman[125] found that polymyxin-sensitive bacteria and their isolated cell walls fixed 4–5 times as much antibiotic as did resistant organisms. However, when sensitive cells were treated with a fluorescent derivative of polymyxin and then fractionated, Newton[108] found that only 10% of the fluorescence was associated with the cell wall and about 90% with a "small particle" fraction which sedimented at 100,000 g. Also, if cell walls and "small particles" were prepared and the walls incubated with the polymyxin derivative, 270 μg. were fixed per milligram dry weight of cell wall (not per milligram of cells, as is stated in the original paper[108]). This could not be removed by repeated washing with distilled water. On adding the "small particles," however, about 90% of the fluorescence was transferred to this fraction in 10 minutes at 30°. In addition, lysozyme treatment of cells "labeled" with fluorescent polymyxin removed the cell wall from the fixed protoplasts without releasing fluorescent material. Supersonic disruption of the fluorescent protoplasts resulted in all the fluorescence being associated with particles sedimenting at 100,000 g.

5. EFFECT OF pH VALUE ON THE STABILITY OF PROTOPLASTS

There are virtually no reports on the effects of the pH value of the suspending medium on the integrity of protoplasts but Gilby[22] found that while intact cells of *Micrococcus lysodeikticus* did not show any leakage of 260 mμ-absorbing material at pH values down to 4.4, protoplasts of the same species were stable only over the range pH 5.5–8.5. They were ex-

TABLE V
Lytic Concentrations of Alcohols[a] for Protoplasts of *Micrococcus lysodeikticus*[b]

Alcohol	Concentration causing lysis of protoplasts suspended in sucrose (1.0 M) and NaCl (0.05 M)
Ethanol	4.0 M
Propanol	1.0 M
n-Butanol	0.25 M
Isoamyl alcohol	0.08 M

[a] These concentrations of alcohols have similar thermodynamic activities.
[b] Data of Gilby.[22]

tremely sensitive immediately below pH 5.5 and were gradually destroyed at pH values above 8.5.

6. Alcohols

When protoplasts of *Micrococcus lysodeikticus* in sucrose (1.0 M) containing NaCl (0.05 M) were treated with a series of alcohols, lysis was caused at concentrations widely differing but at approximately equal thermodynamic activities[22] (Table V).

7. Linoleic Acid and Vitamin D_2

Some investigations on the effects of the unsaturated fatty acid, linoleic acid, on protoplasts have been made by McQuillen and Kodicek (unpublished observations). Kodicek[126] had earlier shown that various unsaturated fatty acids inhibited the growth of *Lactobacillus casei* and that addition of certain steroid compounds in approximately equimolecular amounts could reverse this effect. Substances with vitamin D activity were effective but cholesterol and other sterols were inactive except at high concentrations. He suggested that the site of action might be the osmotic barrier of the cells. (It should be mentioned that steroids do not, so far as is known, occur in bacterial cells and vitamin D is not a known growth factor for any species, so that this lesion caused by fatty acids and cured by vitamin D is a most unnatural form of "rickets.") McQuillen and Kodicek found that linoleic acid (100 μg./ml.) caused visible lysis of protoplasts of *Bacillus megaterium* (250 μg. dry weight/ml.) in a few minutes; a concentration of 10 μg./ml. caused slow lysis over a period of at least an hour; 1 μg./ml. and lower concentrations did not cause lysis, even during 12 hours' incubation. Vitamin D_2 did not have any deleterious effects at 500 μg./ml. However, the effect of 100 μg./ml. of linoleic acid was prevented by adding 250 μg./ml. vitamin D_2, whereas even 500 μg./ml. cholesterol had no protective ac-

TABLE VI

pK Values of *Micrococcus lysodeikticus* and Some of Its Components, Derived from Electrophoretic Mobility Measurements[a]

Material	pK$_a$	pK$_b$
Intact cells	2.1	8.6
Protoplast membranes[b]	3.6	—
Defatted protoplast membranes[b]	3.6	9.95
Lipid fraction of protoplast membranes[b]	0.4	—

[a] Data of Gilby and Seaman, see Gilby.[22]
[b] Methods of preparing these fractions are given by Gilby et al.[137]

tion. Similarly, growth of protoplasts (see Sections VI, L, 1 and 2) was possible in media containing 10 μg./ml. linoleic acid only if vitamin D$_2$ (50 μg./ml.) were present. Low concentrations of linoleic acid also reduced the incorporation of radioactive tracers (C^{14}-labeled glucose, acetate, and glycine) into protoplasts; this effect could be nullified by adding vitamin D$_2$. Similar effects were observed with intact cells of *B. megaterium* and *L. casei*. It was later found that transport reactions were not inhibited but that the subsequent incorporation of amino acids and bases into proteins and nucleic acids was sensitive to linoleic acid.[195]

D. Electrophoretic Studies on Protoplasts

The method of microelectrophoresis has considerable use in investigating the nature of the surfaces of bacteria.[71, 127-132] Gilby[22] has reported experiments carried out in collaboration with G.V.F. Seaman on electrophoresis of *Micrococcus lysodeikticus* and its subfractions. Isolated protoplasts and the protoplast membranes derived from them behaved similarly as to variation of surface density of charge with pH value, etc. Table VI records the pK values of various structures deduced from this electrokinetic study.

Results of measurements carried out in the presence of UO$_2^{++}$ (cf. McQuillen[129]) and Th^{++++} suggested that, in the protoplast membrane and in the isolated protoplast, the lipid is not exposed. This lipid has groups with a pK of 0.4, whereas the protoplasts, membranes, and defatted membranes all have pK values in the region of pH 3.6. It is more likely that the protein component of the protoplast membrane is exposed and that the lipid is underlying, perhaps as the middle layer of a triplex structure.

V. Composition of Protoplasts

A. *Bacillus megaterium*

Since the strict definition of a protoplast is the whole cell minus the cell wall, it is apparent that most of the cellular constituents are present in the protoplast. The whole cytoplasm including the chromatinic body (nuclear

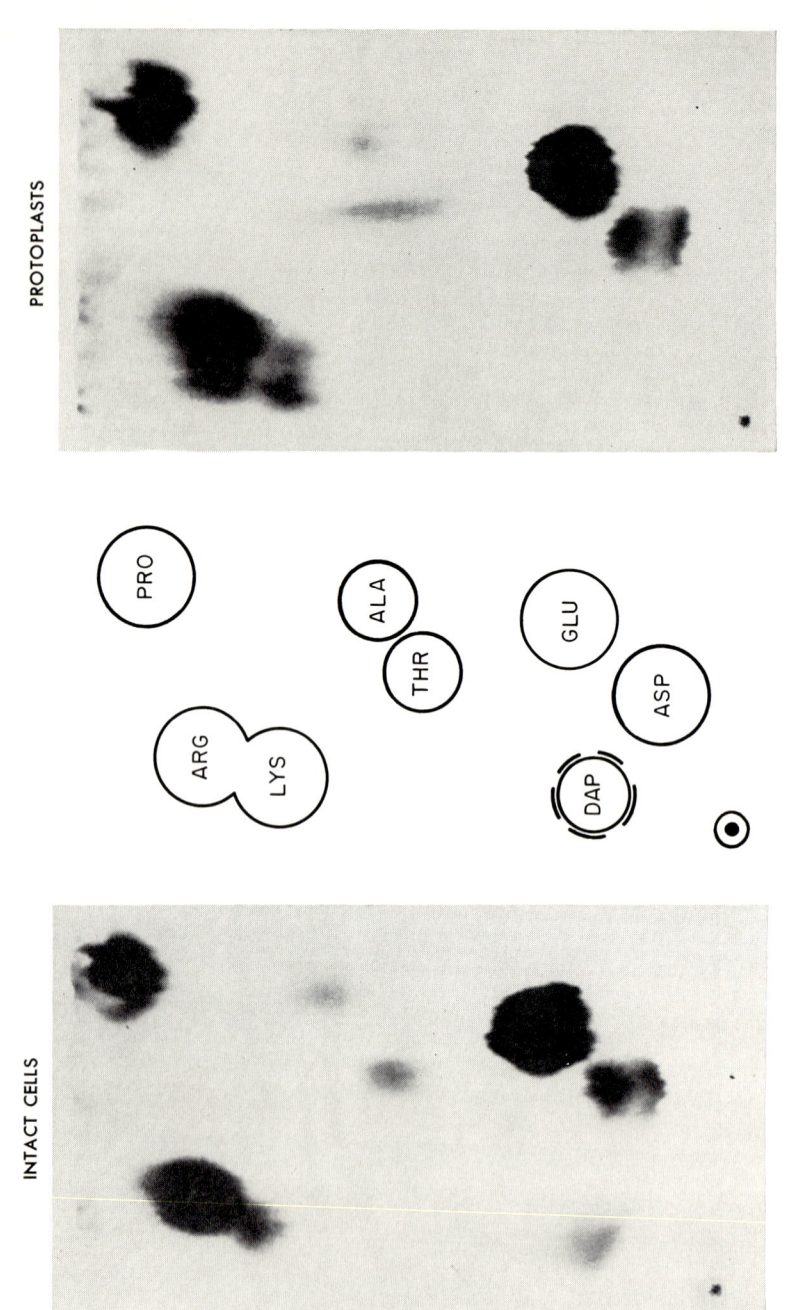

Fig. 9. Radioautograph of amino acid chromatogram of hydrolyzate of trichloroacetic acid-insoluble fractions of intact cells (left) and protoplasts (right) from *Bacillus megaterium* KM. The cells and protoplasts had been allowed to grow in the presence of C^{14}-acetate. Diaminopimelic acid does not occur as a structural component of protoplasts but can be found excreted into the medium (McQuillen, unpublished radioautograph).

apparatus) and granules, particles, inclusions, etc., and the cytoplasmic membrane are recovered in the protoplast. There is no loss, even of small molecular weight cytoplasmic constituents; the only remaining possibility is that between the cytoplasmic membrane and the cell wall there are minor components, which are released into the medium when the wall is removed. Since there is no detectable ultraviolet-absorbing material present in the medium after lysozyme treatment of *Bacillus megaterium*,[13] there can be little or no protein, nucleic acid, or coenzyme lost from the cell during conversion to the protoplast form.

The constituents unique to the cell wall structure are not, of course, found in protoplasts except as transient intermediates. Fig. 9 shows some of the amino acids which become radioactive when intact cells and protoplasts of *Bacillus megaterium* grow in the presence of C^{14}-acetate. The only difference is the complete absence of the diaminopimelic acid (DAP) component from protoplasts.[17]

B. Chemical Composition of "Protoplasts" of *Escherichia coli*, L- Forms of *Proteus vulgaris* and Pleuropneumonia-like Organisms (PPLO)

Weibull[80] estimated the content of diaminopimelic acid (DAP), hexosamine and hexose in the trichloroacetic acid-precipitable fraction of normal *Proteus vulgaris* and the corresponding stable L-forms (see Chapter 7). Table VII shows that the latter contain much less of these substances than do the bacillary forms. It was also shown that practically all of these substances were present in the cell wall fraction; consequently, Weibull[80]

TABLE VII

Analyses of Trichloroacetic Acid-Precipitable Fractions of Normal Cells and Stable L-Forms of *Proteus vulgaris*[a]

Constituent	Content (% by weight) in TCA-precipitable fraction			
	Normal Cells		L forms	
	Medium a	Medium b	Medium a	Medium b*
Hexose	1.1	1.2	0.1	0.2
			0.2	0.1
Hexosamine	1.13	1.25	0.32	0.03
			0.26	0.09
Diaminopimelic acid	0.7	0.7	0.2	0.1
			0.2	0.2

* Media a and b are two different synthetic media which were used.
[a] Data of Weibull.[80]

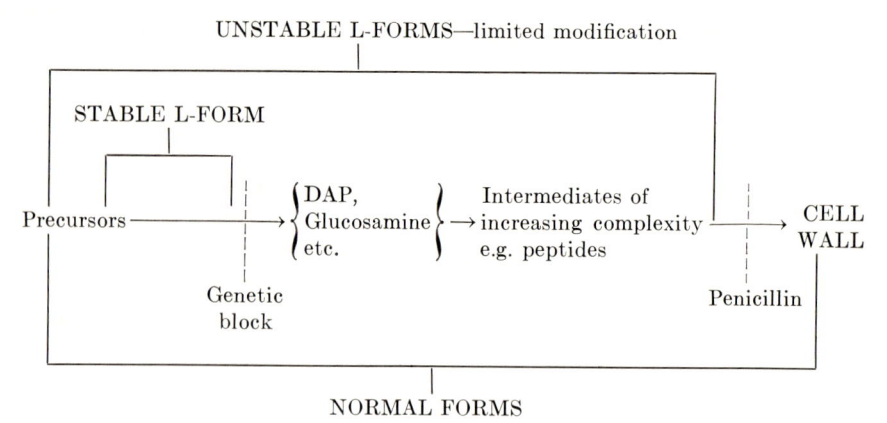

```
                    UNSTABLE L-FORMS—limited modification
    ┌──────────────────────────────┬───────────────────────────────┐
    │ STABLE L-FORM                                                │
    │      ┌──────────┐                                            │
    │      │          ┊   ⎧ DAP,        ⎫                          │
    │ Precursors ─────┼─→ ⎨ Glucosamine ⎬ → Intermediates of       │  →  CELL
    │                 ┊   ⎩ etc.        ⎭   increasing complexity  │     WALL
    │                 ┊                     e.g. peptides          │
    │              Genetic                              Penicillin │
    │               block                                          │
    └──────────────────────────────────────────────────────────────┘
                                    │
                              NORMAL FORMS
```

FIG. 10. Representation of the interrelations between bacillary forms, unstable L-forms and stable L-forms of *Proteus vulgaris*, according to Kandler and Zehender.[78]

concluded that stable L-forms contain much less but not negligible amounts of cell wall material.

Kandler and Zehender,[78] however, reached different conclusions from qualitative analyses of 12 stable L-forms of *Proteus vulgaris*, 6 unstable L-forms (capable of reversion to bacillary forms) and 3 strains of pleuropneumonia-like organisms (PPLO). Neither DAP nor glucosamine was found in the stable L-forms or in the PPLO strains but both substances were present in the unstable L-forms. The authors suggest that the scheme shown in Fig. 10 represents the situation in the various kinds of organism. Unstable L-forms are partly modified in cell wall synthesis; penicillin is postulated to interfere with the last stages of wall synthesis; and stable L-forms are thought to have a genetic block in the synthesis of some essential constituent, such as DAP or glucosamine.

Kandler et al.[240] carried out further analyses of *Proteus vulgaris* in the bacillary, unstable L- and stable L-forms. Stable L-forms did not contain any diaminopimelic acid nor was a cell wall fraction obtainable by the technique used—this involved extracting cells with $1.0\ N$ NaOH at 60°C. for two one-hour periods and then for 12 hours at 37°; the insoluble residue was used as cell wall. Unstable L-forms cultivated in the presence of penicillin were found to contain in their wall fraction substantially *more* diaminopimelic acid, glutamic acid, alanine, glucosamine, and muramic acid than did the walls of the bacillary form. Moreover, the diaminopimelic acid content of whole L-forms was 2 to 3½ times greater than that of whole bacillary forms. These findings are surprising in view of the many other data which suggest that penicillin inhibits the synthesis of the mucopeptide component of bacterial cell walls.

McQuillen[31] found that conversion of *Escherichia coli* to globular forms by growth in the presence of penicillin resulted in preparations which contained much less DAP in their trichloroacetic acid-precipitable fraction. No quantitative determinations were carried out, however, but it is perhaps relevant that similar forms could be produced by growth of a DAP-requiring mutant in the complete absence of this amino acid.[32] Salton and Shafa[235] analyzed penicillin spheroplasts of *Vibrio metchnikovii* and *Salmonella gallinarum* and found the amino sugar and diaminopimelic acid content to be reduced by 30–50% as compared with normal cells. Furthermore, the fraction of cell wall digestible by lysozyme (part of the mucopeptide) was almost completely absent from the penicillin spheroplasts of both species. These results are consistent with the idea that penicillin interferes directly or indirectly with formation of the mucopeptide component of these Gram-negative cell walls. The lipid and polysaccharide components (and probably the protein) were relatively little affected by growth of the organisms in penicillin media.

C. The Protoplast Membrane

1. Preparation

In most cases, dilution of the suspension medium of protoplasts is enough to cause lysis of these fragile structures, release of the bulk of the cytoplasmic contents and the formation of "ghosts"—delicate membranes containing granules and more or less debris (Fig. 25 b, c, d). Weibull[133] prepared protoplast membranes for analysis as follows: cells of *Bacillus megaterium* (10–20 mg./ml.) in sucrose (0.5 M), NaCl (0.01 M), MgCl$_2$ (0.005 M), and lysozyme (100–200 μg./ml.) were incubated until conversion to protoplasts was complete. The preparation was centrifuged and the pellet resuspended in phosphate buffer (0.02 M, pH 7.0). Lysis occurred immediately and "ghosts" were spun down at 15,000 g and washed 3 times with distilled water. A careful study has shown that the number of "ghosts" recoverable after lysis is almost exactly one for each protoplast if Mg^{++} (0.005–0.01 M) is present. If this cation is not added the membranes disintegrate.[118]

Vennes and Gerhardt[134] added EDTA (0.01 M) and deoxyribonuclease (DNase) (10 μg./ml.) to suspensions of protoplasts of *Bacillus megaterium* KM in buffered sucrose and after allowing 30 minutes at 37° for disruption, spun down the protoplast membranes at 12,800 g for 10 minutes. The pellet was resuspended in NaCl (0.85%) which helped to remove small particles and then washed 5 times with distilled water. Delicate membranous structures were obtained.

Storck and Wachsman[135, 136] prepared protoplasts of *Bacillus megaterium* (11 mg./ml.) in polyethylene glycol (10% w./v.), K-phosphate buffer (0.04

M, pH 7.0), MgSO$_4$ (0.002 M) and lysozyme (400 µg./ml.). The protoplasts were spun down at 13,000 g for 5 minutes and resuspended for lysis in K-phosphate buffer (0.04 M, pH 7.0) containing MgSO$_4$ (0.002 M) and some DNase. Protoplast "ghosts" were spun out at 24,000 g for 10 minutes and washed once with the same medium.

Protoplast membranes can also be prepared by direct lysis of lysozyme-sensitive bacteria without intermediate isolation of protoplasts.[137] For example, Gilby, Few and McQuillen treated *Micrococcus lysodeikticus* (10 mg./ml.) in NaCl (0.1 M) with lysozyme (50 µg./ml.) for 45 minutes. A yellowish pellet was spun down at 20,000 g for 30 minutes. Above was a clear viscous layer (nucleic acid) and a clear supernatant. The pellet, contaminated with a little of the viscous material, was dispersed in NaCl (0.05 M) and then washed twice more with saline, being centrifuged for 10 minutes at 20,000 g each time. The orange-brown product consisted of membranes and parts of membranes with very little contaminating material as judged by electron microscopy and analysis[137] (Fig. 25 d).

Mitchell and Moyle[35] "exploded" autolysates of *Staphylococcus aureus* which contained protoplasts and by repeated centrifugation obtained four practically homogeneous fractions: (1) supernatant; (2) light fraction, bright yellow—protoplast membranes; (3) intermediate fraction, white—cell walls; (4) heavy fraction, pale yellow—residual intact cells. Dispersal of pellets was in distilled water and centrifuging was carried out at 3,500 g for 1 hour at 5°. The protoplast membrane fraction had disintegrated into particles on washing but amounted to about 10% of the dry weight of the cells. Earlier, the same workers described a "small particle" fraction derived from mechanically broken cells of *S. aureus* which they believed came from the cytoplasmic membrane. It amounted to 8.6% by weight.[138]

Many other workers have described small particles containing lipoprotein and it is probable that this material is derived, at least in part, from the cytoplasmic membrane.

2. Chemical Composition

It is generally assumed that the osmotic barrier in bacteria is of lipoprotein nature, although there are few direct analyses. Staining reaction often indicate a lipid zone underlying the cell wall in Gram-positive bacteria[139] (Gram-negative organisms contain lipoprotein in their cell wall). Furthermore, the permeability properties of bacteria are consistent with the existence of a lipid membrane.[12, 121]

a. *Bacillus megaterium*. Weibull[14] differentially centrifuged lysates of *Bacillus megaterium* protoplasts and found that the "ghost" fraction came down between 590 and 14,800 g (15 minutes) and amounted to 10% of the dry weight of the whole. Some granules and practically all of the pigmented material, including the cytochromes and a small amount of the ribonucleic acid, were associated with this fraction.

Vennes and Gerhardt[134] could not detect any diphenylamine-reacting material (Dische test for deoxyribose) but claimed that their preparations of protoplast membranes of *Bacillus megaterium* contained 11% ribonucleic acid. This was based solely on the orcinol reaction; no other tests (absorption spectra, purine/pyrimidine chromatography, etc.) were carried out. It seems, therefore, that the only justifiable conclusion is that about 5% of the weight of these preparations was pentose. The isolated membranes were unaffected by ribonuclease although they were disintegrated by treatment with wheat germ lipase.

Weibull[133] has recently studied the lipids of a strain of *Bacillus megaterium* (*Bacillus* M) and found that 55 to 75% of the total cell lipid was present in the protoplast membrane which constituted 15–20% by weight of the vegetative cell. The lipid, which was mainly phosphatidic acid and neutral fat or fatty acid, accounted for 13–21% of the weight of the membrane. Cholesterol, inositol, serine, ethanolamine, choline, and hexoses were virtually absent from all cellular lipids including those from the protoplast membrane. The P content[241] was 1.31–1.58% of which the major part was about equally divided between lipid and protein fractions. The N content was 10.3–10.9%. Protein amounted to 63–69%, lipid to 16–21%, hexoses 1–10%, while nucleic acids were negligible. The only hexose detected was glucose, possibly as glycogen. Proline was not detected and aspartic acid, glutamic acid, and tyrosine were not present in substantial amounts as were all the other protein amino acids. Hydroxyproline was also present in appreciable quantities but the cell wall amino acid, diaminopimelic acid was absent. The presence of the complete cytochrome system was confirmed.

b. Staphylococcus aureus. Mitchell and Moyle[138] showed that the "small particle" fraction of *Staphylococcus aureus* contained 41% protein and 22.5% lipid. The protein had a high content of glycine, alanine, and glutamic acid and the lipid had N = 1.3% and P = 1.85%.[121]

3. The Protoplast Membrane of *Micrococcus lysodeikticus*

a. Yield and General Properties. Purified preparations of protoplast membranes from *Micrococcus lysodeikticus* have been made by Gilby *et al.*[137] and have been analyzed in somewhat greater detail than those from other organisms. Several independent batches of *M. lysodeikticus* have been processed separately and the results of analyses are reasonably consistent (Table VIII). The membrane fraction (see Section V, C, 1) amounted to 8.5% by weight of the whole cell and was 28% lipid, 50% protein, and 15–20% carbohydrate. The total N was 8.3% and total P, 1.16%.

b. Lipid. The lipid was extracted by ether after refluxing with methanol

TABLE VIII
Analysis of Protoplast Membrane Fraction of *Micrococcus lysodeikticus*[a]

Batch no.	Yield of membrane[b]	Total P[c]	Total N[c]	Lipid[c]
4	8.65	1.16	8.38	28.1
5	8.75	1.15	8.20	27.6
6	8.45	1.18	8.49	28.2
Mean	8.62	1.16	8.36	28.0

[a] Data of Gilby et al.[137]
[b] Expressed as percentage of bacterial dry weight.
[c] Expressed as percentage of dry weight of membrane fraction.

to disrupt lipoprotein complexes. It contained P (3.0%), but virtually no N, and was mainly phospholipid. Sugars, choline, ethanolamine, and serine were not detected and there was only a small content of inositol. Hydrolysis with 6 N HCl was necessary to liberate phosphoglycerol, which was not found after treatment with methanolic KOH. This was taken to mean that the phospholipid was largely polyphosphatidic acid, perhaps similar to cardiolipin. The ratio of double bonds to P was 0.18/1. The lipid had an absorption spectrum with peaks at 416, 440, and 471 mμ, a pattern characteristic of carotenoids. On the basis of the extinction coefficient of lutein, it was calculated that the membrane lipid might contain 0.25% carotenoid. Gilby and Few[242] place a revised estimate at a minimum value of 0.14% of the dry weight of the membrane. Stanier (personal communication) mentioned that carotenoids had been found in the protoplast membrane of a *Sarcina* sp.

c. Protein. Chromatograms of hydrolyzates of defatted membranes showed the presence of the usual protein amino acids and quantitative analysis gave the results shown in Table IX. The relative amounts of individual amino acids were fairly constant from batch to batch. The total α-amino-N of defatted membranes (9.6%), determined directly, agreed closely with that found by summation of the individual amino acids; the N content of the whole membrane was almost entirely accounted for by this protein fraction.

d. Carbohydrate. Chromatography of 2 N H$_2$SO$_4$ and 2 N HCl hydrolyzates of defatted membranes indicated the presence of carbohydrate; this was confirmed by paper electrophoresis. The predominant sugar was found to be mannose and the carbohydrate analyses are given in Table X.

e. General. The cell wall of *Micrococcus lysodeikticus* contains glucose, glucosamine and a small range of amino acids but no lipid, whereas the membrane appears to contain mannose, a complete range of amino acids and a substantial amount of lipid. (Hawthorne[140] had earlier found that mannose was a constituent of the nondialyzable residue after lysozyme

TABLE IX

RELATIVE MOLAR PROPORTION OF AMINO ACIDS IN THE PROTOPLAST
MEMBRANE FRACTION OF *Micrococcus lysodeikticus*[a,b]

Alanine	1.00
Aspartic acid	0.24
Glutamic acid	0.55
Serine + glycine	0.69
Threonine	0.17
Tyrosine	0.05
Lysine	0.37
Arginine	0.21
Proline	0.16
Valine + methionine	0.30
Phenylalanine + leucine + isoleucine	0.60
Methionine sulfoxide	0.02

[a] Data of Gilby et al.[137]
[b] Defatted membranes were hydrolyzed with 6 N HCl at 105° for 15 hours and the amino acids estimated by a ninhydrin method after two-dimensional paper chromatography.

TABLE X

CARBOHYDRATE ANALYSES OF PROTOPLAST MEMBRANES AND CELL WALLS
OF *Micrococcus lysodeikticus*[a]

Material and batch no.	Total sugar (anthrone method)	Reducing sugar (as hexose)	Hexosamine (as glucosamine HCl)
Membrane 4	17.5 (as mannose)	14.3	2.5
Membrane 5	19.7 (as mannose)	16.3	3.0
Membrane 6	19.5 (as mannose)	16.0	2.5
Mean	18.9 (as mannose)	15.5	2.7
Cell wall	10.4 (as glucose)	37	23
Cell wall (Salton[25])	7.5–10 (as glucose)[b]	45	16

[a] Data of Gilby et al.[137]
[b] Glucose estimated by the specific glucose oxidase was 3.5–6.5% (Salton, unpublished observations). Results are expressed as the % of the dry weight of the cell fraction—membrane or wall. The analyses were performed after hydrolysis of either the defatted membrane (which comprises 72% of the whole protoplast membrane) or the cell wall.

digestion of *M. lysodeikticus*.) Thus the two structures are strikingly different in composition. The analyses account for 93–97% of the weight of the protoplast membrane; the only surprise is the presence of the carbohydrate component. It may be that the orcinol-positive material found by Vennes and Gerhardt[134] in the membrane of *Bacillus megaterium* is part of a carbo-

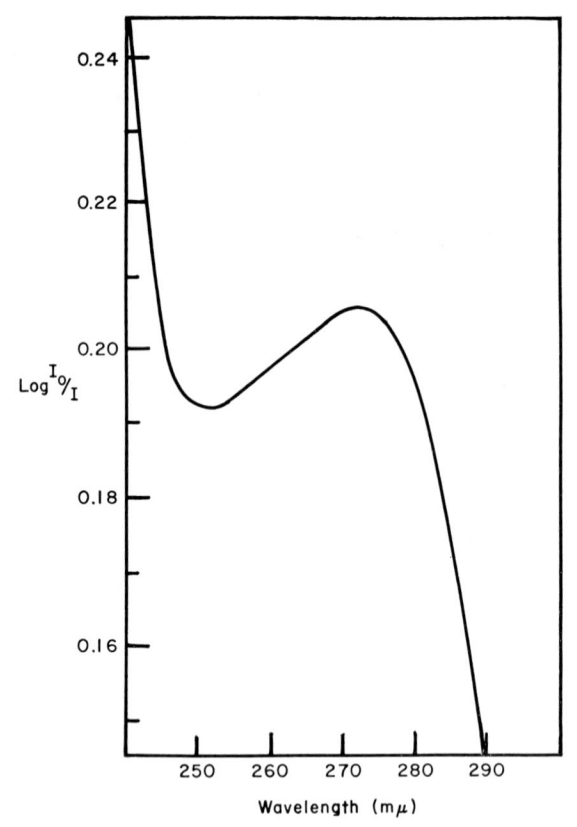

Fig. 11. Absorption spectrum of defatted protoplast membrane fraction of *Micrococcus lysodeikticus* (Gilby et al.[137]).

hydrate fraction, rather than of ribonucleic acid. The ultraviolet absorption spectrum of hydrolyzates of defatted membranes of *M. lysodeikticus* is shown in Fig. 11. Neither it nor paper chromatograms showed evidence of more than trace amounts of purines or pyrimidines.

4. ENZYME ACTIVITIES ASSOCIATED WITH PROTOPLAST MEMBRANES

Weibull[14] recorded that the difference spectrum of protoplast "ghosts" (suspended in 70% glycerol) before and after reduction with hydrosulfite showed maxima at 530, 558 and 600 mμ which correspond to the absorption bands for whole cells of *Bacillus megaterium*.[141] This indicates that the complete cytochrome system is associated with the protoplast membrane. Mitchell and Moyle[121] found 90% of the total cytochrome system, the acid phosphatase, and succinic dehydrogenase, as well as lactic dehydrogenase activity (cytochrome b-linked) were present in the "small particle" fraction believed to have originated from the protoplast membrane of *Staphylococcus aureus*. The cytochromes had maxima at 528, 558, and 604 mμ. The authors suggest that the membrane may be concerned in phosphate transport and that the occurrence of cytochrome-linked enzymes "may be connected with the fact that they are the last members of the enzymic chain and deal with end-products of metabolism, namely lactic, succinic and formic acids which must be specifically carried through the plasma membrane."

Storck and Wachsman[135, 136] also found enzyme activity associated with the protoplast membrane. Using *Bacillus megaterium*, they showed that all of the succinic, DL-lactic, and α-ketoglutaric dehydrogenase systems were present in the protoplast membrane fraction, as was about half of the L-malic dehydrogenase activity.

D. Other Components of Protoplasts

1. Deoxyribonucleic Acid

Several authors have found that even after lysis of protoplasts of *Bacillus megaterium*, the deoxyribonucleic acid (DNA) can readily be centrifuged down. Weibull[14] first pointed out that, of all the macromolecular components released on lysis, the DNA was most readily sedimented in 0.15 M NaCl. He suggested that it may exist as a gel in this medium. McQuillen and Kodicek (unpublished observations) attempted to use estimations of the DNA released by treatment of protoplasts with linoleic acid as a measure of the degree of lysis but found that even when lysis was visibly complete, the entire DNA was readily sedimentable. In view of later work by Spiegelman *et al.*[142] (see Section V, F, 2), it is probable that the DNA is present in a "nuclear body" which either within the protoplast membrane or after release can easily be spun down.

2. Granules

Granules, inclusions, or particles of various sizes are also found within the protoplast and can sometimes be seen with the phase contrast microscope. The granules of *Bacillus megaterium* are lipid, consisting of polymerized β-hydroxybutyric acid, according to Lemoigne *et al.*[106] They can be isolated by differential centrifugation and are soluble in warm alkali and in warm chloroform.[14] They can be stained with Sudan black B. The amount of lipid varies greatly, depending on the strain of organism and on the cultural conditions. *Bacillus* M, a strain of *B. megaterium* much used by Tomcsik and whose lipids were recently studied by Weibull,[133] is reported to contain the β-hydroxybutyric acid lipid granules only rarely.

E. Antigenic Properties of Protoplasts

Tomcsik, who has made great contributions to our knowledge of bacterial anatomy by elegant immunological methods, began to investigate the antigenic relationships of protoplasts and intact cells in 1954[143] and has written fuller accounts,[16, 144] and has also reviewed the use of antibodies as indicators for bacterial surface structures.[34] Since Chapter 10 also deals with this subject, the findings of Tomcsik and his co-workers will only be summarized here.

Of 55 strains of *Bacillus megaterium* examined, 38 distinct serological types have been identified on the basis of capsular and cell wall reactions.[144] The cell wall contains thermostable, mucopolysaccharide antigens but the protoplasts possess thermolabile, protein antigens. Intact cells react with antisera produced against protoplasts probably by virtue of the flagella which both possess. The protoplast membrane and protoplasm of different strains of *B. megaterium* possess another common antigen, which is absent from the cell wall and is not exposed in intact cells (Fig. 12). In addition, 53 of the 55 strains gave a capsular reaction with *Bacillus anthracis* polypeptide antiserum.

Perhaps the most important finding for our present concern is that antisera produced against isolated cell walls did not react with protoplasts. This implies that protoplasts do not possess cell wall antigens exposed in their surface. It would be useful to apply such antigenic analysis to so-called "protoplasts" of Gram-negative bacteria.

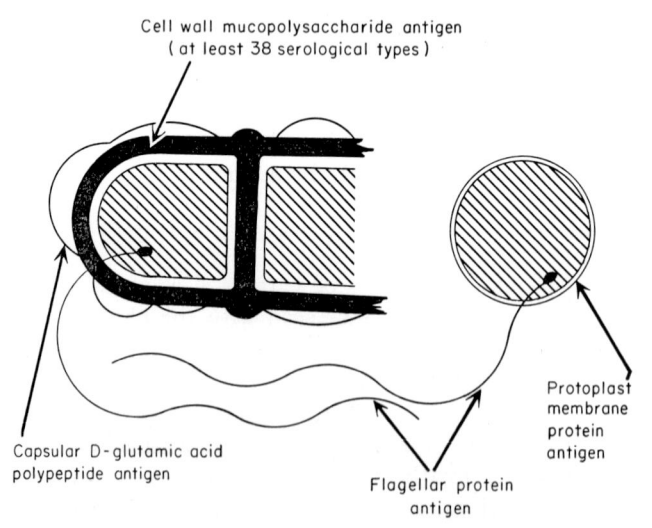

Fig. 12. Representation of antigenic structure of cells and protoplasts of *Bacillus megaterium*. Based on results of Tomcsik and colleagues[16, 34, 143, 144] and Vennes and Gerhardt.[134]

TABLE XI
SEROLOGICAL REACTIONS OF *Bacillus megaterium* STRUCTURES[a]

Antiserum against	Titer of sera against[b]			
	Whole cells	Cell walls	Protoplasts	Protoplast membranes
Whole cells	20,480	2560	1280	1280
Cell walls	5120	5120	<10	<10
Protoplasts	80	<10	1280	1280
Protoplast membranes	40	<10	1280	1280

[a] Data of Vennes and Gerhardt.[134]
[b] Titers determined by 50% end-point method of complement fixation. Lysozyme and other controls were negative.

TABLE XII
CROSS ADSORPTION OF ANTISERA TO COMPONENTS OF *Bacillus megaterium*[a]

Antiserum to whole cells adsorbed with	Titer of sera against[b]		
	Whole cells	Cell walls	Protoplast membranes
Whole cells	<20	1280	1280
Cell walls	<20	<20	1280
—	20,480	2560	1280

Antiserum to cells walls adsorbed with	Titer of sera against		
	Whole cells	Cell walls	Protoplast membranes
Whole cells	<20	1280	<20
Cell walls	<20	<20	<20
—	2560	5120	<20

[a] Data of Vennes and Gerhardt.[134]
[b] Titers determined by complement fixation.

Vennes and Gerhardt[134] have also published information on immunological comparisons of cells, protoplasts, cell walls, and protoplast membranes of *Bacillus megaterium* (Table XI). They have also carried out cross adsorption experiments (Table XII).

Their conclusions, like those of Tomcsik, were that cell wall antigens do not occur in the surface of protoplasts or in preparations of protoplast membranes. Nor do membrane antigens occur in the cell wall. Antiserum to protoplasts did not cross react with intact cells—this is in contradistinction to Baumann-Grace and Tomcsik's[144] finding and may be because flagella were absent from Vennes and Gerhardt's protoplasts.

F. Effects of Enzymes on Protoplasts

The effects of enzymic reaction on protoplasts have been studied from two points of view. Firstly in order to gain information about the constitution of the protoplast and, secondly, in attempts to study cellular components necessary for biochemical activities such as protein synthesis and bacteriophage multiplication. The term "enzymic resolution" has been used in connection with the latter approach.

1. Action of Ribonuclease and Deoxyribonuclease

Lester[46] and Beljanski[47] treated their lysozyme "lysates" of *Micrococcus lysodeikticus* with ribonuclease (RNase) and deoxyribonuclease (DNase). While the latter enzyme somewhat stimulated the incorporation of radioactive amino acids, RNase was a potent inhibitor. This was taken to mean that RNA is concerned in the incorporation. However, at that time the formation of protoplasts by lysozyme action was not realized and the possibility that RNase acted on such a structure rather than on material released by lysis was not considered. Brenner[42] studied the effects of crystalline DNase (Worthington) and crystalline RNase (Armour) on lysozyme-produced protoplasts of *Bacillus megaterium* KM in connection with experiments on the growth of bacteriophage. Neither enzyme affected the yield of phage from intact cells, nor did DNase (10 μg./ml.) affect protoplasts or the phage yield from them. RNase (50 μg./ml.), however, abolished phage growth by causing lysis of the protoplasts—not by specific removal of the substrate RNA. Bridoux and Hanotier[145] described inhibition of glycine incorporation by, and lysis of, protoplasts of *M. lysodeikticus* caused by RNase (33 μg./ml.). Fig. 13 shows the changes in turbidity of a suspension of *B. megaterium* during conversion to protoplasts by lysozyme and subsequently during lysis by addition of RNase.

Nečas[243] found that the development of many yeast "plasmatic droplets" ("protoplasts") was prevented by RNase which also destroyed volutin granules. He suggested that these may contain RNA.

Shortly afterward, Landman and Spiegelman[43] reported that protoplasts of the same strain KM of *Bacillus megaterium*, if suspended in phosphate (0.32 M) instead of sucrose (0.2 M), would lose "80–90% of their RNA with no significant loss of DNA or protein" by treatment with RNase (1000 μg./ml.). DNase (200 μg./ml.) either selectively removed 20–50% of the DNA or (in other preparations) removed this amount of DNA and about 30% of the RNA, in each case without loss of protein. Trypsin (Worthington, twice recrystallised, 100 μg./ml.) and lipase (Nutritional Biochemical Company, 100 μg,/ml.) caused physical destruction of the protoplasts. None of these enzymes had any deleterious effect on the enzyme-forming ability of intact cells under similar conditions.

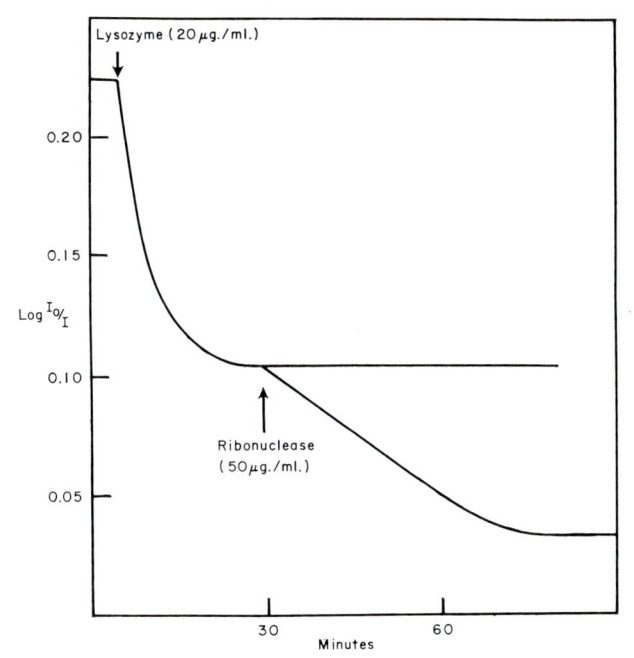

Fig. 13. Effect of lysozyme (20 μg./ml.) and ribonuclease (50 μg./ml.) on turbidity of suspensions of *Bacillus megaterium* KM in buffered sucrose. Data of Brenner.[42]

Subsequently Brenner (unpublished observations) confirmed that RNase resulted in lysis of protoplasts suspended in sucrose-stabilized media but not in phosphate-stabilized media. However, he found that trypsin (crystalline, 250 μg./ml.) did *not* cause lysis of protoplasts but markedly reduced phage yields from them.

Spiegelman[146, 147] has carried out further work on the resolution of protoplasts. His procedure was to add RNase or DNase to the suspension medium of cells in hypertonic phosphate medium containing casein digest (1%) and lysozyme (200 μg./ml.), and to incubate at 30° for 45 minutes with shaking. After centrifuging, the pellet was analyzed and compared with controls to which no nuclease had been added (Table XIII). This procedure led to removal of variable amounts of RNA or DNA, or both, but if protoplasts were first formed and then suspended in a medium containing succinate (0.5 M) as stabilizer, enzymic resolution was impossible (Table XIV).

2. Action of Lipase on Protoplasts of *Bacillus megaterium*

From the point of view of bacterial structure, the action of lipase on protoplasts is most interesting. Spiegelman *et al.*[142] found that lipase digestion of protoplasts of *Bacillus megaterium* led to the liberation of struc-

TABLE XIII
Effect of Ribonuclease and Deoxyribonuclease on Protoplasts of Bacillus megaterium[a]

Enzyme	Percentage removal[b]	
	DNA	RNA
DNase 400 µg./ml.	97	0
	87	0
	41	4
	99	13
	59	46
	39	42
RNase 500 µg./ml.	0	34
	0	72
	16	52
	21	33
	58	78

[a] Data of Spiegelman.[147]
[b] Lysozyme treatment of *Bacillus megaterium* was carried out in the presence of DNase, RNase, or neither (control). Amounts of DNA and RNA determined on pellet sedimentable after 45 minutes. Percentage removal is based on difference between amounts found in control and in nuclease-treated samples.

TABLE XIV
Effect of Nucleases on Preformed Protoplasts of Bacillus megaterium[a,b]

Treatment	Percentage recovery after treatment		
	DNA	RNA	Protein
Suspension in succinate (0.5 M)	95	97	95
RNase 400 µg./ml.	91	89	105
DNase 400 µg./ml.	100	98	102
NaCl (1.0 M)	87	81	80

[a] Data of Spiegelman.[147]
[b] Protoplasts of *B. megaterium* were analyzed and samples suspended in succinate (0.5 M) and subjected to RNase, DNase, or NaCl for 30 minutes at 30° with appropriate ionic supplements. Sedimentable material (5 minutes at 8000 r.p.m.) was analyzed.

tures which could be sedimented at 10,000 g in 5 minutes and which were believed to be the "nuclear bodies" of the organism. They were about 1 µ in diameter and appeared to be composed of chromatinic material (presumably DNA) wrapped around a nonchromatinic core. The chromatin was invisible by phase contrast microscopy but the core could be seen by dark phase contrast. Electron micrographs of preparations were also presented. Density gradient and immunological methods indicated that the "nuclear bodies" were distinct from residues of protoplast membranes. Purified preparations were analyzed and found to contain DNA, RNA, and protein in the ratio of 1:1:3. This exciting development is dealt with more fully in the chapter on the nucleus (Chapter 2).

3. ACTION OF *Bacillus cereus* ENZYME PREPARATION

Norris[148] described an enzyme system associated with cultures of *Bacillus cereus* which has the ability to digest away practically the whole protoplast, leaving the cell wall apparently untouched. It was active in a wide range of *Bacillus* spp. and although there are no reports of its action on isolated protoplasts, there is every reason to think it would destroy them. The enzyme preparations have gelatinase and lecithinase activity but the role of these, if any, in the dissolution of the protoplast is not known.

VI. Physiology and Biochemistry of Protoplasts

A. RESPIRATION

Weibull[13] compared the endogenous respiration and glucose-oxidizing abilities of intact cells, protoplasts, and osmotically lysed protoplasts of *Bacillus megaterium*. The Q_{O_2} values were 68, 66, and 3, respectively, for endogenous respiration and 120, 128, and 27 (uncorrected), for glucose oxidation. Mg^{++} (0.005 M) was reported to cause strong inhibition of respiration of protoplasts.[118] Crawford and McQuillen (unpublished observations) also found protoplasts and intact cells to have similar respiratory activities but did not observe inhibition by Mg^{++}. The Q_{O_2} of protoplasts oxidizing glucose was 110 in the absence and 115 in the presence of Mg^{++} (0.005 M), the corresponding endogenous values being 5 and 7.

Wiame *et al.*[23] found that protoplasts of *Bacillus subtilis* could oxidize glucose at a constant rate for at least 3 hours but that osmotic lysis resulted in complete loss of activity. Beljanski[47] and Bridoux and Hanotier[145] demonstrated that lysozyme-treated *Micrococcus lysodeikticus* could also respire. The former found that DNase (approximately 200 μg./ml.) had a small activating effect on glucose oxidation, while RNase (approximately 5 μg./ml.) had no effect. Bridoux and Hanotier,[145] on the other hand, using less concentrated suspensions, found gradual inhibition of respiration by RNase (33 μg./ml.). Since RNase has been shown to promote lysis of protoplasts under certain conditions it is probable that the inhibition was caused by disintegration of these structures.

B. Incorporation of Radioactive Tracers

1. *Micrococcus lysodeikticus*

Lester,[46] Beljanski,[47] and Bridoux and Hanotier[145] have studied incorporation of C^{14}-labeled amino acids into the protein fraction of lysozyme-treated *Micrococcus lysodeikticus*. Lester[46] first showed that if "lysis" occurred in a thick suspension of cells (25 mg./ml.) in sucrose (0.64 M), the product could still incorporate leucine, whereas if enzyme treatment were carried out in dilute buffer this was not so. It is probable that all three groups of workers used preparations which contained both protoplasts and lysed protoplasts since a concentration of greater than molar sucrose is required to stabilize protoplasts of *M. lysodeikticus*.[21, 22] Lester[46] also found that DNase (5 μg./ml.) enhanced leucine incorporation while RNase (200 μg./ml.) abolished it.

An account of studies of the action of ionic detergents on *Micrococcus lysodeikticus* and its protoplasts was given by Gilby and Few.[239]

Beljanski[47] made similar observations using C^{14}-glycine and, in addition, found that ATP (0.01 M) inhibited incorporation. Both workers suggested that their results indicated that RNA is involved in protein synthesis. However, it is doubtful if this conclusion is justified. The highest rates of incorporation in these experiments were in the range 0.001–0.0025 μM of amino acid/mg. protein/hour. This is 200 to 500 times slower than is to be expected for a bacterial cell growing with a mean generation time of one hour and is about one-hundredth of that which can be demonstrated in protoplasts of *Bacillus megaterium*.[20] Secondly, no protein synthesis was demonstrated and Gale and Folkes[149] have shown that a single amino acid can be incorporated into bacterial protein by reactions which do not involve net protein synthesis. Thirdly, the effect of enzymes on protoplasts is not simply to remove their homologous substrates. It is unlikely that under most conditions enzymes can penetrate into protoplasts, since small molecular weight components, such as nucleotides and sucrose, do not readily pass in or out. Finally, Brenner[42] has subsequently shown that RNase can cause lysis of protoplasts suspended in sucrose solutions.

Bridoux and Hanotier[145] observed incorporation of glycine into the protein fraction of lysozyme-treated *Micrococcus lysodeikticus* and found that if preparations (20 mg./ml.) were centrifuged at 13,000 g for 10 minutes, the supernatant fraction, "SP," was 10 times as active at incorporating glycine as was the whole "lysate." It is not known in what way it differs from the sedimented fraction. The maximum rates of incorporation were again low—about 0.001 μM/mg./hour for the whole lysate and 0.01 μM/mg./hour for the "SP" fraction. RNase (33 μg./ml.) abolished glycine incorporation almost instantaneously in some cases but only after a lag in others. The authors reported that glycine incorporation was more sensitive to RNase than was respiration and that DNase (7 μg./ml.) was sometimes stimulatory.

Oparin et al.[245] demonstrated incorporation of radioactive amino acids by protoplasts of *Micrococcus lysodeikticus* and net increase (up to 12%) in protein. Most remarkably, 2,4-dinitrophenol (0.001 M) and azide (0.01 M) were reported to reduce incorporation of labeled glycine by 90% without affecting the increase in protein-N. This increase, however, only occurred in protoplasts maintained in a suitable concentration of sucrose; reduction of the concentration led to lysis and inactivation.

A study of the uptake of lysine by intact cells and protoplasts of *Micrococcus lysodeikticus* has been made by Britt and Gerhardt.[236, 237] The whole cells contained considerably more lysine but the additional amount was found to be located in the bacterial cell wall. The uptake of lysine into the internal pool (which can be liberated by hot water extraction) was similar in cells and protoplasts. Isolated cell walls, however, were able to bind an amount of lysine greater than that in the internal pool and large enough to account for the difference in total amount in cells as compared with protoplasts.

Abrams[203] prepared "protoplasts" of *Streptococcus faecalis* 9790 using lysozyme and suspended them in solutions of sucrose or other nonpenetrating sugars. In the presence of glucose and K^+ the "protoplasts" swelled as the glucose was metabolized. Excessive quantities of glucose led to lysis but with lower amounts the "protoplasts" contracted to their original volume after all the glucose had been consumed. Rb^+ could replace K^+ but Li^+ and NH_4^+ were ineffective and Na^+ antagonized K^+ without impairing glycolysis. It was suggested that in the presence of metabolizable substrates (glucose) and K^+ there occurred an active transport of the sucrose into the "protoplasts."

2. *Bacillus megaterium*

More detailed incorporation studies were made by McQuillen,[17, 20] using *Bacillus megaterium*. The strain KM was trained to grow in a glucose/NH_3/salts medium and the synthetic abilities of cells, protoplasts, and osmotically lysed protoplasts were compared. Lysozyme treatment was carried out in sucrose- or phosphate-stabilized media and protoplasts were spun down and washed before use. Suspension densities were kept low, since the incorporation per unit mass of preparation was found to be greater if this were done. Rates of incorporation of C^{14}-labeled substrates were comparable in protoplasts and in intact cells that were capable of growing under the conditions used. For example, glycine incorporation could occur at a rate of 0.2 μM/mg. protein/hour and this rate could be maintained for several

TABLE XV

Incorporation of C^{14} into Major Fractions of
Protoplasts of *Bacillus megaterium*[a,b]

C^{14}-substrate	Fraction[c]		
	Lipid	Nucleic acid	Protein
Glucose (random)	+++	+++	+++
$CH_3C^{14}OOH$	+++	+	+++
$C^{14}H_3COOH$	+++	+	+++
$CH_2NH_2C^{14}OOH$		+++	+++
Aspartic acid (random)		+++	+++
Uracil (random)		+++	
Thymine (random)		—	

[a] Data of McQuillen.[17]

[b] Protoplasts were incubated in aerated medium containing glucose, NH_3, and inorganic salts with sucrose as stabilizer. Intact cells of *B. megaterium* showed the same pattern of incorporation of radioactivity from the added labeled substrate.

[c] Symbols +++, +, and — indicate strong, weak, and no labeling, respectively.

hours. Since *B. megaterium* is a strict aerobe and is unable to carry out syntheses anaerobically, it was necessary to devise a method of aeration which would cause minimum damage to the fragile protoplasts. Bubbling or vigorous shaking is highly deleterious, but rocking of tubes or even Roux bottles is a satisfactory procedure, as is gentle shaking. C^{14}-labeled substrates which have been used include glucose, acetate, glycine, aspartic acid, glutamic acid, adenine, uracil, and thymine. Table XV lists the major cell fractions into which radioactivity passes and Tables XVI and XVII indicate the specific end products which are labeled from each substrate.

There is almost exactly parallel behavior in intact cells and protoplasts—with the outstanding exception that the latter do not incorporate radioactivity into diaminopimelic acid which is found in the protein fraction of the intact cells. This amino acid, a constituent of the cell wall, does not in fact occur at all in the trichloroacetic acid-precipitable fraction of protoplasts. However, by using radioactive aspartic acid which is a good precursor of DAP, it was found that protoplasts do synthesize DAP but instead of incorporating it into cell wall structure, excrete it or a derivative into the medium from which it can be isolated.[31]

The pathways of synthesis of amino acids and purine and pyrimidine bases are probably similar to those in other organisms; the tracer studies on protoplasts are consistent with this view.[17, 20] The pattern of incorpora-

TABLE XVI
INCORPORATION OF C^{14} INTO AMINO ACIDS OF THE TRICHLOROACETIC ACID-PRECIPITABLE FRACTION OF CELLS AND PROTOPLASTS OF *Bacillus megaterium*[a]

Amino acid	\multicolumn{8}{c}{C^{14}-substrate[b,c]}
	Glucose		Acetate		Aspartic acid		Glycine	
	KM	PP	KM	PP	KM	PP	KM	PP
Alanine	+	+	+	+	+	+	—	—
Aspartic acid	+	+	+	+	+	+	—	—
Glutamic acid	+	+	+	+	+	+	(+)	(±)
Glycine	+	+	—	—	—	—	+	+
Leucine(s)	+	+	+	+	+	+	—	—
Lysine	+	+	+	+	+	+	—	—
Proline	+	+	+	+	+	+	(+)	—
Serine	+	+	—	—	—	—	(±)	(±)
Threonine	+	+	+	+	+	+	—	—
Diaminopimelic acid	+	—	+	—	+	—	(+)	—

[a] Data of McQuillen.[17]
[b] Symbols +, (+), and — indicate strong, weak, and no labeling, respectively; (±) indicates occasional traces of activity.
[c] Intact cells (KM) and protoplasts (PP) incubated as described in Table XV.

TABLE XVII
INCORPORATION OF C^{14} INTO NUCLEIC ACID CONSTITUENTS OF CELLS AND PROTOPLASTS OF *Bacillus megaterium*[a]

Nucleic acid constituent	\multicolumn{10}{c}{C^{14}-substrate[b,c]}
	Glycine		Acetate		Aspartic acid		Uracil		Thymine	
	KM	PP	KM	PP	KM	PP	KM	PP	KM	PP
Adenine	+	+			(+)	(+)	—	—	—	—
Guanine	+	+			(+)	(+)	—	—	—	—
Cytidylic acid	—	—	+	+	+	+	+	+	—	—
Uridylic acid	—	—	+	+	+	+	+	+	—	—
Thymidylic acid	—	—			+	+	+	+	—	—

[a] Data of McQuillen.[17]
[b] Symbols +, (+), and — indicate strong, weak, and no labeling, respectively.
[c] Intact cells (KM) and protoplasts (PP) incubated as described in Table XV.

tion of acetate carbon into amino acids and pyrimidines is the same as that found in *Escherichia coli* by McQuillen and Roberts,[150] with the exception that alanine was also labeled. Aspartic acid is a major precursor of pyrimidine carbon and glycine, of the purines. Thymine was not used either by intact cells or protoplasts but this is not uncommon.

3. Inhibitors

Beljanski[47] found that ATP (0.01 M), azide (0.005 M) and 2:4 dinitrophenol (0.001 M) inhibited glycine incorporation into protein of lysozyme-treated *Micrococcus lysodeikticus*. Dinitrophenol also inhibited incorporation into both nucleic acid and protein fractions of protoplasts of *Bacillus megaterium*.[20] Uranyl chloride is a reagent found by McQuillen[129] to be strongly bound to the cell walls of various species of bacteria. At concentrations up to 100 μM it had no effect on the incorporation of glycine into the protein fraction of protoplasts, but it did reduce incorporation into the nucleic acid fraction (Fig. 14). The reason for this is not known.[17, 20]

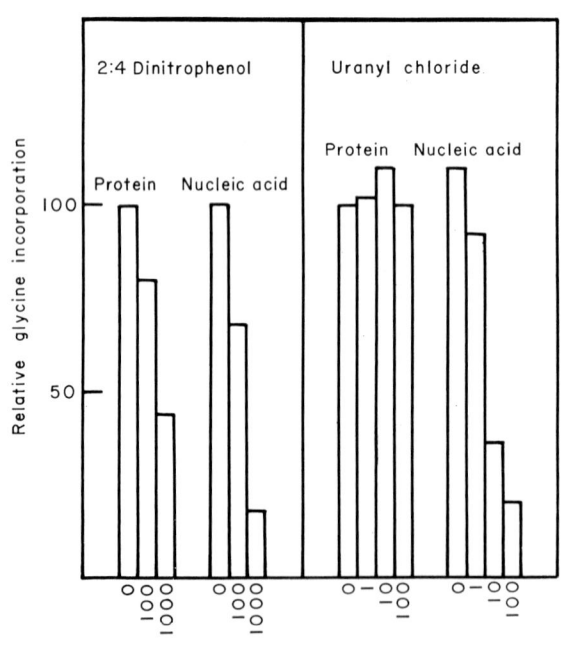

Fig. 14. Inhibition of incorporation of C^{14} from glycine into protein and nucleic acid fractions of protoplasts of *Bacillus megaterium* KM. Results expressed relative to 100 for control in absence of inhibitor. Data of McQuillen.[20]

The diversity, rate, and extent of incorporation of carbon from a wide variety of substrates make it apparent that *de novo* synthesis of most of the cellular constituents can occur in isolated protoplasts. Moreover, these syntheses can occur at rates comparable to those observed in intact, growing cells.

C. Synthesis of Enzymes by Protoplasts

1. Constitutive Enzymes

Little work appears to have been done on the ability of protoplasts to synthesize constitutive enzymes but the results of the tracer experiments mentioned in the previous section suggest that since protoplasts can make proteins, lipids, and nucleic acids about as rapidly as growing cells, it is likely that this involves enzyme synthesis rather than elaboration of inactive, nonspecific products.

McQuillen[17] found that succinic dehydrogenase activity of protoplast preparations increased during incubation in a medium containing glucose and amino acids with sucrose as stabilizer. The increase occurred at about the same rate as in intact cells treated similarly. Fivefold increases were measured over a period of some hours. This enzyme is associated with the protoplast membrane[135, 136] and there is no doubt but that this structure can increase many times in amount during suitable incubation[17, 41] (see Sections VI, L, 1 and 2).

2. Inducible Enzymes

Three independent demonstrations of inducible enzyme synthesis in protoplasts were made in 1955.[23, 43, 151] Wiame et al.[23] studied arabokinase synthesis in *Bacillus subtilis*. Manometric experiments were made with protoplasts in the absence of substrate, in the presence of glucose, and in the presence of L (+)-arabinose. Endogenous and glucose respiration rates were more or less constant for 3 hours but in the presence of arabinose a gradual increase in the rate of oxidation occurred. Arabokinase activity of extracts of protoplasts after varying periods of incubation were also measured (Table XVIII).

Landman and Spiegelman[43] formed protoplasts of *Bacillus megaterium* in phosphate-stabilized media (0.3–0.5 M) and showed that they could be induced by lactose (0.06 M) to synthesize β-galactosidase. Casein hydrolyzate (20 mg./ml.), $MnCl_2$ (100 μM), hexose diphosphate, HDP, (6 mg./ml.), and adenosine triphosphate, ATP, (1 mg./ml.) were also present in the induction medium and of these "the hydrolyzate, HDP and lactose are mandatory."[43] Fig. 15 compares the time course of enzyme synthesis in cells and protoplasts.

TABLE XVIII
Synthesis of the Inducible Enzyme, Arabokinase, by Protoplasts of Bacillus subtilis[a,b]

Substrate	Rate of CO_2 production (μl./hr.) after preincubation for		
	0 hr.	1 hr.	2 hr.
ATP (10 μmole)	24	19	28
ATP (10 μmole) + arabinose (20 μmole)	46	106	220
Arabokinase activity	22	87	192

[a] Data of Wiame et al.[23]

[b] Protoplasts were shaken with arabinose (0.02 M) for various periods of time in a solution containing NaCl (0.5 M), $(NH_4)_2SO_4$ (0.005 M), and yeast extract (0.1%). Enzyme activity was determined on extracts of lyophilized preparations.

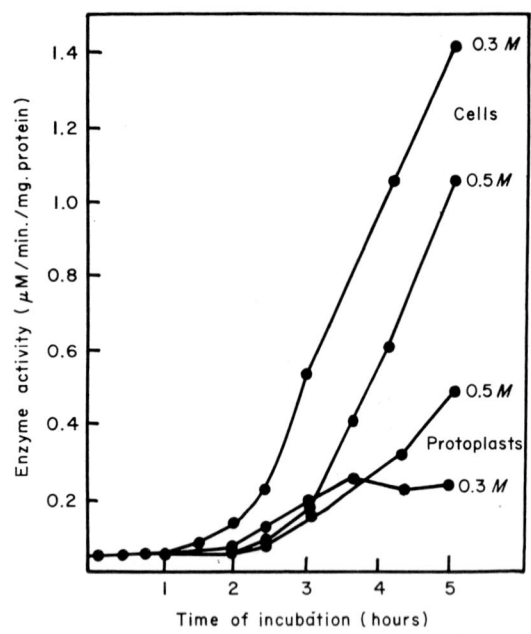

Fig. 15. Induced formation of β-galactosidase in intact cells and protoplasts of Bacillus megaterium KM suspended in media containing 0.3 M and 0.5 M phosphate. Data of Landman and Spiegelman.[43]

The same workers found that HDP was essential for synthesis of the enzyme by intact cells suspended in 0.5 M but not in 0.05 M phosphate. These curious requirements were not found by McQuillen[17] or by Crawford and McQuillen (unpublished observations) who showed that protoplasts in phosphate- or sucrose-stabilized media could be induced by galactose to form β-galactosidase. Glucose (100 μg./ml.) was used as energy source and addition of amino acids or peptone enhanced enzyme formation. No Mn^{++}, ATP, or HDP was added. Dinitrophenol (0.001 M), chlortetracycline (1 μg./ml.) and puromycin (25 μg./ml.) inhibited adaptation.[17] Rather surprisingly, UO$_2^{++}$ (0.001 M) had no effect on the induction, although it flocculated the protoplasts, inhibited glycine incorporation into nucleic acid at much lower concentrations (Fig. 14) and reduced phage growth in protoplasts by 80% or more (see Section VI, G, 1).

Trypsin (100 μg./ml.) and lipase (100 μg./ml.) also prevented enzyme formation but this was because incubation of protoplasts for an hour with either of these enzymes resulted in lysis.[43] RNase can, under certain conditions, cause lysis but Spiegelman[146, 147] reported that under other conditions both it and DNase could specifically remove nucleic acids. The ability of protoplasts to form β-galactosidase was compared before and after "resolution" with nucleases. The RNase or DNase was added together with the lysozyme (see Section V, F, 1). Treatment which resulted in removal of 97% of the DNA but no RNA increased enzyme-forming ability more than 4-fold; removal of 30% or more of the RNA reduced enzyme formation. In one experiment, 99% of the DNA and 13% of the RNA were removed by DNase treatment without altering the amount of β-galactosidase subsequently synthesized.

However, this "removal" of DNA must not be taken to imply removal of the fragments formed by DNase action since 60% could still be found within the protoplasts after enzyme resolution. This type of experiment was later abandoned for various reasons and the capabilities of osmotically shocked protoplasts were investigated.[147]

D. Capabilities of Osmotically Shocked Protoplasts

McQuillen[17] found that lysates of protoplasts produced by diluting the suspension medium gave consistently negative results in incorporation studies, enzyme synthesis, etc., but, on occasion, lysis in a stabilizing medium gave preparations which retained wide synthetic activities. It was also found (unpublished observations) that if protoplast preparations were diluted to induce lysis and then centrifuged, the pellet had ability to incorporate amino acids. This was retained after washing but could be considerably reduced by adding back the original supernatant. The activity was also reduced by treatment with RNase. The washed lysate pellets incorporated radioactivity from a single labeled amino acid in the presence or absence of 17 unlabeled amino acids. Initially, glucose, ATP, and HDP were all added but it was found that omission of the ATP and HDP increased the incorporation several-fold and that further omission of glucose did not cause any reduction. (The preparations contained lipid granules as well as protoplast membranes and these may have acted as energy source.) However, it was difficult to get consistently reproducible results.

Crathorn and Hunter[152] have recently carried out similar experiments. Hunter *et al.*[153] studied incorporation of C^{14}-glycine by protoplasts of *Bacillus megaterium* and found that after incubation periods of 5 minutes or longer, the protein released on osmotic lysis had a higher specific activity than that which could be sedimented. Incubation for only 1 minute, however, gave the converse result. Similar observations were made with intact cells, that is, if cells which had incorporated glycine were treated with lysozyme to remove their walls and the resulting protoplasts were then lysed, the sedimentable protein had a higher specific activity for very short incubation periods than had the soluble cytoplasmic protein. Since this sedimentable fraction contained the protoplast membrane it was reported that "initial observable incorporation of C^{14}-glycine into proteins occurs in this case in the cytoplasmic membrane." Later, it was found[152] that an isolated membrane fraction could incorporate radioactive amino acids at a rate equal to that for the same fraction of intact protoplasts, when suspended in a medium containing glucose and inorganic salts. However, whereas the radioactivity in the membrane fraction of intact cells increased very rapidly for about a minute and then much more slowly, the incorporation into isolated membrane fractions continued at a linear rate for 2 hours. Further, if isolated membrane fractions which had been labeled in this way were washed and incubated with the soluble cytoplasmic fraction of protoplasts, there was transfer of radioactivity to the cytoplasmic protein. Since the latter never became labeled in any tests not involving the presence of the membrane fraction, it was concluded that the primary site of protein synthesis in *B. megaterium* is associated with the protoplast membrane, a suggestion which had been made earlier.[17]

Spiegelman[146, 147] has made extensive studies of osmotically shocked protoplasts of *Bacillus megaterium*, following DNA, RNA, protein content and enzyme-forming ability. He first showed that the extent of dilution is highly critical—1:3 and 1:4 dilutions gave best retention of biochemical capabilities, whereas 1:5 dilution often gave inactive preparations. Such shocked preparations were sometimes more active than the original protoplasts in synthesizing β-galactosidase, if assayed on a protein-N basis. Osmotically shocked protoplasts were also found to be readily amenable to enzymic resolution; again it was shown that DNA removal had little effect on enzyme synthesis, whereas RNA was apparently essential. These

TABLE XIX

Synthesis of Nucleic Acids and Protein by Osmotically Shocked Protoplasts of *Bacillus megaterium*[a, b]

Material	Amounts after 3 hr. incubation as % of initial levels		
	DNA	RNA	Protein
Protoplasts	510	600	550
Resolved, osmotically shocked protoplasts	420	120	80
	720	210	140
	1200	520	460
	3000	410	350

[a] Data of Spiegelman.[147]

[b] Protoplasts prepared in succinate (0.5 M), diluted 1:4 with water, spun immediately, and pellet resuspended, after washing, in RM (resolving medium) containing succinate (0.5 M), phosphate (0.1 M), and a complete mixture of amino acids. (This treatment did not degrade intact protoplasts but reduced acid-soluble and -insoluble DNA in osmotically shocked preparations to less than 1%, while not altering protein or RNA.) Protoplasts and resolved, shocked preparations were incubated for 3 hours in succinate (0.5 M), KCl (0.1 M), casein digest (10 mg./ml.), HDP (6 mg./ml.), MgCl$_2$ (0.02 M), and MnCl$_2$ (0.001 M). Analyses carried out on acid-insoluble fraction.

preparations were able to synthesize RNA and DNA as well as protein (Table XIX).

The full experimental details of Spiegelman's work were not available at the time of writing but it appears that "the procedures for removing DNA ... were clearly effective, there being no detectable DNA left in the pellet either as small or large fragments." Nevertheless, it is reported "that the rate of DNA synthesis can exceed by a factor of 10 the maximal rate attained by a culture growing logarithmically in 2 per cent peptone." The synthesis began after a lag of 40 to 90 minutes, proceeded linearly for more than 5 hours and must have been at the expense of succinate and amino acids. When preparations of osmotically shocked protoplasts were treated with RNase or molar NaCl, all synthetic activity was lost. The explanation of these dramatic findings is eagerly awaited.

There are further indications that an important locus of protein and nucleic acid synthesis may be associated with the cytoplasmic membrane of bacteria.[17, 243a] Butler, Crathorn, and Hunter[244] investigated amino acid incorporation by intact cells, protoplasts and fragmented protoplasts of *Bacillus megaterium*. Their results suggested that the protoplast membrane was an important site of amino acid fixation and that soluble protein could be produced by a process involving transfer from membrane to cytoplasm.

Nomura et al.[246, 247] reported that lysozyme treated *Bacillus subtilis* could form amylase even under conditions such that the protoplasts were completely lysed. Addition of polyvinyl sulfate and a protease inhibitor prepared from potato extract were necessary to maintain this activity and it was thought to depend on the presence of RNA. The polyvinyl sulfate may have acted by inhibiting RNase present in the preparations. However, the rate of glycine incorporation by such lysates was only about $\frac{1}{500}$ of that to be expected in a growing culture.

E. Inducible Enzyme Formation in "Protoplasts" of *Escherichia coli*

The "protoplasts" of *Escherichia coli*, formed by growth in the presence of penicillin, are capable of synthesizing inducible enzymes. Lederberg[59] showed that β-galactosidase formation could be induced by lactose; McQuillen and Sheinin[253] have demonstrated induction of both α- and β-galactosidase activity in the globular forms of *E. coli*, at rates comparable with those of intact cells. The same inducers are active in cells and spheroplasts. The β-galactosidase is a soluble enzyme and the activity of preparations increases on lysis, but the α-galactosidase activity is abolished if lysis occurs.

Some inducible enzyme systems (including the β-galactosidase of *Escherichia coli*) involve an inducible transporting system for the substrate, as well as an inducible enzyme for metabolizing the substrate. The former system has been called a "permease,"[154] although some people prefer the term "translocase,"[155] and others dislike both. The concentrating mechanism might well be expected to be present in the osmotic barrier of the cell, i.e., the cytoplasmic membrane. Rickenberg[156] found that both induced cells and "protoplasts" (lysozyme/EDTA) of *E. coli* had similar abilities to concentrate lactose; hence the "galactosidepermease"[157] is present in "protoplasts" as well as in intact cells.

Sistrom[238] used lysozyme and EDTA to prepare spheroplasts of *Escherichia coli* both induced to form β-galactosidase and uninduced. The osmotic behavior of these spheroplasts was consistent with the supposition that the bulk of the intracellular galactosides accumulated by the galactoside-permease of the induced form was present as the free solute and thus osmotically active. Lactose accumulated to the extent of 110 mg./gm. dry weight of cells and β-thiomethyl galactoside to 40 mg./gm.

Spheroplasts of *Escherichia coli* formed by lysozyme and EDTA were shown by Otsuji and Takagi[248] to be able to make tryptophanase even after lysis. DNA was also formed.

Rogers[251] investigated the synthesis of ornithine transcarbamylase by

Escherichia coli preparations. Whereas intact cells could use NH_3, spheroplasts formed by penicillin or by lysozyme ± EDTA required a full range of amino acids for formation of the enzyme. The ability of *E. coli* to make β-galactosidase after growth in penicillin was studied by Hurwitz et al.[252] Samples for assay of enzyme-forming capacity were taken after various periods of growth, washed and resuspended in cold NaCl (1% w./v.) in a tight-fitting glass homogenizer before being incubated in the induction medium. During growth in penicillin the cells became nonviable and osmotically fragile, ultimately becoming spheroplasts; their enzyme-forming ability decreased to a minimum of about 10% (at 30 minutes) but then rose again to its initial value (at 50 minutes) before declining once more. Spheroplast formation occurred during the period when the activity was rapidly rising and was accompanied by massive synthesis of DNA—an 8-fold increase between the 20th and the 60th minute. Such a remarkable rate of DNA synthesis (except in phage-infected systems) has not been explained.

A fraction of *Alcaligenes faecalis* which contained fragments of the cell wall and/or membrane was reported by Beljanski and Ochoa[249] to incorporate amino acids and to increase in net protein (up to 20%). A soluble "incorporation enzyme" was purified 800-fold and this enhanced amino acid incorporation. Increases in protein were also found when amino acids were omitted and also when neither amino acids nor the enzyme was added.

Another striking report of protein and nucleic acid synthesis by a wall/membrane preparation comes from Spiegelman's laboratory.[250] Penicillin-spheroplasts of *Escherichia coli* were lysed and fractionated into soluble protein, ribosomal particles, and a cell wall/cytoplasmic membrane fraction. The latter on suitable supplementation was able to incorporate amino acids and to synthesise RNA-like polynucleotide. The protein content could increase by 15% and the polynucleotide by 1000% in 3 hours. The same preparation was reported to be able to form an inducible enzyme.

F. Nitrogen Fixation and Oxidations by "Protoplasts" of *Azotobacter vinelandii*

P. W. Wilson and his colleagues (José and Pengra[158] and unpublished observations) have shown with N^{15} that "protoplasts" formed by lysozyme treatment of *Azotobacter vinelandii* in the presence of citrate (see Section II, C, 2) can fix nitrogen. It was also shown that succinate and malate oxidation, sparked by acetate, occurred with little or no lag in contrast to the behavior of intact cells. The $Q_{O_2}^N$ values for "protoplasts" were similar to those for cell-free preparations. However, acetate oxidation appeared to be impaired in "protoplasts." These results are the first in which a difference in permeability between intact cells and "protoplasts" might be postulated.

It is clear that protoplasts and "protoplasts" have capabilities of synthesis closely similar to those of intact cells under parallel conditions but that modifications to protoplasts can lead to great changes in their relative abilities.

G. Growth of Bacteriophages in Protoplasts

Protoplasts of *Bacillus megaterium*[17, 39, 40, 42] and *B. subtilis*[159] retain the ability to support the growth of bacteriophages. The initial interaction between virus and cell occurs as a highly specific reaction between the tail tip of the phage and a receptor site in the wall of the bacterium. It is not surprising, therefore, that protoplasts do not interact with phage particles.[13] A. Pirie,[160] in 1940, demonstrated that lysozyme treatment of heat-killed *B. megaterium* abolished the ability to adsorb phage and released any already adsorbed. This release was parallel to the hydrolysis of the bacterial polysaccharide by the lysozyme and she suggested that the two might be interrelated. After adsorption of the phage on the cell wall, the DNA passes into the bacterium leaving most of the phage protein as an empty coat still attached to the cell wall. This protein can be removed (by a Waring Blendor) without impairing the ability of the bacterium/phage DNA system to continue the normal course of intracellular development, lysis of the bacteria and liberation of progeny phage—up to several hundred from each infected cell. Fig. 16 represents diagrammatically the probable sequence of events.

1. Growth of Virulent Bacteriophage in Protoplasts of *Bacillus megaterium*

No success has been achieved in attempts to infect protoplasts with phage but if cells are infected and then converted to protoplasts, development of mature phage can still occur. Brenner and Stent[40] infected cells of *Bacillus megaterium* KM with phage in peptone medium, centrifuged and washed the cells, and suspended them in buffered sucrose. One part was treated with lysozyme and, when protoplast formation was complete, both this and the remainder were diluted with sucrose-buffer-peptone and incubated at 25°. After a 90-minute latent period, the intact cell preparation released phage; ultimately the average burst size was 230. The initial counts of the protoplast sample were about 1% of the input because plating during the "eclipse period" results in lysis of protoplasts which at this stage do not contain any mature phage. Eventually a titer 14 times the input value was reached. A single-burst experiment showed that most of the infected protoplasts produced phage, rather than that a small proportion produced large bursts (Table XX).

The relatively low burst size (14 per protoplast compared with 230 per

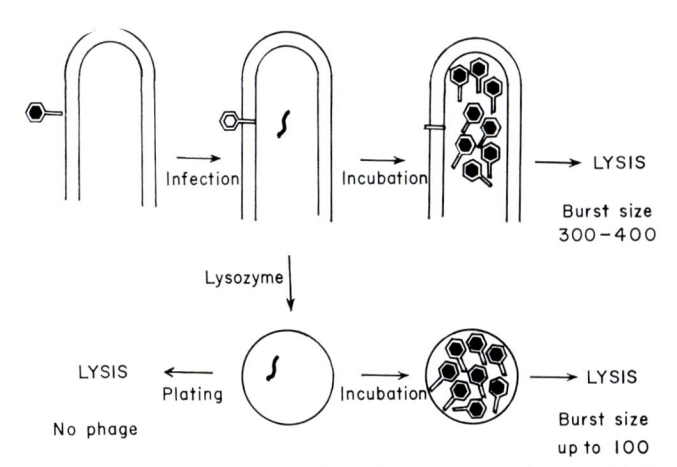

Fig. 16. Representation of current hypotheses concerning multiplication of virulent bacteriophage, together with implications of conversion of cells of *Bacillus megaterium* KM to protoplasts (McQuillen[17]).

TABLE XX

Single-Burst Experiment on Bacteriophage-Infected Protoplasts of *Bacillus megaterium*[a]

No. of plaques	No. of tubes[b]
0	24
1	5
2	2
3	4
5	1
12	2
23	1
25	1

[a] Data of Brenner and Stent.[40]
[b] Forty tubes containing an average of one infected protoplast per tube were incubated and the whole contents of each tube plated to determine the number of infective centers.

Fraction of tubes without burst	$24/40 = 0.60$
Average number of bursts per tube	$-\ln 0.60 = 0.51$
Average number of infected protoplasts per tube	1.0
Fraction of protoplasts yielding bursts	$0.51/1.0 = 0.51$

intact cell) was not always found. Brenner,[42] using an aerated asparagine-phosphate-sucrose medium, obtained bursts of 45 from intact cells and 30 from protoplasts. With these preparations DNase (10 μg./ml.) had no effect but RNase (50 μg./ml.) abolished phage growth by causing lysis of the protoplasts. UO_2^{++} also inhibited production of bacteriophage in protoplasts (Brenner, unpublished observations); this may be related to McQuillen's finding that UO_2^{++} inhibited glycine incorporation into nucleic acid[17, 20] (see Section VI, B, 3 and Fig. 14).

The same strain KM of *Bacillus megaterium*, trained to grow on glucose/NH_3/salts,[20] was used by Salton and McQuillen.[39] Since this medium was unsuitable for adsorption of phage, this step was carried out in peptone, a sample of infected cells was converted to protoplasts by lysozyme and half of this sample was lysed. Phage growth in cells, protoplasts, and lysed protoplasts was compared in the glucose/NH_3/salts medium made hypertonic with sucrose. Fig. 17 shows one-step growth curves with high and low multiplicities of infection. The eclipse period is evident in the

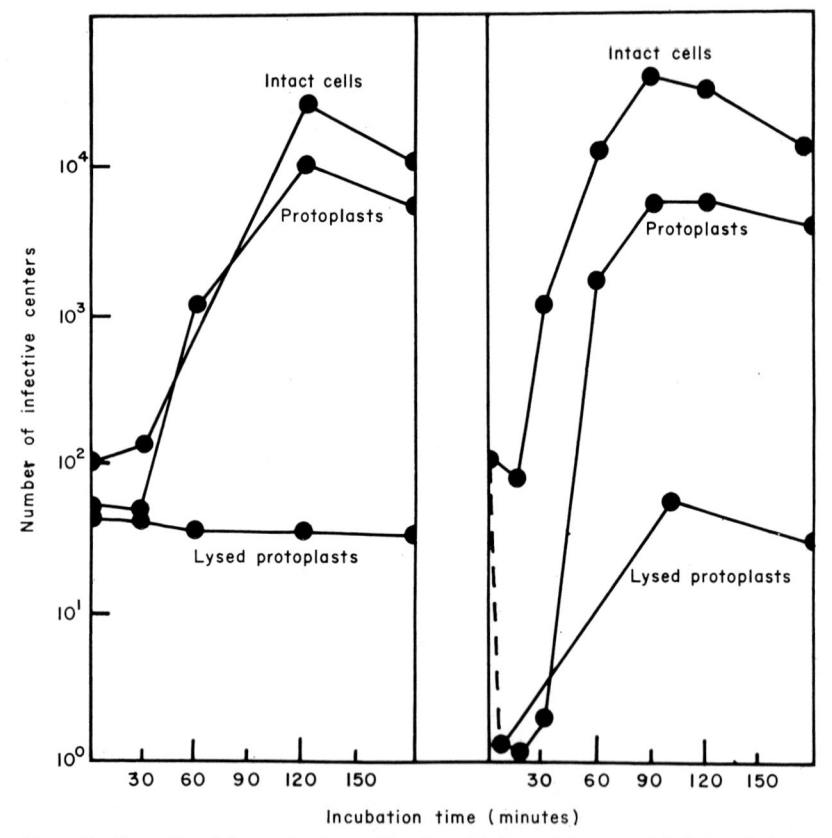

FIG. 17. Growth of bacteriophage C after high and low multiplicity infection of *Bacillus megaterium* KM. Whole cells were infected before washing and conversion to protoplasts. Note "eclipse period," which lasts for about 30 minutes in low multiplicity experiment. Data of Salton and McQuillen.[39]

latter case; in each case the yield of phage from protoplasts is about 30% of that from cells, while the lysed protoplast control shows fewer infective centers at the end of the experiment than at the beginning. The increase in bacteriophage (up to 100 per infected protoplast) occurred in media in which the only sources of carbon and nitrogen were glucose, NH_3, and the sucrose stabilizer.

2. Growth of Bacteriophage in Protoplasts of Lysogenic *Bacillus megaterium*

Lysogenic bacteria are believed to carry "prophage" as a genetic component;[161] cultures of such organisms contain bacteriophage derived from lysis of a small proportion of cells in which prophage has developed spontaneously into many mature phage. Various agents, such as ultraviolet light, certain reducing agents, and hydrogen peroxide, may cause all or most of the cells in a lysogenic culture to be "induced" to produce mature phage and to lyse. The course of events is represented diagrammatically in Fig. 18.

Synthetic media which maintained the lysogenic nature of the strain 899(1) of *Bacillus megaterium* and conferred a condition of "aptitude" for induction[161] were developed by McQuillen and Salton.[44] Using hydrogen peroxide as inducer, they tried without success to induce protoplasts.[39] However, when intact cells were treated with H_2O_2 (67 μM) while growing

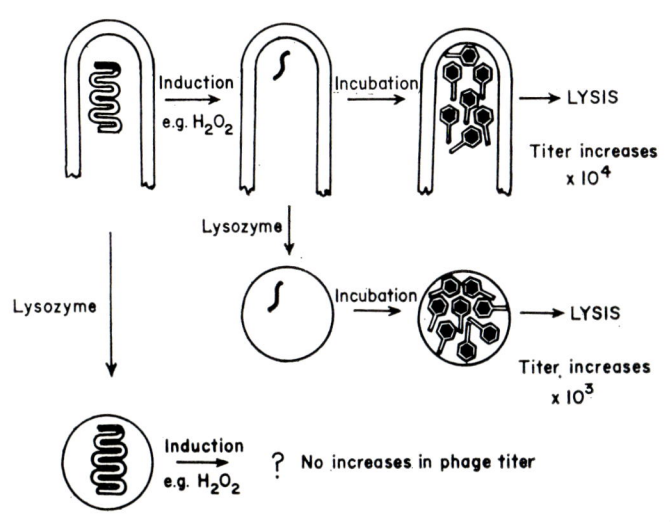

Fig. 18. Representation of current hypotheses concerning induction of a lysogenic culture, together with implications of conversion of cells of *Bacillus megaterium* 899(1) to protoplasts (McQuillen[17]).

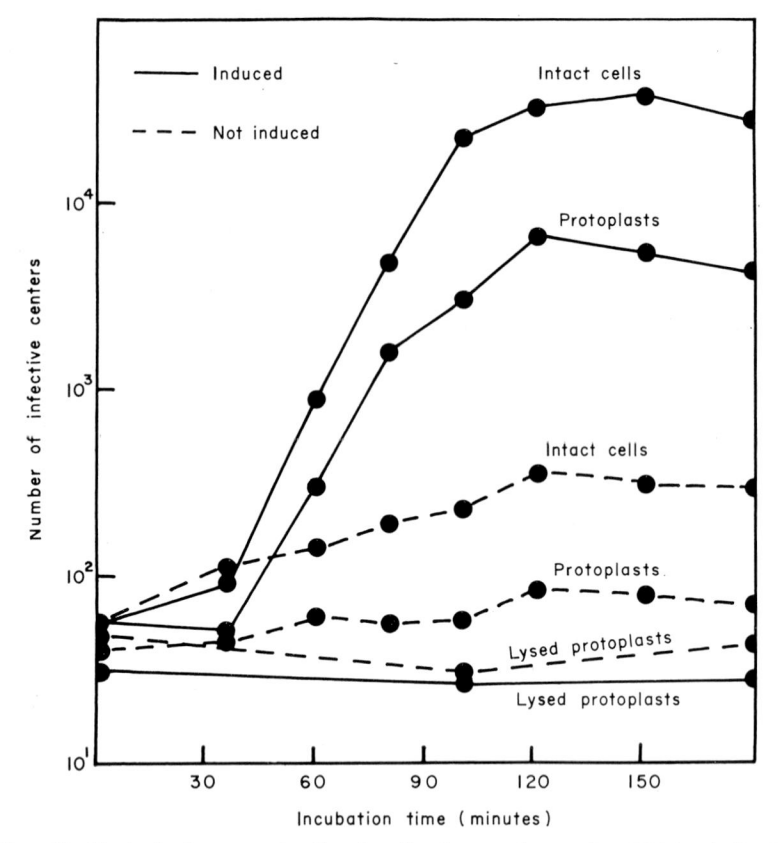

Fig. 19. Bacteriophage production by the lysogenic strain 899(1) of *Bacillus megaterium*, with and without induction by H_2O_2 (67 μM). Protoplasts formed 15 minutes after induction. Data of Salton and McQuillen.[39]

exponentially for 15 minutes in a glucose-sucrose-amino acid medium and then converted to protoplasts, these would go on to give rise to mature bacteriophage on further incubation. Induced and noninduced preparations of intact cells, protoplasts, and lysed protoplasts were studied (Fig. 19).

The phage titer increased 100- to 1000-fold in the samples of protoplasts derived from induced cells; final values were about one-seventh of those found with induced intact cells. Again, lysed preparations were inactive. Failure to induce protoplasts directly (using H_2O_2 concentrations between 16.7 μM and 167 μM) was probably due to unsuitable experimental conditions, rather than to inherent disability on the part of protoplasts.

H. Bacteriophage Experiments with "Protoplasts" of *Escherichia coli*

1. Growth of Bacteriophage in "Protoplasts" Derived from Infected Cells

Spherical, osmotically sensitive forms of *Escherichia coli*, whether produced by lysozyme treatment at pH 9,[58] or by lysozyme and EDTA,[33, 61, 107] are capable of supporting the growth of bacteriophage. Zinder and Arndt[58] infected *E. coli* B in saline containing tryptophan (100 μg./ml.) with phage T4 (multiplicity 0.01) and, after 5 minutes, added lysozyme and tris buffer, pH 9.0. Samples of infected cells, "protoplasts," and lysed "protoplasts" were incubated in P broth (see Section IV, A, 3). The yields of infective centers after 90 minutes are shown in Table XXI. "Protoplasts" gave a final titer about as large as did whole cells, but the time 0 titer indicated recovery of 24 % of the input, as compared with about 1 % for the *Bacillus megaterium* system.[39, 40, 42] This probably indicates the greater survival properties of "protoplasts" of *E. coli* when plated on solid media. The lysed preparations gave only small numbers of plaques—a final recovery of only 3.4 % of the input.

Fraser and Mahler[33, 61] studied growth of T3 using lysozyme/EDTA for treating *Escherichia coli* B. Cells were infected (multiplicity 0.18), washed, converted to "protoplasts", and then incubated in sucrose-broth. Figure 20 shows a one-step growth curve for such a system and Table XXII compares the final yields from cells, "protoplasts", and lysed "protoplasts".

Single-burst experiments showed that the lower average yield from "pro-

TABLE XXI

Growth of Bacteriophage T4 in "Protoplasts" of Preinfected *Escherichia coli* B[a, b]

System	No. of infective centers	
	Time, 0	Time, 90 minutes
Cells	100	6000
"Protoplasts"	24	4400
Lysed "protoplasts"	0.86	3.4

[a] Data of Zinder and Arndt.[58]

[b] *E. coli* B infected in saline/tryptophan (100 μg./ml.) with T4 (multiplicity 0.01). After 5 minutes, one part converted to "protoplasts" by lysozyme at pH 9.0 in Tris buffer. Infected cells, "protoplasts," and lysed "protoplasts" incubated in P broth. Number of infective centers expressed relative to 100 for the initial count with intact cells.

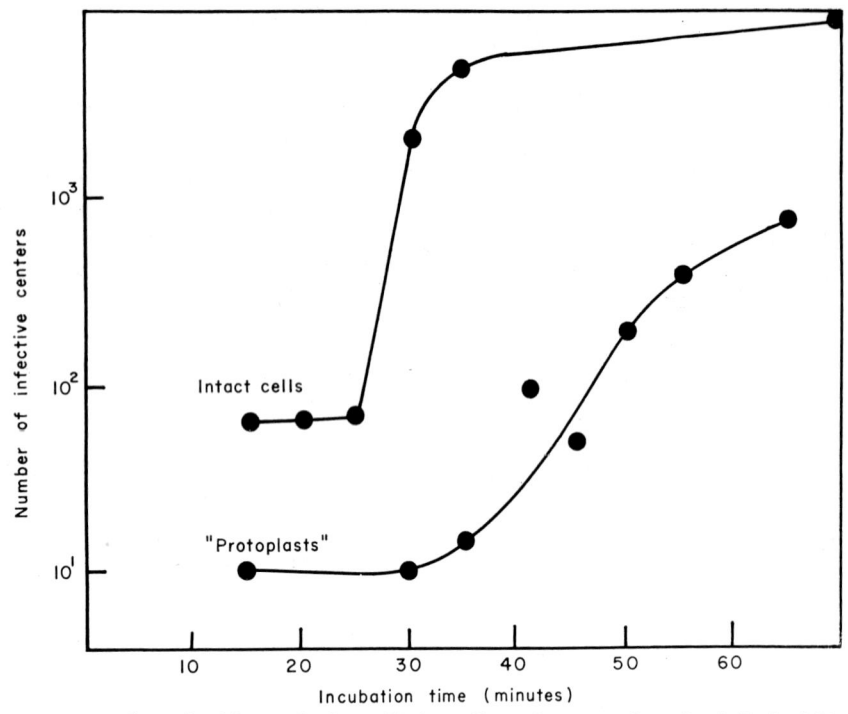

FIG. 20. Growth of bacteriophage T3 in cells and "protoplasts" of *Escherichia coli* B. Cells infected before treatment with lysozyme/EDTA to convert them to spherical forms. Data of Mahler and Fraser.[33]

TABLE XXII

Growth of Bacteriophage T3 in "Protoplasts" of Preinfected *Escherichia coli* B[a, b]

System	No. of infective centers	
	Time, 0	Time, 65 minutes
Cells	100	14300
"Protoplasts"	14	1140
Lysed "protoplasts"	14	214

[a] Data of Mahler and Fraser.[33]
[b] *E. coli* B infected with T3 (multiplicity 0.18), washed, and converted to "protoplasts" by lysozyme/EDTA. Infected cells, "protoplasts," and lysed "protoplasts" incubated in sucrose-broth. Number of infective centers expressed relative to 100 for the initial count with intact cells.

toplasts" was not due to normal bursts from a small proportion of the population. The burst size was fairly constant and most of the infected "protoplasts" produced phage.

2. Reported Infection of "Protoplasts" with Bacteriophage

None of the Gram-positive protoplasts has been shown to be infectable with phage or to adsorb phage. The situation with "protoplasts" of *Escherichia coli* is confused. The suggestion that "apparent infection by T3 subsequent to the transformation of cells into protoplasts"[33] could occur has been modified (Fraser, unpublished observations). Similar preparations made by Fraser *et al.*[107] could not be infected with T2. McQuillen (unpublished observations) found no interaction between T2 and penicillin-formed "protoplasts" of *E. coli* B, nor did Zinder and Arndt[58] and Spizizen[162] using lysozyme "protoplasts." However, Spiegelman (unpublished observations) found that "protoplasts" of *E. coli* prepared by both lysozyme and penicillin methods "adsorb T2 as well as intact cells and furthermore, they grow virus quite well and every protoplast is infectable. The burst size tends to be, however, suppressed as compared with intact cells in normal medium but not when compared to intact cells held under hypertonic conditions." Kellenberger (unpublished observations) found no growth of T2 in "protoplasts" prepared by phage action, nor did they adsorb T4. Lederberg and St. Clair[192] reported that penicillin-spheroplasts of *Escherichia coli* could be infected with T3 and T7 and feebly with T6 but were relatively resistant to T1, T4, T5, and λ-2. Böhme and Taubeneck[255-257] found that penicillin-induced "large bodies" (spheroplasts) of *Proteus mirabilis* could be infected with bacteriophage and responded in exactly the same way as did normal rod-shaped organisms.

At present it cannot be said definitively whether or not the "protoplasts" of *Escherichia coli* interact with intact bacteriophage. There is evidence that heated, osmotically shocked phage[162] and urea-treated phage[107] can infect lysozyme-treated *E. coli* and there is no proof that the cell wall is entirely removed by the procedures used for making "protoplasts." It may be, therefore, that apparently successful infection of protoplasts is due to damaged phage or residual cell wall material or to a combination of both. Salton[254] reviewed work on the development of bacteriophages in protoplasts *stricto sensu* of various species.

3. Effect of Nucleases on Development of Phage in "Protoplasts"

Fraser and Mahler[61] found that RNase and DNase caused lysis of "protoplasts" in some media but not in others; they chose to use a sucrose-broth and D medium (containing phosphate buffer, salts, glycerol, amino acids, and a trace of gelatin). When *Escherichia coli* B which had been infected with T3 was washed, treated with lysozyme and EDTA, and then suspended in either of these media, the "protoplasts" were stable to DNase for 1 hour or more. RNase in sucrose-broth caused a fall in turbidity to 70% in 20 minutes and to 43% in 1 hour. In D medium reduction to 90% occurred in 1 hour. Phage growth was studied in preinfected "protoplasts" and a burst size of 2 to 3 was observed (6 to 10 when corrected for unabsorbed and aborted phage). Exposure to nucleases for 5 minutes after conversion to "protoplasts" resulted in lower final yields, the extent of the reduction depending on the enzyme (RNase or DNase) and the medium (Table XXIII).

Experiments were also performed in which the enzymes were present throughout the entire growth period (Table XXIV). RNase consistently reduced phage yields but DNase action was found to be more variable. These variations and the general variability in response of both "protoplasts" and true protoplasts to the action of enzymes may be due to alterations in permeability induced by changing the nature of the suspension medium. Thomas[163] speculated on the possibility that "competence"

TABLE XXIII

Effect of Brief Exposure of "Protoplasts" from T3-Infected *Escherichia coli* B to Ribonuclease and Deoxyribonuclease[a,b]

"Protoplast" pretreatment	Enzyme treatment	Final titer
Sucrose-broth	5 min. in sucrose-broth	
	No enzyme	100
	DNase 500 µg./ml.	11
	RNase 500 µg./ml.	85
D medium	5 min. in sucrose-broth	
	No enzyme	43
	DNase 500 µg./ml.	31
	RNase 500 µg./ml.	8.5
Sucrose-broth	10 min. in sucrose-broth	
	No enzyme	100
	DNase 500 µg./ml.	17
	RNase 500 µg./ml.	19
D medium lacking Mg^{++}	10 min. in D medium lacking Mg^{++}	
	No enzyme	40
	DNase 500 µg./ml.	27
	RNase 500 µg./ml.	12

[a] Data of Fraser and Mahler.[61]

[b] *E. coli* B infected with T3, converted by lysozyme/EDTA to "protoplasts," suspended in "pretreatment" medium, immediately diluted 10 times into enzyme treatment system. After 5 minutes at room temperature, diluted 100 times into sucrose-broth and titrated at beginning and end of incubation period (60 or 75 minutes) at 30°.

TABLE XXIV

Effect of Exposure of "Protoplasts" of T3-Infected *Escherichia coli* B to Nucleases during the Whole Growth Period[a,b]

"Protoplast" pretreatment medium	Enzyme treatment and growth medium	Enzyme	Final titer
Sucrose-broth	Sucrose-broth	None	100
		DNase	10
		RNase	10
		Both	7.7
D medium	Sucrose-broth	None	117
		DNase	27
		RNase	9.8
		Both	11.3
D medium	D medium	None	133
		DNase	75
		RNase	33
		Both	46

[a] Data of Fraser and Mahler.[61]
[b] RNase and DNase used at 500 μg./ml. Conditions as in Table XXIII, except that after 5 minutes in enzyme treatment system, dilutions were made into same medium still containing enzyme(s).

for transformation of pneumococci by DNA preparations might be associated with a permeable, protoplast-like phase, but he found no evidence that this was the case.

4. Infection of "Protoplasts" with Damaged Bacteriophage

a. Heated, Osmotically Shocked Bacteriophage. Spizizen[162] reported apparent infection of "protoplasts" of resistant strains and even of other species by heated, osmotically shocked preparations of phage. He used T2r$^+$ which in the intact state would not infect "protoplasts" of *Escherichia coli* B formed by lysozyme/EDTA treatment. Fortyfold dilution of suspensions in NaCl (3.0 M) with gelatin solution (1 % w./v.) reduced the infectivity against whole cells by more than 99 %. Preparations were also heated to 71° for 15 minutes either before or after being shocked. Samples which gave a titer of less than 10^2/ml. when plated on whole cells could increase to 10^8 or 10^9/ml. after 60 minutes' incubation with "protoplasts" of *E. coli* B or of the resistant strain B/2. After a lag the increase occurred more or less exponentially, but could be inhibited by EDTA (0.01 M), NaCl (0.1 M), cyanide, arsenite and arsenate, or by absence of casein hydrolyzate or nutrient broth. Pretreatment of heated, shocked phage with trypsin, DNase (with *or* without citrate), or ultraviolet light, markedly reduced activity, as did 2 minutes' treatment in a Waring Blendor or a

Mickle disintegrator. The enzymes, however, were not inhibitory if added 10 minutes after mixing the damaged phage with the "protoplasts."

Unheated, shocked phage could also increase in titer on incubation with "protoplasts," but differed in being insensitive to trypsin and DNase. However, addition of either of these enzymes or RNase to the phage/"protoplast" system prevented increase in titer. Spizizen[162] suggested that the enzyme action might be on the host rather than on the phage.

The titer of heated, shocked phage also increased when incubation was carried out with "protoplasts" of certain other species and, conversely, there was no increase with some strains of *Escherichia coli* which would support growth when intact cells were infected with intact phage (Table XXV). Spizizen pointed out that these experiments do not establish with certainty the synthesis of new phage particles by replication in these systems, since in most cases much less virus was ultimately recovered than was put in. He has also tried to obtain infection of "protoplasts" with phage DNA. Preparations were made from T2 by phenol extraction[164] and also by the acetate-duponol method,[165] but when used at 1 mg./ml. these DNA preparations were unable to cause infection.

TABLE XXV

Effect of Incubation with "Protoplasts" on Titer of Heated, Shocked Bacteriophage T2[a,b]

Organism	Sensitivity of intact cells to intact T2 (10^4/ml.)	Titer of T2/ml. after 60 min. incubation of heated, shocked T2 with "protoplasts"
Escherichia coli B	+	$> 10^8$
E. coli B/2	0	$> 10^8$
E. coli K 12	+	5×10^3
E. coli Crookes	+	1.5×10^5
Anaerogenic *E. coli* (WR-2)	0	10^4
Atypical *Klebsiella* (WR-3)	0	5×10^3
Providence group paracolon (WR-6)	0	1.3×10^4
Aerobacter aerogenes 417	0	2×10^6
Proteus vulgaris (WR-1)	0	0
Bacillus megaterium 899	0	0
Pseudomonas saccharophila	0	0

[a] Data of Spizizen.[162]

[b] Phage T2 (10^{12}/ml.) heated 15 min. at 71°, incubated 5 min. in NaCl (3.0 M), rapidly diluted 40-fold with gelatin solution (1.0% w/v). Titer less than 10^2/ml. "Protoplasts" prepared in sucrose (0.5 M), tris (0.01 M, pH 9.0), containing lysozyme (12.5 μg./ml.). After 5 min., preparations spun and pellet resuspended in sucrose (0.25 M) containing ½ concentration nutrient broth and mixed with heated, shocked phage.

b. *Urea-Treated Phage.* Fraser *et al.*[107] also degraded phage T2 and obtained material which in association with "protoplasts" of *Escherichia coli* increased in titer. Phage was diluted into urea (8 M) in saline (0.1 M) of final pH value 8.2. The titer fell from 5×10^{10}/ml. to 5×10^2/ml. in 3 minutes; after 1 hour the preparation was dialyzed overnight against saline (0.1 M). When this "T2-DNA preparation," equivalent to 2.5×10^{11} phage particles, was added to 2.5×10^8 "protoplasts" of *E. coli* B, the plaque count after 10 minutes adsorption was 6.5×10^5 but after 75 minutes' incubation at 37° in broth containing bovine serum albumin, the titer had increased to 6×10^7. The T2-DNA and the "protoplasts," when plated separately on *E. coli* B at time 0 or time 75 minutes, gave no plaques. These results seem to imply that phage DNA can infect "protoplasts" and cause formation of mature bacteriophage, albeit very inefficiently. In no experiment was more than 0.1 % of the number of input phage equivalents finally recovered and in some cases it was much less than this.

A careful investigation seemed to rule out the possibility that the DNA preparations contained intact T2 masked (by DNA or traces of urea), or reversibly inhibited (but able to be revivified by "protoplasts"), or reversibly dissociated (and able to be reconstituted by "protoplasts"). It was also found that T2-DNA preparations could infect and multiply in lysozyme/EDTA preparations of the T2-resistant strain B/2. DNA, equivalent to 2.5×10^{11} phage particles, mixed with 2.5×10^9 B/2 "protoplasts" gave an initial titer of 800; after 2 hours at 37° this had increased to 2.7×10^5 when *E. coli* B was used as indicator. No plaques were formed when the systems were plated on *E. coli* B/2. The same number of B/2 "protoplasts" treated with 4.35×10^5 intact T2 phage particles gave titers of 8.6×10^5 and 9.0×10^5 at times 0 and 2 hours, respectively, when plated on *E. coli* B, but no plaques at all on *E. coli* B/2. The appearance of infective centers immediately after mixing T2-DNA with "protoplasts" contrasts with the low titers found during the eclipse period in preinfected protoplasts.[33, 39, 40, 42] It is, however, probably due to the incorporation of serum albumin in the medium since this reduces lysis and allows maturation of phage in "protoplasts" plated during the eclipse period.

Results of single burst experiments and infections with T2-DNA preparations varying in dilution from 1:1 to 1:243, suggested that a single particle could infect and that all infected "protoplasts" gave a similar yield of about 200 phage particles. Multiple reactivation or cooperation between degradation products seemed unlikely. Finally, as further evidence that infective material was DNA, it was shown that incubation of the T2-DNA preparation (25 µg./ml.) for 80 minutes at 37° with DNase (10 µg./ml.) reduced the time 0 plaque count on mixing with "protoplasts" by 97 %.[107]

In the experiments just described the time 0 titers were of the order of 10^{-6} of the input phage equivalents of DNA and of the order of 10^{-3} of the input "protoplasts." It is not impossible that part of the DNA preparation consisted of relatively intact but DNase-sensitive phage structures. And since the proportion of viable cells in "protoplast" preparations can be 10^{-3}, it is not impossible that an even higher proportion still retained cell wall material. An alternative explanation to DNA infection of protoplasts is, therefore, damaged phage infection of damaged cells. This in no way detracts from the interest of the findings. Brown and Kozloff[166] have presented evidence that the phage tail tip can be removed from T2 and T4 to uncover enzymes which attack cell walls of *Escherichia coli*. This uncovering normally occurs after adsorption of the phage to a sensitive cell but damage to the tail can render the phage enzyme active against walls of resistant strains such as B/2. This may also be relevant to the findings of Spizizen.[162]

Sekiguchi[258] was able to infect lysozyme- and penicillin-spheroplasts of *Escherichia coli* with urea-disrupted preparations of T2r made by the procedure of Fraser *et al.*[107] However, such preparations were more sensitive to inactivation by trypsin and a proteinase from *Bacillus subtilis* than by DNase. It still seems likely that infection of spheroplasts by "DNA" preparations involves more than just the DNA of the phage.

Fraser (unpublished observations) has been unsuccessful in obtaining phage from the interaction of T3-DNA preparations and "protoplasts" of *Escherichia coli* in experiments conducted similarly to those which were positive with phage T2-DNA.

Spiegelman (unpublished observations) has a system "in which we can demonstrate synthesis of what is presumably virus DNA in lysates of coli protoplasts."

I. Transformation of "Protoplasts"—The Redintegration Phenomenon in *Escherichia coli*

Bacterial transformations are still something of a mystery and it is not clear how a very large molecule like DNA can enter a "competent" cell by a process which leaves the DNA accessible to the action of DNase for as long as 15 minutes and yet does not allow leakage of essential intracellular components. Thomas[163] has already suggested that "competence" might be associated with a protoplast-like phase in pneumococci but was unable to find evidence to support this idea. Recently, however, Chargaff *et al.*[167] have carried out experiments which may represent transformation of "protoplasts" of *Escherichia coli*.

Crude DNA preparations were made from penicillin-produced "protoplasts" of strain W of *Escherichia coli* by NaCl extractions and ethanol precipitations. From a 300 ml. culture one preparation yielded material containing 6.3 mg. DNA, 2.6 mg. RNA, and 2.2 mg. protein; another gave 4.1 mg. DNA, 2.4 mg. RNA, and 2.2 mg. protein. The mutant strain 26-26 of this organism differs from the wild-type, W, in lacking diaminopimelic acid decarboxylase and thus requiring an exogenous supply of lysine. "Protoplasts" of the mutant were prepared by the Lederberg technique[59] in media containing penicillin, with and without human serum albumin (1 %) and crude nucleic acid preparation from strain W (approximately 20 μg. nucleic acid-P per ml.). After conversion to spherical forms (2½ hours) the preparations were washed with sucrose solution ± albumin and shaken at 35° for 30 minutes in a minimal medium containing sucrose ± albumin ± nucleic acid. Then Penassay medium containing sucrose ± albumin ± nucleic acid was added. After shaking at 35° for 2 hours the spheres had reverted to rod-shaped organisms; these were shaken in the same medium for a further 14 hours at room temperature. Samples were washed and plated on minimal medium with and without a lysine supplement.

Table XXVI shows that the presence of the DNA-containing material from the wild-type during the incubations and manipulations increased the proportion of prototrophs about 5-fold in the absence and 20-fold in the presence of albumin. Positive results of this kind were obtained in 6 out of 10 preparations. No results of comparable experiments with DNA prepara-

TABLE XXVI

REDINTEGRATION—INCREASE IN PROPORTION OF PROTOTROPHS ON TREATING "PROTOPLASTS" OF AUXOTROPHIC *Escherichia coli* (26-26) WITH PREPARATIONS CONTAINING DEOXYRIBONUCLEIC ACID FROM THE PARENT STRAIN[a,b]

Additions		Final no. of viable cells/ml. Pennassay medium. Plated		Probable initial nos. of cells used—calculated to same dilution[d]	
Crude DNA from prototrophs[c]	Albumin	+ Lysine	− Lysine	Auxotrophs	Prototrophs
−	−	2.2×10^8	8.0×10^3	3×10^8	3×10^3
−	+	4.0×10^8	1.3×10^4		
+	−	2.0×10^8	4.1×10^4		
+	+	4.3×10^8	2.4×10^5		

[a] Data of Chargaff et al.[167]

[b] See text for experimental details.

[c] Contains DNA, RNA, and protein from strain W (prototrophic).

[d] Based on Lederberg's figure of 2×10^9 cells/ml. for overnight culture of *E. coli* in same medium[59] and Chargaff et al. figure of 0.01 prototrophs per 1000 cells in such cultures.[167]

tions from the mutant itself have yet been reported, nor is it known what the input number of cells was. It appears probable, however, that the absolute number of prototrophs increases in the absence of both albumin and DNA, even supposing that every "protoplast" survives the many manipulations (this is based on a viable count of 2×10^9/ml. found by Lederberg for an overnight culture of *Escherichia coli* growing in the same medium[59]). Addition of albumin further increases this absolute number of prototrophs, as does the DNA preparation and a mixture of the two. Whether all of these effects are due to protection such as is known to be afforded by serum albumin, or whether there is a specific transformation by the wild-type DNA is not known. Chargaff et al.[167] suggest that if the phenomenon is shown to be the rehabilitation of a mutant by specific material from the wild type, then the term "redintegration" might be used.

J. Spore Formation in Protoplasts of *Bacillus megaterium*

Under appropriate conditions, vegetative cells of *Bacillus* species can be "irreversibly committed" to sporogenesis and eventually there is virtually complete conversion to spores.[168] Suspensions of *Bacillus mycoides* shaken in distilled water sporulated after about 10 hours but were irreversibly committed much earlier. Working with *Bacillus megaterium*, Salton[169] obtained inconsistent results in distilled water, but in a medium containing phosphate buffer (0.033 M, pH 7.0), sucrose (7.5% w/v), and Mg^{++} (0.01 M), vegetative cells were committed to sporogenesis after 4½ hours' shaking. When such preparations were treated with lysozyme, the resulting protoplasts could also continue the process and yield spores, although at a low efficiency. Estimating numbers of viable spores by plating on 5% peptone/0.1% starch agar, it was found that cells gave 10^7/ml., lysozyme-treated cells (protoplasts) gave 10^5/ml., and osmotically lysed protoplasts gave only 10^1/ml. The refractile spores could be seen within the protoplast under the phase contrast microscope. Spiegelman (unpublished observations) has confirmed these findings which are illustrated diagrammatically in Fig. 21.

K. Protoplasts Derived from Motile Species

Weibull[13] observed that although flagella were not removed during digestion of the cell wall of *Bacillus megaterium* by lysozyme, the resultant protoplasts were nonmotile. He demonstrated the presence of these organs of motility both by flagellar stains and by electron microscopy. The latter finding has been confirmed by other workers.[17, 23, 41, 108] The spherical forms of *Salmonella typhimurium* lost their motility and failed to agglutinate in homologous antiflagellar serum, according to Lederberg,[59] but electron microscope studies were not made. Similarly, motile strains of

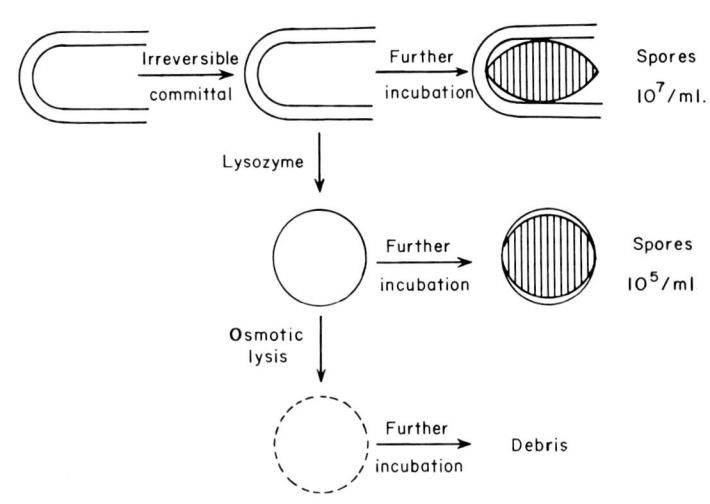

FIG. 21. Representation of sporogenesis of *Bacillus megaterium* and implications of conversion of cells to protoplasts (McQuillen,[17] Salton[169]).

Escherichia coli became nonmotile on treatment with lysozyme at pH 9, but Zinder and Arndt[58] did not investigate the fate of the flagella.

It is true to say that at the present time there is no evidence of any osmotically sensitive, spherical forms being motile, whether or not they bear flagella. If the conversion of actively motile cells of, for example, *Bacillus megaterium* is watched under the microscope, it is seen that motility ceases abruptly when the protoplasts emerge from the last visible remnants of cell wall. Since lysozyme itself does not interfere with motility and has no effect on the motion of lysozyme-insensitive species, it seems that loss of this capacity is bound up with loss of rigidity in the cell wall (McQuillen, unpublished observations). Stocker[169a] suggested that a rigid or semirigid structure may be necessary for the flagella "to exert their thrust upon."

L. Growth of Protoplasts

1. Increase in Dry Weight of Protoplasts of *Bacillus megaterium*

It is six years since Weibull[13] first demonstrated a method of preparing stable protoplasts of *Bacillus megaterium* and reported that they appeared to be incapable of forming colonies on solid media. This situation still obtains and no one has achieved colony formation from megaterium protoplasts. Although they can increase in size in liquid media and come to what appears to be a division stage (see Section L, 2) there is no well-established case of separate protoplasts arising by division.

TABLE XXVII

GROWTH OF PROTOPLASTS OF *Bacillus megaterium*[a, b]

Time of incubation, hours	Dry weight of preparation, mg./100 ml.	
	Intact cells	Protoplasts
0	10.4	8.2
2	23.4	17.2
4	52.4	24.8

[a] Data of McQuillen.[41]

[b] Preparations of intact cells and protoplasts were incubated in a glucose-sucrose-peptone medium and samples removed, treated with formalin, washed, and dried to constant weight.

The size, optical density and dry weight of protoplasts of *Bacillus megaterium* can increase substantially during incubation in suitable media. McQuillen[17, 41] compared the dry weight increase of intact cells and protoplasts suspended in 250 ml. lots of glucose/sucrose/peptone medium in Roux bottles fitted with air vents and gently rocked at 28°. Table XXVII shows that the dry weight of well-washed, formalinized preparations could more than double in 2 hours, and then increase by a similar increment in the next 2 hours. Peptone could be replaced by a mixture of amino acids and probably by the synthetic salts/NH_3 medium which, with glucose as carbon and energy source, could support the growth of the cells of this organism. Phosphate buffer (0.5 M, pH 7.0) could be substituted for sucrose (7.5 to 10%, w./v.) as stabilizer.

An excellent biochemical and cytological study has been made by Fitz-James[259] of the growth of protoplasts of *Bacillus megaterium* both in slide culture and in aerated liquid media. Increases in dry weight, protein, RNA, DNA, and lipid were followed. Increase in mass was linear with time rather than exponential but no division was observed and protoplasts 10 to 20 times as large as the initial ones were obtained after some hours' growth. DNA increased at the same rate as in intact cells but RNA formation was slower so that the RNA/DNA ratio gradually decreased. A 10-fold increase in DNA during 4 hours' incubation was noted and was consistent with the increase in the number of chromatinic bodies in the protoplasts. Lipid-P increased during growth at a rate equivalent to the increase in surface area as would be expected if the lipid were largely present in the protoplast membrane. After prolonged aeration, considerable amounts of fibrous material accumulated in the culture medium. This had also been observed by the present author. It was shown by Fitz-James to be of phospholipoprotein nature and it may be related to the cytoplasmic membrane.

Since protoplasts are inherently fragile, any prolonged experiments are complicated by the occurrence of lysis to a greater or lesser degree and it must be taken that increases in weight (or protein or any other component) are less than maximal. It is not known, for instance, whether individual protoplasts increase exponentially in weight as do cultures of cells. If they do, then the measured increases in weight, which are often linear with time, may be so because of lysis of part of the population. Certainly the diameter of individual protoplasts may double during only a few hours' incubation; this implies an increase of 4 times in surface area and 8 times in volume and probably in dry weight.

2. Division of Protoplasts of *Bacillus megaterium*

When incubation is prolonged to about 6 hours at 28°, it is sometimes found that among the enlarged spherical protoplasts there are some which have a small protuberance. As incubation is continued the proportion of the "buds" increases, as does their size. McQuillen[17, 41] has obtained preparations in which 50% of the surviving protoplasts were of this form (Figs. 23 a, b, c; 24). Eventually, dumbbell forms consisting of two spheres of equal size joined by a short bridge may be seen. The probable sequence of forms is shown in Fig. 22. Brenner (unpublished observations) has confirmed the occurrence of dumbbell forms.

The dumbbell forms are still osmotically sensitive, lysing on dilution of the medium, and are still surrounded by a membrane. This can be seen as a faint "ghost" in the phase contrast microscope and, after fixation, in the electron microscope.[17, 41] Preparations of membranes from suspensions

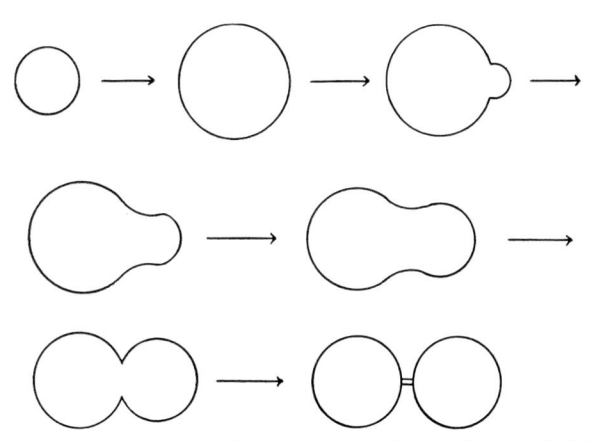

Fig. 22. Representation of possible sequence of growth and division of protoplasts of *Bacillus megaterium* KM on continued incubation in a suitable medium (McQuillen[17, 41]).

which have been incubated for a long time sometimes contain dumbbell-shaped "ghosts" as well as spherical ones, suggesting that the membrane may be responsible for the assumption of this shape. In view of the fact that the protoplast membrane of *Micrococcus lysodeikticus* contains about 20% carbohydrate, mainly mannose, it is possible that as growth proceeds a modification can occur such that the carbohydrate component (which may be a mannan like that found in the yeast cell envelope[170]) can impart a nonspherical form. More likely, perhaps, is the notion that the "bud" occurs at a part of the membrane which is weaker than the remainder. There is a little evidence that in dumbbells there exists a cross membrane dividing it into two separate spheres. On two or three occasions such forms have been seen to undergo spontaneous lysis in such a manner that one sphere became a "ghost" but not the other. It is hard to see how this could happen if a single membrane surrounded the whole dumbbell.

3. Growth of "Protoplasts" of Gram-Negative Species

The spherical forms of *Escherichia coli*, induced by incubation in hypertonic growth medium containing penicillin[31, 32, 59] or deficient in diaminopimelic acid,[31, 32] can enlarge considerably in volume. Lederberg[59] mentioned an increase of 50% in turbidity during transformation to spheres; his diagram illustrates a rod of axial ratio 3.3:1 developing into a sphere of diameter equal to the length of the rod. This involves an increase in volume of nearly 10-fold. Increases of up to 40-fold have been recorded by McQuillen,[32] both for penicillin forms and DAP-less forms of *E. coli* (Figs. 23 d; 26; 27). Similar enlargement occurs with the globular forms of *Proteus vulgaris*.[75] There are no reports of division of these forms produced by penicillin. On continued incubation most of the spheres lyse; some become very granular, and development of large bodies and L-forms may eventually result. It is outside the province of the writer to discuss this further but see Chapter 7, and references 75, 77, and 79.

A short note by Jeynes[98] describes the properties of spherical forms of various species of *Vibrio* and *Salmonella* induced by growth in liquid media containing glycine (3% w./v.). The vegetative rods "lost their cell walls in a few hours and a dense growth of protoplasts was obtained in 18–24 hours." These forms could be propagated for long periods without reversion and it was reported that "division is controlled by a growing point granule consisting of lipoprotein in association with Feulgen-positive nuclear material."[98]

Lark[222] studied the growth of penicillin-spheroplasts of *Alcaligenes faecalis* and found that division did not occur in synthetic medium containing the antibiotic but that prolonged exponential growth and division occurred in tryptone-penicillin medium. It was suggested that a constituent of the tryptone was essential for division of the globular forms. Lark and Lark[260] also showed that growth in penicillin-tryptone led to a conversion to the globular form (spheroplasts) and to a fairly sharp change in the exponential rate of synthesis of protein, RNA, and DNA. The suggestion that the rate of RNA and protein synthesis was reduced *before* that of DNA does not appear valid on examination of the published data but it is well known that the RNA/DNA ratio in large bodies, and L-forms, etc. is lower than that in the corresponding rod forms.

M. Reversion of Spherical Forms to Normal Morphology

1. Protoplasts of *Bacillus megaterium*

The protoplasts of Gram-positive species appear to have all the synthetic abilities of the whole cells from which they were derived, with the exception of the ability to make a new wall. However, they can form diaminopimelic acid, a substance which is a constituent of the mucopolysaccharide cell wall of the vegetative cell but which is absent from the structural make-up of the protoplast. This amino acid, or a derivative, is synthesized and excreted into the surrounding medium by protoplasts (McQuillen[31] and unpublished observations). There are other compounds such as muramic acid[26, 28, 171-173] which are probably uniquely present in bacterial cell walls, but there is no reason to think that protoplasts are incapable of their synthesis or that this is the cause of their inability to make cell walls. Washed protoplasts of *Bacillus megaterium* have been incubated in a variety of media supplemented with isolated cell wall fraction (prepared by Dr. M. R. J. Salton), hydrolyzed cell wall, and such substances as diaminopimelic acid and glucosamine, but in no case has cell wall formation been provoked (McQuillen, unpublished observations).

Three possible explanations remain as to why protoplasts do not make new walls. Firstly, cell wall-synthesizing enzymes may exist between the cell wall proper and the protoplast membrane and may be lost when the cell wall is removed by lysozyme. This possibility is remote, since no proteins have been detected in the supernatant after conversion of cells to protoplasts. Secondly, it may be that enough lysozyme remains associated with the protoplast membrane to degrade any new wall material as fast as it is laid down. This again seems unlikely, since in antigenic tests on protoplasts of *Bacillus megaterium*, Vennes and Gerhardt[134] found that controls with lysozyme were negative. Lastly, it may be that a primer or "starter" of existing cell wall must be present before more can be added, just as phosphorylase[174] and polynucleotide phosphorylase[175, 176] require oligosaccharide and oligonucleotide primers before additional units can be polymerized.

Attempts have also been made to see whether or not protoplasts of *Bacillus megaterium* would revert to bacillary form on solid media. Washed protoplasts were plated on solid, semisolid, and sloppy agar containing various supplements, including hypertonic sucrose solution and peptone. Although individual spheres increased greatly in size—often more than in liquid media—there was never detected any reversion or even any "budding" on these media (McQuillen, unpublished observations).

Preparations have also been studied in microdrop cultures hanging from cover slips and immersed in liquid paraffin. Time lapse photomicrography enabled the fate of individual protoplasts to be followed. One such preparation was observed for 20 hours (McQuillen, unpublished observations). Initially about 60 spherical protoplasts were visible in a single field of view. After about 2 hours at 28° their movements had become so reduced that it was possible to recognize individuals by their position relative to their neighbors. After 6–8 hours the diameter of the spheres had increased to nearly double, i.e., the volume had increased nearly 8-fold, and some had developed small buds. After 12 hours the majority still survived, although about 5 had been seen to burst. At this time nearly half of the population were no longer spherical. By 20 hours somewhat less than half of the initial bodies were visible, only a few remained spherical, most were very gross and distorted and a few had formed tangled masses of tubular growth (Fig. 28).

It is, unfortunately, not known whether or not this represents reversion to bacillary form and cell wall formation. If reversion did occur, it was a relatively slow process, since a single intact rod could produce a vast progeny within 6 hours under the conditions used. It would be useful to carry out both chemical and immunochemical analyses of growing preparations of protoplasts to determine if there is a gradual recovery of the ability to synthesize walls. It is possible, always, that a small proportion of cells retains a small amount of cell wall material during lysozyme treatment and that it is this minority which subsequently reverts.

2. Spherical Forms of Gram-Negative Species

Lederberg[59] reported that of the spherical forms of *Escherichia coli* produced by growth in penicillin media, up to 50% were capable of reverting to rods when transferred to penicillin-free medium containing sucrose and were able to form colonies when plated on agar containing 20% sucrose.

Lederberg and St. Clair[192] dealt at length with observations of the growth of penicillin-spheroplasts of *Escherichia coli* in sucrose-agar to yield L-colonies. Conversion of rod-shaped forms to spheroplasts and reversion of these to normal rods was readily followed in sucrose-broth preparations but no division was ever observed on incubation of spheroplasts in liquid media nor were L-colonies formed on agar surfaces. L-colonies were subcultured as many as 20 passages (equivalent to a cumulative increase of 10^{20}) but no stabilized L-forms were obtained.

The "protoplasts" of *Escherichia coli* formed by lysozyme treatment at pH 9 have been incubated in P broth by Zinder and Arndt[58] (see Section IV, A, 3). For 4 hours there was no change in their appearance; then the majority began to shrink and degenerate. However, about 10% of them increased in volume, formed a vacuole, and then developed a protrusion, which ultimately became a rod capable of normal division. In P broth just solidified with agar, between 2 and 20% of a population of "protoplasts" would form colonies.

Thus, in these experiments a fraction of both lysozyme-formed and penicillin-formed spherical bodies were capable of reversion to rods and of forming colonies. However, Fraser (unpublished observations) has never found reversion of globular forms of *Escherichia coli* produced by the action of lysozyme and EDTA. Nor has he observed division of these forms.

Sinkovics[261] who earlier[180] observed "protoplast"-like forms in senescent cultures, has since reported that *Escherichia coli* cultures aged for several months yielded small units (about 350 mμ) capable of regeneration. Fusion of two or more of these small bodies (which had been filtered through collodion) appeared to precede the development of a new wall structure and no growth or division occurred prior to this fusion. Addition of washed cells of *Sarcina lutea* was said to assist regeneration of the *E. coli* fragments perhaps by allowing adsorption to the surface of the cells.

The variability of response of individual members of these "protoplast" populations and the seemingly contradictory reports concerning reversion to bacillary form, ability to form colonies, reaction with bacteriophage, and so on, make it probable that the term "protoplast" is being used for a heterogeneous collection of forms. It may well be that the variations in response to different treatments are indications of variable amounts of residual cell wall components in the spherical forms. Perhaps a technique such as microelectrophoresis might be used to gain information as to the degree of inhomogeneity, since it gives quantitative results on the surface properties of individual members of a population.[71, 115, 128, 177]

Jeynes[98] claimed that reversion to vegetative forms from the spherical forms induced by growth of *Salmonella* spp. and *Vibrio* spp. in media containing glycine can "be obtained under suitable conditions without difficulty."

The "protoplast"-like forms of *Neurospora crassa* were found by Emerson and Emerson[209] to be able to divide by "a simple pinching in two by a furrowing process or sometimes by a process resembling the budding of yeasts."

Reversion to hyphal-type growth occurred on transfer to media lacking the hemicellulase or snail enzymes used for formation of "protoplasts."

N. Conjugation of Protoplasts

Stähelin[4] reported that spherical bodies, produced by plasmoptysis of *Bacillus anthracis*, could fuse under appropriate conditions—two could coalesce to form a larger sphere if they were placed in hypotonic medium or if the temperature were maintained at more than 40°. No other reports of fusion of protoplasts have appeared but it is now known that recombination in bacteria depends upon conjugation of intact bacterial cells. Lederberg[178] and Anderson *et al.*[179] have demonstrated that conjugation of sexually compatible strains of *Escherichia coli* involves temporary fusion followed by separation. It would seem likely that surface constituents play a dominant role in this process.

Spheroplasts of *Escherichia coli* were found by Lederberg and St. Clair[192] to be capable of conjugation if compatible strains ($F^+ \times F^-$) were used. Either one or both parents could be in the spheroplast form. No fusion of sexually incompatible genotypes was observed. Hagiwara[262] also found that with lysozyme-EDTA spheroplasts of *E. coli* transfer of characters only occurred from F^+ to F^-. Since the spherical forms of Gram-negative bacteria may well retain much of their cell wall structure, it is likely that the successful recombination experiments may arise from the presence of these components. Should this be so, and should the experiments of Chargaff *et al.*[167] turn out to involve transformation of "protoplasts" by DNA, further advances might be made in our knowledge of how genetic information is transferred.

O. Natural Occurrence of Protoplast-like Forms

McQuillen[20] reported that infection of *Bacillus megaterium* with phage can lead to protoplast-like forms and Zinder and Arndt[58] made similar observations with *Escherichia coli* treated with phage ghosts. McQuillen[20] also noted that the margin of old colonies may contain protoplast-like forms. Jeynes[98] and Sinkovics[180] have also drawn attention to the occurrence of protoplast-like structures in senescent cultures; the former found such forms in pond and other natural waters.

Although they are essentially fragile, it is possible that such spherical forms may have survival value under certain conditions: for instance, in the presence of bacteriophage which can infect only when the appropriate receptor site is present in the cell wall. However, to be of value such properties must be manifested by a structure capable of reverting to normal form or capable of continued growth in a stabilized new form. As has already been discussed, the evidence on reversion of Gram-positive protoplasts is almost completely negative so far, but Gram-negative species can often either become stabilized in some different mode or revert to bacillary form when the globular forms are transferred to different environments.

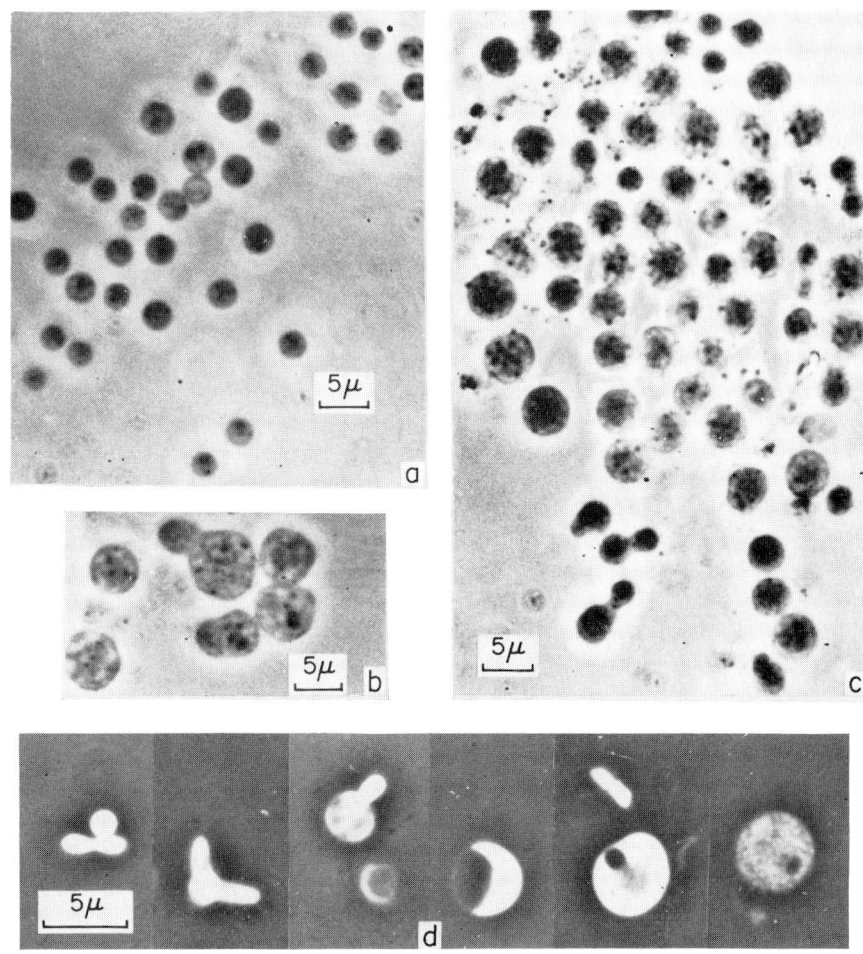

Fig. 23. (a, b, c) Protoplasts of *Bacillus megaterium* KM before (a) and after (b and c) incubation for 8 hours (phase contrast). (d) *Escherichia coli* 173-25 growing in absence of diaminopimelic acid (anoptral phase contrast).

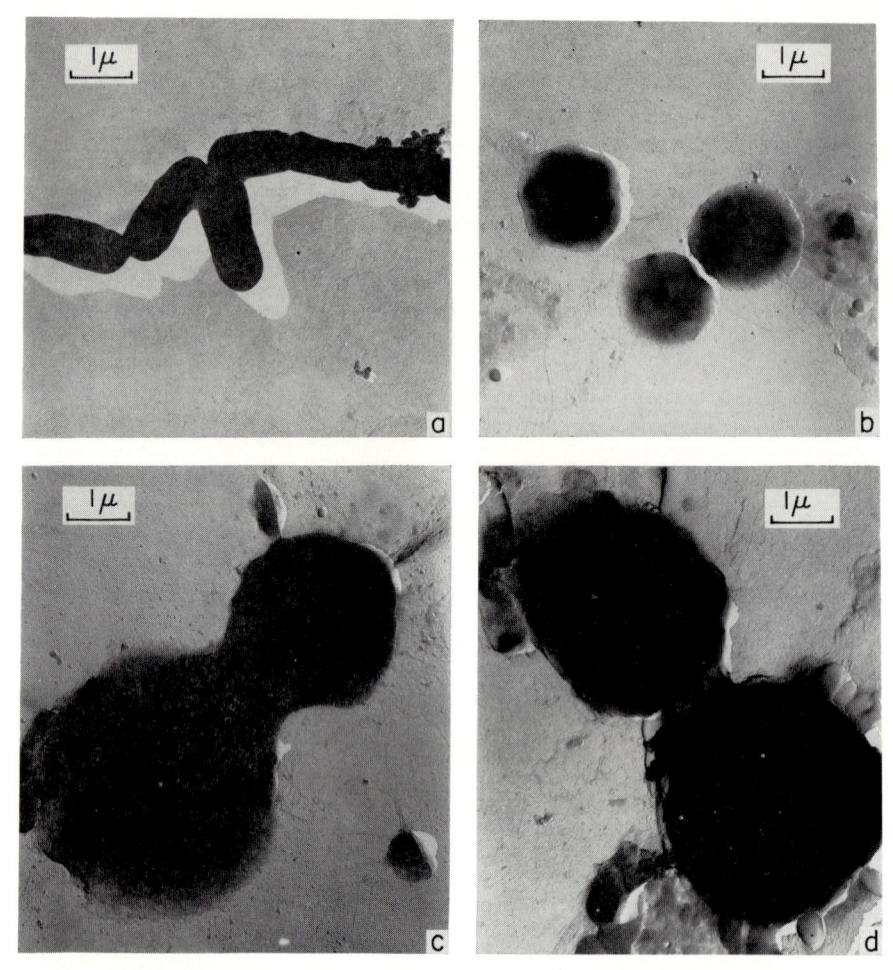

FIG. 24. *Bacillus megaterium* KM (a) Whole cells. (b) Protoplasts at beginning of incubation. (c and d) Protoplasts dividing after 9 hours' incubation (electron micrographs).

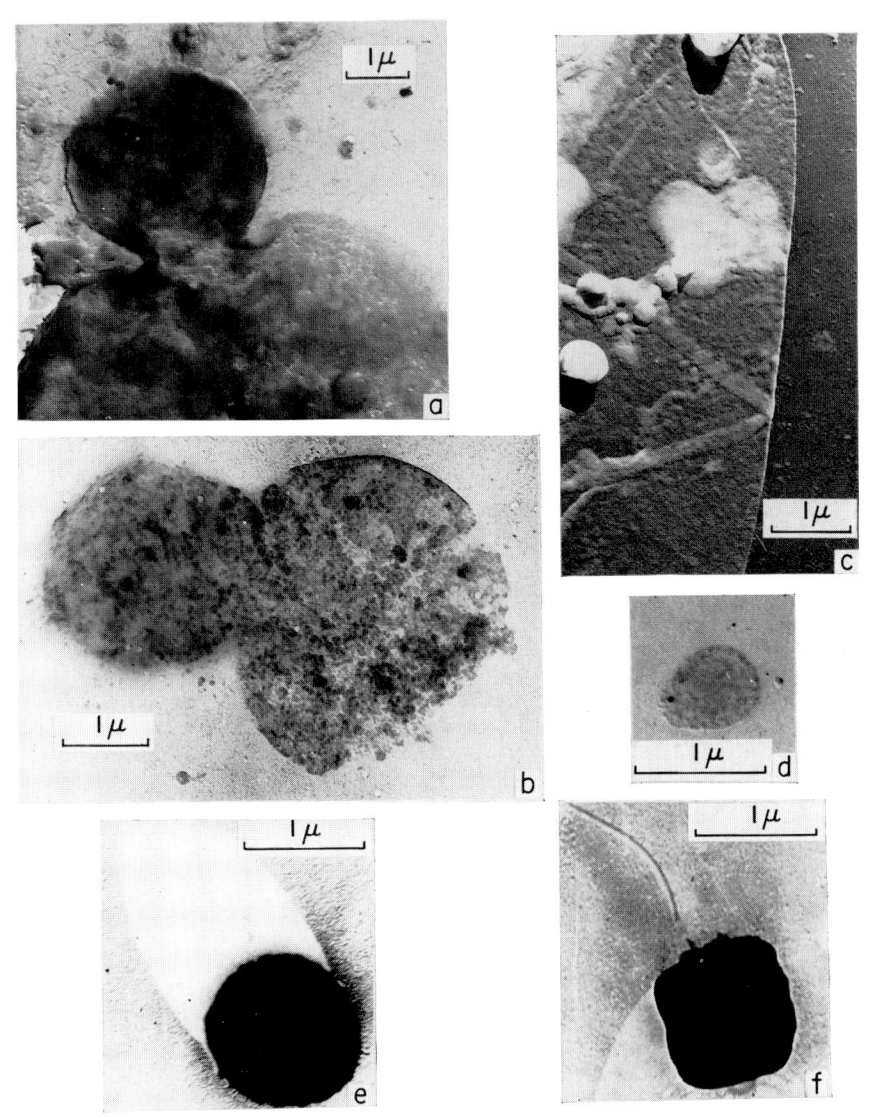

Fig. 25. *Bacillus megaterium* KM (a, b, c, e, and f) and *Micrococcus lysodeikticus* (d) (electron micrographs). (a) "Bud" of greatly enlarged protoplast. (b) "Ghost" of dividing protoplast. (c) Part of protoplast membrane. (d) Entire protoplast membrane of *M. lysodeikticus*. (e and f) Protoplasts from normal and polymyxin-treated cells.

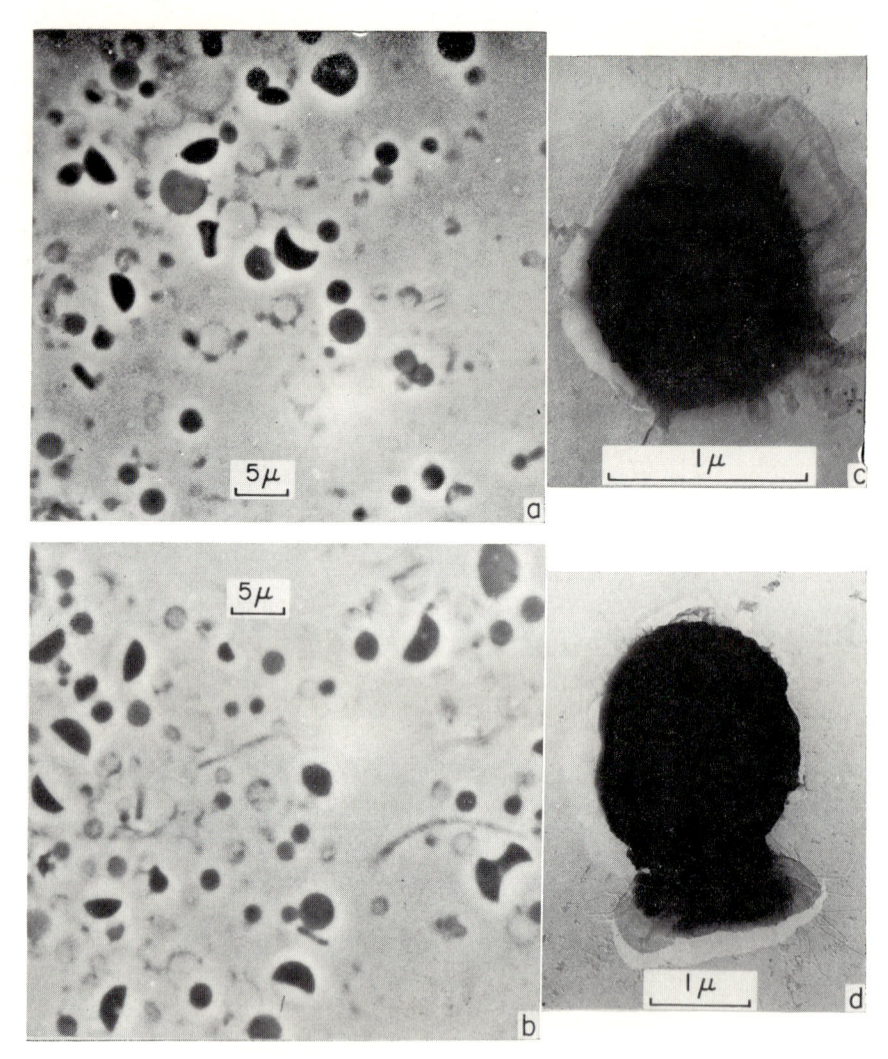

FIG. 26. *Escherichia coli* 173-25 grown in presence of penicillin. (a and b) Phase contrast. (c and d) Electron micrographs.

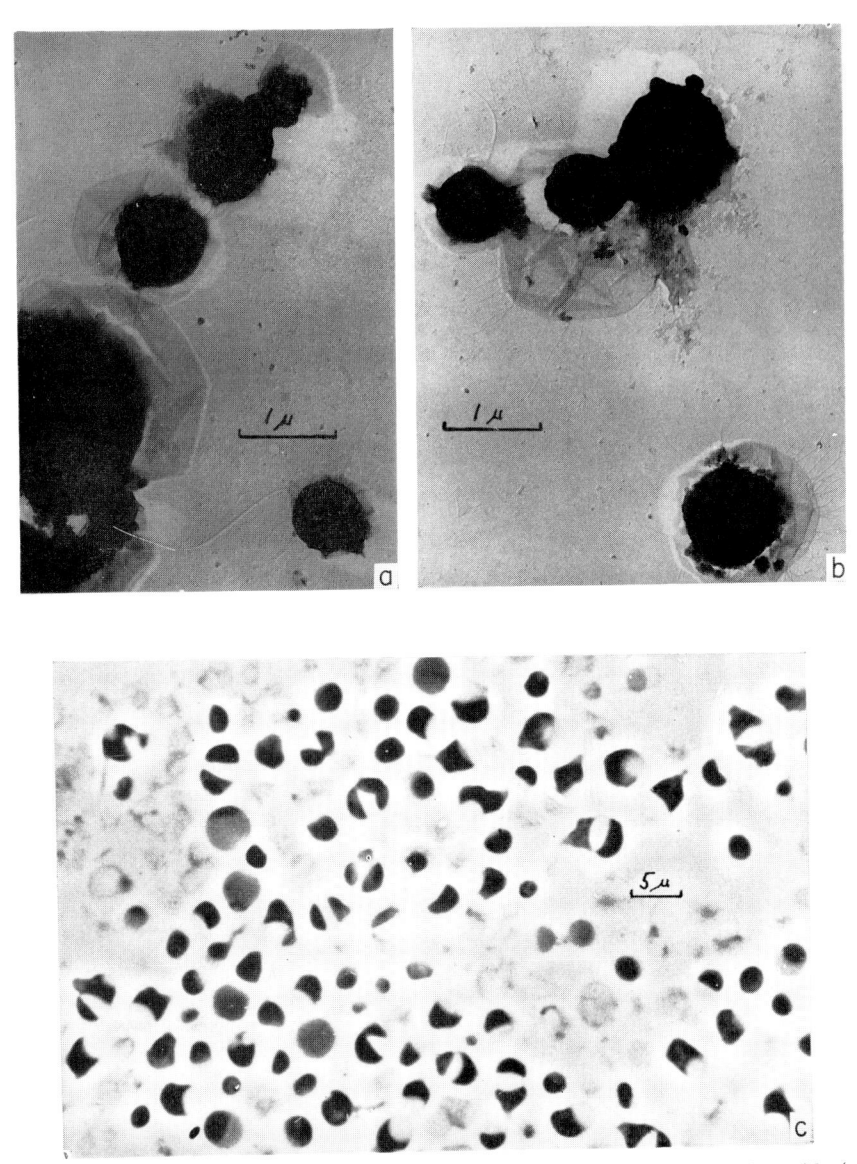

FIG. 27. *Escherichia coli* 173-25 grown in absence of diaminopimelic acid. (a and b) Electron micrographs. (c) Phase contrast.

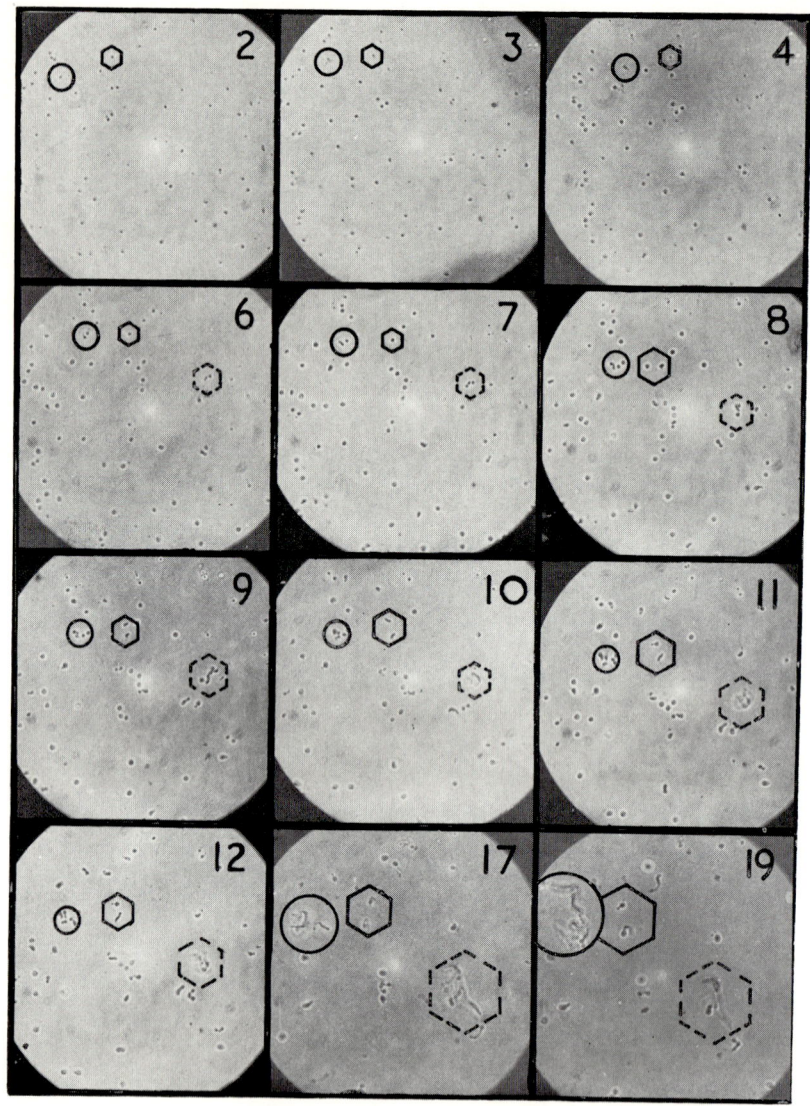

FIG. 28. Serial phase contrast photomicrographs of a single field of a microdrop of protoplasts of *Bacillus megaterium* KM. The numbers indicate hours of incubation at 28°.

REFERENCES

[1] A. Fischer, *Z. Hyg. Infektionskrankh.* **35**, 1 (1900).
[2] A. Fischer, *Ber. deut. botan. Ges.* **24**, 55 (1906).
[3] H. Stähelin, *Schweiz. Z. allgem. Pathol. u. Bakteriol.* **16**, 892 (1953).
[4] H. Stähelin, *Schweiz. Z. allgem. Pathol. u. Bakteriol.* **17**, 296 (1954).
[5] C. F. Robinow and R. G. E. Murray, *Exptl. Cell Research* **4**, 390 (1953).
[6] J. R. G. Bradfield, *Symposium Soc. Gen. Microbiol.* **6**, 296 (1956).
[7] G. B. Chapman and J. Hillier, *J. Bacteriol.* **66**, 362 (1953).
[8] S. G. Tomlin and J. W. May, *Australian J. Exptl. Biol. Med. Sci.* **33**, 249 (1955).
[9] R. G. E. Murray, *Can. J. Microbiol.* **3**, 531 (1957).
[10] A. Birch-Andersen, O. Maaløe, and F. S. Sjöstrand, *Biochim. et Biophys. Acta* **12**, 395 (1953).
[11] G. B. Chapman and A. J. Kroll, *J. Bacteriol.* **73**, 63 (1957).
[12] P. Mitchell and J. Moyle, *Symposium Soc. Gen. Microbiol.* **6**, 150 (1956).
[13] C. Weibull, *J. Bacteriol.* **66**, 688 (1953).
[14] C. Weibull, *J. Bacteriol.* **66**, 696 (1953).
[15] C. Weibull, *Symposium Soc. Gen. Microbiol.* **6**, 111 (1956).
[16] J. Tomcsik and J. B. Baumann-Grace, *Verhandl. naturforsch. Ges. Basel* **67**, 218 (1956).
[17] K. McQuillen, *Symposium Soc. Gen. Microbiol.* **6**, 127 (1956).
[18] L. Colobert and J. Lenoir, *Ann. inst. Pasteur* **92**, 74 (1957).
[19] J. Tomcsik and S. Guex-Holzer, *Schweiz. Z. allgem. Pathol. u. Bakteriol.* **15**, 517 (1952).
[20] K. McQuillen, *Biochim. et Biophys. Acta* **17**, 382 (1955).
[21] P. Mitchell and J. Moyle, *J. Gen. Microbiol.* **15**, 512 (1956).
[22] A. R. Gilby, "The Physical Chemistry of Bacterial Protoplasts." Ph.D. Dissertation, University of Cambridge, England, 1957.
[23] J. M. Wiame, R. Storck, and E. Vanderwinkel, *Biochim. et Biophys. Acta* **18**, 353 (1955).
[24] M. R. J. Salton, *Bacteriol. Revs.* **21**, 82 (1957).
[25] M. R. J. Salton, *Symposium Soc. Gen. Microbiol.* **6**, 81 (1956).
[26] C. S. Cummins and H. Harris, *J. Gen. Microbiol.* **14**, 583 (1956).
[27] C. S. Cummins, *Intern. Rev. Cytol.* **5**, 25 (1956).
[28] E. Work, *Nature* **179**, 841 (1957).
[29] L. A. Epstein and E. Chain, *Brit. J. Exptl. Pathol.* **21**, 339 (1940).
[30] W. Weidel, and J. Primosigh, *J. Gen. Microbiol.* **18**, 513 (1958).
[31] K. McQuillen, *J. Gen. Microbiol.* **18**, 498 (1958).
[32] K. McQuillen, *Biochim. et Biophys. Acta* **27**, 410 (1958).
[33] H. R. Mahler and D. Fraser, *Biochim. et Biophys. Acta* **22**, 197 (1956).
[34] J. Tomcsik, *Ann. Rev. Microbiol.* **10**, 213 (1956).
[35] P. Mitchell and J. Moyle, *J. Gen. Microbiol.* **16**, 184 (1957).
[36] G. D. Shockman, J. J. Kolb, and G. Toennies, *J. Biol. Chem.* **230**, 961 (1958).
[37] G. Alderton, B. H. Ward, and H. L. Fevold, *J. Biol. Chem.* **157**, 43 (1945).
[38] H. L. Fevold and G. Alderton, in "Biochemical Preparations" (H. E. Carter ed.), Vol. 1, p. 67. Wiley, New York, 1949.
[39] M. R. J. Salton and K. McQuillen, *Biochim. et Biophys. Acta* **17**, 465 (1955).
[40] S. Brenner and G. S. Stent, *Biochim. et Biophys. Acta* **17**, 473 (1955).
[41] K. McQuillen, *Biochim. et Biophys. Acta* **18**, 458 (1955).
[42] S. Brenner, *Biochim. et Biophys. Acta* **18**, 531 (1955).
[43] O. E. Landman and S. Spiegelman, *Proc. Natl. Acad. Sci. U. S.* **41**, 698 (1955).

[44] K. McQuillen and M. R. J. Salton, *Biochim. et Biophys. Acta* **16**, 596 (1955).
[45] A. Fleming, *Proc. Roy. Soc.* **B93**, 306 (1922).
[46] R. L. Lester, *J. Am. Chem. Soc.* **75**, 5448 (1953).
[47] M. Beljanski, *Biochim. et Biophys. Acta* **15**, 425 (1954).
[48] E. A. Grula and S. E. Hartsell, *J. Bacteriol.* **68**, 171 (1954).
[49] M. H. Richmond, *J. Gen. Microbiol.* **16**, iv (1957).
[50] M. H. Richmond, "The Production of a Lysozyme-like Enzyme by a Strain of *Bacillus subtilis*." Ph.D. Dissertation, University of Cambridge, England, 1958.
[51] M. Nomura and J. Hosoda, *J. Bacteriol.* **72**, 573 (1956).
[52] M. Nomura and J. Hosoda, *Nature* **177**, 1037 (1956).
[53] R. E. Strange and F. A. Dark, *J. Gen. Microbiol.* **16**, 236 (1957).
[54] R. E. Strange and F. A. Dark, *J. Gen. Microbiol.* **17**, 525 (1957).
[55] F. A. Dark and R. E. Strange, *Nature* **180**, 759 (1957).
[56] P. Mitchell and J. Moyle, *Nature* **178**, 993 (1956).
[57] O. Nakamura, *Z. Immunitätsforsch.* **38**, 425 (1923).
[58] N. D. Zinder and W. F. Arndt, *Proc. Natl. Acad. Sci. U. S.* **42**, 586 (1956).
[59] J. Lederberg, *Proc. Natl. Acad. Sci. U. S.* **42**, 574 (1956).
[60] R. Repaske, *Biochim. et Biophys. Acta* **22**, 189 (1956).
[60a] R. Repaske, *Biochim. et Biophys. Acta* **30**, 225 (1958).
[61] D. Fraser and H. R. Mahler, *Arch. Biochem. Biophys.* **69**, 166 (1957).
[62] D. Fraser and E. A. Jerrel, *J. Biol. Chem.* **205**, 291 (1953).
[63] T. Amano, Y. Seki, K. Fujikawa, S. Kashiba, T. Morioka, and S. Ichikawa, *Med. J. Osaka Univ.* **7**, 245 (1956).
[64] J. G. Hirsch, *J. Exptl. Med.* **103**, 598 (1956).
[65] W. Weidel and J. Primosigh, *Z. Naturforsch.* **12b**, 421 (1957).
[66] M. G. Harrington, "The Action of Antibiotics on *Bacterium coli*." Ph.D. Dissertation, University College, Cork, National University of Ireland, Eire, 1954.
[67] L. S. Prestidge and A. B. Pardee, *J. Bacteriol.* **74**, 48 (1957).
[68] A. D. Gardner, *Nature* **146**, 837 (1940).
[69] J. P. Duguid, *Edinburgh Med. J.* **53**, 401 (1946).
[70] E. F. Gale and E. S. Taylor, *J. Gen. Microbiol.* **1**, 314 (1947).
[71] K. McQuillen, *Biochim. et Biophys. Acta* **6**, 534 (1951).
[72] P. D. Cooper, *Bacteriol. Revs.* **20**, 28 (1956).
[73] J. T. Park, *J. Biol. Chem.* **194**, 877, 885, and 987 (1952).
[74] J. T. Park and J. L. Strominger, *Science* **125**, 99 (1957).
[75] K. Liebermeister and E. Kellenberger, *Z. Naturforsch.* **11b**, 200 (1956).
[76] F. E. Hahn and J. Ciak, *Science* **125**, 119 (1957).
[77] E. Kellenberger, K. Liebermeister, and V. Bonifas, *Z. Naturforsch.* **11b**, 206 (1956).
[78] O. Kandler and C. Zehender, *Z. Naturforsch.* **12b**, 725 (1957).
[79] O. Kandler and G. Kandler, *Z. Naturforsch.* **11b**, 252 (1956).
[80] C. Weibull, *Acta Pathol. Microbiol. Scand.* **42**, 324 (1958).
[81] E. Work, *Biochem. J.* **49**, 17 (1951).
[82] B. D. Davis, *Nature* **169**, 534 (1952).
[83] P. Meadow and E. Work, *Biochem. J.* **64**, 11P (1956).
[84] P. Meadow, D. S. Hoare, and E. Work, *Biochem. J.* **66**, 270 (1957).
[85] L. E. Rhuland, *J. Bacteriol.* **73**, 778 (1957).
[86] N. Bauman and B. D. Davis, *Science* **126**, 170 (1957).
[87] B. D. Davis, *Proc. Natl. Acad. Sci. U. S.* **35**, 1 (1949).
[88] B. D. Davis and E. S. Mingioli, *J. Bacteriol.* **60**, 17 (1950).
[89] G. Toennies and D. L. Gallant, *J. Biol. Chem.* **177**, 831 (1949).

[90] G. Toennies and D. L. Gallant, *Growth* **13**, 21 (1949).
[91] G. Toennies and G. D. Shockman, *Arch. Biochem. Biophys.* **45**, 447 (1953).
[92] G. D. Shockman, J. J. Kolb, and G. Toennies, *Federation Proc.* **16**, 247 (1957).
[93] G. D. Shockman, J. J. Kolb, and G. Toennies, *Bacteriol. Proc. (Soc. Am. Bacteriologists)* **73**, P72, p.131 (1957).
[94] J. T. Park, *Symposium Soc. Gen. Microbiol.* **8**, 49 (1958).
[95] E. S. Maculla and P. B. Cowles, *Science* **107**, 376 (1948).
[96] M. Rubio-Huertos and P. R. Desjardins, *Microbiol. Españ.* **9**, 375 (1956).
[97] P. Pease, *J. Gen. Microbiol.* **14**, 672 (1956).
[98] M. H. Jeynes, *Nature* **180**, 867 (1957).
[99] O. Nečas, *Folia Biol. (Prague)* **1**, 19 (1955).
[100] O. Nečas, *Folia Biol. (Prague)* **1**, 104 (1955).
[101] O. Nečas, *Folia Biol. (Prague)* **2**, 29 (1956).
[102] O. Nečas, *Nature* **177**, 898 (1956).
[103] J. Giaja, *Compt. rend. soc. biol.* **86**, 708 (1922).
[104] A. A. Eddy and D. H. Williamson, *Nature* **179**, 1252 (1957).
[105] C. F. Robinow, *Symposium Soc. Gen. Microbiol.* **6**, 181 (1956).
[106] M. Lemoigne, B. Delaporte, and M. Croson, *Ann. inst. Pasteur* **70**, 224 (1944).
[107] D. Fraser, H. R. Mahler, A. L. Shug, and C. A. Thomas, Jr., *Proc. Natl. Acad. Sci. U.S.* **43**, 939 (1957).
[108] B. A. Newton, *J. Gen. Microbiol.* **12**, 226 (1955).
[109] P. Gerhardt, J. W. Vennes, and E. M. Britt, *J. Bacteriol.* **72**, 721 (1956).
[110] T. Amano, K. Kato, K. Okada, Y. Tamatani, and Y. Higashi, *Med. J. Osaka Univ.* **7**, 217 (1956).
[111] C. Weibull, *Exptl. Cell Research* **9**, 139 (1955).
[112] C. Weibull and K. G. Thorsson, in "Electron Microscopy: Proceedings of the Stockholm Conference, September, 1956" (F. Sjöstrand and J. Rhodin, eds.), p. 266. Academic Press, New York, 1957.
[113] G. Kellenberger and E. Kellenberger, *Schweiz. Z. allgem. Pathol. u. Bakteriol.* **15**, 225 (1952).
[114] E. Kellenberger, *Z. wiss. Mikroskop.* **60**, 408 (1952).
[115] C. C. Brinton, A. Buzzell, and M. A. Lauffer, *Biochim. et Biophys. Acta* **15**, 533 (1954).
[116] J. P. Duguid, I. W. Smith, G. Dempster, and P. N. Edmunds, *J. Pathol. Bacteriol.* **70**, 335 (1955).
[117] J. P. Duguid and R. R. Gillies, *J. Gen. Microbiol.* **16**, vi (1956).
[118] C. Weibull, *Exptl. Cell Research* **10**, 214 (1956).
[119] T. Amano, K. Kato, and R. Shimizu, *Med. J. Osaka Univ.* **3**, 292 (1952).
[120] C. Weibull, *Exptl. Cell Research* **9**, 294 (1955).
[121] P. Mitchell and J. Moyle, *Discussions Faraday Soc.* **21**, 258 (1956).
[122] J. Tomcsik, *Proc. Soc. Exptl Biol. Med.* **89**, 459 (1955).
[123] A. R. Gilby and A. V. Few, *Nature* **179**, 422 (1957).
[124] B. A. Newton, *Bacteriol. Revs.* **20**, 14 (1956).
[125] A. V. Few and J. H. Schulman, *J. Gen. Microbiol.* **9**, 454 (1953).
[126] E. Kodicek, in "Biochemical Problems of Lipids" (G. Popják and E. Le Breton, eds.), p. 401. Butterworths, London, 1956.
[127] H. A. Abramson, L. S. Moyer, and M. H. Gorin, "Electrophoresis of Proteins and the Chemistry of Cell Surfaces." Reinhold, New York, 1942.
[128] K. McQuillen, *Biochim. et Biophys. Acta* **5**, 463 (1950).
[129] K. McQuillen, *Biochim. et Biophys. Acta* **6**, 66 (1950).
[130] K. McQuillen, *Biochim. et Biophys. Acta* **7**, 54 (1951).

[131] A. M. James and P. J. Barry, *Biochim. et Biophys. Acta* **15,** 186 (1954).
[132] J. H. B. Lowick and A. M. James, *Biochim. et Biophys. Acta* **17,** 424 (1955).
[133] C. Weibull, *Acta Chem. Scand.* **11,** 881 (1957).
[134] J. W. Vennes and P. Gerhardt, *Science* **124,** 535 (1956).
[135] R. Storck and J. T. Wachsman, *Biochem. J.* **66,** 19P (1957).
[136] R. Storck and J. T. Wachsman, *J. Bacteriol.* **73,** 784 (1957).
[137] A. R. Gilby, A. V. Few, and K. McQuillen, *Biochim. et Biophys. Acta* **29,** 21 (1958).
[138] P. Mitchell and J. Moyle, *J. Gen. Microbiol.* **5,** 981 (1951).
[139] K. L. Burdon, *J. Bacteriol.* **52,** 665 (1946).
[140] J. R. Hawthorne, *Biochim. et Biophys. Acta* **6,** 94 (1950).
[141] D. Keilin, *Nature* **133,** 290 (1934).
[142] S. Spiegelman, A. I. Aronson, and P. C. Fitz-James, *J. Bacteriol.* **75,** 102 (1958).
[143] J. Tomcsik and S. Guex-Holzer, *Experientia* **10,** 484 (1954).
[144] J. B. Baumann-Grace and J. Tomcsik, *J. Gen. Microbiol.* **17,** 227 (1957).
[145] M. Bridoux and J. Hanotier, *Biochim. et Biophys. Acta* **22,** 103 (1956).
[146] S. Spiegelman, *in* "Enzymes: Units of Biological Structure and Function" (O. H. Gaebler, ed.), p. 67. Academic Press, New York, 1956.
[147] S. Spiegelman, *in* "Chemical Basis of Heredity" (W. D. McElroy and B. Glass, eds.), p. 232. Johns Hopkins Press, Baltimore, Maryland, 1957.
[148] J. R. Norris, *J. Gen. Microbiol.* **16,** 1 (1957).
[149] E. F. Gale and J. P. Folkes, *Biochem. J.* **55,** 521 (1953).
[150] K. McQuillen and R. B. Roberts, *J. Biol. Chem.* **207,** 81 (1954).
[151] K. McQuillen, *J. Gen. Microbiol.* **13,** iv (1955).
[152] A. R. Crathorn and G. D. Hunter, *Biochem. J.* **68,** 4P (1958).
[153] G. D. Hunter, A. R. Crathorn, and J. A. V. Butler, *Nature* **180,** 383 (1957).
[154] G. N. Cohen and J. Monod, *Bacteriol. Revs.* **21,** 169 (1957).
[155] P. Mitchell, *Nature* **180,** 134 (1957).
[156] H. V. Rickenberg, *Biochim. et Biophys. Acta* **25,** 206 (1957).
[157] H. V. Rickenberg, G. N. Cohen, G. Buttin, and J. Monod, *Ann. inst. Pasteur* **91,** 829 (1956).
[158] A. G. Jose and R. M. Pengra, *Bacteriol. Proc. (Soc. Am. Bacteriologists)* **73;** P76, p. 132 (1957).
[159] W. Mutsaars, *Ann. inst. Pasteur* **89,** 166 (1955).
[160] A. Pirie, *Brit. J. Exptl. Pathol.* **21,** 125 (1940).
[161] A. Lwoff, *Bacteriol. Revs.* **17,** 269 (1953).
[162] J. Spizizen, *Proc. Natl. Acad. Sci. U. S.* **43,** 694 (1957).
[163] R. Thomas, *Biochem. J.* **66,** 38P (1957).
[164] A. Gierer and G. Schramm, *Nature* **177,** 702 (1957).
[165] V. L. Mayers and J. Spizizen, *J. Biol. Chem.* **210,** 877 (1954).
[166] D. D. Brown and L. M. Kozloff, *J. Biol. Chem.* **225,** 1 (1957).
[167] E. Chargaff, H. M. Schulman, and H. S. Shapiro, *Nature* **180,** 851 (1957).
[168] W. A. Hardwick and J. W. Foster, *J. Gen. Physiol.* **35,** 907 (1952).
[169] M. R. J. Salton, *J. Gen. Microbiol.* **13,** iv (1955).
[169a] B. A. D. Stocker, *Symposium Soc. Gen. Microbiol.* **6,** 19 (1956).
[170] D. H. Northcote and R. W. Horne, *Biochem. J.* **51,** 232 (1952).
[171] R. E. Strange and J. F. Powell, *Biochem. J.* **58,** 80 (1954).
[172] R. E. Strange, *Biochem. J.* **64,** 23P (1956).
[173] L. H. Kent, *Biochem. J.* **67,** 5P (1957).
[174] G. T. Cori and C. F. Cori, *J. Biol. Chem.* **131,** 397 (1939).
[175] S. Mii and S. Ochoa, *Biochim. et Biophys. Acta* **26,** 445 (1957).

[176] M. F. Singer, L. A. Heppel, and R. J. Hilmoe, *Biochim. et Biophys. Acta* **26**, 447 (1957).
[177] J. H. B. Lowick and A. M. James, *Biochem. J.* **65**, 431 (1957).
[178] J. Lederberg, *J. Bacteriol.* **71**, 497 (1956).
[179] T. F. Anderson, E. L. Wollman, and F. Jacob, *Ann. inst. Pasteur* **93**, 450 (1957).
[180] J. Sinkovics, *Acta Microbiol. Acad. Sci. Hung.* **4**, 59 (1957).
[181] E. Kellenberger and A. Ryter, *J. Biophys. Biochem. Cytol.* **4**, 323 (1958).
[182] A. E. Vatter and R. S. Wolfe, *J. Bacteriol.* **75**, 480 (1958).
[183] Y. Tokuyasu and E. Yamada, *J. Biophys. Biochem. Cytol.* **5**, 123 (1959).
[184] C. M. F. Hale, *Exptl. Cell Research* **12**, 657 (1957).
[185] J. Baddiley, J. G. Buchanan, and B. Carss, *Biochim. et Biophys. Acta* **27**, 220 (1958).
[186] J. J. Armstrong, J. Baddiley, J. G. Buchanan, and B. Carss, *Nature* **181**, 1692 (1958).
[187] J. J. Armstrong, J. Baddiley, J. G. Buchanan, B. Carss, and G. R. Greenberg, *J. Chem. Soc.* p. 4344 (1958).
[188] A. Abrams, *J. Biol. Chem.* **230**, 949 (1958).
[189] W. Brumfitt, A. C. Wardlaw, and J. T. Park, *Nature* **181**, 1783 (1958).
[190] P. J. O'Brien and F. Zilliken, *Biochim. et Biophys. Acta* **31**, 543 (1959).
[191] C. Weibull, *Ann. Rev. Microbiol.* **12**, 1 (1958).
[192] J. Lederberg and J. St. Clair, *J. Bacteriol.* **75**, 143 (1958).
[193] S. Madoff and L. Dienes, *J. Bacteriol.* **76**, 245 (1958).
[194] S. Brenner, F. A. Dark, P. Gerhardt, M. H. Jeynes, O. Kandler, E. Kellenberger, E. Klieneberger-Nobel, K. McQuillen, M. Rubio-Huertos, M. R. J. Salton, R. E. Strange, J. Tomcsik, and C. Weibull, *Nature* **181**, 1713 (1958).
[195] K. McQuillen, in *Colloquia of 4th Internat. Cong. Biochem.* (Vienna, 1958).
[196] M. R. J. Salton, *J. Gen. Microbiol.* **18**, 481 (1958).
[197] L. Colobert, *Compt. rend. soc. biol.* **151**, 114 (1957).
[198] L. Colobert, *Bull. Soc. Chim. Biol.* **39**, Suppl. 1, 135 (1957).
[199] L. Colobert, *Ann. inst. Pasteur* **95**, 156 (1958).
[200] L. Colobert, *Compt. rend. soc. biol.* **115**, 1904 (1957).
[201] M. C. Karunairatnam, J. Spizizen, and H. Gest, *Biochim. et Biophys. Acta* **29**, 649 (1958).
[202] A. Sohler, A. H. Romano, and W. J. Nickerson, *J. Bacteriol.* **75**, 283 (1958).
[203] A. Abrams, *J. Biol. Chem.* **234**, 383 (1959).
[204] H. D. Slade and W. C. Slamp, *Bacteriol. Proc.* (Soc. Am. Bacteriologists) **75**, P38, p. 108 (1958).
[205] H. Gooder and W. R. Maxted, *Nature* **182**, 808 (1958).
[206] K. Horikoshi and K. Sakaguchi, *J. Gen. Appl. Microbiol.* **4**, 1 (1958).
[207] K. Horikoshi and S. Iida, *Nature* **181**, 917 (1958).
[208] K. Horikoshi and S. Iida, *Nature* **183**, 186 (1959).
[209] S. Emerson and M. R. Emerson, *Proc. Natl. Acad. Sci. U.S.* **44**, 668 (1958).
[210] J. S. Murphy, *Virology* **4**, 563 (1957).
[211] J. Panijel and J. Huppert, *Compt. rend.* **245**, 240 (1957).
[212] J. Panijel, *Compt. rend.* **246**, 1776 (1958).
[213] W. F. Carey, W. Spilman, and L. S. Baron, *J. Bacteriol.* **74**, 543 (1957).
[214] A. Bolle and E. Kellenberger, *Schweiz. Z. allgem. Pathol. u. Bakteriol.* **21**, 714 (1958).
[215] G. Koch and W. J. Dreyer, *Virology* **6**, 291 (1958).
[216] F. Jacob and C. R. Fuerst, *J. Gen. Microbiol.* **18**, 518 (1958).
[217] J. M. Gebicki and A. M. James, *Nature* **182**, 725 (1958).

[218] G. Falcone and F. Graziosi, *Giornale di Microbiologia* **3**, 269 (1957).
[219] J. T. Park, *Biochem. J.* **70**, 2P (1958).
[220] R. E. Trucco and A. B. Pardee, *J. Biol. Chem.* **230**, 435 (1958).
[221] O. E. Landman, R. A. Altenbern, and H. S. Ginoza, *J. Bacteriol.* **75**, 567 (1958).
[222] K. G. Lark, *Canad. J. Microbiol.* **4**, 165 (1958).
[223] K. G. Lark, *Canad. J. Microbiol.* **4**, 179 (1958).
[224] K. Crawford, cited in "Biochemistry of some Peptide and Steroid Antibiotics" by E. P. Abraham. John Wiley and Sons, Inc., New York, 1957.
[225] J. Ciak and F. E. Hahn, *Bacteriol. Proc.* (Soc. Am. Bacteriologists) **75**, P40, p. 108 (1958).
[226] F. E. Hahn, unpublished observation (1958).
[227] J. T. Holden and J. Holman, *J. Bacteriol.* **73**, 592 (1957).
[228] C. Lark and K. G. Lark, *Bacteriol. Proc.* (Soc. Am. Bacteriologists) **75**, P39, p. 108 (1958).
[229] M. Welsch, *Schweiz. Z. allgem. Pathol. u. Bakteriol.* **21**, 741 (1958).
[230] P. Gerhardt, unpublished observation.
[231] P. Mandel, M. Sensenbrenner, and G. Vincendon, *Nature* **182**, 674 (1958).
[232] I. Takahashi and N. E. Gibbons, *Canad. J. Microbiol.* **5**, 25 (1959).
[233] P. C. Fitz-James, *J. Bacteriol.* **75**, 369 (1958).
[234] K. G. Thorsson and C. Weibull, *J. Ultrastructure* **1**, 412 (1958).
[235] M. R. J. Salton and F. Shafa, *Nature* **181**, 1321 (1958).
[236] E. M. Britt and P. Gerhardt, *J. Bacteriol.* **76**, 281 (1958).
[237] E. M. Britt and P. Gerhardt, *J. Bacteriol.* **76**, 288 (1958).
[238] W. R. Sistrom, *Biochim. et Biophys. Acta* **29**, 579 (1958).
[239] A. R. Gilby and A. V. Few, *2nd Internat. Cong. Surface Activity.* "Solid-Liquid Interface and Cell-Water Interface" p. 262. Butterworth's Scientific Publications, London, (1958).
[240] O. Kandler, A. Hund, and C. Zehender, *Nature* **181**, 572 (1958).
[241] C. Weibull and L. Bergström, *Biochim. et Biophys. Acta* **30**, 340 (1958).
[242] A. R. Gilby and A. V. Few, *Nature* **182**, 55 (1958).
[243] O. Nečas, *Exptl. Cell Research* **14**, 216 (1958).
[243a] R. B. Roberts, K. McQuillen, and I. Z. Roberts, *Ann. Rev. Microbiol.* **13**, in press.
[244] J. A. V. Butler, A. R. Crathorn, and G. D. Hunter, *Biochem. J.* **69**, 544 (1958).
[245] A. I. Oparin, N. S. Gelman, and G. A. Deborin, *Arch. Biochem. Biophys.* **69**, 582 (1957).
[246] M. Nomura, J. Hosoda, and S. Nishimura, *J. Biochem. (Tokyo)* **44**, 863 (1957).
[247] M. Nomura, J. Hosoda, and S. Nishimura, *Biochim. et Biophys. Acta* **29**, 161 (1958).
[248] N. Otsuji and Y. Takagi, *Proc. 7th Symposium on Nucleic Acids*, p. 6 (1958).
[249] M. Beljanski and S. Ochoa, *Proc. Natl. Acad. Sci. U.S.* **44**, 494 (1958).
[250] S. Spiegelman, in "Recent Progress in Microbiology," Symposia held at VII International Congress of Microbiology. Almqvist and Wiksell, Stockholm, 1959.
[251] P. Rogers, *Federation Proc.* **17**, 298 (1958).
[252] C. Hurwitz, J. M. Reiner, and J. V. Landau, *J. Bacteriol.* **76**, 612 (1958).
[253] R. Sheinin and K. McQuillen, *Biochim. et Biophys. Acta* **31**, 72 (1959).
[254] M. R. J. Salton, in "Ciba Foundation Symposium on the Nature of Viruses," p. 263. J. & A. Churchill, London, 1956.
[255] H. Böhme and U. Taubeneck, *Naturwissenschaften* **12**, 296 (1958).

[256] U. Taubeneck and H. Böhme, Z. Naturforsch. **13b**, 471 (1958).
[257] H. Böhme and U. Taubeneck, Proc. Intern. Cong. Genetics, Montreal, 1958.
[258] M. Sekiguchi, Virology **6**, 777 (1958).
[259] P. C. Fitz-James, J. Biophys. Biochem. Cytol. **4**, 257 (1958).
[260] C. Lark and K. G. Lark, J. Bacteriol. **76**, 666 (1958).
[261] J. Sinkovics, Nature **181**, 566 (1958).
[262] A. Hagiwara, Nature **182**, 456 (1958).

Acknowledgment

The author is grateful to many people for information in advance of publication and for permission to quote unpublished results.

He also wishes to thank Mr. R. W. Horne of the Cavendish Laboratory, University of Cambridge, who made the electron micrographs.

The following diagrams and photographs have appeared in publications of the Society for General Microbiology and the copyright owners are thanked for permission to reproduce them: Figures 16, 18, 21, and 22 from McQuillen[17] and Figure 25, e and f, from Newton.[108]

The following photographs have appeared in Biochimica et Biophysica Acta and the copyright owners are thanked for permission to reproduce them: Figures 23 a, b, and c; 24 a, b, and d; 25 b from McQuillen[41] and Figures 26 c and 27 from McQuillen.[32]

Chapter 7

L-Forms of Bacteria

E. KLIENEBERGER-NOBEL

I. Introduction.. 361
II. The Discovery of the L-Form... 362
III. Definition of L-Form.. 363
IV. Appearance of Growth on Solid and in Liquid Media................ 363
V. Production of L-Form... 365
VI. Microscopic Demonstration of L-Form............................... 367
VII. Morphology of L-Form.. 368
    A. Cultures in the Stage of Transformation........................... 368
    B. Development and Morphology of L-Form......................... 370
VIII. Properties of L-Form... 371
    A. Metabolism and Chemical Constitution........................... 371
    B. Serology of L-Form... 372
    C. Pathogenicity of L-Form... 373
IX. The Similarities of L-Forms and Pleuropneumonia-like Organisms........ 374
    A. Properties of PPLO.. 374
    B. Similarities between L-Forms and PPLO......................... 375
    C. Differences between L-Forms and PPLO......................... 376
X. Electron Microscopic Demonstration of L-Forms of Bacteria and of PPLO. 377
XI. L-Forms and Protoplasts... 381
XII. Summary and Conclusions... 382
    References... 383

I. Introduction

Bacteria usually appear in the rigid form of a bacillus, a coccus, or a vibrio; but ever since high power microscopic lenses have been focused on them bacteriologists have observed that they also produce aberrant forms, which vary from dwarf cells to giant bodies and may be quite irregular in their outline. The aberrant forms occur in cultures freshly isolated from pathological material, and may also be found in old laboratory cultures. They are rare in frequently transplanted strains. In the early literature they were often described as degenerate and designated "involution forms." Some workers were stimulated by the fantastic appearance of the aberrant or pleomorphic forms and conceived theories that attributed special significance to them, as reviewed by Klieneberger.[1] For example, Kuhn[2] propagated the idea that all bacterial cultures were contaminated with or lived in symbiosis with a parasite of protozoan nature. He designed a special fixing and staining technique, later modified and widely used by Klieneberger.[3] He stained the nuclear structures in the aberrant forms at a time when the Feulgen reaction had just been invented and bacteria were supposed not to contain nuclear bodies.

Almquist,[4] Enderlein,[5] Mellon,[6] Hadley,[7] and others believed that bacteria went through life cycles and produced minute reproductive "gonidia" and large "zygotes" formed by fusion or copulation. Löhnis,[8] who collected and published numerous photographs of pleomorphic bacteria, tried to show that they can exist in a protoplasmic state in which they coalesce and form a "symplasma." These theories were probably inspired by the expectation that, like higher organisms, bacteria possessed some means of regenerating their genetic material. Of course, recombination, induced mutation, and transduction, discovered in recent years, were not then thought of.

Many workers found that conditions of culture were often responsible for the production of the pleomorphic forms; for instance, high concentrations of common salt,[9,10] lithium chloride,[7] and low temperature[11] induced bacteria to produce aberrant forms which were not destined to die but grew and multiplied when suitable conditions were provided.

II. The Discovery of the L-Form

In 1935 Klieneberger[12] reported that she had successfully separated the aberrant, soft, protoplasmic forms from the bacilli in cultures of *Streptobacillus moniliformis*. She succeeded in keeping this protoplasmic growth in pure culture and designated it "L-form" in contrast to the bacillary form (L stands for Lister Institute).

Streptobacillus moniliformis is the organism causing rat bite fever in man.[13] It is carried in the nasopharynx of tame and wild rats[14] and it may occur in their lungs when they are suffering from pulmonary disease. This organism shares with a few other bacteria, for example, *Fusiformis necrophorus* (syn. *Bacteroides*) and a small Gram-negative Listeria-like bacterium causing disease in South African rodents, the ability to produce L-form colonies spontaneously. When these organisms were sown on suitable nutrient agar two types of colonies appeared, one consisting of protoplasmic elements only, and the other of a mixture of protoplasmic and bacterial cells. It was found that the colonies consisting of protoplasmic bodies had an appearance which distinguished them from the parent strain colonies. Such colonies sometimes bred true from first isolation, but at other times a series of consecutive isolations was required to produce a stable strain. A strain which was stable on solid medium was not necessarily stable in liquid medium and then further passages on the solid medium were required to produce an L culture which no longer reverted to bacillary phase.

The author possesses L-form cultures which have been maintained for ten to twenty years and have been propagated in liquid, on semisolid and on solid media without changing their colony type or morphology.

When the author discovered the L-form she was struck by the similarity of its colonies and of its morphology to that of the organisms of the so-called pleuropneumonia group,[15] and indeed thought that a pleuropneumonia-like organism (PPLO) existed in close symbiosis with the bacilli of *S. moniliformis*. However, other workers, foremost among them Dienes[16-18] showed that the soft bodies are derived from the bacilli; it is generally admitted today that the small and large soft elements found in bacterial cultures are derived from the bacterium and may transform into L-form.

With the discovery of penicillin, interest in the L-form was greatly aroused because Dienes and his collaborators showed that under its influence many bacteria produce soft bodies from which L cultures can be obtained. Penicillin-induced L cultures have been reported from strains of *Salmonella*, *Shigella*, *Flavobacterium*, *Haemophilus*, *Bacillus coli*, *B. proteus*, and even from Gram-positive spore-forming bacteria. Recently, Dienes and Sharp[19, 20] have added a streptococcus to the list of L-form-producing bacteria.

III. Definition of L-Form

Numerous articles on L-forms published up to 1955 have been cited in a number of reviews.[21-25] In this literature the name "L-form or L-phase" is often used in rather a loose way and any aberrant form observed in bacterial cultures is described as "an L-form" or an "L-element." The designation was first used to describe a type of growth, not single elements. Confusion could be avoided if the name L-form were only applied to the stable growth which consists of soft protoplasmic elements without defined morphology, which no longer possesses rigid bacterial forms nor reverts to them, which can be propagated indefinitely, and which has a characteristic colony on solid medium independent of the bacterial species from which the growth was derived. The term is used in this sense in this article; it was agreed to by the members of the "Réunion sur les Formes L et sur les Formes évolutives des bactéries" ("L-forms" of bacteria[26]).

Probably a biochemical distinction exists between the L-form and the unstable "transition" form which is its predecessor; recent studies by Weibull[27] point in this direction.

IV. Appearance of Growth on Solid and in Liquid Media

When L-form strains are sown on a suitable solid medium they form round colonies, characterized by a dark center and a lighter peripheral zone that has a conspicuous lacelike pattern. When they grow densely the individual centers can always be distinguished, but the peripheral parts may join together and the whole growth then resembles a number of eggs which have been fried in one pan. The lacelike pattern of the peripheral growth is caused by excreted metabolic products condensed in "oily drop-

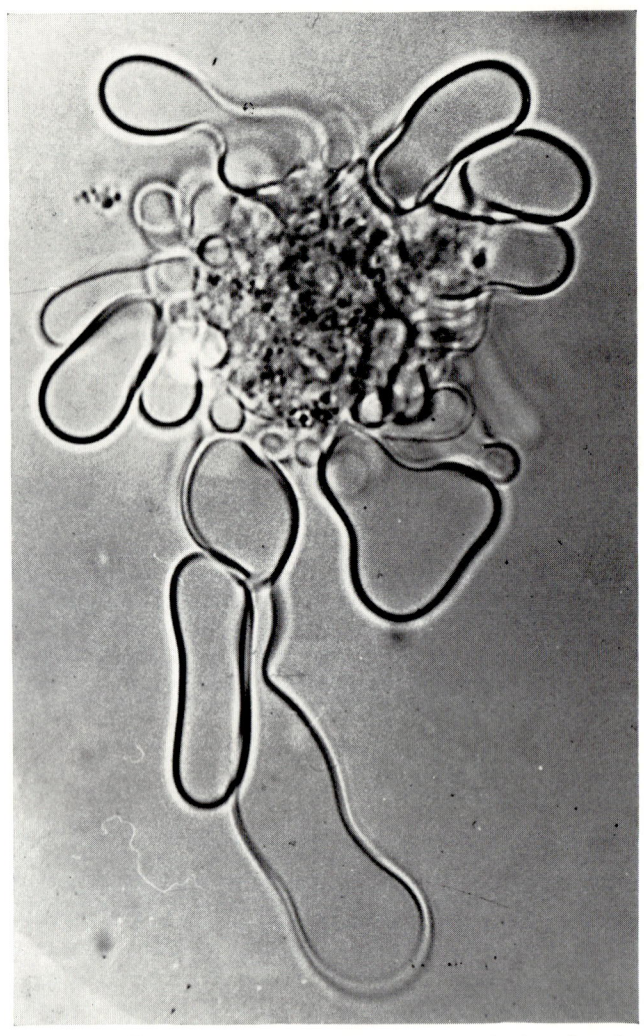

Fig. 1. L-form of *Streptobacillus moniliformis*, grown in hanging drop culture, alive. Note cholesterol droplets. (Magnification: × 1800)

lets" (myelin structures). In the L-form of *S. moniliformis* these droplets are shown to contain cholesterol (Fig. 1[28]). The dense central part of the L-colony is the result of growth into the depth of the medium. It is surprising that L colonies derived from a *Streptobacillus moniliformis*, a *Fusobacterium necrophorus*, a *Salmonella* sp., a *Proteus vulgaris*, or a *Streptococcus* are morphologically indistinguishable from each other; biochemical and serological studies may, however, distinguish them even if their parentage is not known.

In liquid media L-form growth is usually granular in appearance and may adhere to the test tube walls or sink to the bottom; individual granules vary, from small aggregates which are just visible, to clumps the size of a pin's head. The *Proteus vulgaris* L-form is exceptional in forming a pellicle on the surface of liquid medium.[29] All L-strains grow very well in semisolid medium and both facultative and obligate anaerobic L cultures can be maintained to advantage in Brewer's medium enriched with serum.

V. Production of L-Form

When the author first isolated L-form she incubated a serum broth culture of *S. moniliformis* for a number of days and inoculated serum agar plates with a drop from this broth on 5 successive days. After 1 day's incubation of the broth only a few L-type colonies appeared, but after 2, 3, 4, and 5 days of incubation an increasing number of L-type colonies and a decreasing number of bacillary colonies grew. Thus, aging had favored the development of L-like colonies in *S. moniliformis*. By a process of repeatedly cutting out these L-type colonies and transferring them to new plates L-form cultures were obtained. Subsequently it was found that incubation at room temperature resulted in the formation of a larger number of soft protoplasmic bodies, from which L-form culture could be obtained more easily. Temperatures higher than 37°C. had no stimulating effect.

With the advent of penicillin L-cultures were obtained from many organisms that did not usually develop soft protoplasmic forms.[30] Dienes used this antibiotic freely in order to induce L-form production in a variety of different bacteria, and many other workers followed in his wake. However, there are other substances that induce L-like growth and eventually L-form in various organisms. Thus, antiserum and complement were used by Dienes et al.[31] to produce L-form from *Salmonella typhi*. Poetschke[32] used normal sera to obtain L-like colonies from *Corynebacterium* sp.; bovine sera were most potent and anaerobic conditions favored the transformation. Dienes and Zamecnik[33] studied the effect of amino acids on *Salmonella typhimurium* and *Haemophilus influenzae*; L-form was induced by glycine, methionine, and phenylalanine, but glycine was the most effective of the three. Medill and O'Kane[34] described chemically defined amino acid media which support rapid growth of Proteus bacteria as well as its L-form. They found that L-form grew better on these chemically defined media solidified with agar than on more complex media which they describe as "natural media." Their explanation is that L-form is very sensitive to inhibitors present in the various constituents of bacteriological media, such as yeast extract, hydrolyzed casein, and peptones. The effect of serum or plasma on L-form development is due to a detoxification of the "natural media" rather than to supplying nutrients. This would at the same time explain why such enormous amounts of serum are required to allow L-form to grow on ordinary bacteriological media. Abrams[35] cultivated L-form in a casamino acid penicillin broth without serum. Lorkiewicz[36] grew L-form without additional serum on a nutrient medium containing 0.4 to 1.2 % of charcoal for detoxification. Similarly, Pollock[37] showed that charcoal can replace serum in media supporting the growth of *Haemophilus influenzae.*

Dienes and Sharp[19] found that group A hemolytic streptococci produced L-form in penicillin-containing media when the common salt concentration was higher than 1.5 %. The sodium chloride could be replaced by salts not toxic for the streptococci, such as potassium chloride, disodium hydrogenphosphate, ammonium chloride, calcium chloride, and magnesium chloride. Salts toxic for the streptococci, such as the compounds of lithium, mercury, copper, and cadmium, induced swellings of the bacteria but not the production of the L-form. A high salt concentration ("4 times physiological") was also used by Grasset and Bonifas[38] for the induction of globular forms in *Klebsiella pneumoniae.* High salt concentration activates transformation toward L-form but is not generally effective in producing the stabilized form.

Lavillaureix[39] used mineral salts in a penicillin-containing medium in attempts to obtain L-forms; De Gregorio and Terranova[40] used both a complex and a chemically defined medium for production of L-form and obtained the best results with the complex medium. They covered their inoculum with a layer of agar medium and when transformation had occurred continued this "sandwich growth method." The usefulness of this method was fully confirmed by the author. Nelles[41] sums up a number of the factors which lead to the establishment of stable L-forms, and the present author has also proved their importance:

L-forms are more easily obtained from freshly isolated strains than from old laboratory cultures. The bacteria should be in the logarithmic phase but must not yet have reached the culmination point; for *Proteus* and *Salmonella* spp. a broth culture is best after 3 hours' incubation at 37°C.

Certain amino acids, salts, and antibiotics, such as penicillin and related substances, serve as inducers. The best concentrations of penicillin for *Proteus* and *Salmonella* strains are 100 to 150 microns per milliliter. Higher concentrations speed up the transformation of the bacteria but soon lead to lysis of the swollen forms. Low concentrations allow the globular bodies to revert to bacilli.

A soft agar medium is favorable. The addition of peptone to the medium is recommended, but the various brands differ in their effect. Nelles considers the addition of serum (20 %) in particular, horse serum, necessary. It is certainly a useful addition. Anaerobic conditions are preferable to aerobic ones. With these methods Nelles obtained a stable L-culture from a *Salmonella breslau* in 8 passages on penicillin-containing medium, from a *Proteus vulgaris* in five, from a *Salmonenella typhi* in four, and from a *Salmonella paratyphi B* in two passages.

To obtain growth in liquid media Nelles cut out a piece of agar on which the L-culture had grown well, transferred it to a broth tube and thoroughly squashed it with a glass rod; the growth developed as a granular sediment.

Thus a number of factors may initiate the transformation of bacteria into L-form, for example, the presence in the media of salts, amino acids, sera, antibiotics, etc. The suitable concentrations have to be found by trial and a procedure suitable for every organism cannot be given. In most cases, but not invariably, frequent transfers of the transformed growth on the initiating medium are required for the establishment of L-form.

VI. Microscopic Demonstration of L-Form

L-form growth consists of soft protoplasmic forms and therefore cannot be demonstrated in the usual bacteriological smear preparation. The elements are broken up by drying and fixing, and only a cloudy material is seen, containing no formed elements. Such a preparation is useful only for the demonstration of the purity of the culture. All authors who have demonstrated L-form microscopically have preserved the original structure of the growth by making preparations of colonies *in toto*.[21-23]

For the demonstration of unfixed colonies, growth on agar is cut out and placed on a coverslip with the colonies toward the coverslip. A slide is used as carrier and the sides of the agar piece are sealed to the coverslip and slide with paraffin wax. This simple preparation can be studied microscopically. It shows the honeycomb structure of the colony and the sizes and shapes of the elements. If, as recommended by Dienes, a drop of stain (methylene blue or azure in alcohol) has been dried on the coverslip the colonies take up the stain quickly and appear in greater contrast on the unstained agar background.

The agar fixation technique[15] is still very useful for the demonstration of L colonies and of the bacterial growth in transformation. Growth on agar is placed on a coverslip, or better still a piece of agar carrying a suitably diluted inoculum is placed on a sterile coverslip and incubated in a petri dish fitted with sterile filter paper. For fixation the coverslip with adhering agar piece is placed in a block dish or watch glass and covered with Bouin's fixative; the fixation is slow as the fixative has to diffuse through the agar; the process should be allowed to continue for 8 to 18 hours. Then the agar piece is peeled off with the tip of a scalpel. The coverslip with adhering growth is thoroughly washed and stained in dilute Giemsa solution (1:50) for at least 5 hours. The stained coverslip is mounted either in water, or in Canada balsam after dehydration in acetone-xylol mixtures.[42]

A stained precipitate is formed where the agar is in direct contact with the fixative; in examining the preparation its edges should therefore be avoided but inside this precipitate the colonies and their elements show up brilliantly.

A modification of this method can be applied to the electron microscope study of the L-form.[43, 44] A piece of nutrient agar carried on a slide is covered with a Formvar solution (0.3% in chloroform); when the slide is tilted the solution flows off quickly, leaving a thin Formvar film. A diluted suspension of the organism is then prepared in a buffer or Tyrode-Ringer solution and small droplets are placed with a finely drawn out pipette on the surface of the film. The slides are incubated in a moist chamber for periods desired. The films carrying the growth are then floated on formalin, fixed for 2 to 3 minutes, refloated on distilled water for washing, placed on coverslips, and dried. They are then stained, dehydrated, and mounted as before or mounted on a grid for the electron microscope.

VII. Morphology of L-Form

A. Cultures in the Stage of Transformation

Numerous papers have been published on the appearance of cultures transforming into "large bodies" (Fig. 2). These bodies are formed by the swelling of bacteria at the end or in the middle, by the oozing out of the contents of the bacterium through a hole in the membrane, or by the transformation of the whole bacterium into a big element; they have the important property, which distinguishes them from L-form, that they can revert to the bacterial form. The reversion occurs in a very characteristic way (Fig. 3) by extensions which resemble pseudopodia.[1, 45-49] Recently it

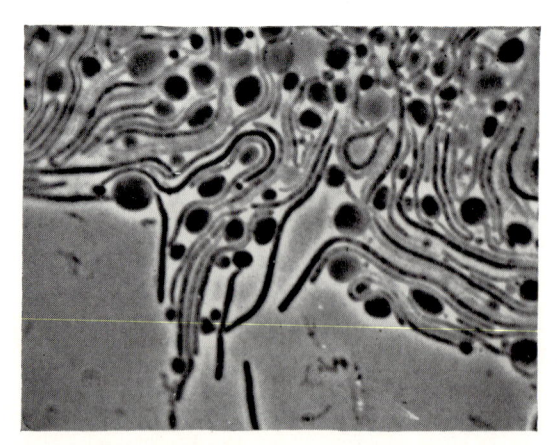

Fig. 2. "Swarming edge" of *Proteus vulgaris* kept in the cold (4° C.) overnight. Many "large bodies" have been produced. Phase contrast. (Magnification: × 1200)

FIG. 3. The "swarming edge" of *Proteus vulgaris* which, as seen in Fig. 2, had produced "large bodies" was further incubated for several hours at 37° C. The reversion into bacillary forms has set in. Agar fixation, stained with Giemsa solution. (Magnification: × 1700)

was shown by Lederberg[50] that penicillin-induced "protoplasts" reproduce bacteria in an identical way. Eventually, the bacteria are set free by division. Thus fantastic configurations are sometimes produced, which have been likened to antlers. The phenomenon has been illustrated by time-lapse photography and by cinematography.[51] As the large bodies reproduce bacilli it is not surprising that Stempen[52] was able to show by staining methods and by phase-contrast microscopy that they can possess cell walls. The development into large bodies (Fig. 2) and their *in vitro* and *in vivo* reversion to bacterial cells (Fig. 3) have been frequently described in recent years.[38, 53-63] Winkler[25] has a good description of the swarming of *Proteus* spp. cells developing aberrant forms under the influence of penicillin: the cells form giant globules at their ends or in the region of their transverse septa. These bodies, which at first have an even density, either lyse completely or become granulated and produce vacuoles; the outlines of the globules disappear slowly until a heap of granules has formed in their place. Winkler draws attention to the numerous nuclear structures often contained in penicillin-induced giant forms, an observation also made by the author.

Opinions are divided about multiplication of the large bodies. Some authors think that they are destined to disintegration and death; others are convinced that they multiply; their actual fate is difficult to observe microscopically. Microscopic and filtration evidence strongly suggest to the author that they reproduce by the formation of small bodies which increase and grow into large forms. Otherwise it would hardly be possible to understand how colony formation is brought about.

B. Development and Morphology of L-Form

The elements of L-form cultures vary in size from miniature cocci to leucocytes; how these elements are related is not yet known. Tulasne is the major exponent of the theory that big forms divide into a large number of small, filterable forms, "les formes naines" and that each dwarf form grows into a big element. In dark-field illumination and sometimes in stained preparations the large forms certainly appear granulated. Yet they never appear to release these granules, nor do we find released masses of larger granules which continue to grow. However, among the large, extremely flat elements grown on the Formvar film lighter and darker spots are seen and it is tempting to interpret the darker parts as centers of new growth which may develop into small coccal forms. Coccal elements indeed occur singly and in small groups (Fig. 4); they are doubtless the small bodies which grow into bigger forms. The observation that very small elements are produced in the L cultures, although not in large numbers, is in excellent agreement with the filtration experiments (see later) and with some electron microscope studies of L-form carried out by the author in collaboration with Cuckow.[43, 44] Further observations show that L-form can multiply by the division of large forms into a number of smaller parts which grow up individually; at the edges of colonies thinly spread forms with a frayed border can often be seen (Fig. 5). The fringes of the border subdivide into separate islets and grow into new protoplasmic bodies.

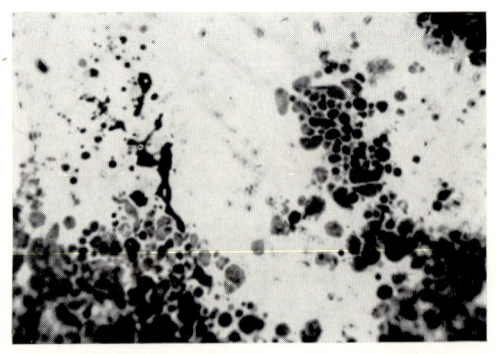

Fig. 4. An L-form of *Proteus vulgaris* grown on the Formvar film as described. Many very small coccal forms can be seen. Formalin fixation, stained with Giemsa solution. (Magnification: × 1700)

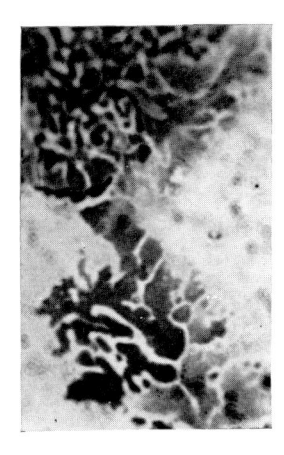

FIG. 5. An L-form of *Proteus vulgaris* grown on the Formvar film as described. Note feathery edge dividing into small forms. Formalin fixation, stained with Giemsa solution. (Magnification: × 1200)

It was mentioned above that little is known about the nuclear structures of L-form, but structures of various shapes (granules, filaments, rings) occur in the transition stage and in the L-form.[64] Their relation to the nuclear structures of bacteria is dubious. The cytoplasm may stain very deeply in some of the younger elements; in others it is very thin and stains faintly. Old elements which disintegrate usually take little stain. We can assume that there is an outer membrane, but it must be very soft as it allows the soft bodies to spread widely and flatly in all directions when growing. In the developing bodies vacuoles can be discerned which vary in number and size, but can grow into fairly large droplets on which the protoplasmic body may sit in the shape of a crescent moon. It is probable that these vacuoles are filled with metabolic products; in the L-form of *S. moniliformis* it is cholesterol (see p. 364). These droplets apparently coalesce and produce the astounding myelin structures which form interesting patterns in preparations of living organisms and appear as holes in the fixed and stained ones. Attention has been drawn by various authors to the production of similar lipid-like material when rich media are incubated in the uninoculated stage.[15, 65]

VIII. Properties of L-Form

A. Metabolism and Chemical Constitution

The metabolism of the L-form may be the same as that of the parent bacterium in some respects, and different in others. Thus, an L-form derived from an anaerobic culture is itself an obligate anaerobe and L-strains derived from facultative anaerobes grow well under both conditions, aerobic and anaerobic. But, on the other hand, the L-form is uninhibited by high concentrations of such substances as penicillin and sulfonamides, which are lethal for the bacterial phase; the L strain of *F. necrophorus* in

Brewer's medium plus serum produces gas like the parent strain, although more slowly and less abundantly.

Little is known about the chemical composition of L-forms. Vendrely and Tulasne[66] found that the content of lipids is much greater in the L-form (24.3 % of the dry weight) than in the bacterial phase (8 %), and that the ratio of deoxyribonucleic acids to ribonucleic acids is considerably higher in the L-form than in the bacterial form. Kandler *et al.*[67] found that the L-form of *Proteus vulgaris* oxidizes the same substrates (carbohydrates, organic acids, amino acids, and alcohols) as the bacterial phase. Minck[68] showed that *Proteus vulgaris* and its L-form behaved identically in regard to sugar fermentation, production of hydrogen sulfide, indole, catalase, and reductase.

The L-form of *S. moniliformis*[28] contains a large amount (30 % of the dry weight) of cholesterol. This is not synthesized but set free from the serum used to enrich the medium. The L-form contains little of the material found in bacterial cell wall preparations. The chemistry of the cell wall of various bacteria has been studied by Work and Dewy[69] and Work.[70] A large amount of hexosamine and neutral sugars is present in the walls and also a new amino acid, diaminopimelic acid. Chemical analysis of the trichloroacetic acid precipitable part of the bacterial phase and the L-form of *Proteus vulgaris* carried out by Weibull[27] showed that the bacteria contain large amounts of the cell wall material, the L-form (which therefore is mainly protoplasmic in nature) a very small amount, and the protoplasts of *Bacillus megaterium* none at all. Kandler and Zehender[71] have recently studied the content of cell wall material in *Proteus vulgaris*, in its transition form and in its L-form. Their results are at variance with those of Weibull, for they found no diaminopimelic acid in the L-form strains but appreciable amounts in the transition form. This latter finding is in good agreement with that of Stempen who was able to demonstrate cell wall in transition culture. Although there is some disagreement between Weibull's and Kandler's results we can conclude that the L-form possesses either very little or no cell wall material; the amount may even vary from one L culture to the other and give rise to the observed discrepancies.

B. Serology of L-Form

Serological analysis reveals a close antigenic relationship between the bacterial and the L-form of one strain.[72] The two forms differ in certain respects which can be explained by their different surfaces. Tulasne[23] and Minck[73] reported that the L-forms of *Proteus vulgaris* and of *Vibrio cholerae* do not possess H antigen. This is not surprising as they have no flagella. The author[74] prepared antisera from *S. moniliformis* and its L-form. She found that a bacillary serum absorbed with bacilli had lost all its agglutinins When it was absorbed with an L-form antigen it had lost its L agglutinins yet was still able to agglutinate the bacilli although to a lower titer.

By absorption of an L antiserum with either the bacillary or the L antigen all agglutinins were removed. This shows that the bacilli contain an antigen not present in the L-form. From her experiments with *Proteus* sp. and its L-form von Prittwitz und Gaffron[75] concluded that the L-form possessed O and not H antigen.

In regard to other effects of antisera Dienes et al.[72] reported that typhoid antiserum had an L-form-activating effect on typhoid bacilli in the presence of complement. Edward and Fitzgerald[76] described the inhibition of growth of L-form of *Proteus vulgaris* by an L-form antiserum which had no effect on the parent strain. Von Prittwitz und Gaffron[65] found that addition of homologous antiserum to the media had no inhibitory effect on the L and bacterial cultures of a *Proteus vulgaris*. Klieneberger found that the L-form of *S. moniliformis* grew well on media containing the homologous serum. The observed differences in the inhibitory effect of homologous antiserum on L-form growth may be the result of different nutritional conditions during the tests and might be resolved if experiments of this kind were carried out under identical conditions.

C. Pathogenicity of L-Form

The author[77] found that mice could not be infected by any of the conventional routes by the L-form of a *S. moniliformis* strain which itself was pathogenic. Freundt[78] showed that L-form of *S. moniliformis* was apathogenic. When he used an unstable culture the infection established itself after the bacillary form developed in the peritoneal cavity. He could inhibit the outbreak of the disease by an appropriate administration of penicillin. By vaccination with L1 (L-form of *S. moniliformis*) he was not able to protect mice against *S. moniliformis* infection. Hannoun et al.[79] showed that a "granular form" of a "*Streptococcus viridans sanguis*" inoculated into the yolk sac of the chick embryo could, after a period of delay, produce lesions in which the normal form of the bacterium again was present.

Schnauder[80] infected mice orally with *Salmonella typhimurium* transition form. A delayed lethal infection occurred and the parent bacterial form was isolated from the feces and the heart blood of the mice. Silberstein[81] showed that the L-form of *Salmonella typhimurium* was nonpathogenic when injected intraperitoneally into white mice. However, a transition culture of *Proteus vulgaris* reverted in the second mouse passage and proved lethal.

Thus it appears that the L-forms of the cultures so far mentioned are not pathogenic, whereas compared with the bacterial form transition cultures ("unstable L-form") are feebly pathogenic.

Quite different results were obtained with vibrios by Tulasne and his school at Strasbourg. Minck[73, 82] injected cholera vibrios, protoplasmic transformation form, and L-form into white mice intraperitoneally. The bacterial form killed the mice within 24 hours. They died from an acute peritonitis; vibrios were present in the peritoneal exudate. When L-form was injected many animals died within 2 to 3 days from peritonitis. When protoplasmic unstable form was injected the mice survived and they were not resistant to the normal vibrio form.

Lavillaureix,[83] and Tulasne and Lavillaureix[84, 85] studied a vibrio isolated from water in Egypt. Both alive and heat-killed, it was lethal for white mice when injected by the intraperitoneal route. L-form was also pathogenic by every route used. The animals died with congested organs and L-organism was isolated from the peritoneal cavity. Even heat-killed L-form and extracts from L-form were lethal.

These interesting studies show that bacteria, such as vibrios, which produce an endotoxin are almost equally pathogenic in their L-form and their bacterial form whereas their transition form is apathogenic, contrary to the findings with all other bacillary and L-forms studied in this way. Further work with a wider variety of bacteria should bring to light interesting information on the pathogenicity of various L-forms.

IX. The Similarities of L-Forms and Pleuropneumonia-Like Organisms

A. Properties of PPLO

The organism of pleuropneumonia of cattle was discovered in 1898 and described as a filterable virus. The smallest elements of this organism are indeed of a size that would justify its classification with the viruses. However, its growth on artificial media shows that it is not a virus. The elements of the organism are readily deformed and have no rigid cell walls; Kandler and Zehender[71] have recently shown that they possess no cell wall substances. The elements of the culture are very pleomorphic and vary greatly in size. The smallest elements are 100–150 mμ in diameter. The diameter of the bigger forms can measure several microns. The organism of bovine pleuropneumonia is regarded as the prototype of a whole class of organisms designated "pleuropneumonia-like organisms" (PPLO). Edward and Freundt[86] suggested the new name of "Mycoplasmataceae," but this has not yet been generally accepted. Numerous different species or varieties are known today. Some are pathogenic for animals, for instance, the organisms causing pleuropneumonia in cattle and in goats agalactia in sheep and goats, respiratory disease in chickens, polyarthritis and bronchopneumonia in rats, and pneumonia in mice. Others found in sewage and soil appear to be saprophytic. Three varieties are so far known to occur in man: (1) the type commonly found on human genitals;[87] (2) the type found in fusospirillary gangrenous lesions by Ruiter and Wentholt;[88-90] and (3) the type which is a frequent inhabitant of the human mouth.[87] All these different organisms are very similar in regard to their morphology and the appearance of their colonies. The colonies are relatively small (between 10 and 500 microns in diameter) with a dark center and a lighter peripheral zone.

B. Similarities between L-Forms and PPLO

When the author first saw the colonies of L-form she thought she was dealing with a PPLO; there is a great likeness between the colonies of the two types of organisms. In both the elements in the centers of the colonies embed themselves in the agar, whereas they grow on the agar surface in the peripheral part. In both the outer zone may have a lacelike pattern (Figs. 6 and 7) although as a rule the PPLO colonies are more delicately marked than the L-colonies and are of a more finely granular appearance. In both types of organisms the single elements are readily deformed and lack rigid cell walls; therefore they are very pleomorphic in appearance and vary widely in their sizes.

The organisms of both groups are rather exacting in their nutritional

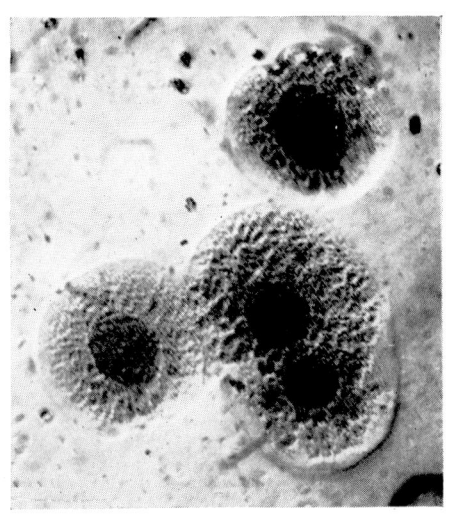

FIG. 6. L-colonies of *Proteus vulgaris*. Note the dark center and the lacy peripheral part. (Magnification: × 54)

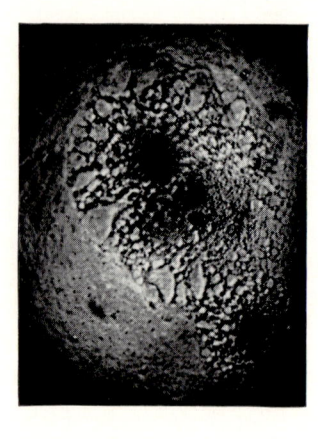

Fig. 7. Colonies of a human genital PPLO which has grown an exceptionally lacy peripheral part. (Magnification: × 140)

requirements and most of them need serum. They are both resistant to high concentrations of penicillin and sulfonamides. Owing to their lack of a rigid cell wall they burst and die in an environment of low osmotic pressure, such as distilled water or saline.

Both have relatively small elements which pass through filters that retain bacteria. These similarities have led to the conception that PPLO are in fact L-forms of bacteria the origin of which can no longer be traced.[23, 91, 92] There are, however, a number of outstanding differences between the two, some of which suggest to the author that they belong to different classes of organisms.

C. Differences between L-Forms and PPLO

1. Media, Growth, and Isolation

The media required differ with the species in both groups. It can, however, be said that as a rule the L-cultures grow on a relatively simple medium. Some of the pathogenic PPLO are most exacting; they require serum and additional growth factors present in yeast extract and in filtrates of bacterial broth cultures. Thus, PPLO require cholesterol whereas the L-form does not.[86]

L-form cultures grow reluctantly and in small colonies when first isolated; the longer an L-culture is subcultured the easier it grows until it reaches an optimum, which is then maintained. At this stage the colonies are visible to the naked eye and in the liquid they grow—as described—in fairly large clumps. In contrast, PPLO do not alter in their growth characteristics on subculture. The colonies remain small compared with L-colonies and in the liquid medium they grow with a very slight turbidity or in granules which can only be detected by means of a hand lens.

L-form cultures have rarely been found in pathological material or in nature. The author once cultured the L-form of a *Streptobacillus moniliformis* from a lung lesion of a rat suffering from bronchiectasis; the streptobacillus was present simultaneously. Dolman and Chang[93] cultivated an apparent L-form (it reverted later) of *S. moniliformis* from the blood of a child that had been ill with rat bite fever and had been treated with penicillin several weeks previously. These are the only cases known in which L-form cultures have been isolated directly from an animal or human. In all other cases the authors have obtained their L-cultures from laboratory strains which, as reported, produced them either spontaneously or by stimulation. In contrast, PPLO are widely spread and have been isolated from sewage and soil, from animals and humans on numerous occasions, and can be cultured from suitable material whenever desired.

2. COMPARATIVE FILTRATION EXPERIMENTS

The main and fundamental difference between L-form and PPLO is the size of their smallest units. This can be demonstrated by filtration experiments and electron microscopy; both these methods have been used by the author. Filtration experiments[94] determined the number of particles passing sintered glass and gradocol filters of different average pore diameter, and defined the size of their smallest units.

Four different PPLO cultures (organism of agalactia, of pleuropneumonia, of "rolling disease" in mice, and a sewage organism) and four different L-form cultures (L-forms of two different *S. moniliformis* strains, and L-form of *F. necrophorus* and of *B. proteus*) were used in the experiments.

More PPLO elements than L-form elements passed both types of filters.[95] When the sediment of the Jena glass filtrate of PPLO (agalactia organism) was centrifuged at 1750 g for 1 hour there was a visible sediment; an electron micrograph of the PPLO elements in this sediment is shown in Fig. 8. There was no visible sediment when an L-form filtrate was centrifuged in the same way and no elements were seen in an electron micrograph of the sediment.

The end point of filtration, using gradocol membranes, was much lower for PPLO than for L-forms; using Elford's[96] formula it was found that the diameter of the smallest particles of the L-form was almost twice as large as those of the PPLO. As far as a comparison of the different types of PPLO and of L-forms can be regarded as representing the groups, it is clear that the two belong to different size classes.

X. Electron Microscopic Demonstration of L-Forms of Bacteria and of PPLO

The methods used for the electron micrographs have been described by Klieneberger-Nobel and Cuckow[43] and by Cuckow and Klieneberger-

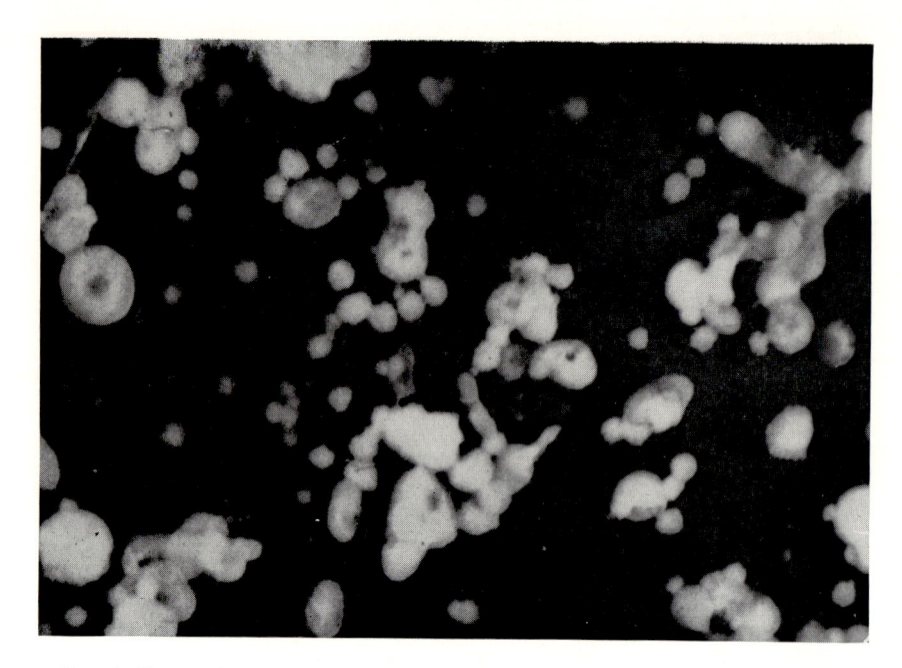

Fig. 8. Suspension of organism of agalactia of sheep filtered through Jena glass filter; the filtrate was spun and the visible sediment was resuspended in Tyrode-Ringer solution. The elements of this suspension are shown in the illustration. The "minimal reproductive units" and elements up to a diameter of 350 mμ are seen (Magnification: $\times$ 20,000; 1 μ = 2 cm.)

Nobel.[44] It was essential to suspend the organisms in a liquid of such osmotic qualities that their shapes and their viability are preserved; a Tyrode-Ringer suspension fulfilled these requirements.

Figure 9 shows a microcolony of an L-form of *S. moniliformis* which had grown on the Formvar film for 5 hours. It has a number of flat elements, some fairly large, some small, which seem interconnected by protoplasmic extensions. In two of the elements vacuoles can be seen. Some of the elements are very electron-opaque, others less so. Usually the opacity is less pronounced at the edges. In the middle of the colony a large body has disintegrated. Traces of protoplasm at the edges of the growth indicate the position of cholesterol droplets present before fixation.

Figure 10 shows a microcolony of Laidlaw's A organism from sewage (a saprophytic PPLO), which had grown on the membrane for 15 hours. The larger elements which are 0.3 to 1 microns in diameter have varying densities. Around the major part of the growth a large number of very small particles can be seen which have a diameter of approximately 100 to 150 mμ. These represent a uniform phase of small elements which are doubt-

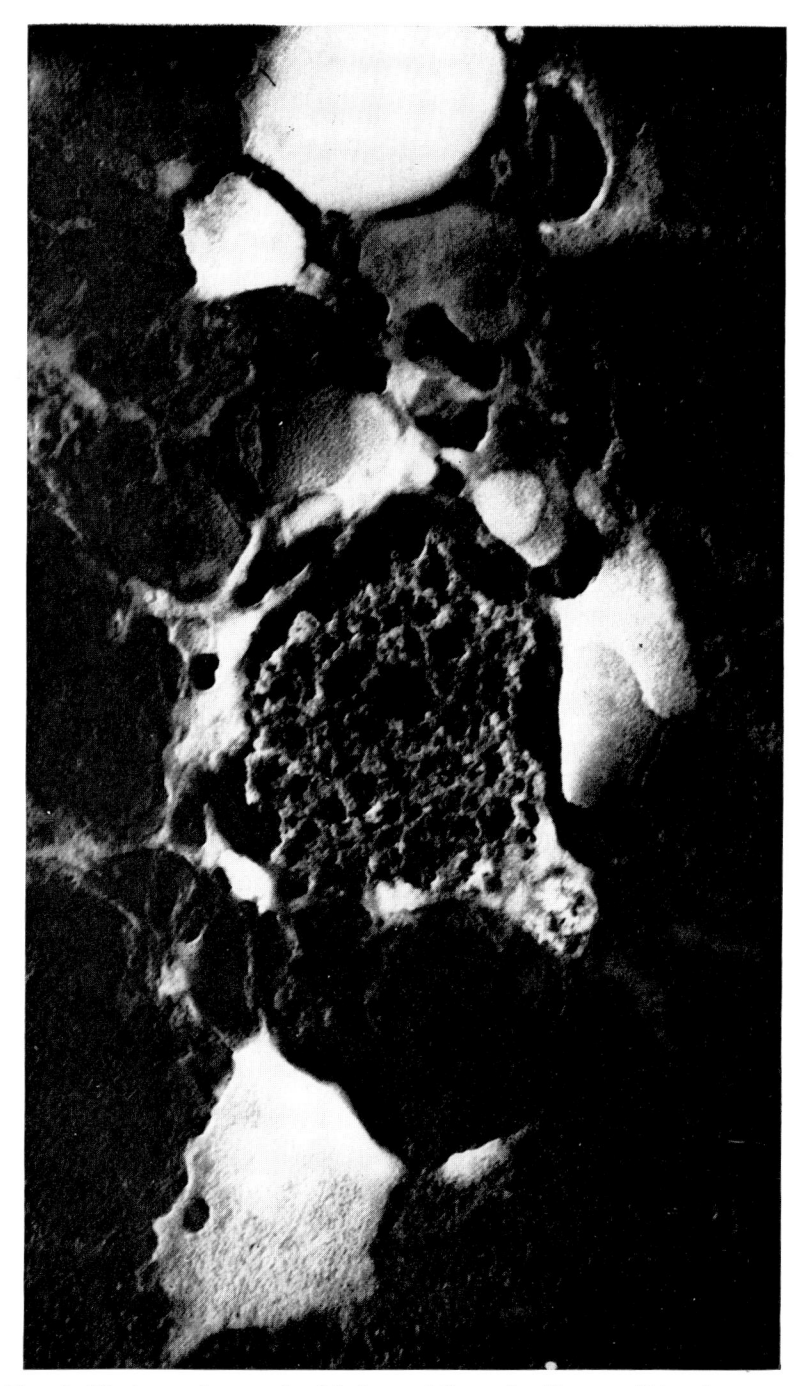

Fig. 9. Electron micrograph of L-form of *Streptobacillus moniliformis* grown on Formvar film for 5 hours. (Magnification: × 20,000; 1 μ = 2 cm.)

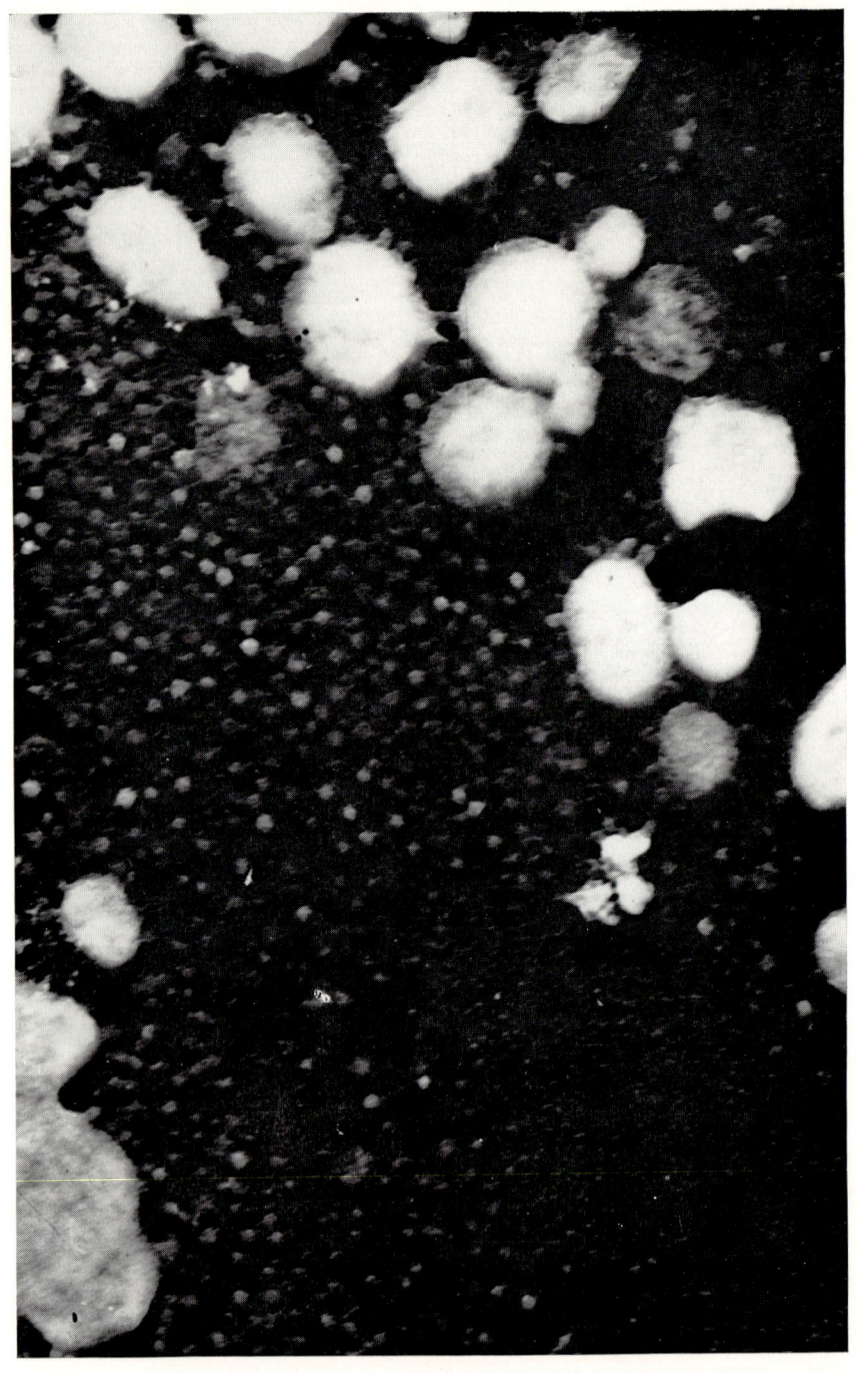

Fig. 10. Electron micrograph of a PPLO (Laidlaw's A organism from sewage) grown on Formvar film for 15 hours. (Magnification: ×20,000; 1 μ = 2 cm.)

less identical with the "minimal reproductive units." The electron micrographs shown are representative illustrations of the two kinds of organisms and they confirm the filtration experiments. Both filtration and electron microscopy demonstrate that L-forms produce elements of various sizes, the smallest of which have a diameter of about 0.3 microns and occur only in small numbers. In contrast, the larger bodies in cultures of PPLO always pass through a phase in which they are resolved into small elements; at any given moment during the free growth of a culture the small elements are far more numerous than they are in cultures of L-form. Moreover, the elements are smaller, of a size similar to that of the elementary bodies of the poxviruses.

Kellenberger et al.[97] formed similar conclusions about the dimensions and filterability of L-cultures but their results with PPLO (Laidlaw A) do not agree with those reported above. However, the results found by Elford[96] for PPLO, in his classic experiments as long ago as 1938, are similar to those of Klieneberger-Nobel[94] and of Cuckow and Klieneberger-Nobel.[44]

The differences between the PPLO and the L-forms revealed by filtration and electron microscopy seem so fundamental that the two groups of organisms cannot possibly be regarded as closely related. They must be considered as two different types of organisms: the PPLO, a particular class of microbes and the L-cultures, a form of life which bacteria are able to assume. Their striking similarities are due to the readiness with which their shape is changed, probably as a consequence of the absence of a cell wall.

XI. L-Forms and Protoplasts

When *Micrococcus lysodeikticus* or *Bacillus megaterium* is incubated in the presence of lysozyme, the cell walls are broken down, resulting in the clearing of the bacterial suspension.[98, 99] If, however, the medium has a high osmotic pressure, the action of lysozyme does not result in lysis of the cells but merely of the bacterial cell wall, leaving the contents intact; these assume a spherical shape and are known as protoplasts.[100] The protoplasts can assimilate, increase their substance, and even produce division forms;[101] but they do not revert to bacteria or give rise to a new protoplast population.

Under suitable conditions lysozyme affects the gram-negative organisms giving rise to a similar structure.[102, 103] Lederberg[104] was able to obtain protoplasmic elements, which he called protoplasts, by the action of penicillin on *B. coli*.

The protoplasmic bodies differ from protoplasts obtained from gram-positive organisms by lysozyme in that they can revert to bacteria in the absence of penicillin and also are able to reproduce themselves under suitable conditions in the presence of penicillin. Using Lederberg's technique we have obtained the protoplasmic bodies from a strain of *Proteus vulgaris* and have been able to maintain them in continuous subculture; we have found that after 10 passages these protoplasmic elements no longer require the presence of penicillin to prevent their reversion into bacteria. Therefore this protoplasmic growth now appears to be typical L-form.

The nature of the change from bacteria to the L-form is obscure. It occurs gradually and at first reversion takes place in the absence of penicillin in a matter of hours. After further subculturing in the presence of penicillin, reversion on penicillinless medium is delayed; finally an agar-stable L-type culture is obtained which eventually gives rise to typical L-form. Mutation and segregation have been suggested as operating factors. If is difficult, however, to interpret the experimental results on the basis of gene mutations.

So far L-forms have not been produced from protoplasts which have been obtained by the action of lysozyme on either gram-positive or gram-negative bacteria. It seems likely, therefore, that the lysozyme-induced protoplasts differ from the protoplasmic bodies produced under the influence of penicillin. It is probable that the differences can be correlated with the mechanism of their production; lysozyme treatment gives rise to protoplasts by the dissolution of the bacterial cell wall, whereas penicillin has been shown to interfere with the synthesis of the bacterial cell wall.[105]

XII. Summary and Conclusions

1. An L-form of a bacterium is defined as a growth of protoplasmic elements which lack a rigid cell wall and do not revert to the parent bacterium under any conditions; the colony form of the growth is characterized by a dark center and a lighter periphery; the growth in liquid media is as a rule in the form of clumps.

2. L-form can arise spontaneously but is usually produced by stimulation, for example, by the presence in the medium of penicillin, some amino acids, antisera, or some salts.

3. L-form arises usually by means of a gradual process: a number of passages in the presence of the stimulating substances are necessary before the transition stage, which is unstable, develops into L-form, which is stable.

4. The metabolism of the L-form is similar to that of the parent bacterium but differs in certain respects, such as the resistance of L-form to penicillin and sulfonamides.

5. The chemical composition of the L-form is quantitatively but not qualitatively different from that of the parent bacterium, with the exception of cell wall material which is lacking in the L-form.

6. L-form and parent bacterium are not identical in their serological behavior, for instance, L-form of *Proteus* sp. has no H antigens but possesses the same O antigen as the parent strain.

7. The L-form of pathogens is completely apathogenic in certain groups of organisms (*Salmonella typhimurium, Proteus vulgaris, Streptobacillus moniliformis*, etc.). It possesses a degree of pathogenicity in others (vibrios).

8. The organisms of the pleuropneumonia group and the L-forms of bacteria have characteristics in common, such as their colony type and the lack of a rigid cell wall. They show, however, differences in their filterability and their electron microscopic appearance, the smallest elements (about 300 mμ diameter) found in L-forms are twice as large as those of PPLO (about 150 mμ).

9. Protoplasts, penicillin-induced protoplasmic bodies (Lederberg's "protoplasts"), and L-forms are not identical. Protoplasts produced by the action of lysozyme on gram-positive organisms have no cell wall material and cannot produce a protoplast culture or revert to bacilli; the L-type transition form, which is identical with the penicillin-induced protoplasmic bodies of Lederberg's, probably possesses an appreciable amount of cell wall material and can reproduce itself or revert to bacilli; the L-form reproduces itself, does not revert, and contains either a small amount of cell wall material (Weibull) or none (Kandler).

Since the L-form can multiply without the presence of penicillin (with and without serum) and can be produced fairly easily, it can be assumed that it would lend itself admirably as a tool for biochemical studies concerned with the properties and activities of the bacterial cytoplasm deprived of its cell wall.

It would further be of value to explore on a large scale the pathogenicity of the L-form of various organisms not only from a theoretical point of view but also from the viewpoint of possible protection against the bacterial form by L-form vaccination.

The example of the vibrios justifies the speculation that in some yet unexplored bacteria L-form may play a role in the etiology of disease.

References

[1] E. Klieneberger, *Ergeb. Hyg. Bakteriol. Immunitätsforsch. u. Exptl. Therap.* **11**, 499 (1930).

[2] Ph. Kuhn and K. Sternberg, *Zentr. Bakteriol. Parasitenk. Abt. I. Orig.* **121**, 113 (1931).

[3] E. Klieneberger and J. Smiles, *J. Hyg.* **42**, 110 (1942).

[4] E. Almquist, *Zentr. Bakteriol. Parasitenk. Abt. I. Orig.* **37**, 18 (1904).

[5] G. Enderlein, "Bakteriencyclogenie." de Gruyter, Berlin, 1925.

[6] R. R. Mellon, *Am. J. Med. Sci.* **159**, 874 (1920).

[7] Ph. Hadley, *J. Infectious Diseases* **40**, 1 (1926).

[8] F. Löhnis, *Mem. Natl. Acad. Sci. U. S.* **16**, 2nd Memoir (1921).

[9] N. Gamaleia, "Elemente der allgemeinen Bakteriologie." A. Hirschwald, Berlin, 1900.
[10] M. Maassen, *Arb. kaiserl. Gesundh.* **21**, 383 (1904).
[11] E. Almquist, *Zentr. Bakteriol. Parasitenk. Abt. I. Orig.* **45**, 491 (1908).
[12] E. Klieneberger, *J. Pathol. Bacteriol.* **11**, 93 (1935).
[13] T. Mc P. Brown and J. C. Nunemaker, *Bull. Johns Hopkins Hosp.* **70**, 201 (1942).
[14] W. I. Strangeways, *J. Pathol. Bacteriol.* **37**, 45 (1933).
[15] E. Klieneberger-Nobel, *Biol. Revs. Cambridge Phil. Soc.* **29**, 154 (1954).
[16] L. Dienes, *J. Infectious Diseases* **65**, 24 (1939).
[17] L. Dienes, *Proc. Soc. Exptl. Biol. Med.* **42**, 636 (1939).
[18] L. Dienes, *Proc. Soc. Exptl. Biol. Med.* **43**, 703 (1940).
[19] L. Dienes and J. T. Sharp, *Bacteriol. Proc. (Soc. Am. Bacteriologists)* **55**, 49 (1955).
[20] J. T. Sharp, *Proc. Soc. Exptl. Biol. Med.* **87**, 94 (1954).
[21] E. Klieneberger-Nobel, *Bacteriol. Revs.* **15**, 77 (1951).
[22] L. Dienes and H. J. Weinberger, *Bacteriol. Revs.* **15**, 245 (1951).
[23] R. Tulasne, *Rev. immunol.* **15**, 223 (1951).
[24] R. Tulasne, *Biol. méd. (Paris)* **44**, 1 (1955).
[25] A. Winkler, "*Die Bakterienzelle.*" Fischer, Stuttgart, 1956.
[26] "L Forms" of Bacteria, *Nature* **179**, 461 (1957).
[27] C. Weibull, *Acta Pathol. Microbiol. Scand.* **42**, 324 (1958).
[28] S. M. Partridge and E. Klieneberger, *J. Pathol. Bacteriol.* **52**, 219 (1941).
[29] L. Dienes, *J. Bacteriol.* **66**, 274 (1953).
[30] J. P. Duguid, *Edinburgh Med. J.* **53**, 401 (1946).
[31] L. Dienes, H. J. Weinberger, and S. Madoff, *Proc. Soc. Exptl. Biol. Med.* **75**, 409 (1950).
[32] G. Poetschke, *Intern. Congr. Microbiol., 6th Congr., Rome 1953. Rept. Proc.* **1**, (1) 142 (1955).
[33] L. Dienes and P. C. Zamecnik, *J. Bacteriol.* **64**, 770 (1952).
[34] M. A. Medill and D. J. O'Kane, *J. Bacteriol.* **68**, 530 (1954).
[35] R. Y. Abrams, *J. Bacteriol.* **70**, 251 (1955).
[36] Z. Lorkiewicz, *Acta Microbiol. Polon.* **6**, 3 (1957).
[37] M. R. Pollock, *Brit. J. Exptl. Path.* **28**, 295 (1947).
[38] E. Grasset and V. Bonifas, *Schweiz. Z. allgem. Pathol. u. Bakteriol.* **18**, 1074 (1955).
[39] J. Lavillaureix, *Ann. inst. Pasteur* **90**, 376 (1956).
[40] P. De Gregorio and T. Terranova, *Experientia* **13**, 317 (1957).
[41] A. Nelles, *Zentr. Bakteriol. Parasitenk. Abt. I. Orig.* **164**, 78 (1955).
[42] E. Klieneberger-Nobel, *Quart. J. Microscop. Sci.* **91**, 340 (1950).
[43] E. Klieneberger-Nobel and F. W. Cuckow, *J. Gen. Microbiol.* **12**, 95 (1955).
[44] F. W. Cuckow and E. Klieneberger-Nobel, *J. Gen. Microbiol.* **13**, 149 (1955).
[45] L. Dienes and W. E. Smith, *J. Bacteriol.* **48**, 125 (1944).
[46] R. Tulasne, *Intern. Congr. Microbiol. 6th Congr. Rept. Proc. Rome 1953* **1**,(1) 144 (1955).
[47] J. von Prittwitz und Gaffron, *Naturwissenschaften* **40**, 590 (1953).
[48] M. A. Medill and W. G. Hutchinson, *J. Bacteriol.* **68**, 89 (1954).
[49] U. Taubeneck and M. R. Muller, *Zentr. Bakteriol. Parasitenk. Abt. I. Orig.* **163**, 309 (1955).
[50] J. Lederberg, *J. Bacteriol.* **73**, 144 (1957).
[51] R. J. V. Pulvertaft, *J. Pathol. Bacteriol.* **65**, 175 (1953).
[52] H. Stempen, *J. Bacteriol.* **70**, 177 (1955).
[53] A. Eisenstark, C. B. Ward, Jr., and T. S. Kyle, *J. Bacteriol.* **60**, 525 (1950).

[54] L. Carrère, J. Roux, and J. Mandin, *Compt. rend. soc. biol.* **148**, 2050 (1954).
[55] M. R. Huertos, E. Kuster, and W. Flag, *Zentr. Bakteriol. Parasitenk. Abt. I. Orig.* **165**, 514 (1956).
[56] H. Schellenberg, *Zentr. Bakteriol. Parasitenk. Abt. I. Orig.* **161**, 433 (1954).
[57] G. Pontieri, *Zentr. Bakteriol. Parasitenk. Abt. I. Orig.* **165**, 524 (1956).
[58] W. Höpken and K. Bartmann, *Zentr. Bakteriol. Parasitenk. Abt. I. Orig.* **162**, 372 (1955).
[59] W. Höpken and K. Bartmann, *Zentr. Bakteriol. Parasitenk. Abt. I. Orig.* **165**, 514 (1956).
[60] K. Bartmann and W. Höpken, *Zentr. Bakteriol. Parasitenk. Abt. I. Orig.* **166**, 30 (1956).
[61] J. von Prittwitz und Gaffron, *Arch. Hyg. u. Bakteriol.* **140**, 151 (1956).
[62] K. Liebermeister and E. Kellenberger, *Z. Naturforsch.* **11b**, 200 (1956).
[63] V. Sundman and C. R. Hackman, *Nature* **179**, 149 (1957).
[64] H. Schellenberg, *Zentr. Bakteriol. Parasitenk. Abt. I. Orig.* **161**, 425 (1954).
[65] J. von Prittwitz und Gaffron, *Zentr. Bakteriol. Parasitenk. Abt. I. Orig.* **163**, 313 (1955).
[66] R. Vendrely and R. Tulasne, *Nature* **171**, 262 (1953).
[67] O. Kandler, C. Zehender, and F. Muller, *Arch. Microbiol.* **24**, 219 (1956).
[68] R. Minck, *Compt. rend.* **234**, 764 (1952).
[69] E. Work and D. L. Dewy, *J. Gen. Microbiol.* **9**, 394 (1953).
[70] E. Work, *Nature* **179**, 841 (1957).
[71] O. Kandler and C. Zehender, *Z. Naturforsch.* **12b**, 125 (1957).
[72] L. Dienes, H. J. Weinberger, and S. Madoff, *J. Bacteriol.* **59**, 755 (1950).
[73] R. Minck, *Schweiz. Z. allgem. Pathol. u. Bakteriol.* **14**, 595 (1951).
[74] E. Klieneberger, *J. Hyg.* **42**, 485 (1942).
[75] J. von Prittwitz und Gaffron, *Naturwissenschaften* **42**, 113 (1955).
[76] D. G. FF. Edward and W. A. Fitzgerald, *J. Pathol. Bacteriol.* **68**, 23 (1954).
[77] E. Klieneberger, *J. Hyg.* **38**, 458 (1938).
[78] E. A. Freundt, *Acta. Pathol. Microbiol. Scand.* **38**, 246 (1956).
[79] C. Hannoun, J. Vigoroux, F. Levaditi, and O. Nazimoff, *Ann. inst. Pasteur* **92**, 231 (1957).
[80] G. Schnauder, *Z. Hyg. Infektionskrankh.* **141**, 404 (1955).
[81] J. Silberstein, *Schweiz. Z. Pathol. u. Bakteriol.* **16**, 741 (1953).
[82] R. Minck, *Compt. rend.* **231**, 386 (1950).
[83] J. Lavillaureix, *Compt. rend.* **239**, 1155 (1954).
[84] R. Tulasne and J. Lavillaureix, *Compt. rend. soc. biol.* **148**, 2080 (1954).
[85] R. Tulasne and J. Lavillaureix, *Compt. rend. soc. biol.* **149**, 178 (1955).
[86] D. G. FF. Edward and E. A. Freundt, *J. Gen. Microbiol.* **14**, 197 (1956).
[87] C. S. Nicol and D. G. FF. Edward, *Brit. J. Venereal Diseases* **29**, 141 (1953).
[88] M. Ruiter and H. M. M. Wentholt, *J. Invest. Dermatol.* **15**, 301 (1950).
[89] M. Ruiter and H. M. M. Wentholt, *J. Invest. Dermatol.* **18**, 313 (1952).
[90] M. Ruiter and H. M. M. Wentholt, *Acta Dermato-Venereol.* **33**, 123 (1953).
[91] R. G. Wittler, S. G. Cary, and R. B. Lindberg, *J. Gen. Microbiol.* **14**, 763 (1956).
[92] C. R. Amies and S. A. Jones, *Can. J. Microbiol.* **3**, 579 (1957).
[93] C. E. Dolman and H. Chang, *Can. J. Public Health* **42**, 73 (1951).
[94] E. Klieneberger-Nobel, *Zentr. Bakteriol. Parasitenk. Abt. I. Orig.* **165**, 329 (1956).
[95] A. A. Miles and S. S. Misra, *J. Hyg.* **38**, 732 (1938).
[96] W. J. Elford, *in* "Handbuch der Virusforschung" (R. Doerr and C. Hallauer, eds.), p. 126. Springer, Vienna, 1938.

[97] E. Kellenberger, K. Liebermeister, and V. Bonifas, *Z. Naturforsch.* **11b,** 206 (1956).
[98] C. Weibull, *J. Bacteriol.* **66,** 688 (1953).
[99] M. R. J. Salton, *Bacteriol. Revs.* **21,** 82 (1957).
[100] J. Tomcsik and S. Guex-Holzer, *Schweiz. Z. allgem. Pathol. u. Bakteriol.* **15,** 517 (1952).
[101] K. McQuillen, *Symposium Soc. Gen. Microbiol.* **6,** 127. Cambridge.
[102] R. Repaske, *Biochim. et Biophys. Acta* **22,** 189 (1956).
[103] N. D. Zinder and W. F. Arndt, *Proc. Natl. Acad. Sci. U. S.* **42,** 586 (1956).
[104] J. Lederberg, *Proc. Nat. Acad. Sci. U. S.* **42,** 574 (1956).
[105] J. T. Park and J. L. Strominger, *Science* **125,** 99 (1957).

CHAPTER 8

Bacterial Viruses—Structure and Function

THOMAS F. ANDERSON

	page
I. Introduction	387
A. Sources of Bacteriophages	388
B. Mutants of Bacteriophages and Bacteria—Host Range	388
C. The Life Cycles of Virulent Bacteriophages	390
D. Lysogeny and the Life Cycles of Temperate Bacteriophages	392
II. Structure of Bacteriophage Particles	394
A. Gross Morphology	394
B. Properties of the Head Membrane	397
C. Properties of the Phage Nucleic Acids (DNA)	399
D. Minor Components in Phage Heads—Neutralization of the Nucleic Acids	403
E. The Proteins of Phage Tails	403
III. Relation of Structure to Function—Mechanism of Infection	408
A. Adsorption	408
B. Injection of Phage DNA	409
C. Effects of Infection on Properties of the Host	410
IV. Importance of Bacteriophages in Bacteriology	411
A. Typing Bacteria with Bacteriophages	411
B. Bacteriophages in the Classification of Bacteria	411
References	412

I. Introduction

Bacteriophages are viruses that infect bacteria. Thus to a bacteriologist the study of bacteriophages or bacterial viruses is a branch of the broader subject of the pathology of bacteria which includes among other subjects malnutrition, poisoning, and radiation sickness. Because their action is highly specific the bacteriophages are useful to epidemiologists for typing bacteria. Other bacteriologists with interests in genetics are finding the bacteriophages important because many bacteria inherit the capacity for producing certain bacteriophages. The chromosome of the phage seems to reside on a definite locus of the host's chromosome and multiplies whenever the host's chromosome does. These phages can also transport bacterial genes from one bacterium to another. To bacteriologists interested in viruses, the bacteriophages are of interest in themselves, for of all viruses, those infecting bacteria are the most easily studied under controlled conditions.

In this chapter we shall be concerned primarily with the morphological properties of bacteriophages and the mechanisms with which they infect their host bacteria. First, however, we will describe the natural history of bacteriophages: their discovery and life cycles.

A. Sources of Bacteriophages

Bacteriophages or bacterial viruses were discovered over forty years ago by Twort[1] and by d'Herelle.[2] The former noted the existence of "nibbled" bacterial colonies growing on nutrient agar. He reported that the nibbling was caused by an agent that could be transferred indefinitely from culture to culture. d'Herelle observed holes or "plaques" on bacterial smears and showed that the agent causing them was "particulate" in nature since the number of plaques observed on a smear was proportional to the amount of the lytic material spread on the smear. Both investigators showed that the transmittable agent was so small that it readily passed through filters that held back bacteria. In the early days one spoke of "the bacteriophage" as though there were only one type of agent. As we shall see, there are many kinds of bacteriophages.

Modern experimental methods used in working with bacteriophages have been described by Adams.[3] Bacteriophages may usually be found in nature wherever their host bacteria are present and are being destroyed in large numbers. Thus samples of feces and sewage are sources for phages active against enteric bacteria; pus is a source of phages active against staphylococci; soil is a source for phages active against soil bacteria, etc. To isolate the phage from such samples the bacteria may first be removed by filtration or they may be killed by adding chloroform to the sample without affecting the phages in it. Then when the filtrate is added to a growing culture of the prospective host, the activity of virulent phages is manifest by lysis of the cells: eventual clearing if the culture is in the liquid, or the appearance of plaques or holes in the smear, if the culture is spread on nutrient agar.

B. Mutants of Bacteriophages and Bacteria—Host Range

Often more than one type of plaque may be recognizable on a smear. The phage preparation then represents a mixture from which pure lines of phage may be isolated by picking material from well-isolated plaques. Each plaque contains from 10^7 to 10^9 plaque-forming particles as determined by suspending an individual plaque in broth, spreading appropriate dilutions with host bacteria on nutrient agar, and counting the plaques that appear after incubation. Most of the plaques so obtained appear identical to the parent plaque that was picked, but a few may be different and, if they breed true in subculture, represent the offspring of mutant forms (called "plaque-type mutants") of the bacteriophage which initiated the original plaque.

The host bacteria are usually capable of mutating to resistance to a given phage. Such resistant mutants may easily be isolated on nutrient agar by spreading some 10^9 bacteria with a large number (say 10^8) of plaque-forming particles. The plaques then merge to give confluent lysis, except for a few colonies of resistant bacteria. By means of fluctuation tests on parallel cultures grown from small inocula, Luria and Delbrück[4] have shown that these resistant bacteria arise spontaneously during the growth of the bacteria and are not caused or induced by the phage.

The phage, however, can in turn mutate to activity against these otherwise resistant bacteria. The great majority of phage particles in a preparation fails to form plaques on them, but again if a large number of plaque-forming particles (say 10^{10}) is spread with the resistant bacteria a few plaques may appear on the otherwise resistant bacterial smear. These plaques represent "host-range mutants" of the phage. They too arise spontaneously without the intervention of the mutant bacteria.[5] The mutant bacteria may mutate a second time—this time to resistance to the host-range phage mutant. And so by alternate mutations of bacteria and phage, from a single plaque and a single colony of sensitive bacteria, one may obtain a set of mutant bacteria and mutant phages, each bacterial strain characterized by a range of sensitivity to the phages and each phage by its plaque type and spectrum of activity against the bacteria.

The most intensive studies of host range have been made by Demerec and Fano[6] with a set of seven bacteriophages, arbitrarily labeled T1 to T7, and each active against the single strain B of *Escherichia coli*. They are listed in Table I together with the sizes of the plaques they form. As shown in column 6, there are two ways in which *E. coli* strain B, may mu-

TABLE I

BACTERIOPHAGES ACTIVE AGAINST STRAINS OF *Escherichia coli*

Bacterio-phage	Plaque size	Serological relatives	Dimensions of Particle Head, A.	Dimensions of Particle Tail, A.	Principal resistant mutants of B
T1	Medium	Alone	400	1600 × 100	B/1, tr; B/1,5
T2	Small	T4, T6	900 × 600	1000 × 200	B/2
T4	Small	T2, T6	900 × 600	1000 × 200	B/4,3; B/4,3,7,pr
T6	Small	T2, T4	900 × 600	1000 × 200	B/6
T3	Large	T7	450	100 × 100	B/3,4; B/3,4,7,pr
T7	Large	T3	450	100 × 100	B/7,3,4,pr
T5	Small	Alone	650	1700 × 100	B/5,1
λ	Large	—	600	1000 × 100	Host: *E. coli* K12
φ X 174	Large	S13	300	—	Host: *E. coli* C122

tate to become resistant to T1; one mutant, B/1, tr, requires tryptophan for growth, but is sensitive to the other T phages; the other mutant, B/1,5, while requiring no new growth factors, is resistant to T5 as well as to T1. The mutant B/5,1, isolated in the presence of T5, is identical to the mutant B/1,5 isolated in the presence of T1. Such "cross-resistance tests" carried out for each of the seven bacteriophages show that each phage is different from the others with the exception of T3 and T4. However, T3 and T4 are markedly different in plaque size, and as we shall see later, they differ in other properties including the morphology and serology of the infectious particle (columns 3, 4, and 5 of Table I). These T phages are termed "virulent" bacteriophages because bacteria infected with them are killed and lysed.

C. The Life Cycles of Virulent Bacteriophages

The life cycle of a virulent bacteriophage can be illustrated in the following way.[7] If a sample of a virulent bacteriophage is added to a growing aerated broth culture of susceptible bacteria and the mixture is plated at intervals on a smear of susceptible bacteria, the number of plaques obtained follows a curve like curve A shown in Fig. 1. The number of plaques remains constant for a time called the "latent period," which is charac-

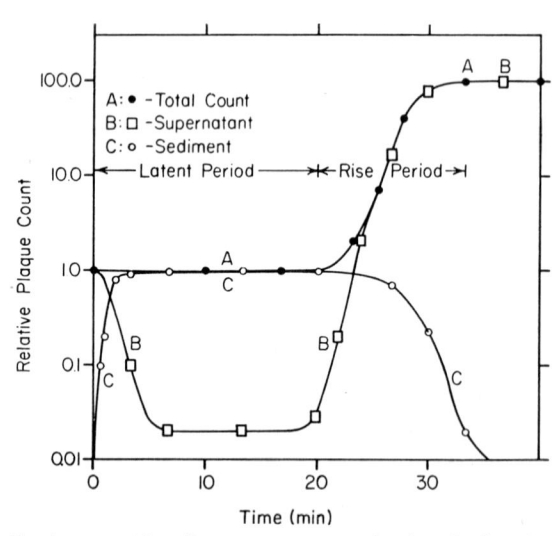

Fig. 1. Idealized curves showing one-step growth of a virulent bacteriophage on susceptible bacteria. The relative plaque counts (ordinates) are plotted on a logarithmic scale against the time after mixing a bacteriophage suspension with host bacteria. Curve A shows the total plaque counts. Curve B shows the plaque counts of supernatants and curve C those of sediments of aliquots taken at intervals and centrifuged to throw down the bacteria.

teristic of the phage and the conditions of culture. At the end of the latent period the plaque count rises suddenly during a "rise period" to a new stationary level some 20 to 100 times the original titer. Sometime during the latent and rise periods the bacteriophage evidently has multiplied 20- to 100-fold.

For the interpretation of such a "one-step growth curve" the roles of free phage and bacteria must be separated. To do this, aliquots of the incubating mixture are chilled at intervals and centrifuged to throw down the bacteria. The supernatant and resuspended sediments are then assayed separately. One then obtains curves like B and C of Fig. 1. A study of these curves will show that initially the phage activity is not sedimented with the bacteria because the infectious agent is much smaller than the bacteria. However, in the first few minutes after mixing, the phage activity originally in the supernatant (curve B) becomes adsorbed onto the sensitive bacteria (curve C). During the remainder of the latent period each of the infected bacteria forms a single plaque when plated; even though a bacterium may contain many phage particles these are localized in the agar and together form but a single plaque. However, at the end of the latent period the bacteria liberate phage activity in a form that is not sedimented at low speeds for the new high activity now appears in the supernatant.

To see what happens to the bacteria during such an experiment one may examine the living preparation under the microscope. The bacteria appear more or less normal through most of the latent period. However, near the end of the latent period they swell slowly, then one by one during the rise period the bacteria disappear by lysis. Evidently in the process of lysis the organisms liberate the 20 to 100 daughter phage particles they had contained within them.

From this experiment the life cycle of a virulent bacteriophage may be seen to consist of three stages: (1) adsorption of the infectious particle on the host; (2) multiplication within the host cell; (3) liberation of mature daughter particles by lysis of the host. Each liberated daughter particle is free to infect further susceptible bacteria and form a plaque. As we shall see in later chapters, the results of more refined experiments permit one to divide each of these stages into steps and in some cases even to visualize how the steps merge into each other in a continuous process.

1. Mechanism of Plaque Formation

We can now appreciate how plaques are produced by a localized epidemic of the virus disease: a single infected bacterium in a smear of susceptible cells lyses and liberates daughter phage particles which then infect other bacteria in the neighborhood; these lyse, in turn, liberating more infectious particles and so on until, by diffusion of daughter particles, the epidemic spreads to visible proportions and produces a local zone of transparent lysed cells and debris in the otherwise healthy bacterial smear. In its terminal stages the growth of a plaque is checked by nutritional and other natural limitations to the growth of the bacteria, but most of the factors that determine the appearance of plaques of various phage mutants (size, halos, degree of cloudiness, etc.) are poorly understood.

2. Lysis of Liquid Cultures

The infection of a single bacterium in a young broth culture of susceptible organisms may lead to an epidemic that spreads principally by convection of phage particles and infected bacteria throughout the vessel and may lead to lysis of all the susceptible bacteria. Such epidemics, invading the bacterial vats on which large industrial operations depend, can be of major economic concern. There are two remedies: to take precautions against such infections, and to develop strains of bacteria that are resistant to the phages known to be present in the plant.

Soon after the discovery of bacteriophages it was thought that the bacteriophages might prove invaluable to medicine in curing bacterial infections. However, few if any well-controlled experiments have been reported that prove their value in therapy. Presumably, the presence of resistant strains of bacteria and the development of antibodies against the phage prevents this. On the other hand, d'Herelle[8] maintained that all natural recovery from bacterial infections is due to bacteriophage action rather than antibody production. Perhaps the truth lies somewhere between these points of view.

D. Lysogeny and the Life Cycles of Temperate Bacteriophages[9, 10]

Infection of bacteria by the *virulent* phages, just discussed, almost invariably leads to proliferation and liberation of daughter phage particles from the infected cell. We turn now to the *"temperate"* phages which do not always kill the bacteria they infect. Their genetic material has the capacity ultimately to enter into what might be termed "genetic symbiosis" with the host.

Many strains of bacteria called "lysogenic" bacteria carry in their genetic make-up the capacity to produce one or more bacteriophages active against other sensitive or "indicator" strains. If a few such lysogenic bacteria are incubated with a smear of the indicator strain, plaques will appear surrounding each colony of lysogenic bacteria. (These plaques are not to be confused with the halos around colonies of colicinogenic bacteria on smears of colicine-sensitive bacteria. The colicines[11] are antibiotic substances which cannot be propagated on bacteria). Although the phage is adsorbed on bacteria of the lysogenic strain from which it originated, the adsorbed phage does not multiply or lyse bacteria that are lysogenic for the same or closely related strain of phage. However, nonlysogenic bacterial strains may be isolated from the parent lysogenic strain. These "cured" bacteria are sensitive. When plated on a smear of such nonlysogenic derivatives, the phage forms more or less turbid plaques containing many or only a few minute colonies. When isolated, these colonies turn out to be of two types; (1) nonlysogenic—they represent mutants resistant to the temperate phage; (2) lysogenic for the phage—lysogenic bacteria are resistant to the phage they carry.

Evidently the life histories of the so-called "temperate" bacteriophages are, superficially at least, more complex than those of the virulent bacteriophages. Besides an abortive infection in which the ability to produce phage is lost, the adsorption of a temperate bacteriophage on a sensitive bacterium may produce either one of two responses. One response leads directly to production of new phage particles as already described. This "lytic response" is shown schematically on the left of the diagram of Fig. 2. Superficially, at least, this lytic response of the temperate phage-host complex seems to differ in no essential way from that of the virulent phage-host complex. The alternative response is the lysogenic one shown on the right side of Fig. 2. Here, the infectious principle, which we shall see is phage nucleic acid, may exist in the bacterial cytoplasm for some time. Eventually,

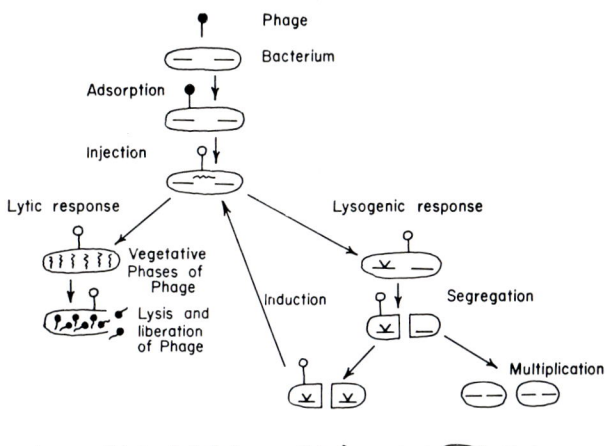

●— phage particle; ○— shell of phage particle; ⸨ gonophage; ⊂⊃ bacterium;
— genetic system of non-lysogenic bacterium; ∨ prophage; ⩔ genetic system of lysogenic bacterium.

FIG. 2. A schematic diagram illustrating the adsorption of a temperate bacteriophage, the injection of its DNA, and the alternative lytic and lysogenic responses of the complex that results.

however, it becomes incorporated as "prophage" into the genetic system of the bacterium, where it multiplies in step with the genetic system of the host to make up a lysogenic clone.[12] Depending on the strains of bacteria and phage, the proportion of complexes giving the lytic response is known to depend on many factors including the multiplicity of infection, the temperature, and the physiological state of the bacteria.

No mature infectious phage particles can be found within lysogenic bacteria, as they can in the latter part of the latent period during a lytic response.[13] However, occasionally a bacterium in a lysogenic clone spontaneously passes to the vegetative stage to give the lytic reaction with phage production. The phage particles spontaneously produced give rise to the small amount of free phage found in broth cultures of lysogenic bacteria. Spontaneously produced phage particles also initiate the production of the plaque around each lysogenic colony placed on a smear of sensitive bacteria. Bacteria lysogenic for some phages can also be "induced" to pass to the lytic state by treatment with mutagenic agents (ultraviolet irradiation or nitrogen mustards) or with carcinogenic chemicals. Almost all the bacteria in such a culture can be induced to lyse and liberate phage in this way. A later chapter, in Volume V, of this treatise will cover the details of these interesting and important phenomena.

II. Structure of Bacteriophage Particles

The complex structures of bacteriophages, like those of other biological entities, are of interest from many points of view. Their structures imply, on the one hand, the manner in which they are fashioned by nature—approaching molecular dimensions, as they do, it is felt by many that the solution to the problem of specific biological synthesis may be found by studying the mechanism of their formation. At the same time, phage structures reflect the function each element has to play when the virus particle encounters a susceptible host bacterium. The phages are big enough to be examined individually in the electron microscope, so that changes in the structure of an individual after an encounter with a host bacterium can be seen. On the other hand, a purified phage preparation presents a static, homogeneous population that, unlike metabolizing physiological systems, holds still for significant chemical analyses based on counted particles. Thus the bacteriophages provide for study a set of elementary biological structures in a form suitable both for the minute examination of individuals and for gross chemical analyses of huge but essentially homogeneous populations.

A. Gross Morphology

With the aid of the electron microscope most of the bacteriophages are seen to consist of two easily recognizable components: a more or less round or polyhedral-shaped "head" to which is usually attached a relatively thin appendage called a "tail." The particles of any one type are remarkably uniform in size and shape. The dimensions of the phages in the T-set are indicated in Table I. T1 has a roughly spherical head (diameter $\cong$ 400 A.) to which a thin (100 A.) and relatively long (1600 A.) tail is attached. It is not inactivated by drying. The even-number phages T2, T4, and T6 have identical morphologies with heads some 900 A. long and 600 A. wide that appear to have the shapes of hexagonal bipyramids (Fig. 3B) when care is taken to avoid drying artifacts in preparing specimens for study.[14] Their tails are well-defined structures about 1000 A. long and 200 A. wide and frequently exhibit strands at the slightly thickened tips. T3 and T7 are small (450 A. diameter) polyhedra with barely discernible nubs of tails.[15] T5 has a large roughly spherical head (diameter $\cong$ 650 A.) with a hexagonal silhouette. The tail of T5 is long and thin (1700 × 100 A.). Other bacteriophages have morphologies analogous to those of the T-set—well-defined head and tail structures with two known exceptions: (1) the phage φX 174 is only 300 A. in diameter;[16] (2) a bacillus-shaped phage, reported by Kottmann,[17] which has not been studied further. As far as structure goes, the temperate phages like λ (Fig. 4) differ in no obvious way from the virulent ones.[18]

It might be well to list some of the criteria that indicate that these remarkable structures are really the bacteriophages.[19] (1) They are present in phage suspensions in numbers that equal (within experimental error) the numbers of plaque-forming particles. (2) The structure of both the head and tail is characteristic of the strain of phage. (The exceptions here prove the rule: phages like T2, T4, and T6, which have the same morphology, are closely related serologically and genetically.[20]) (3) Under conditions suitable for adsorption (as determined by biological criteria) the particles are seen to be absorbed specifically by the ends of their tails to sensitive bacteria (Figs. 3C and 4) and under unsuitable conditions they are not adsorbed.[14, 21] (4) At the end of the latent period, particles having the same morphology as the infecting ones are liberated in appropriate numbers from infected bacteria.[22] (5) They are specifically aggregated and inactivated by homologous antiphage sera but not by heterologous sera or by antibacterial sera. (6) Agents which destroy phage activity usually have visible effects on the particles.

Chemically, these particles are made up primarily of protein and deoxyribose nucleic acid (DNA). The intact particle is not attacked by nucleases. The protein and nucleic acid can be separated from each other, however, by a number of techniques that put the nucleic acid into solution. In the following sections we shall discuss some of the properties of the morphologically distinguishable parts: the head protein, the nucleic acid, and the tail protein.

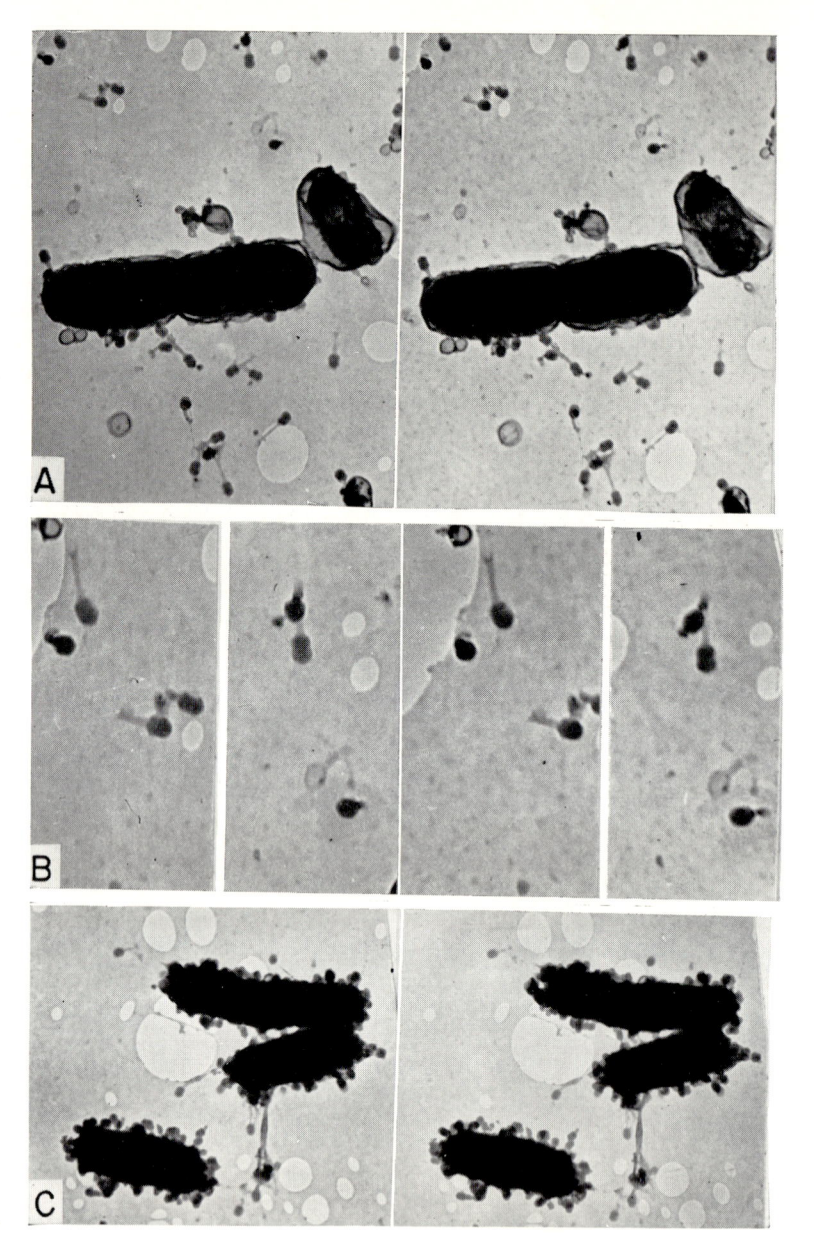

Fig. 3. Stereoscopic electron micrographs showing: (A) a mixture of *Escherichia coli* and the bacteriophage T4; (B) the morphology of T4 particles from (A) at higher magnification; and (C) a similar mixture of B and T4 but with the cofactor for adsorption, L-tryptophan, added. Note that in (A) very few tryptophan-requiring T4 particles are adsorbed on host bacteria while in (C), with L-tryptophan added, the bacteria are covered with particles adhering by their tails like pins in pin cushions. The hexagonal morphologies of the heads of T4 particles are well brought out in (B) when the micrographs are viewed stereoscopically. [From T. F. Anderson, *Am. Naturalist* **86,** 91 (1952).] (Magnifications: (A) × 22,000; (B) × 43,000; (C) × 22,000.)

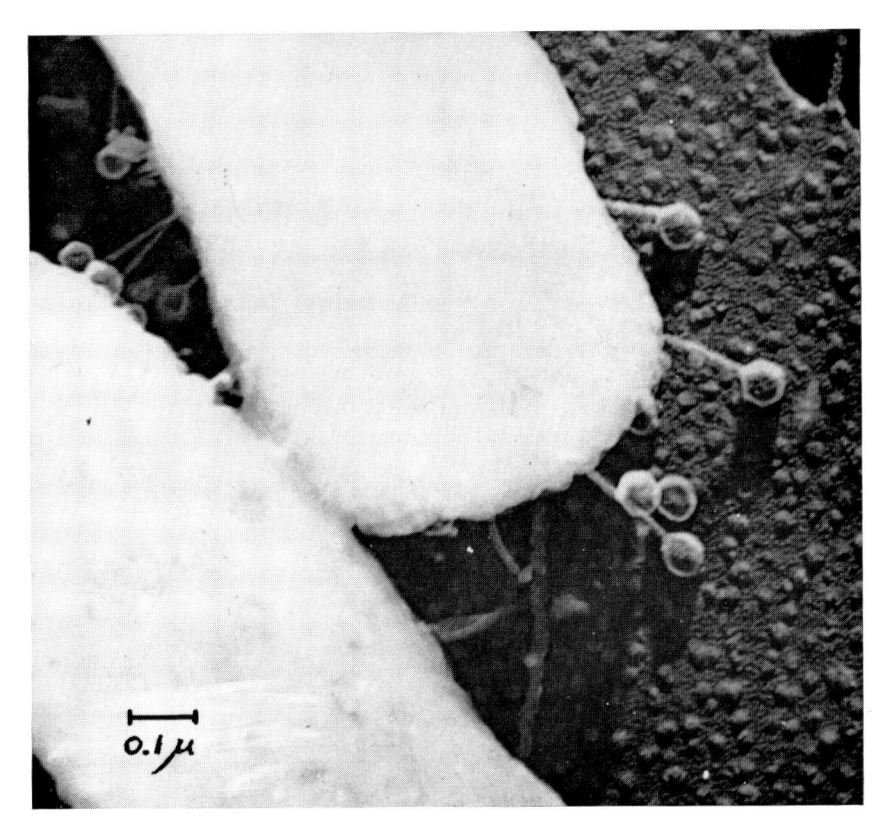

Fig. 4. The specific adsorption of λ bacteriophage particles to a conjugating F-bacterium (*Escherichia coli* K 12) shown in the upper part of the field. None of the λ particles is adsorbed on its mate, an Hfr/λ bacterium, seen in the lower left part of the field. Note that the phage particles adhere to the host bacteria by the tips of their tails. Although the heads are already empty, injection of DNA having already occurred when the specimen was prepared, the head membranes still retain their polyhedral shape. (From T. F. Anderson, E. L. Wollman, and F. Jacob, *Ann. inst. Pasteur* **93**, 450 (1957).) (Electron micrograph 24.I.56 E₃ × 95,000)

B. Properties of the Head Membrane

1. Sensitivity to Osmotic Shock[23]

When the bacteriophages T2, T4, or T6 are placed in concentrated salt solutions and then rapidly diluted with water, the turbidity drops, the solution becomes viscous, and the viable phage count drops to 1% or less of its initial activity. Since such effects are not produced if the dilution is carried out slowly, the term "osmotic *shock*" has been applied to the phenomenon.[24] The effect is nonspecific: osmotic shock can be produced by rapid dilution of T-even phage in organic solutes such as glycerol as long

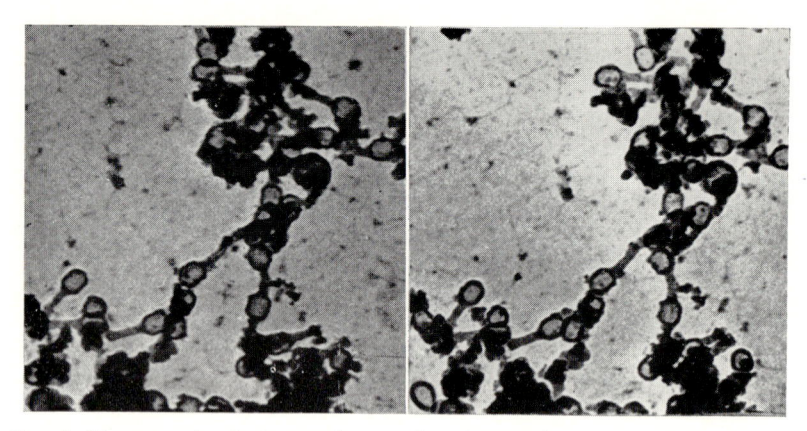

FIG. 5. Stereoscopic electron micrographs of osmotically shocked T4 showing "ghosts" of phage particles with empty head membranes that tend to retain the polyhedral shapes of the original particles.

as the decrement in osmotic pressure is at least some 60 atmospheres. Electron micrographs of shocked phage show particles having empty heads (Fig. 5) but with morphologies otherwise similar to the intact phage. These "ghosts" of phage particles can be purified by centrifugation. They prove to consist of protein,[25] while the nucleic acid which had made up some 40% of the intact phage particles remains in the supernatant. This is the proof that the nucleic acid had been contained in the proteinaceous head membrane of the T-even phage particle.

To explain the phenomena associated with osmotic shock it is supposed that the head membrane is somewhat more permeable to water than it is to most solutes. On slow dilution the solute and solvent can exchange without causing an appreciable rise in internal pressure. However, on sudden dilution of particles containing solute, the water rushes into the heads more rapidly than the solute can escape. Then if the resulting diffusion pressure is some 60 atmospheres or more the head membrane ruptures to release the DNA into the medium. As yet no holes have been seen in the 100 A.-thick head membranes of phage ghosts that have released their DNA, but neither can holes be seen in ghosts of erythrocytes that have released their hemoglobin. Holes in membranes seem to be able to open under pressure and then close when the pressure is released.

Concentrated sucrose inactivates T-even particles directly to produce ghosts, whether the subsequent dilution is rapid or slow. Presumably the head membranes of normal particles are essentially impermeable to sucrose and are ruptured by collapse under the external pressure. This concept receives support from the observation that some strains of T6 exist at high temperature (50° C.) in forms that cannot be shocked by dilution from salt solutions but which can be shocked by dilution from sucrose. It is thought that the enlarged pores in the membranes of these high-temperature forms are quite permeable to salts and only slightly permeable to sucrose; as a result, the particles at high temperature can be shocked by the latter solute but not by the former which, at higher temperature, pass almost as freely as water through them.

Attempts to disrupt the T-odd bacteriophages by osmotic shock have been unsuccessful. However, in a medium low in Ca^{++} or at elevated temperatures T5 particles release their DNA into the medium. The resulting ghosts of T5 particles have the morphologies of native T5 but with empty heads; it is inferred that the DNA had been in the heads.[26] Supposedly Ca^{++} is essential to the integrity of the head membrane of T5. It is interesting in this connection that Ca^{++} is also essential to the successful early stages in the infection by T5 of *E. coli*, strain B, and to the production of mature virulent T5 particles as well.

2. Antigenic Properties

Antigenically the protein(s) of the heads of T-even particles is distinct from those of the tail.[27] When T-even phages are allowed to infect a bacterium, to develop, and lyse the host in the presence of proflavine, only the head protein is formed. Antisera prepared against these head membranes fix complement in the presence of whole T-even phage and cause the particles to aggregate in star-shaped head-to-head clumps. Such antisera do not inactivate or neutralize whole T-even phage, however, suggesting that, even though the head membrane may be essential to the formation of mature phage particles, it does not play an active role in the adsorption or infection of the host bacteria. The role of the head membrane seems instead to be as a coat or envelope containing and protecting the virus nucleic acid.

C. Properties of Phage Nucleic Acids (DNA)

The nucleic acids of all the phages so far analyzed are of the deoxyribose type. As in the DNA's of other systems, the amount of the base *adenine* is equivalent to the *thymine*; and *guanine* is equivalent to *cytosine*—or in the case of the T-even phages, to 5-hydroxymethyl cytosine.[28,29] Also, as for other DNA's, the high molecular weight, the high viscosity and the X-ray diffraction pattern[30] suggest a structure of the Watson-Crick type,[31] i.e., a double helix of complementary chains, indicated schematically in Fig. 6. The individual chains are thus held together by phosphate diester linkage between the number 3-carbon of the deoxyribose of one nucleoside and the number 5-carbon of the deoxyribose of the adjacent nucleoside. Hydrogen bonds hold each base of one chain to its complementary base in

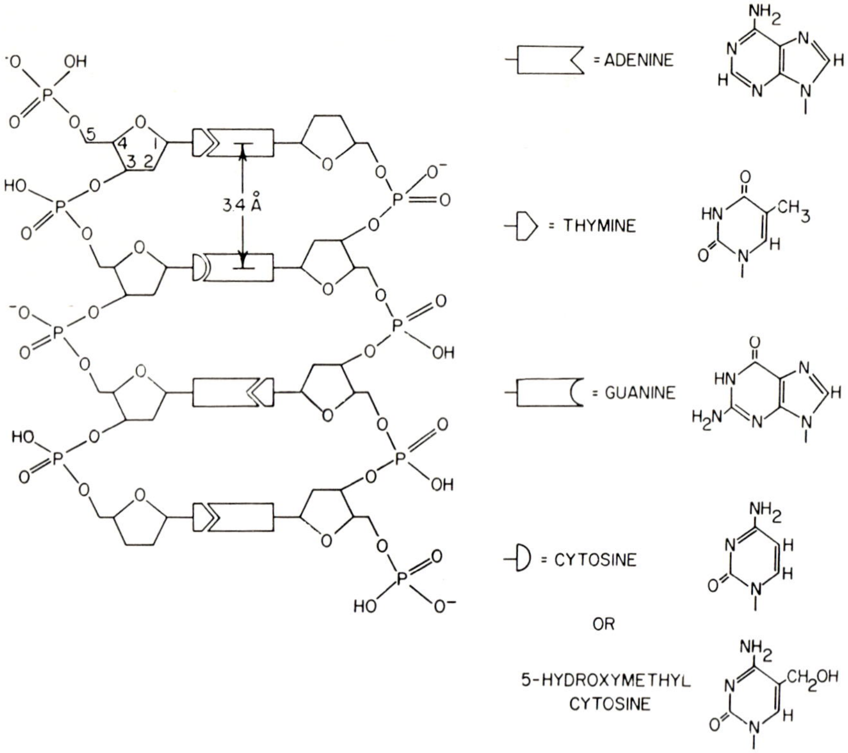

FIG. 6. A schematic drawing of the organization of DNA showing how steric fitting of complementary bases holds the double helix together and how each chain is linked together by phosphate diesters linking the 3-carbon atom of each deoxyribose to the 5-carbon atom of the next deoxyribose. Note that the sequence on one chain is the reverse of that on the other. On the right are given the structures of the various bases with the position of the bond connecting a nitrogen atom of each base to the 1-carbon of its deoxyribose. In the actual structure each base lies perpendicular to the axis of the double helix formed by the sugar-phosphate-sugar links. The spacing between layers of bases is 3.4 A. and there are thousands of bases in each chain.

the other to form a rigid helical structure. The genetic information carried by DNA has been suggested to reside in a code expressed by the variable sequence of the four bases in the chains, the actual sequence in one chain determining the sequence in the other.[32]

1. Nucleic Acids of T2, T4, and T6

The DNA of the T-even phages is unique in two respects: (1) as mentioned above, it contains 5-hydroxymethyl cytosine instead of cytosine, and (2) it contains glucose bound to the 5-hydroxymethyl group.[33] In the case of T4 all the hydroxymethyl groups are substituted with glucose, while in the case of T2 only some 77 % is substituted.[34] If, however, T2 is grown with T4 in the same bacterium, both the T2 and T4 particles liberated have all their hydroxymethyl groups linked to glucose.[35] The modified T2, labeled $\overline{T2}$, is endowed with the high efficiency of plaque formation on certain strains of bacteria that is characteristic of T4. This appears to be the first case in which a chemical structure of an isolated genetic material has been associated with a functional property.

As we shall see in Volume III, Chapter 7, the qualitative difference of the nucleic acids of T-even phages from host DNA permits one to study a number of unique aspects of the synthesis of phage DNA in infected host cells.

2. Nucleic Acids of Temperate Bacteriophages

For the bacterial DNA's all the reported ratios, adenine/thymine and guanine/cytosine, are unity. The base composition of different bacteria, however, varies over a wide range. Among different species the ratio (adenine + thymine)/(guanine + cytosine) varies from 0.4 to 2.7[36] and seems to be highly characteristic of the bacterial group.

The base composition of the DNA of the temperate bacteriophage DNA might be expected to correspond with that of the host, or at least to the composition of the part of the host's chromosome it occupies as prophage in lysogenic bacteria. The composition of temperate phage DNA does not seem as yet to have been determined.

Certain temperate bacteriophages have the property of incorporating genetic markers from the host's chromosome into their complement of DNA. These markers can then be transmitted to another bacterium in a phenomenon called "transduction." Certain transducing bacteriophages seem to pick up markers from the host chromosome at random, whereas others behave in a more directed fashion. For example, in *Escherichia coli*, strain K12, λ phage behaves as though closely linked to the Gal (galactose-fermenting) markers. Thus λ lysogenic K12 transduces only the Gal marker, as though the λ prophage took along only the Gal region of the host chromosome when lysis was induced. Transducing bacteriophage particles seldom make the transduced bacterium lysogenic, as though there were not enough room in the phage for both their own DNA and the transduced piece of the host's chromosome.

3. Structural Organization of Nucleic Acids

The manner in which the DNA is packed in heads of phage particles is an interesting problem. For example, each T-even particle contains about 5.3×10^5 atoms of phosphorus in its DNA. Assuming with Watson and Crick[31] that the DNA is in the form of a double helix with 3.4 A. between the layers of bases, one obtains $½ \times 5.3 \times 10^5 \times 3.4 = 900,000$ A. = 90 μ for the total length of DNA in the head of each particle. Since the head is only some 900 A. long and 600 A. wide, the DNA must obviously be folded, bent, or broken some 1000 times.

a. Sizes of Nucleic Acid Chains. The sizes of the DNA chains have been estimated by Levinthal and Thomas[37] using a very ingenious method. They grew T-even phages in media containing a certain proportion of radioactive phosphorus atoms (P^{32}) so that the phage DNA contained this same proportion of P^{32} atoms. The "hot" T-even preparations were placed in photographic emulsions of a type used for studying nuclear reactions and after an appropriate interval the emulsions were developed. The disintegration of each P^{32} atom produces a high energy electron which creates a track of silver grains in the emulsion. Levinthal and Thomas found that "hot" phage particles produced stars in the emulsion: the number of stars corresponded to the number of particles and the average number of tracks per star was proportional to the number of P^{32} atoms that had disintegrated in the emulsion before it was developed. To determine the size of DNA molecules the "hot" phage was osmotically shocked. With shocked phage the same number of large stars was obtained in the emulsion, but the average number of tracks per star was reduced to about 40% of the controls with intact "hot" phage. This result indicates that some 60% of the DNA of each particle exists in small fragments that are dispersed by osmotic shock while the remaining 40% exists in a form that is not broken up by osmotic shock. The length of the long piece would therefore be about 36 μ (360,000 A.) and in the particle would contain 400 or more bends. The long piece seems to be the carrier of phage inheritance. The function of the smaller fragments is uncertain.

b. The Nucleic Acid of φX 174. The nucleic acid of φX 174, the smallest bacteriophage examined so far, consists of only 5500 nucleotides.[16] According to the Watson-Crick model it would be $½ \times 5500 \times 3.4 = 9300$ A. long and would need at least 30 bends to fit into a particle whose diameter is only 300 A. This DNA seems to differ from that of the T-even phages in at least two respects however: (a) its density is greater than either the even-numbered phages or calf-thymus DNA, and (b) it shows an "abnormal" radiosensitivity.

c. Decay of P^{32} Incorporated into Nucleic Acid: "Suicide." As the P^{32} incorporated into the DNA of phage particles decays to sulfur, the particles lose their infectivity, i.e., "commit suicide" at a rate proportional to the P^{32} content. An average of some 10 P^{32} disintegrations inactivates a T-even particle's infectivity. This indicates that only about $1/10$ of the phosphorus atoms are essential to the preservation and transmission of the information necessary to successful infection, i.e., to the production of viable phage in an infected bacterium. With φX 174, on the other hand, almost every P^{32} disintegration leads to inactivation of the particle.[38, 39] This indicates that in φX 174 almost every phosphorus atom is essential to successful infection of a host bacterium and production of daughter phage particles.

These results suggest that the DNA of φX 174 has novel features in either structure and/or function. One possibility is that φX 174 might contain a higher proportion of essential DNA than T-even phages. On the other hand, the φX 174 DNA may be single-stranded, in which case, lacking the redundancy of a double helix, its content of information would be more sensitive to P^{32} decay than the double-stranded Watson-Crick helix.[39a]

D. Minor Components in Phage Heads—Neutralization of the Nucleic Acids

In most biological systems the strongly acidic nucleic acids are neutralized by associated basic substances. Phage nucleic acid seems to be no exception.

Osmotic shock of the T-even phages liberates small amounts of at least three substances besides nucleic acid. One is a peptide reported by Hershey.[40] The other two are strongly basic substances that might neutralize some of the acid groups of the DNA. For each mole of phosphate in the T4 DNA there is released 0.2 mole of the diamine, putrescine ($NH_2CH_2CH_2$-$CH_2CH_2NH_2$), and 0.04 mole of the polyamine, spermidine ($NH_2CH_2CH_2$-CH_2NH—$CH_2CH_2CH_2CH_2NH_2$).[41] The amino groups on these bases could neutralize about half the strongly acidic phosphate groups of the phage nucleic acid.

In addition, osmotic shock of T2 is reported to release a strongly basic protein that may neutralize more of the phosphate groups in the head. This protein is antigenically distinct from the other proteins of T2. It seems to be specific for T2 since osmotically shocked T4 and T6 react only very weakly with antibodies against this protein.[42]

E. The Proteins of Phage Tails

The structures of phage tails are complex and most remarkable in the ways they function during the infectious process.

1. Localization of Adsorption Sites

The sites responsible for the specific adsorption of phage particles on their host bacteria are located at the tip of the tail. This is shown by electron microscopy of the adsorption of phage T4 on its host (Fig. 3): in the presence of L-tryptophan, of which five or more molecules are required for the activation of each particle,[43] many particles can be seen adhering to host bacteria by the tips of their tails, while in control preparations made without this cofactor, particles are not adsorbed.[14] All phages studied so far adsorb to host bacteria by their tails.

The protein "ghosts" of the bacteriophages T2, T4, and T6 prepared by osmotic shock have some of the properties of whole phage particles. They are adsorbed onto susceptible bacteria and kill them without the production of new phage.[44] The nature of this killing action of the phage protein is not well understood. Ghosts of the temperate bacteriophages are also adsorbed on their hosts, but do not kill them.

2. The Number of Adsorption Sites—Phenotypic Mixing of T2 and T4

In the T-even phages there appear to be two or more adsorption sites per particle, as suggested by the following observations.[45] When T2 and T4 are grown together in a single bacterium there are many types of infectious particles produced by a process called "phenotypic mixing." Besides producing standard T2 and T4 particles, this cross produces particles with T2 genetics that will adsorb only on B/2 to produce T2. These particles evidently have the adsorption sites of T4 but the genetic make-up of T2. The cross also produces particles with T2 genetics that will adsorb on and infect *either* B/2 or B/4 to produce T2 daughter particles. They therefore must have two kinds of sites: T4 sites responsible for adsorption on B/2, and T2 sites responsible for adsorption on B/4. Similarly there are particles with T4 genetics from the T2 × T4 cross that will adsorb on and infect only B/4, and others with T4 genetics that will adsorb on and infect either B/2 or B/4. The latter must therefore have had adsorption sites characteristic of both T2 and T4 particles while the former had sites characteristic of T2 particles. It would appear that the materials responsible for specific adsorption of daughter particles are produced in a pool in the infected bacterium and are randomly associated with the genetic material of phage particles during their maturation. It is inferred that two or more host-range specific sites are incorporated into each particle.

Phage-neutralizing antibodies evidently react with the adsorption sites since neutralized phage is not adsorbed on host bacteria and neutralizing antibody causes tail-to-tail agglutination of the particles.[27] Furthermore, the specificity of neutralizing antibodies against the related phages T2 and T4 is inseparable from the host range of the phages as it relates to their adsorption on bacteria.[46]

3. Contractile Properties of T-Even Tails

The tail structures of T-even phages also have some very interesting contractile properties. When treated with a number of reagents including cyanide complexes of cadmium or zinc,[47] hydroxyl amine, a mixture of hydrogen peroxide and ethanol, or with host cell walls,[48] a sheath in the

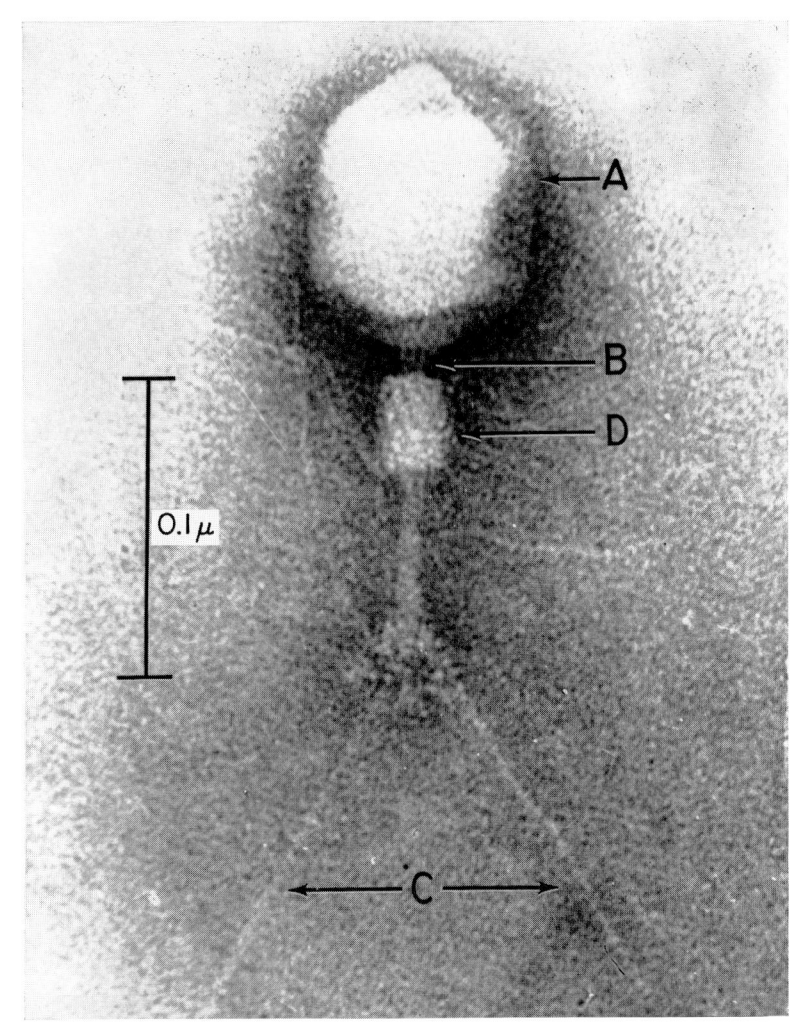

Fig. 7. An electron micrograph of a T-even phage particle that had been treated with H_2O_2 to cause the sheath of the tail to contract. This specimen as well as those shown in Figs. 8 and 9 have been prepared by adding phosphotungstic acid to the preparations and allowing them to dry. Since the phosphotungstic acid has a much higher density than the organic material of which the virus is composed, the organic material appears bright in the positive prints shown, while the phosphotungstic acid appears dark. In this figure one can distinguish: the head (A), its attachment to one end of the core (B), the tail fibers attached to the other end of the core (C), and the contracted sheath which surrounds the core (D). On the sheath, diagonal striations can barely be seen making an angle of about 30° to its axis. The author is greatly indebted to S. Brenner, G. Streisinger, R. W. Horne, S. P. Champe, L. Barnett, and S. Benzer[48a] for making this and the following two figures available before publication.[48a] (Magnification: ×400,000)

tail contracts to expose a central core or spike about 100 A. in diameter that projects about 500 A. The contraction is thought to be related to the breaking of thiol ester bonds. There are some 120 molecules of adenosine triphosphate (ATP) plus deoxy ATP tightly bound to the tail protein of intact phage, whereas in phage particles that have reacted with host cell walls 70% of these energy-rich nucleotides are hydrolyzed to adenosine

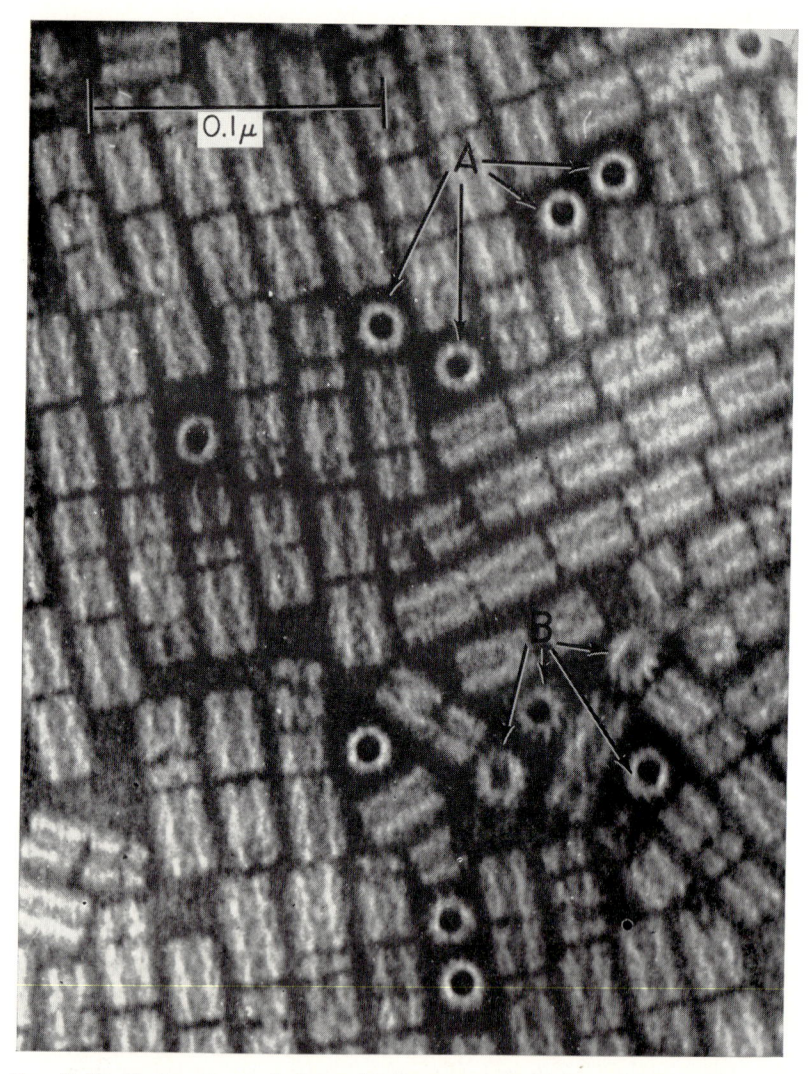

FIG. 8. Purified contracted phage sheaths, most of which are lying on their sides. A few, however, are standing squarely on their ends so the hollow centers can be clearly seen (A). Others are inclined so that the helical channels and projections on their surfaces are seen in relief (B).[48a] (Magnification: ×400,000)

diphosphate (ADP), deoxy ADP, and inorganic phosphate.[49] This breakdown is inhibited by sodium versenate. The tails of the T-even phages thus resemble in some respects the myosin from muscle of higher organisms. In phage tails, however, some of the nucleotide is deoxy ATP instead of the ordinary ATP which usually serves as an energy-rich compound in higher organisms. Since the chemical difference between deoxy ATP and ATP is small, it is interesting to speculate that the deoxy ATP was incorporated "by mistake" from the precurser pool for phage DNA.

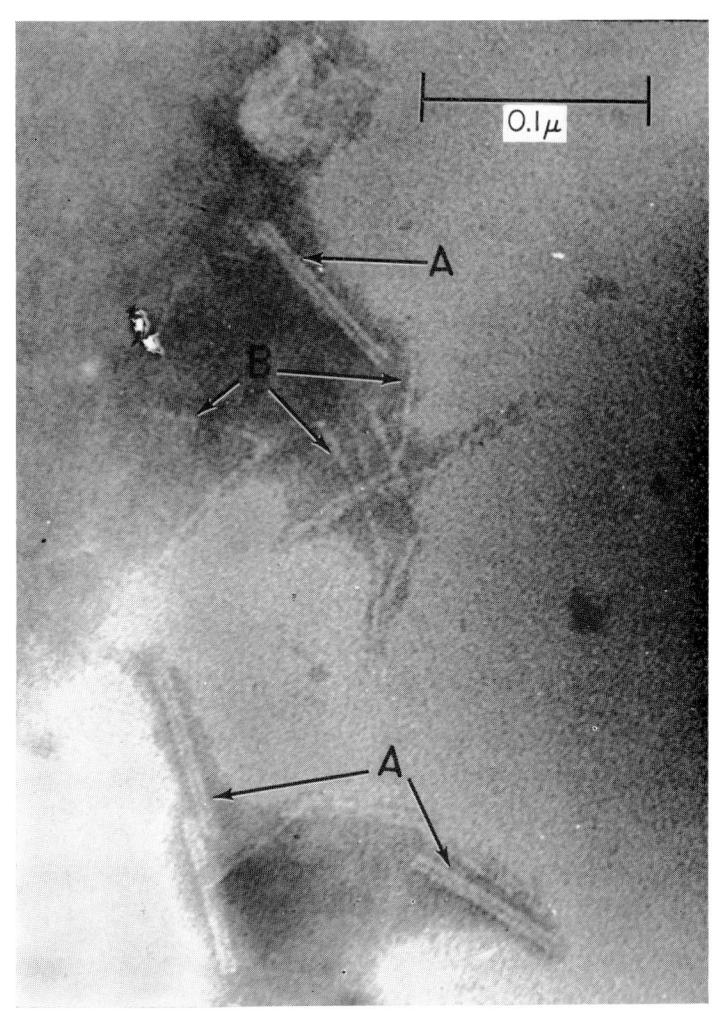

Fig. 9. A mixture of cores (A) and tail fibers (B). The holes running through the cores are filled with phosphotungstic acid and so appear dark. The 1300 A long tail fibers have characteristic kinks near their centers.[48a] (Magnification: ×300,000)

In high resolution electron microscopy[48a] the fine structure of H_2O_2-treated phage (Fig. 7) shows the head attached to the thin core which in turn is surrounded by the contracted sheath. At the other end of the core are attached a number of fine tail fibers by which the phage is thought to be specifically attached to the host. After purification, the contracted sheaths show the remarkable structure shown in Fig. 8, looking like cylindrical beads when viewed from the side and like tiny worm gears when viewed at an angle slightly inclined to the axis of the cylinder. Cores and tail fibers are shown separately in Fig. 9. The cores appear to be hollow cylinders, 800 A. long, and 80 A. in diameter with the holes running through their centers just big enough (25 A.) for DNA to pass through. The 1300-A. long tail fibers appear to be about 25 A. in diameter and to have kinks near their centers that give them a characteristic "V" shape.

III. Relation of Structure to Function—Mechanism of Infection

A. Adsorption

1. Kinetics

When a suspension of bacteriophage is mixed with a suspension of susceptible host bacteria and the bacteria are sedimented by centrifugation at intervals, the free virus titer in the supernatant (T) falls at a rate proportional to the bacterial concentration (B):

$$-\frac{d(T)}{dt} = k_1 (B) (T)$$

The value of the adsorption rate constant k_1 for many phages approaches but does not exceed the value calculated for the rate of diffusion of the bacteriophage particles to the bacterial surface. The rate of adsorption is therefore considered to be diffusion-limited; thus under optimal conditions a phage particle adsorbs to the host bacterium on almost every contact.

Adsorption occurs in at least two steps.[50] The initial step, which is rapid even at 0° C., involves a reversible, presumably electrostatic association of the phage with the host bacterium. At higher temperatures the first step is followed by a second irreversible step which definitely commits the particle to an attack on the bacterium with which it is associated.[51] Presumably the second step involves the specific steric fitting of the adsorption sites on the particle to receptor sites on the host.

2. Receptor Sites on Bacterial Surfaces

The receptor sites for many phage strains exist on the rigid cell wall of the host, for many phages are readily adsorbed on ghosts of receptive, susceptible bacteria from which the cytoplasm and its membrane have been removed by sonic vibration or by autolysis and enzymic digestion. Since a single bacterium can adsorb a hundred or more phage particles of any one kind, the cell wall of a bacterium must be covered with hundreds of receptor sites for each of the phages to which it is sensitive.

Receptor sites for T5 have been purified and proven to consist of lipoglycoprotein particles in the form of spheres about 200 A. in diameter. These spheres attach specifically to the tails of T5 particles whereupon the particles are inactivated and lose their DNA.[52] Each receptor site can react with only one T5 particle and vice versa. Similar particles can be isolated from ghosts of the T5-resistant strain B/1,5, but they do not react with T5 particles. No chemical differences were demonstrated between particles from B and B/1,5, which suggests that their biological activities depend on minor differences in composition or in structure; the serological properties of the two are different.[53]

3. Digestion of the Cell Wall

Shortly after adsorption, bacteriophages like T2 and T4 begin to digest the material of the cell wall,[54] as is clearly visible in the electron microscope and by chemical analysis, which shows a large proportion of the cystine to be removed. The phenomenon of early lysis of bacteria (lysis from without), induced by adsorbing large numbers of phages on them, may be due to the rapid digestion of cell walls by many particles. It also seems likely that injection occurs through a hole made in the cell wall.

B. Injection of Phage DNA

The infection process appears to occur by an injection of the phage DNA into the cytoplasm of the bacterium. Using tracer techniques, Hershey and Chase[55] found, within experimental error for the T-even phages, that the protein shell of the phage remains on the outside, where it can be shorn off in a blender without affecting the production of daughter phage particles inside the infected bacterium. The particles so obtained have empty heads and short, thick tails[55a] like those seen after treatment of free particles with reagents like $Cd(CN)_2$ or H_2O_2 + alcohol but without the spike seen after the latter treatments. Apparently contact with the host causes the "muscle" in the tail to contract. In accord with Hershey and Chase's tracer results, the heads have lost their DNA and are empty. It is as though the T-even phage protein were a tiny disposable syringe with a built-in muscle for delivering the DNA specifically to susceptible host bacteria.[56]

The mechanism of injection of DNA by the other bacteriophages is not well understood.

The minimum time required for adsorption and injection of T4 DNA into the bacterium (where the virus activity is resistant to osmotic shock) is of the order of 15 seconds with an average time of some 3 minutes at 33° C.; and of the order of 7 minutes at 15° C.[21]

It should be pointed out that before the phage DNA gets into host protoplasm two cell layers must be penetrated: the rigid cell wall on which the particle is adsorbed, and the cytoplasmic membrane.[57] As yet, it is not clear whether the phage particle delivers its DNA directly into the protoplasm or into the space between the cell wall and the cytoplasmic membrane. If the DNA were delivered into the space between the two layers it would probably be able to infect the bacterium, for osmotically shocked phage can penetrate and infect host protoplasts from which the cell wall has been removed.[58] Such disrupted phage cannot infect intact bacteria. On the other hand, intact phage particles cannot infect protoplasts, for without receptor sites the phage cannot be adsorbed and inject its DNA.

To summarize, normal, intact phage seems designed to infect its host by adsorption on the cell wall and injection of its DNA. If the cell wall is removed from the host, the normal interactions cannot occur; but then the steps can be bypassed by disrupted phage whose DNA, being liberated from the phage envelope, can enter and infect the protoplasts of its normal host. Such disrupted phage can even infect protoplasts of many bacterial strains that normally lack receptor sites and are therefore as whole cells resistant to the intact phage.

If the bacterium is already infected by one of the T-even phages or by T5, a peculiar phenomenon occurs when the same or another phage is added 5 minutes later. Then the DNA of this so-called "suprainfecting" phage is depolymerized and liberated into the medium.[59] This remarkable phenomenon may owe its origin to the presence of a deoxyribonuclease brought into activity by the phage having priority in the host. Previous infection of the host with T1, T3, or T7 does not stimulate this mechanism, but nevertheless bacteria already infected with these phages cannot be suprainfected with other phages.

C. Effects of Infection on Properties of the Host

A number of remarkable cytological changes follow infection of host bacteria with phage. Infected bacteria soon become leaky and liberate a number of substances into the medium as though their membranes had become permeable to intracellular materials.[60] After infection with T-even phages light scattering by the bacteria decreases markedly and then increases again before lysis occurs.[61] Also the "nuclear equivalents" of bacteria are broken down and dispersed after infection.[62] Concomitant with these phenomena, the respiration of T-even-infected bacteria remains at the level reached at the time of infection rather than increasing logarithmically as in normal growing cultures.[63] Infected bacteria therefore do not synthesize more respiratory enzymes, nor are they able to synthesize adaptive enzymes.[64] Infection therefore seems to halt the syntheses of bacterial substances, in preparation for the diversion of syntheses to the production of substances characteristic of the infecting phage. The details of these and other phenomena associated with phage proliferation will be dealt with in later chapters.

IV. Importance of Bacteriophages in Bacteriology

Studies of obligate parasites reap a double harvest: not only do they increase knowledge of the parasite itself but, by their very nature, such studies increase our knowledge of the host. The study of the bacteriophages has been no exception.

This chapter has shown how the bacteriophages have served as powerful tools for discovering such bacterial properties as genetic mutability and the nature of bacterial cell walls. Many more properties uncovered by research on bacteriophage will be dealt with in later chapters. Here we will briefly discuss a few of the more general aspects.

A. Typing Bacteria with Bacteriophages

The bacteriophages are invaluable in practical bacteriology, particularly in epidemiological studies. The typing of pathogenic strains of enteric bacteria is an example.[65] The bacteriophages have two great advantages over older methods of typing: (a) they are remarkably stable and (b) they reveal differences between bacterial cultures that cannot be brought out by known serological methods. The use of bacteriophages has uncovered 33 types of Vi *Salmonella typhi* and doubtless more types will be discovered as the work progresses. They are equally useful for typing *S. paratyphi* B and staphylococci.

Furthermore, many bacterial strains are characterized by the types of bacteriophages they carry in lysogenic form. For example, pathogenic strains of *Corynebacterium diphtheriae* are lysogenic for the bacteriophage β.[66]

B. Bacteriophages in the Classification of Bacteria

The classification of bacteria poses many difficulties because, in the absence of sexual mechanisms in most species, it is impossible to perform the classic biological tests of intraspecies fertility. Antigenic cross reactions can show misleading relationships—as in the case of Proteus OX-19 and typhus rickettsia. Furthermore, by selecting successive mutants of an organism, its chemical, fermentative, or even morphological characteristics can be made to mimic those of organisms classified in an entirely different group. Bacteriophages are useful for the classification of bacteria; for a particular phage will multiply on a limited range of closely related bacteria in which it finds both the receptor spots for adsorption and the protoplasmic elements necessary for the proliferation of its DNA and the maturation of its particles.[67] Also, in the case of the temperate bacteriophages at least, the phage DNA appears to find a region on the host's chromosome that is so nearly homologous to a region on its own that its DNA can perch there and reduplicate whenever the host's DNA does so. Two bacterial strains that can become lysogenic for the same bacteriophage must therefore be rather closely related. The absence of such a common property, however, provides no evidence for a lack of relatedness, for a single bacterial mutation can destroy a bacterium's capability of becoming lysogenic for a given phage.

Another area in which phages can be useful in showing relatedness (but not unrelatedness) is to be found in the phenomenon of transduction, by which a phage particle maturing in one bacterium is able to incorporate a part of the host's chromosome and transmit it to another bacterium where it is incorporated into the latter's genome.[68, 69] Two bacterial strains that can give expression to the same genes would seem to be rather closely related. The bacteriophages thus seem to provide substitutes for the classic tests of sexual fertility. Their usefulness in classification would be increased if large collections of type phages could be made available for study in the same way type cultures of bacteria are.

It may well be that in nature transduction between asexual bacteria is the primary means of genetic communication between them and provides opportunities for a partial but effective exchange of genetic material that is akin to a sexual process.

REFERENCES

[1] F. W. Twort, *Lancet* **2**, 1241 (1915).

[2] F. d'Herelle, *Compt. rend.* **165**, 373 (1917).

[3] M. H. Adams, in "Methods in Medical Research" (J. H. Comroe, ed.), Vol. 2 Chapter 1, p. 1. Year Book, Chicago, Illinois, 1950.

[4] S. E. Luria and M. Delbrück, *Genetics* **28**, 491 (1943).

[5] S. E. Luria, *Genetics* **30**, 89 (1945).

[6] M. Demerec and U. Fano, *Genetics* **30**, 119 (1945).

[7] E. L. Ellis and M. Delbrück, *J. Gen. Physiol.* **22**, 365 (1939).

[8] F. d'Herelle, "The Bacteriophage and its Behavior." Williams & Wilkins, Baltimore, Maryland 1926.

[9] A. Lwoff, *Bacteriol. Revs.* **17**, 269 (1953).

[10] G. Bertani, *Advances in Virus Research* **5**, 151 (1958).

[11] P. Fredericq, *Ann. Rev. Microbiol.* **11**, 7 (1957).

[12] S. E. Luria, D. K. Fraser, J. N. Adams, and J. W. Burrous, *Cold Spring Harbor Symposia Quant. Biol.* **23**, 71 (1958).

[13] T. F. Anderson and A. H. Doermann, *J. Gen. Physiol.* **35**, 657 (1952).

[14] T. F. Anderson, *Am. Naturalist* **86**, 91 (1952).

[15] D. K. Fraser and R. C. Williams, *J. Bacteriol.* **65**, 167 (1953).
[16] R. L. Sinsheimer, *Eng. and Sci.* **21**, 21 (1958).
[17] V. Kottmann, *Arch. ges. Virusförsch.* **2**, 388 (1942).
[18] T. F. Anderson, E. L. Wollman, and F. Jacob, *Ann. inst. Pasteur* **93**, 450 (1957).
[19] S. E. Luria and T. F. Anderson, *Proc. Natl. Acad. Sci. U.S.* **28**, 127 (1942).
[20] M. H. Adams, *J. Bacteriol.* **64**, 387 (1952).
[21] T. F. Anderson, *Cold Spring Harbor Symposia Quant. Biol.* **18**, 197 (1953).
[22] S. E. Luria, M. Delbrück, and T. F. Anderson, *J. Bacteriol.* **46**, 57 (1943).
[23] T. F. Anderson, C. Rappaport, and N. A. Muscatine, *Ann. inst. Pasteur* **84**, 5 (1953).
[24] T. F. Anderson, *J. Appl. Phys.* **21**, 70 (1950).
[25] R. M. Herriott, *J. Bacteriol.* **61**, 752 (1951).
[26] K. G. Lark and M. Adams, *Cold Spring Harbor Symposia Quant. Biol.* **18**, 171 (1953).
[27] F. Lanni and Y. T. Lanni, *Cold Spring Harbor Symposia Quant. Biol.* **18**, 159 (1953).
[28] G. R. Wyatt and S. S. Cohen, *Biochem. J.* **55**, 774 (1953).
[29] G. R. Wyatt, *Cold Spring Harbor Symposia Quant. Biol.* **18**, 133 (1953).
[30] M. H. F. Wilkins, A. R. Stokes, and H. R. Wilson, *Nature* **171**, 738 (1953).
[31] J. D. Watson and F. H. C. Crick, *Cold Spring Harbor Symposia Quant. Biol.* **18**, 123 (1953).
[32] G. Gamow, A. Rich, and M. Ycas, *Advances in Biol. and Med. Phys.* **4**, 23 (1956).
[33] R. L. Sinsheimer, *Science* **120**, 551 (1954).
[34] R. L. Sinsheimer, *Proc. Natl. Acad. Sci. U.S.* **42**, 502 (1956).
[35] G. Streisinger and J. J. Weigle, *Proc. Natl. Acad. Sci. U.S.* **42**, 504 (1956).
[36] Ki Yong Lee, R. Wahl, and E. Barbu, *Ann. inst. Pasteur* **91**, 212 (1956).
[37] C. Levinthal and C. A. Thomas, Jr., *Biochim. et Biophys. Acta* **23**, 453 (1947).
[38] I. Tessman, E. S. Tessman, and G. S. Stent, *Virology* **4**, 209 (1957).
[39] I. Tessman, *Biophys. Soc. Program and Abstr. 1958 Meeting* p. 42 (1958).
[39a] G. S. Stent and C. R. Fuerst, *Advances in Biol. and Med. Phys.* **7**, (in press).
[40] A. D. Hershey, *Virology* **4**, 237 (1957).
[41] B. N. Ames, D. T. Dubin, and S. M. Rosenthal, *Federation Proc.* **17**, 181 (1958).
[42] L. Levine, J. L. Barlow, and H. Van Vunakis, *Virology* **6**, 702 (1958).
[43] T. F. Anderson, *J. Bacteriol.* **55**, 637 (1948).
[44] R. M. Herriott and J. L. Barlow, *J. Gen. Physiol.* **41**, 307 (1957).
[45] G. Streisinger, *Virology* **2**, 388 (1956).
[46] G. Streisinger, *Virology* **2**, 377 (1956).
[47] L. M. Kozloff and M. Lute, *J. Biol. Chem.* **228**, 511 (1957).
[48] E. Kellenberger and W. Arber, *Z. Naturförsch.* **10b**, 698 (1955).
[48a] S. Brenner, G. Streisinger, R. W. Horne, S. P. Champe, L. Barnett, and S. Benzer, *J. Mol. Biol.* **1**, (in press).
[49] L. M. Kozloff, *Federation Proc.* **17**, 257 (1958).
[50] T. F. Anderson, *Botan. Rev.* **15**, 464 (1949).
[51] T. T. Puck, *Cold Spring Harbor Symposia Quant. Biol.* **18**, 149 (1953).
[52] W. Weidel and E. Kellenberger, *Biochim. et Biophys. Acta* **17**, 1 (1955).
[53] W. Weidel and G. Koch, *Z. Naturförsch.* **10b**, 694 (1955).
[54] G. Koch and W. Weidel, *Z. Naturförsch.* **11b**, 345 (1956).
[55] A. D. Hershey and M. Chase, *J. Gen. Physiol.* **36**, 39 (1952).
[55a] C. Levinthal and H. W. Fisher, *Cold Spring Harbor Symposia Quant. Biol.* **18**, 29 (1953).
[56] A. D. Hershey, *Cold Spring Harbor Symposia Quant. Biol.* **18**, 135 (1953).

[57] E. Kellenberger and A. Ryter, *J. Biophys. Biochem Cytol.* **4**, 323 (1958).
[58] J. Spizizen, *Proc. Natl. Acad. Sci. U.S.* **43**, 694 (1957).
[59] R. G. French, S. M. Lesley, A. F. Graham, and C. E. Van Rooyen, *Can. J. Med. Sci.* **29**, 144 (1951).
[60] C. D. Prater, Ph.D. Thesis, University of Pennsylvania, Philadelphia, Pennsylvania (1951).
[61] A. H. Doermann, *J. Bacteriol.* **55**, 257 (1948).
[62] S. E. Luria and M. L. Human, *J. Bacteriol.* **59**, 551 (1950).
[63] S. S. Cohen and T. F. Anderson, *J. Exptl. Med.* **84**, 511 (1946).
[64] J. Monod and E. Wollman, *Ann. inst. Pasteur* **73**, 937 (1947).
[65] A. Felix, *Bull. World Health Organization* **13**, 109 (1955).
[66] V. J. Freeman, *J. Bacteriol.* **61**, 675 (1951).
[67] B. A. D. Stocker, *J. Gen. Microbiol.* **12**, 375 (1955).
[68] N. D. Zinder and J. Lederberg, *J. Bacteriol.* **64**, 679 (1952).
[69] N. D. Zinder, *Cold Spring Harbor Symposia Quant. Biol.* **18**, 261 (1953).

Chapter 9

Antigenic Analysis of Cell Structure

E. S. Lennox

I. Introduction	415
A. Useful Properties of Antisera	415
B. Kinds of Serological Reactions	416
C. Problems Explorable with Serological Techniques	417
II. Preparation of Antisera	417
A. Choice of the Animal	418
B. Collection, Storage, and Fractionation of Sera	418
C. Preparation of the Antigen for Injection	419
D. Injection Routes, Schedules, and Dosages	420
E. Test Bleedings and Terminal Bleedings	422
III. Quantitative Methods of Using Antisera	423
A. Reactions in Liquids	423
B. Reactions in Gels	424
IV. Applications of Serological Techniques to Problems of Bacteriology	426
A. Serum-Gel Techniques	426
B. Uses of Antisera to Localize Cellular Components	426
C. Uses of Antisera to Determine Chemical Basis of Cross Reaction	428
D. Antibodies to Enzymes and Proteins Related to Them	429
E. Antibodies to Bacteriophage	438
References	439

I. Introduction

A. Useful Properties of Antisera

The immune response of several animals provides a way of preparing specific reagents useful for the identification, enumeration and quantitative determination of various bacterial components.

These reagents, the specific antisera, may be used not only with soluble material, but also for particulate material such as whole cells, cell walls, and protoplasts. The usefulness of these reagents lies in (a) their high degree of specificity, (b) their sensitivity, (c) their relative ease of preparation, and (d) the range of materials for which they may be prepared.

The visualization of a reaction between an antiserum and the agent that elicited it can be made in several ways. These methods are first classified regarding their occurrence in liquid or in solid media.

B. Kinds of Serological Reactions

1. Reactions in Liquids[1]

These occur between antibody and either soluble or particulate antigens. Soluble antigens, or antibodies directed to them, are usually assayed by methods which involve determination of the quantity of specific precipitate by any of a number of different methods for protein determination. With particulate antigen, e.g., whole cells, the reaction of the antigen with antiserum is assayed by the appearance of flocculation which alters the settling pattern of this particulate antigen.

The test for a soluble antigen can be converted into an agglutination type reaction by methods which couple the antigen to the surface of particles, e.g., sheep erythrocytes,[2,3] which are then agglutinated by the specific antiserum. Such techniques gain in sensitivity, lose in specificity and in quantitation. Agglutination reactions are usually run as end-point reactions which are not always easy to quantitate in terms of weight of antibody.

One should also note another reaction in liquid, namely, staining with fluorescent antibodies.[4] In this technique a fluorescent compound (usually fluorescein isocyanate) is coupled to the specific gamma globulin with little loss in activity of the gamma globulin. The attachment of antibody then stains the antigen with fluorescent material visible in a microscope especially equipped for this work. The method is of course most useful for particulate antigens. As a specific stain for cell wall material it becomes an extremely useful method for classification of bacteria once a group of fluorescent reference sera are obtained. When cells larger than bacteria are used the method is useful as a method of localizing a specific antigen within the cell.*

2. Reactions in Solid Media.[5]

These reactions are usually restricted to soluble antigens. Antigen, antibody, or both are allowed to diffuse into a solid medium—usually an agar gel. Reaction of antigen with antibody is seen as an opaque band of precipitation in the solid medium. Quantitative assays of antigen or antibody depend upon quantitative densitometry of the zones of precipitation.† For complex mixtures, the advantages of these methods over liquid reactions are several. (a) Visualization as a set of precipitin bands of a multi-component antigen mixture is easy. (b) One may easily ascertain that two

---

* Fluorescent reagents and microscopy equipment are available from Scientific Products, 1210 Leon Place, Evanston, Illinois.

† An instrument for this purpose is available from American Instrument Co., Silver Springs, Maryland.

preparations contain antigens in common which either cross react or are identical. (c) A fairly clear idea of the molecular weight of an antigen[5, 6] (or at least its diffusion coefficient) may be obtained from measurements of positions and intensities of the bands of precipitate. (d) Less information is required about the concentration of antibody in an antiserum and also about the concentration of antigen components than would be necessary for quantitative reactions in liquid.

Details of the use of these methods in studying bacteria will be given below.

C. Problems Explorable with Serological Techniques

It should be emphasized at the outset that a material can be detected by serological methods insofar as it or an antigenically similar substance can elicit the formation of antibodies in a suitable animal. Some substances may be antigenic in one animal but not in another. Also, one must remember that identity of two components by serological methods implies a different criterion from other methods such as ultracentrifugation, electrophoresis, chromatography, etc., and is a method supplementary to these.

Some of the problems of bacteriology explorable with the techniques of serology are as follows: (a) Is a given protein (or polysaccharide) preparation homogenous? (b) What are the kinetics of formation of a certain protein? (c) Two enzyme activities are exhibited in the same preparation—are they associated with the same protein? (d) Two strains of bacteria are to be compared—are their cell constituents identical by serological criteria? (e) Two enzymes with similar activities are isolated from the same or different strains of bacteria—are these activities associated with the same antigenic entity? (f) A bacterial strain mutates to lose a certain enzymic activity—is this due to formation of an altered protein, enzymically inactive, cross reacting serologically, or to the lack of formation of a protein similar to the enzyme? We shall ignore in most of what follows the uses of serology in bacteriological classification since this has been adequately discussed in many places.[7] For the same reason toxin-antitoxin reactions,[8, 9] which have indeed formed the foundations of serology, will be only briefly discussed. The discussion will concern mainly those uses of serological techniques which will prove useful to the bacteriologist concerned with problems of genetics, of synthesis of new cell materials, and of localization of antigenic components within cells.

II. Preparation of Antisera

Since the utility of a serological method starts with well-prepared and assayed antiserum, one must deal first with the problems of preparing and testing antisera. These problems are well discussed in the excellent article of Cohn; therefore only some simple rules are included here. In the matter of injecting animals for antiserum production, the novice is provided in the literature with a wealth of diverse recipes and finds himself extremely confused in deciding the bare essentials. One must admit at the outset in the matter of immunization procedures, that an insufficient number of tightly controlled experiments have been done to know which of the parameters usually varied are important. The following comments are addressed to those investigators, relatively unskilled in the methods of serology, who want some simple recipes, and who want to know what kind of problems lend themselves to the special techniques of serology.

A. Choice of the Animal

For ease of handling, for general responsiveness to a wide variety of antigens, and for ease of handling of the sera, the chicken and the rabbit are the animals of choice. From each of these animals one can draw enough blood to yield 50 to 60 ml. of serum on terminal bleeding or can draw repeatedly enough blood (about 50 ml.) to yield about 25 ml. of serum. In case only small amounts of sera are needed and immunization against many different antigen preparations is required, the smaller animals, such as mice, rats, and guinea pigs, may be preferred.

In choosing an animal one should keep in mind that a preparation poorly antigenic in one species may elicit good response in another. Furthermore, since one animal of a given species may respond poorly, even on repeated injection, while another may respond well, it is well to immunize several animals at the same time.

B. Collection, Storage, and Fractionation of Sera

This subject, discussed very adequately in the works of Cohn,[10] Kabat and Mayer,[11] and Boyd,[7] will be summarized briefly here. The animal is fasted for about 18 hours before bleeding. Blood is generally collected and allowed to clot in 40-ml. centrifuge tubes. After incubation at 37°C. for about 1 hr., the clots are separated from the wall of the tube and the contents allowed to stand in the icebox overnight. The serum should be centrifuged and poured into a fresh tube to free from the clot and then re-centrifuged to free from the remaining cells. Centrifugation at about 8×10^3 g for 15 minutes is sufficient. For most purposes, serum can be handled without sterile precautions and then sterilized by Seitz filtration. Sterile serum can be stored at 4°C. indefinitely. For some purposes merthiolate to a concentration of 1 part in 10^4 can be added as a preservative. Merthiolate is, however, extremely difficult to remove (except by dilution); therefore one must be cautious in its addition if the serum is to be used with living cells. Serum will also keep well when frozen.

Fractionation of sera may be desirable for some purposes, e.g., for concentration of a weak serum or to free gamma globulin from other, possibly interfering, components. Several fractionation methods have been sucessfully used. Probably the easiest is that of Kekwick,[12] which uses Na_2SO_4 at room temperature, or one of the methods using $(NH_4)_2SO_4$.[13] Alcohol fractionation[14] gives a cleaner preparation, but if one is interested only in concentrating the serum it is probably not worth the added effort needed for careful control of temperature, pH, and ionic strength. These methods will give about 70% recovery of the antibody activity. The gamma-globulin fractions recovered by these methods are dialyzed against 0.5 to 0.9% NaCl, their volume measured, and convenient volumes lyophilyzed and stored in sealed ampoules.

C. Preparation of the Antigen for Injection

The methods of antigen preparation vary depending upon whether a soluble or particulate antigen is to be used. Here again an enormous variety of methods have been used successfully. Only a few, which have proved useful in other laboratories and in our own, will be included here.

1. Particulate Antigens

Cell wall antigens can be made from whole cells by washing them several times in buffered saline and resuspending to an optical density of about 1 in a 1-cm. cell at 520 mμ. Several precautions should be observed in preparing cell suspensions for use as antigens. Cells grown in liquid synthetic media are generally preferable because meat extracts and agar frequently contain antigenic substances. The concern with antigenicity of media ingredients arises from the desire to be sure that the antibodies elicited are indeed a response to factors intrinsic to the cells themselves. The question of the age of the culture, i.e., cells harvested in the exponential phase of growth or after onset of stationary phase depends, of course, upon the cell component to which one wants to immunize. In general, for anti-cellwall serum the cells are harvested not too long after stationary phase or near the end of the exponential phase. With some cells, such as the pneumococcus, careful control of age and pH must be kept or the cells lyse and fill the suspension with debris. Cell preparations after growth and washing are usually killed by heating or with formalin (see Boyd,[7] p. 638). Such antigen preparations kept refrigerated are usually stable at least for the course of injection. Some cell preparations, e.g., pneumococcus, may serve as a good antigen (in the rabbit) only so long as the preparation continues to show a Gram-positive staining reaction. It is well, therefore, to check staining characteristics and microscopic appearance of cell preparations.

2. Soluble Antigens

Of the several methods for preparing soluble proteins for injection, the most widely used and most generally successful are those involving adsorption of the material to an insoluble material, such as aluminum hydroxide or phosphate, or involving incorporation into a stable water-oil emulsion.

An alum-precipitated antigen may be prepared by the following recipe: Add to a protein solution (about 0.5% concentration) a volume of 10% $Al_2(SO_4)_3 \cdot K_2SO_4 \cdot 24H_2O$ to give a weight of alum equal to 2.5 × the weight of the protein. Then bring the pH to 7.0 by the addition of $0.1N$ NaOH to form a precipitate which binds the antigen. One should check the supernatant to ascertain that all the protein is bound. If not, more alum may be added and precipitated as before. This suspension is to be injected in quantities discussed below. There are many variants of the method of preparing the antigen suspension; whether they differ in any essential aspect is unclear.

The second generally useful method of preparing protein antigens for injection is that of Freund (see reference 10). A solution of antigen is prepared in a stable water-oil emulsion containing heat-killed mycobacterium to stimulate antibody formation. The recipe is as follows: 8.5 ml. Bayol F* containing 10 mg./100 ml. heat-killed mycobacterium; 1.5 ml. Arlacel A† plus 10 ml. of a 1 to 2% solution of antigen in buffer. The emulsion is best prepared in a high-speed homogenizer (e.g., Virtis 45). The oil and emulsifier (Arlacel A) are added to the homogenizer cup and the antigen solution added slowly as the speed of the motor is increased. The cup should be cooled during the process and the motor run at top speed for only brief intervals (10 sec.). The emulsion is ready when a drop on water shows no tendency to spread. Such preparations are stable for months in the icebox. The Bayol F-mycobacterium suspension may be sterilized by autoclaving as may the emulsifier Arlacel A. *Mycobacterium tuberculosis* or *butyricum* are the strains generally used. Mixing is accomplished by grinding the dried bacteria with oil in a mortar. This mixture can be kept indefinitely in the icebox.‡

D. Injection Routes, Schedules, and Dosages

The variety of recipes for administering antigen to an animal is again bewildering to the occasional user of serological techniques, and again the reason for this multiplicity is the same—not all the variables involved have been adequately studied.

* Stanco Distributors, New York.
† Atlas Powder Co., Wilmington, Delaware.
‡ An oil-mycobacterium mixture can be purchased from Difco Laboratories, Detroit, Michigan.

The recipes given here seem to us simple and have given sera of good titer. They are a distillate of advice from many sources (see especially reference 10).

1. Routes of Injection

The antigen in various forms, such as alum precipitates, solutions, and virus suspensions, may be injected by several routes, e.g., intravenously, intraperitoneally, into the foot pads, or subcutaneously. Oil-water emulsions are injected intraperitoneally or subcutaneously in the rabbit, under the scapula. It is the practice to vary the route of injection for the solutions and suspensions. This may prove necessary for antigen preparations which are toxic. Toxic effects are often reduced by beginning the immunization schedule with subcutaneous injections for the first few days. Other routes of injection should follow.

2. Schedules of Injection

An injection schedule which usually gives good results is: alum precipitates, solutions, and suspensions, once daily Monday through Thursday; no injections for balance of week; oil-water emulsions once a week. In general, immunization continues for about 4 weeks, a test is made, and either the animal is bled several times over the period of a week or is bled out by carotid cannulation. With animals which have not responded well within a 4-week period, the course of immunization should not be continued, for this generally leads to less specific and weak antisera. An animal which has responded well may be bled several times and saved for further injection and bleeding. Often several months after the first course of injection a very small dose of antigen will yield a relatively high amount of antibody, although this may depend on the particular animal and the particular antigen. In general, animals are bled out after the first course of injections.

3. Dosages of Antigen

Recommended dosages are almost as varied as injection schedules. Here one compromises between wasting precious antigen by injecting more than necessary and wasting it by injecting too little to elicit a response. Bacterial antisera can be prepared by injecting a suspension containing about 10^9 organism/ml. (see reference 7, p. 639 and reference 11, p. 544). The first day one injects a few tenths of a milliliter and increases it to as much as several milliliters in the final week of injection. For bacterial viruses, using a preparation purified by low- and high-speed centrifugation (about $3 \times 10^3\ g$ for 15 min., then $2 \times 10^4\ g$ for 1 hr.), one can prepare excellent antisera by injection for 4 weeks with 4×10^{11} virus particles per injection. We have had excellent results (inactivation constants $K \geqslant 1000$, see Section IV, E) for the bacteriophages T2, T5, and PLT22, by injecting 0.1 ml. of a suspension of 10^{12} particles/ml. into each hind foot pad and 0.2 ml. intravenously for 4 weeks. We do not have a sufficient number of experiments to know whether foot pad injections could be omitted, using only the intravenous route with the same total dosage. We do know that this injection schedule and dosage used in foot pads alone gives sera with almost as high titer. The total number of viruses per injection cannot be decreased very much without sacrificing serum titer. A similar injection schedule with 10^{10} particles per injection will give usable titers ($K = 100$ to 500). Bacterial virus suspensions, no matter how well purified, may contain cell material some of which is toxic. It is well, therefore, to start the course of immunization with subcutaneous injections.

Alum-precipitated protein solutions using either rabbits or chickens give good antisera when 20 to 30 mg. per animal[7, 10, 11] is injected over the course of the injection period. For a complex mixture of proteins one would give much more (ca. 100 mg.) in order to get response to minor components of the mixture. For response to minor components one immunizes for a shorter time (e.g., 3 weeks) with a larger quantity of antigen. To obtain antibodies principally against the major component one immunizes with a few milligrams of protein spread over 6 to 8 weeks.[10]

With antigens prepared by the Freund technique one injection of 10 to 15 mg. protein may suffice to give a good antiserum in a few weeks. If the serum is too weak at this time, a second injection should be given and bleeding made after 3 or 4 more weeks.

More details of injection schedules, dosages, and their variants may be found in references 7, 10, and 11.

E. Test Bleedings and Terminal Bleedings

Recommended methods of bleeding animals are given in many publications.[7, 10, 11] Nothing, however, can replace a demonstration of bleeding techniques by someone skilled in their execution. A good animal board (for example, the Latapie type, available from most scientific supply houses) is worth a lot of descriptions of how to hold animals properly. The firmness with which such a board holds rabbits or guinea pigs, during carotid bleeding or cardiac punctures, is invaluable. Chickens can be held with sandbags on the wings flattened against a table. Terminal bleeding of rabbits or chickens is most easily accomplished from the carotid artery using polyethylene tubing as the cannula. One can collect almost as much blood from a rabbit by cardiac puncture. For test bleedings, the ear veins of the rabbit, and the veins on the underside of the wing in the chicken are easiest.

Methods of testing sera for precipitating antibody are described in reference 10. A good serum contains several hundred gammas of precipitating antibody nitrogen/milliliter. Much weaker sera will be useful, however, especially for reactions in solid media. A good antibacteriophage serum should have an inactivation constant (K) (see Section IV, E) of several hundred to a thousand min.$^{-1}$, although much lower titers are usable. Some bacterial viruses (e.g., *Escherichia coli* phage T1) seem poorly antigenic in the rabbit and give $K \leqslant 50$.

Antiflagellar sera may show an end point at a dilution as high as $1:10^5$. Agglutinating antibodies to somatic antigen have end point dilutions as high as $1:10^3$ to $1:10^4$. Normal sera occasionally have agglutinin titers of about $1:10^1$ against many bacteria, therefore the serum titer must be much higher to be useful. A preliminary bleeding before the course of injection is essential. The correspondence between agglutinin titer and antibody concentration seems to depend on the particular antigen and antiserum (see, for example, reference 11, Chapter 3). The author is not aware of measurements correlating neutralizing antibody concentration with a bacteriophage inactivation constant.

III. Quantitative Methods of Using Antisera

An antiserum for which the antibody content is accurately determined is a specific quantitative reagent for measuring soluble antigen. There are several distinct methods of measuring quantity of antibody, and many variants of each. The more useful of these will be discussed here.

A. Reaction in Liquids

Measurement of neutralizing antibody is different from measurement of precipitating antibody. So also are those methods which depend upon staining, changes in optical properties of cell surfaces, flagella inactivation, and agglutination.

Precipitating antibody is measured primarily by the quantitative precipitin reaction in which accurately measured quantities of an antigen are added to an antiserum, precipitation allowed to occur, the precipitates washed and measured in one of many ways of determining protein. The method of choice against which all others should be calibrated is the Kjeldahl determination. Details of these methods are found in references 10 and 11. Several useful modifications of these techniques should be mentioned here.

The Conway modification, one of the most useful of these, employs room temperature distillation of ammonia from one part of a closed vessel to a second part containing acid in excess. The amount of ammonia absorbed by the acid is determined by titration.*, † Quantities of nitrogen down to a few micrograms can be determined by these methods.

With the Kjeldahl nitrogen determination as standard, it is possible to use many other methods for determining the extent of antigen antibody reaction. For example, the turbidity can be followed, using a Beckman D. U. spectrophotometer at 520 mμ or the Photronreflectometer.[15]‡ In the first case one measures increase in optical density; in the second, light scattering at 90° to the incident beam.

A very useful extension of the sensitivity range of the precipitin reaction, down to millimicrograms of nitrogen, was made by Glick et al.,[16] by using dye binding to the specific precipitate.

The prime utility of the precipitin reaction is in its precision. As indicated above it may be used to measure small quantities of antigenic material in the presence of contaminating substances. It is also useful for giving quantitative data on the degree of antigenic relatedness of two materials. The method loses its usefulness with mixtures of several components, which react with different antibodies, for it becomes difficult to interpret the overlapping precipitin curves. In such cases the advantages of the gel procedure, which shows multiantigen-antibody interactions as multibanded zones, predominate.

B. Reactions in Gels

The utility and limitations of precipitin reactions in gels are well discussed in the articles of Oudin[5] and of Munoz.[17] We shall just call attention here to some of the simpler, more useful techniques and emphasize their application to problems of bacteriology.

There are several ways of performing gel reactions, all of which use diffusion of antigen or antibody or both into agar§ gels where their interaction is visualized as a zone of turbidity where precipitation occurs. The following features account for gel reactions being the preferred method of analyzing complex mixtures. (1) Diffusion of the reactants allows adjustment to antigen-antibody equivalence for a wide range of starting concentrations. (2) Reactions of different components of an antigenic complex

---

* Special tubes in which digestion, diffusion, and titration can all be done are manufactured by the Microchemical Specialties Co., Berkeley, California.

† A microburette for titration can be purchased from Emil Greiner Co., New York 6, N. Y.

‡ Photronreflectometer is manufactured by American Instrument Co., Silver Springs, Md.

§ A refined agar (Oxoid Ionagar No. 2), suitable for these applications, may be obtained from Consolidated Laboratories, Inc., Chicago Heights, Illinois, or from Oxo Limited, London.

to the several antibodies occur in general at different positions in the gel; thus each component is represented by a separate band of precipitation. (3) The existence of cross-reacting components in different preparations is easily recognized (see below and reference 5). (4) Relatively small quantities of antiserum are necessary. (5) Antigenic products of bacterial growth during colony formation can be visualized by growth on agar containing appropriate antiserum.

Several agar diffusion methods have been used: In the method of Oudin,[5] one-dimensional diffusion in tubes is employed. Antiserum in agar is placed in the bottom of a tube and overlayed with antigen, concentrated as much as possible, in agar or in solution. An essential feature of this method is the constant concentration of antibody in agar throughout each tube and, for comparative purposes, from tube to tube. In this procedure the antigen can be identified, regardless of its concentration, by the intensity of the precipitated band which is determined by the antibody concentration. Thus an antigen can readily be recognized in a complex mixture.

One of the principle objections to the gel diffusion technique, as originally used by Oudin, was the occurrence of false banding caused by slight changes in temperature even for short periods. A one-dimensional diffusion technique, free of the artifacts of the original Oudin method, was published by Oakley and Fulthorpe,[18] and investigated in detail by them and by Preer[19] and Wilson.[20] In this method the antigen and antibody are separated by a portion of gel into which they both diffuse. Wilson[20] has made a detailed comparison of the Oakley-Fulthorpe method with the Oudin method in terms of sensitivity, resolving power, effective reaction time (duration of band persistence), reaction range, i.e., concentration of reactants, and freedom from artifact.

Ouchterlony[5] has devised a two-dimensional gel diffusion method employing a Petri dish with several wells for antigen and for antiserum. These reactants diffuse into the gel, interact, and, at equivalence, form a precipitate. By suitable arrangement of antigen wells around an antibody well, many samples can be examined simultaneously for antigenic components to a single antiserum. One major advantage of Ouchterlony's two-dimensional method over the single dimension procedure using tubes is the ease of recognition of the existence of complex mixtures of cross-reacting components by the fusion of their bands.[5]

An interesting and extremely useful variation of gel methods is the immunoelectrophoresis technique of Grabar and Williams,[21] as modified by Poulik.[22] This procedure first separates a protein mixture by electrophoresis in gel and then allows diffusion into a suitable antiserum.

IV. Applications of Serological Techniques to Problems of Bacteriology

A. Serum-Gel Techniques

Among the earliest applications of gel diffusion antigen-antibody reactions were the studies of Sia and Chung[23] and of Petrie[24] investigating the antigens diffusing from colonies growing on agar. Elek[25, 26] used this technique to distinguish toxigenic bacteria from nontoxin formers. Freeman[27] and Groman,[28] by plating bacteria on agar containing diphtheria antitoxin were able to study the conversion of nontoxigenic strains of *Corynebacterium diphtheriae* to toxin forming ones following infection by a lysogenizing phage. An example of the use of these methods to compare the antigens of one bacterial strain with antiserum prepared to another is the work of Olitzki and Sulitzeanu,[29] comparing extracts of *Haemophilus aegyptius* and *H. influenzae*. Antisera were prepared to the soluble portion of sonic extracts of these strains and gel diffusion tests run against the homologous and heterologous extracts to identify antigens in common by fusion of lines from adjacent wells. The relatedness of these two strains is indicated by the number of common antigens revealed in this manner.

Another method of identifying common antigens in two bacterial extracts is by preparing antiserum to one extract (A) and cross absorbing with the second extract (B). Unabsorbed serum then is put in the well of one Ouchterlony plate and the absorbed serum in a second. Extract A is placed in a well and allowed to diffuse toward the first two wells.[5] Lines that appear with interaction of A and unabsorbed serum, but which are absent or diminished in intensity with the absorbed serum are due to components serologically related in A and B. With this technique, Munoz[17] showed the presence in an extract of rough *H. pertussis* of 3 or 4 antigens in common with a smooth strain, and, moreover, the presence of 3 or 4 antigens in the smooth strain which were absent in the rough strain.

B. Uses of Antisera to Localize Cellular Components[30]

Antisera prepared by injection of whole cells are directed principally against external structures of the cell, i.e., cell wall and flagella. Such sera, used routinely for the classification of bacterial strains according to the antigenic constitution of their walls and of their flagella, form the basis for typing schemes. Distinction between reactions of antisera with cell wall material and with flagella is possible by several means: (1) Boiled cell preparations are devoid of flagellar antigenicity, but retain cell wall antigenicity. (2) Flagella detached from the cell walls by mechanical agitation and recovered by differential centrifugation retain antigenicity. (3) Cell preparations reacting with antisera at high dilution reveal flagella immobilization by antibody.

Cell fractionation techniques allow preparation of antisera to localize other cellular components. Serological comparison of cell wall and cytoplasmic membrane was made,[31, 32] for example, with protoplasts of Gram-positive bacteria prepared by lysozyme treatment. By preparing antisera to whole cells or to protoplasts, the cell wall and protoplast membrane (of *Bacillus megaterium*) were shown to be distinct, both in agglutination and absorption tests, i.e., neither could absorb more than a very small portion of the antibodies obtained by injection of the other.

There is an interesting difference in the degree of reaction of antiprotoplast serum with living bacteria as compared to heat-killed bacteria;[31] the living, flagellated preparation reacts, while heat-inactivated cells do not. This serological result is in accord with the finding of Weibull,[33] that action of lysozyme to protoplasts does not remove flagella which thus can elicit antibodies to react with living, flagellated cells.

Cell walls can be prepared by disruption of cells in a Mickle disintegrator and purification by differential centrifugation.[34-37] Antisera from injection of such preparations react strongly with whole cells and weakly or not at all with protoplasts. Antiserum prepared by injection of whole cells, and absorbed with cell wall preparations no longer agglutinates whole cells; on the other hand, antiserum to cell walls, adsorbed with whole cells, does, to a diminished extent, agglutinate the cell wall preparation.[32] Thus cell wall preparations contain antigens not accessible (to the antibody system) on whole cells.

Injection of whole cells occasionally yields antisera to soluble bacterial components. An example is the light emission system in *Achromobacter fischeri*, which consists of a long chain aldehyde, reduced flavin mononucleotide, oxygen, and a soluble bacterial protein—luciferase. Tsuji and Davis[38] observed the following: (1) Antibodies elicited by injection of whole bacteria agglutinate whole cells and cause a decrease in light emission by them, apparently not by a penetration of antibody into the cells, but only by interfering with O_2 exchange. The original brightness of the agglutinated cells can be restored by dispersing them with shaking. (2) Antisera, elicited by injection of whole cells, cell-free extracts, or partially purified luciferase inhibit light emission in cell-free extracts. Antisera to cell walls, on the other hand, do not.

These results would indicate that injection of a whole cell preparation does stimulate formation of antibodies to internal components, either because the preparation contains lysed cells at the time of injection or the preparation is lysed in the animal.

Antisera can be used to study the relationship of subcellular structures to other cell components. The chromatophores of photosynthetic bacteria, rather large structures, can be isolated from broken cell preparations by differential centrifugation. Antisera, prepared by injecting chromatophore preparations, revealed[39] a polysaccharide component in the chromatophores serologically related to a cell surface component. The chromatophores also contain proteins antigenically related to soluble cell proteins. How essential these components are to the structure remains to be reported.

One should remember that in all these experiments one needs antisera prepared against a variety of cell fractions, the purity of which is established by some other criteria in order to use these sera for localization of cell components.

The materials of cell walls, especially those of the pathogenic organisms, have been studied extensively by serological means. Details of these studies can be found in the books of Boyd[7] and of Dubos.[40] Such investigations were of the type described above, that is, they are concerned with analyses of high molecular weight, antigenic materials, and their localization in cell walls of various organisms.

C. Uses of Antisera to Determine Chemical Basis of Cross Reaction

Another level at which the methods of serology can be useful for determination of structure is in the determination of linkages and types of subunits in some of the polysaccharide antigens. An example is the investigation of the antigenic relationship of the polysaccharides of pneumococcus type XIV to the antigens of human red cells. Antisera prepared by injection of pneumococcus type XIV into a horse agglutinate human red cells of the four major blood types.[41] That the antibodies against the polysaccharides and red cells seemed to be the same was demonstrated by removing the red cell agglutinating antibodies by absorption of the serum with either type XIV polysaccharides or whole type XIV organisms. Conversely, absorption of the serum with purified blood group, blood substance A could remove almost all the type XIV activity.[42]

Clues to the molecular basis of this cross reaction were revealed by further investigation[43] indicating a strong cross reaction with a galactan from cow lung. This cross reaction had been looked for by Heidelberger and Wolfrom on the basis of the knowledge that type XIV polysaccharide contained galactose.

Subsequent investigations have revealed much about the nature of the sites responsible for these cross reactions. Details may be found in references 44, 45, and 46; some of these are summarized below.

It became clear that the sites in common between blood group substances and the type specific polysaccharides were not the ones responsible for blood group antigenicity. This was demonstrated by several main facts: (1) The ability of group A substances to precipitate antibody from horse anti-type XIV serum were independent of their potency or specificity as blood group substances. (2) Hydrolysis of group A substances by dilute HCl (2 hr.

at 100°C., pH 1.5–2.) splits off fucose and some oligosaccharide side chains, leaving a nondialyzable residue which, though having lost blood group activity, shows a marked increase in reactivity with the anti-XIV serum. (3) Treatment of the group A substance with enzyme preparations from snail liver and from *Trichomonas foetus* yielded preparations that lost blood group activity, but which had enhanced reactivity with anti-XIV serum. (4) An enzyme preparation from *Clostridium tertium* destroys A activity, leaves B and O(H) unaffected, but destroys almost completely the reactivity with anti-XIV serum. All of these facts point clearly to a difference in antigenic site between the blood group substances and the pneumococcal polysaccharide. A galactose unit had already been implicated in this cross reaction. Further evidence of the involvement of galactose was furnished by the demonstration that the enzyme preparation from *C. tertium* had β-galactosidase activity and destroyed reactivity of type XIV polysaccharide with its antiserum while releasing galactose.

The exact nature of the galactose-containing group was investigated by assaying inhibition of the precipitin reaction of type XIV polysaccharide and its antiserum by various galactosides. The concentration of the various galactosides needed to cause 50% inhibition of the precipitin-reaction was determined. By this method the compounds galactosyl 1-4-β-N-acetylglucosamine and galactosyl 1-3-β-N-acetylglucosamine were on a molar basis the most effective inhibitors. Therefore, these bases, as terminal groups, are assigned the role of the sites responsible for cross reactivity between the blood group substances and type XIV pneumococcus polysaccharide.

D. Antibodies to Enzymes and Proteins Related to Them

Many enzymes elicit formation of antibodies which react to form precipitate, permitting a serological assay of the amount of enzyme protein and of all other proteins cross reacting with it. In addition to measurement of enzyme antibody interaction by precipitation, one has the additional parameter of enzyme activity, for antibodies are often effective antagonists of enzymic activity.

1. Antibodies as Enzyme Inhibitors

The degree of enzyme inhibition by its antibodies can be measured by the enzyme activity of the precipitate, in the region of antibody excess, or of the soluble complexes, in the region of antigen excess. Some few enzymes, such as the β-galactosidase of *E. coli*, are not at all inhibited by antibody. Most others are inhibited to varying degrees. A plot of residual enzyme activity as a function of antibody added shows a region of approximately linear decrease, then a level of residual activity which changes only very slowly, or not at all, with added antibody. The residual activity for a given enzyme will in general vary from one serum to another.

There are several uses of antienzyme serum:

(1) To inhibit a specific enzyme step in a complex reaction mixture. Krebs and Wright,[47] for example, used an antiserum prepared against purified yeast triose phosphate dehydrogenase to inhibit glycolysis in a cell autolysate.

(2) To ascertain whether two (or more) enzymic activities are assignable to the same protein molecule. Suppose one has purified a protein preparation while following a certain activity (A) and finds that another activity (B) continues to be present through the purification steps. The question will arise whether these two activities are associated with molecules which are identical by serological criteria. An antiserum prepared by injection of the protein preparation will answer this question. To do this, the ratio A/B is measured in the preparation, antibody is added in varying amounts, precipitates (if any) removed by centrifugation, and the supernatants assayed for the ratio A/B. In the simplest case the equivalence points of A and B are not identical, the ratio changes, and A and B clearly belong to antigenically different species. The more difficult case occurs when the equivalence points of A and B are the same, the ratio A/B remaining unchanged as antiserum is added. In this case we would conclude that A and B are associated with the same protein species, but of course may not be at the same site on this protein.

In the first case, where A and B are separable serologically, the possibility that they are cross reacting may be investigated by the technique of agar gels.

These techniques have been successfully applied, for example, to indicate that the lactase and the β-galactosidase of *E. coli* are the same antigenic molecules[48] and that the α-toxin and the lecithinase of clostridia are the same.[49] Other examples and extensions of this technique are referred to in the article by Cohn.[1]

(3) As another specific reagent for studying mechanisms of enzyme inhibition. The excellent article of Cinader[50] reviews the results of such experiments.

An interesting application of the technique described in (2), above, was made by Pollock, who examined the serological relationship of the penicillinases of *Bacillus cereus*.[51] These enzymes had been separated into three types on the basis of the purification procedure and the degree to which they could be neutralized by an antiserum prepared to one of them (α-penicillinase). These classifications are: (a) the α-fraction—the exoenzyme released into the culture fluid by the cell culture and neutralizable to the extent of about 99% by the antiserum; (b) the β-fraction, most of which was removable from the cells by repeated washing and which was neutralized by anti-α-penicillinase; (c) the γ-fraction, the cell-bound enzyme—not neutralizable until freed from the cell and even then incompletely so (about 70%, but variable).

Other experiments, correlated with serological data indicated that the α- and β-penicillinases were identical. On the other hand, the γ-penicillinase had several properties different from the α-enzyme. These properties included activity vs. pH, sensitivity to inactivation by iodine, and solubility in ammonium sulfate solution. In spite of these differences, the close serological relationship, the comparison of substrate specificities, the Michaelis constants, and rough sedimentation rates led the investigators to question whether α- and γ-penicillinase were not the same protein in two different states. Clarification of this point came from experiments by Citri,[52] indicating clearly that a purified preparation of α-penicillinase can under the proper conditions demonstrate the iodine sensitivity and antigenic properties of γ-penicillinase. Treatment of the α-penicillinase with dilute alkali ($N/30$) resulted in a partial loss of activity and an increased sensitivity to iodine. If the exposure to alkali were brief, (10 min.), there was a partial return of activity, which activity again showed the iodine sensitivity of α-penicillinase. If, on the other hand, exposure were prolonged (240 min.) the change was irreversible. Similarity to γ-penicillinase is further indicated by the fact that antiserum which completely inhibits the untreated enzyme inhibits the 10-minute-treated enzyme by 88% and the 240-minute preparation by only 13%. These properties were those of the previously isolated γ-penicillinase. Citri showed, moreover, that α-penicillinase absorbed to glass particles had the iodine sensitivity and serological behavior of γ-penicillinase. After elution from glass, on the other hand, α-penicillinase properties were largely restored.

In other experiments Pollock[53] investigated serological relationships among exocellular penicillinases of various strains of *B. cereus*. The following were shown: (a) In a given bacterial strain (569) the basal enzyme and the inducible enzyme were serologically identical. (b) The enzyme formed by a constitutive mutant of the inducible strain was serologically identical to the inducible enzyme. (c) A derivative of an independent isolate from soil produced a penicillinase serologically related to the one from strain 569, but clearly not identical. Thus two strains of *B. cereus* make serologically related, but distinct, proteins with similar enzymic activity. Details of the serological comparison of these enzymes are to be found in the paper of Pollock.[53]

2. Cross-Reacting Material

The methods of serology are extremely useful in providing answers to certain questions that arise in studying microbial genetics. Some of these questions are: (1) In a mutant strain selected for loss of a certain enzyme activity is no enzyme protein formed or is a related protein with reduced activity formed? (2) After loss of a certain activity and subsequent recovery by mutation, is this recovery always associated with the appearance of the original protein? (3) Are there other proteins in an organism serologically related to the enzyme being studied but without the same activity? (4) If, by some form of genetic recombination, genetic control of synthesis of a protein occurs in a strain that previously lacked it, will the protein in the new strain be identical to the one in the parent strain? Experiments bearing on these questions are discussed below.

There are several ways of answering the first question, i.e., on loss of enzyme activity, does a protein as detected serologically disappear? One method is to prepare an antiserum to the purified enzyme and assay cross-reacting protein in extracts of strains lacking the activity. This is most easily determined by observing if the extracts from the nonenzyme strain adsorb antibodies reacting with the enzyme used as antigen. Another method is to prepare an antiserum to the extract lacking activity and to assay inactivating ability of such an antiserum on that enzyme of the active extract.

Investigation of a number of mutants of *Neurospora crassa* lacking tryptophan synthetase activity[54] revealed the presence in several of proteins related serologically to the enzyme. Antiserum prepared to partially purified extracts of the wild type organism could, in proper concentration, completely neutralize enzyme activity. Mutants were tested for their ability to remove antisynthetase from this serum; and several were found to make cross-reacting material (CRM), which was similar in other ways to the synthetase enzyme, e.g., in solubility properties during fractionation and in being a good antigen to elicit antienzyme formation. Another interesting point found in these investigations was that a temperature-sensitive mutant which at 25°C. produces little active enzyme does produce large amounts of CRM.

Corresponding investigations have been carried out on mutants of *E. coli* lacking tryptophan synthetase activity.[55] As with *N. crassa*, such mutants fall into two groups—those which form a cross-reacting material and those which do not. Investigation of the revertants to tryptophan independence from various negative strains indicated that a suppressor mutation could restore the ability of a strain to form an active protein serologically related to the original one even in the absence of CRM formation.[56] This had not been found to be the case in a small sample studied in *Neurospora*, where it had been found that while most negative mutants were suppressible, one lacking CRM was not.[57]

The tryptophan synthetase system (TSase) in *E. coli* on further examination revealed unexpected complexities. Just as had been shown for the enzyme from *N. crassa*,[58] it was discovered that the coli TSase could catalyze the several reactions thought previously due to several enzymes:[59]

$$\text{Indole} + \text{L-serine} \rightarrow \text{L-tryptophan}$$

$$\text{Indoleglycerolphosphate} \rightleftharpoons \text{indole} + \text{triosephosphate}$$

$$\text{Indoleglycerolphosphate} + \text{L-serine} \rightarrow \text{L-tryptophan} + \text{triosephosphate}$$

Moreover, it was shown that catalysis of each of these reactions is due to a combined form of two proteins A and B which can be separated from each other by chromatography on diethylaminoethyl (DEAE) cellulose. Mutants capable of producing A or B were isolated. Either component was assayed enzymically in the presence of the excess of the other. Serological assays measuring ability to remove antisynthetase from a serum indicate CRM to be an altered form of B, incapable of catalyzing reactions involving serine which would lead to the characterization of a mutant with CRM as one that required tryptophan for growth. The basis of the assignment of CRM as a modified form of B is several fold:[60] B and CRM are precipitated to the same extent even when a serum is used which has part of its antibody removed by absorption with B and CRM; in addition, regarding stability and solubility properties, they are similar.

The β-galactosidase enzyme of *E. coli* and related bacterial strains is one of the best studied immunologically and offers a model of how the techniques of serology can yield information about protein formation.[48, 61] For careful analyses of the techniques used in serological determination of an enzyme and of proteins related to it serologically, one should consult references 1, 48, and 61. The β-galactosidase of *E. coli* is an extremely good antigen in the rabbit. Single injections of 150γ of the purified protein yielded antisera in 4 to 8 weeks, containing 600 to 1600γ precipitating antibody. Since an antiserum will precipitate but not inactivate the enzyme, all serological assays are based on the precipitin reaction, i.e., measuring the amount of enzyme or antibody left in the supernatant after the specific precipitate is removed. An equivalence point G can be defined in the following manner: Increasing amounts of enzyme in a fixed volume are added to a constant volume of antiserum. The specific precipitates are removed by centrifugation and supernatant fluid assayed for enzyme and antibody. A ratio of antigen to antibody is reached where only negligible amounts of either is detectable. This defines the equivalence point G.

A first point of interest is in the use of the equivalence point to ascertain whether the several activities of β-galactosidase preparation are associated with the same protein. The enzyme preparation could be assayed either for lactase activity, i.e., the ability to hydrolyze lactose, or its ability to hydrolyze the chromogenic substrate orthonitrophenyl-β-D-galactoside

(ONPG). The equivalence point for these two activities was shown to be identical. Cohn also ascertained, even starting with a crude extract, that the number of enzyme units added to a constant volume of a given antiserum to reach the equivalence point was the same regardless of the specific activity of the enzyme preparation. This fact is assurance that serological titration of the enzyme is not confused by the presence in the mixture of other precipitating antigen-antibody components.

β-Galactosidase is, in some strains of *E. coli*, an inducible enzyme. Bacteria grown without inducer (usually a β-galactoside) are without enzyme activity. Are they without related protein? A titrated antiserum allows a serological comparison of extracts of induced and noninduced bacteria.

At first glance, the answer seemed very simple. Anti-β-galactosidase gave heavy precipitates with the extracts lacking activity. A completely absorbed serum, one to which enough extract of noninduced cells had been added, so that no further precipitate occurred upon further addition, would still precipitate purified enzyme although not to the same degree as before absorption. This seemed to indicate that absorption with the inactive extract was removing proteins not identical to β-galactosidase. This was of special interest, for it revealed that a new serological entity was formed in the presence of inducer. A careful serological examination revealed a more complex situation than casual serology would have shown—i.e., the inactive extracts contained a protein without β-galactosidase activity, which cross reacted with the enzyme. Absorption of the antiserum with inactive cell extract reduced the antiserum titer by about a factor 10, the titer of the antiserum being defined as the number of enzyme units precipitated by one milliliter of serum.

These cross-reacting antigens were labeled P_z (the inactive protein) and G_z (the enzyme). Were both present in active extracts; was one a precursor of the other? To answer this question required an assay for P_z in the presence of G_z. A quantitative study of an artificial mixture of P_z (from an inactive extract) and G_z (purified enzyme) yielded such an assay. The addition of increasing amounts of such a mixture of P_z and G_z to a constant volume of antiserum revealed five reaction zones (Fig. 1):

1. Antibody excess—all of the P_z, G_z, and part of the antienzyme precipitated; antibody remained in the supernatant.

2. Equivalence point T—all of the P_z, G_z, and antibody precipitated.

3. P_z excess—all G_z, antienzyme, and some P_z precipitated; the balance of P_z was in the supernatant.

4. Equivalence point G—all G_z and antienzyme precipitated; P_z was in the supernatant. At this point P_z and G_z had been separated.

5. G_z and P_z excess—soluble complexes of each with antienzyme were

in the supernatant. The existence of the two equivalence points T and G allowed a serological assay for P_z in the presence of G_z. Defining T_z as the total antigenic material in the mixture, then $T_z = G_z + P_z$. The equivalence point T measures T_z, the equivalence point G measures G_z. The difference $T_z - G_z$ defines P_z.

The precipitation curve of active and inactive extracts with antienzyme is depicted graphically in Fig. 2. Curve 1 represents the precipitation reaction of an extract containing a large amount of enzyme. Between the equivalence points T and G, the total nitrogen precipitated remains constant and neither enzyme nor antienzyme can be detected in the supernatant by enzyme assay or by addition of more enzyme. Addition of anti-

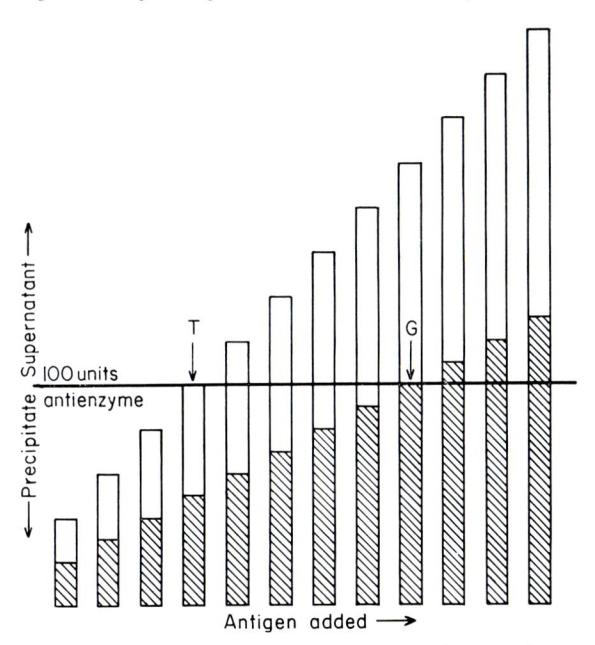

FIG. 1. The reactions of increasing amounts of a mixture of equal parts of P_z (clear part of column) and G_z (striated part of column) with a constant amount of antienzyme. The horizontal line represents 100 units of antienzyme added to each mixture and in relation to their level five zones of reaction are shown. Beginning at the left, the first three columns are in relative antienzyme excess and all P_z and G_z have precipitated. Column 4 is the equivalence point where all P_z, G_z, and antienzyme have precipitated. Columns 5, 6, 7, and 8 show that all of the G_z has combined with the antienzyme, but parts of the P_z remain in the supernatant. Column 9 shows equivalence point G at which all of the G_z and all antienzyme have combined leaving only P_z in the supernatant. The last three columns show soluble complexes of both P_z and G_z in the supernatant fluids. From M. Cohn, *in* "Serological Approaches to Studies of Protein Structure and Metabolism" (W. H. Cole, ed.), p. 38. Rutgers Univ. Press, New Brunswick, New Jersey, 1954.

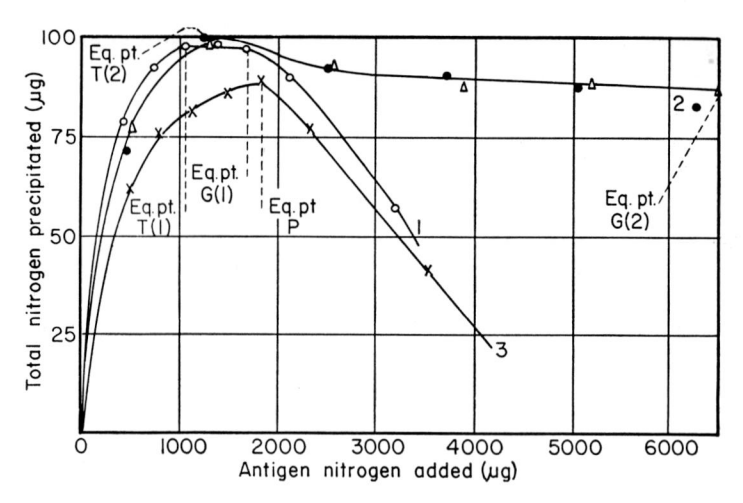

FIG. 2. Precipitated nitrogen vs. increasing amounts of added antigen nitrogen for the reaction of a highly active extract (curve 1), an inactive extract (curve 3), and two different low activity extracts (curve 2). See text for detailed explanation. From M. Cohn, in "Serological Approaches to Studies of Protein Structure and Metabolism" (W. H. Cole, ed.), p. 38. Rutgers Univ. Press, New Brunswick, New Jersey, 1954.

enzyme, however, to the supernatant gives a precipitate indicating the presence of a P_z-like protein. The identity of this protein in the supernatant of an active extract at equivalence point G with the cross-reacting protein in the inactive extracts was demonstrated[48] by comparing their detailed precipitin reactions with antienzyme. An interesting point in connection with the assay of P_z in the presence of G_z is the displacement of P_z from antigen-antibody complexes by G_z. For example, in the titration of an inactive extract with antienzyme even in the region of far P_z excess, G_z precipitates on addition of G_z activity, while addition of more antibody (antienzyme) gives additional P_z precipitation. Such behavior would lead to the confusing appearance of both antigen (P_z excess) and antibody (precipitability of G_z) in the same supernatant. The careful serological investigations of Cohn and Torriani[48] clarified this point.

Search for a cross-reacting material in inactive extracts by using it as an antigen to stimulate production of serum with anti-β-galactosidase activity failed because P_z is not antigenic in the rabbit and this in spite of its close serological relationship to G_z when tested with an anti-G_z serum.

The bacterial strains used in the studies by Cohn and Torriani[48] were primarily *E. coli* ML and a series of mutants selected for their inability to use lactose as a carbon source. All of these strains had the protein P_z. Investigation of P_z and G_z in other Enterobacteriaceae yielded an interesting apparent relationship regarding the presence of these proteins. *Aerobacter aerogenes*, a paracoliform strain, and *Shigella sonnei* all contained a P_z immunologically indistinguishable from that of *E. coli* ML. Moreover, each of these strains either made G_z serologically identical to that of *E. coli* or could mutate readily to form it. Further, the level of P_z fell when an inducer of G_z formation was added to a culture of an inducible strain and, as G_z was formed, seemed to imply a close relationship of the P_z-forming system and the G_z-forming system. The simple assumption of P_z as a precursor of G_z was ruled out by labeling experiments.[62]

This relationship of P_z formation to G_z formation was further explored in a strain of *Shigella dysenteriae*.[63] This strain does not have detectable β-galactosidase and does not seem to mutate to strains able to grow on lactose. Like *Shigella flexneri*, previously studied,[64] it too contained no detectable P_z. Genetic control of lactose utilization could, however, be given to these strains by transduction, or by sexual crossing with *E. coli*. Two main questions could be answered by examining the β-galactosidase system in these *Shigella-E. coli* hybrids: 1. Is the β-galactosidase identical to the one made in *E. coli*? 2. Is P_z also formed?

(1) The β-galactosidases were identical for the criteria tested: comparison of Michaelis constants, activity in the presence of inhibitors, and serology. (2) No detectable P_z was formed even when large amounts of G_z were formed. Thus a separation of P_z formation from G_z formation was effected.

The synthesis of protein under inhibiting growth conditions can be studied by following total activity of some enzyme; serological assays make possible assay of protein related to the enzyme. An example of this is the work by Hamers and Hamers-Casterman[65] on β-galactosidase formation by *E. coli* grown in the presence of thiouracil. The incorporation of this inhibitor into ribonucleic acid (RNA) made it interesting to investigate the effects of such incorporation on protein synthesis. This was studied in a constitutive β-galactosidase former (*E. coli* ML 308) and in an inducible β-galactosidase former (*E. coli* ML 37) in the presence of inducer (isopropyl thio-β-D-galactoside). In either case the growth rate became linear, i.e., there was a linear increase with time of total protein, total RNA, and viable count on agar plates. In both the constitutive and the inducible culture the rate of synthesis of enzyme (measured by activity) was inhibited by 90%. Sonic extracts of these cultures were then examined serologically by measuring the equivalence point for precipitation of activity by antiserum compared to a measurement with the same antiserum in control cultures. In the thiouracil-inhibited inducible strain the number of enzyme units precipitable per ml. of antiserum (at equivalence) was 40 to 50% lower than in the control. With the constitutive mutant the reduction was by 20%. In both cases the implication was that β-galactosidase synthesized in the presence of the RNA inhibitor is not identical to the control, being less active per serological unit.

E. Antibodies to Bacteriophage

The antigenic properties of bacteriophage and the use of antiphage sera as a tool for studying the formation of bacteriophage proteins is well discussed in the book of Adams[66] and in the article by Lanni and Lanni.[67] Here we shall only briefly point out some such applications.

The injection of purified phage preparations into rabbits yield sera which can be assayed by the standard methods of serology including the precipitin reaction (with whole phage particles), complement fixation (with isolated proteins or whole phage proteins), and neutralization of plaque-forming ability. A limitation on the applicability of some of the serological methods is the generally small quantity of material available. In experiments studying phage infection processes the titers seldom exceed 10^9 to 10^{10} particles per ml., which for bacteriophage T2, for example, corresponds to 0.1–1.0 gamma nitrogen per milliliter. For chemical experiments, purified bacteriophage preparations with titers greater than 10^{12} to 10^{13} particles per ml. can be prepared and protein components studied by the techniques of the precipitin reaction.

The kinetics of phage neutralization is generally studied in large antibody excess and in many cases is that of a first-order reaction. It is described by the equation:

$$dP/dt = KP/D$$

or integrating

$$P/P_0 = e^{-Kt/D}$$

where P is the phage titer at any time t, P_0 is the titer at $t = 0$, D is the dilution of the antiserum, and K is the velocity constant of the undiluted serum. P is usually expressed as plaque formers per milliliter, t in minutes, D is dimensionless, therefore K has the dimensions of minutes^{-1}. The K's (measured at 37°C.) for many sera are of the order 100 to 400, although values of 1 to 2×10^3 can be obtained for the phages T2 and T5 (Cohn and Lennox, unpublished). K is temperature-dependent with a Q_{10} of approximately 2. While in general neutralization is a first-order reaction, this is often not the case and may depend on several factors: the particular phage being studied, prior treatment of the phage stock, and the bacterial strain used for phage assay.

There are nonantibody components of serum that inactive bacteriophage T2.[68] In studying neutralization of a bacteriophage one must be careful to assay normal serum controls from the same animals before immunization. Just as one can use the shift in equivalence point of the reaction between a known antigen and antibody as a means of detecting a serologically related antigen, so can one use a change in neutralizing ability (K) of an antiphage serum as an assay for phage-related materials. Such an assay for "serum-blocking power" has been used to study the production of phage-related proteins during the infection process.[69] Such blocking substances were released near the end of the latent period by premature lysis of the infected cells. The blocking material could be separated from whole phage and bacteria by ultrafiltration. Such an assay can of course only detect those proteins involved in phage neutralization. Recent experiments indicate the presence of at least four protein components in phage T2[70] with most of the serum-blocking power being in the tail fibers. The presence of one of these phage components in T2 lysates can be detected by the methods of complement fixation.[69]

Cross reaction of phages with regard to neutralization is another criterion of relatedness. For example, a serum prepared with T2 will also neutralize T4 and T6, both of which will undergo genetic recombination with T2. Cross reaction thus provides a further property to follow in recombination experiments.

References

[1] M. Cohn, in "Methods in Medical Research" (A. C. Corcoran, ed.), p. 301. Year Book Publishers, Chicago, Illinois, 1952.

[2] S. V. Boyden, J. Exptl. Med. **93**, 107 (1951).

[3] A. B. Stavitsky, J. Immunol. **72**, 360 (1954).

[4] A. H. Coons, Ann. Rev. Microbiol. **8**, 333 (1954).

[5] J. Oudin, in "Methods in Medical Research," (A. C. Corcoran, ed.), p. 335. Year Book Publishers, Chicago, Illinois, 1952.

[6] A. C. Allison and J. H. Humphrey, Nature **183**, 1591 (1959).

[7] W. C. Boyd, "Fundamentals of Immunology." Interscience, New York, 1956.

[8] A. M. Pappenheimer, Jr., Advances in Protein Chem. **4**, 123 (1948).

[9] K. Jerne, Acta Pathol. et Microbiol. Scand. Suppl. **87** (1951).

[10] M. Cohn, in "Methods in Medical Research" (A. C. Corcoran, ed.), p. 271. Year Book Publishers, Chicago, Illinois, 1952.

[11] E. A. Kabat and M. M. Mayer, "Experimental Immunochemistry." C. C Thomas, Springfield, Illinois, 1948.

[12] R. A. Kekwick, Biochem. J. **34**, 1248 (1940).

[13] H. N. Eisen and D. Pressman, J. Immunol. **64**, 487 (1950).

[14] H. F. Deutsch, in "Methods in Medical Research" (A. C. Corcoran, ed.), p. 284. Year Book Publishers, Chicago, Illinois, 1952.

[15] A. A. Boyden, in "Serological Approaches to Studies of Protein Structure and Metabolism" (W. H. Cole, ed.), p. 74. Rutgers Univ. Press, New Brunswick, New Jersey, 1954.

[16] D. Glick, R. A. Good, L. J. Greenberg, J. J. Eddy, and N. K. Day, Science **128**, 1625 (1958).

[17] J. Munoz, in "Serological Approaches to Studies of Protein Structure and Metabolism" (W. H. Cole, ed.), p. 55, Rutgers Univ. Press, New Brunswick, New Jersey, 1954.

[18] C. L. Oakley and A. J. Fulthorpe, *J. Pathol. Bacteriol.* **65,** 49 (1953).
[19] J. R. Preer, *J. Immunol.* **77,** 52 (1956).
[20] M. W. Wilson, *J. Immunol.* **81,** 317 (1958).
[21] C. A. Williams and P. Grabar, *J. Immunol.* **74,** 158 (1956).
[22] M. D. Poulik, *J. Immunol.* **82,** 502 (1959).
[23] R. H. Sia and S. F. Chung, *Proc. Soc. Exptl. Biol. Med.* **29,** 792 (1932).
[24] G. F. Petrie, *Brit. J. Exptl. Pathol.* **13,** 380 (1932).
[25] S. D. Elek, *Brit. Med. J.* 493 (1948).
[26] S. D. Elek, *J. Clin. Pathol.* **2,** 250 (1949).
[27] V. J. Freeman, *J. Bacteriol.* **61,** 675 (1951).
[28] N. B. Groman, *J. Bacteriol.* **66,** 184 (1953).
[29] A. L. Olitzki and A. Sulitzeanu, *J. Bacteriol.* **77,** 264 (1959).
[30] J. Tomcsik, *Ann. Rev. Microbiol.* **10,** 213 (1956).
[31] J. Tomcsik and S. Guex-Holzer, *Experientia* **10,** 484 (1954).
[32] J. W. Vennes and P. Gerhardt, *Science* **124,** 535 (1956).
[33] C. Weibull, *J. Bacteriol.* **66,** 688 (1953).
[34] M. R. J. Salton and R. W. Horne, *Biochem. et Biophys. Acta* **7,** 177 (1951).
[35] M. R. J. Salton, *Biochim. et Biophys. Acta* **8,** 510 (1952).
[36] M. R. J. Salton, *Biochim. et Biophys. Acta* **9,** 334 (1952).
[37] M. R. J. Salton, *Biochim. et Biophys. Acta* **10,** 512 (1953).
[38] F. I. Tsuji and D. L. Davis, *J. Immunol.* **81,** 242 (1958).
[39] J. W. Newton, *Bacteriol. Proc.* 56 (1959).
[40] R. J. Dubos, *in* "The Bacterial Cell in Its Relation to Problems of Virulence, Immunity and Chemotherapy." Harvard Univ. Press, Cambridge, Massachusetts, 1945.
[41] M. Finland and E. C. Curnen, *J. Immunol.* **38,** 457 (1940).
[42] P. B. Beeson and W. F. Goebel, *J. Exptl. Med.* **70,** 239 (1939).
[43] M. Heidelberger and M. L. Wolfrom, *Federation Proc.* **13,** 496 (1954).
[44] E. A. Kabat, "Blood Group Substances." Academic Press, New York, 1956.
[45] P. Z. Allen and E. A. Kabat, *J. Immunol.* **82,** 340 (1959).
[46] P. Z. Allen and E. A. Kabat, *J. Immunol.* **82,** 358 (1959).
[47] E. G. Krebs and R. R. Wright, *J. Biol. Chem.* **192,** 555 (1951).
[48] M. Cohn and A. M. Torriani, *J. Immunol.* **69,** 471 (1952).
[49] M. G. MacFarlane and B. C. J. G. Knight, *Biochem. J.* **35,** 884 (1941).
[50] B. Cinader, *Ann. Rev. Microbiol.* **11,** 371 (1957).
[51] M. R. Pollock, *J. Gen. Microbiol.* **14,** 90 (1956).
[52] N. Citri, *Biochim. et Biophys. Acta* **27,** 277 (1958).
[53] M. R. Pollock, *J. Gen. Microbiol.* **15,** 154 (1956).
[54] S. R. Suskind, *in* "The Chemical Basis of Heredity" (W. D. McElroy and B. Glass, eds.), p. 123. John Hopkins Press, Baltimore, Maryland, 1957.
[55] P. Lerner and C. Yanofsky, *J. Bacteriol.* **74,** 494 (1957).
[56] C. Yanofsky, *Science* **128,** 843 (1958).
[57] C. Yanofsky, *in* "Enzymes: Units of Biological Structure and Function" (O. H. Gaebler, ed.), p. 147. Academic Press, New York, 1956.
[58] C. Yanofsky and M. Rachmeler, *Biochim. et Biophys. Acta* **28,** 640 (1958).
[59] I. P. Crawford and C. Yanofsky, *Proc. Natl. Acad. Sci. U. S.* **44,** 1161 (1958).
[60] C. Yanofsky and J. Stadler, *Proc. Natl. Acad. Sci. U. S.* **44,** 245 (1958).
[61] M. Cohn, *in* "Serological Approaches to Studies of Protein Structure and Metabolism" (W. H. Cole, ed.), p. 38. Rutgers Univ. Press, New Brunswick, New Jersey, 1954.
[62] D. S. Hogness, M. Cohn, and J. Monod, *Biochim. et Biophys. Acta* **16,** 99 (1955).

[63] M. Cohn, E. Lennox, and S. Spiegelman, *Biochim. et Biophys. Acta,* In press.
[64] M. Cohn, and A. M. Torriani, *Biochim. et Biophys. Acta* **10,** 280 (1953).
[65] R. Hamers and C. Hamers-Casterman, *Biochim. et Biophys. Acta* **33,** 269 (1959).
[66] M. H. Adams, "Bacteriophages." Interscience, New York, 1959.
[67] F. Lanni and Y. T. Lanni, *Cold Spring Harbor Symposia Quant. Biol.* **18,** 159 (1953).
[68] H. Van Vunakis, J. L. Barlow, and L. Levine, *Proc. Natl. Acad. Sci. U. S.* **42,** 391 (1956).
[69] R. I. De Mars, S. E. Luria, H. Fisher, and C. Levinthal, *Ann. Inst. Pasteur* **84,** 113 (1953).
[70] G. S. Stent, *Nature* **182,** 1769 (1958).

CHAPTER 10

Localization of Enzymes in Bacteria

ALLEN G. MARR

I. Introduction.. 443
II. Direct Cytochemistry.. 444
III. Analytical Morphology... 446
    A. Methods of Disrupting Bacterial Cells............................. 447
    B. Fractionation of the Extracts..................................... 450
    C. Properties of the Centrifugal Fractions........................... 452
    D. Separation of Submicroscopic Particles............................ 456
    E. Location in Cell.. 458
IV. Pigments of Photosynthetic Bacteria.................................... 461
V. Endospores... 464
    References.. 465

I. Introduction

The study of the localization of enzymes or, more broadly, of the biochemical cytology of bacteria derives its impetus from the analysis of the cells of higher plants and animals, particularly the liver cell. The technique of differential centrifugation of homogenates developed by Claude[1, 2] and adopted by numerous enzymologists has produced a vast literature on the biochemical activities of cell fractions.[3-5] The wedding of cytology and biochemistry resulted from the identification of the mitochondrion (isolated many years earlier by Bensley and Hoerr[6]) as the component of Claude's "large granules" which contained the respiratory enzymes.[7, 8] More recently the development of methods of fixation, embedding, and sectioning of material for high-resolution electron micrographs, pioneered by Palade,[9] has permitted a detailed comparison of the structures in centrifugal fractions with the structures in the cell.[10] The extension of such studies to bacteria is a severe test of the generality of the conclusions based on analysis of larger cells. The small size of a bacterium, which approaches the dimensions of the chondriosomes of larger cells, may demand a gross alteration of biochemical cytology.

Two broad methods are available for determining the location of a substance within the cell. One of these is direct cytochemistry; usually a specific chromogenic reagent is applied and the location within the cell of the colored product is determined with a microscope; in rare instances a substance may have sufficiently distinct physical properties to permit its direct determination, e.g., sulfur by refractive index or nucleic acid by its extinction at 2600 A. The second method is analytical morphology, the isolation in bulk of a certain organelle from a large population of cells. Both methods have distinct advantages and liabilities. Direct cytochemistry, particularly when applied to cells as small as bacteria, suffers from the lack of specificity and sensitivity and from the difficulty of avoiding optical artifacts. Analytical morphology has the obvious advantage of allowing conventional methods of chemical analysis with no problem of sensitivity, but suffers the difficulty of isolating a given organelle (or its fragments) from the remainder of the cell and of identifying the isolated substance cytologically.

II. Direct Cytochemistry

An enzyme can rarely be identified by its extrinsic physical and chemical properties (cytochromes are an obvious exception), but rather by the catalysis of a reaction in which a substrate is converted into a product. Even though the product may have distinctive physical properties which permit its determination by microscopy, the location within the cell of a product of an enzymic reaction bears no necessary relationship to the location of the enzyme which catalyzed the reaction. The most common method of determining the localization of an enzyme is to use a colorless substrate which is converted by the enzyme in question to a colored product. For example, many bacteria are capable of oxidizing the Nadi reagent (α-naphthol and dimethylparaphenylenediamine) to indophenol blue, a reaction presumably mediated by the terminal oxidase system. After the application of the reagents, the location of the indophenol blue is determined microscopically. Unfortunately the location of the colored product cannot be expected to reveal the location of the enzyme. The product can diffuse rapidly, particularly over the short distances imposed by the small dimensions of a bacterial cell, and accumulate in any lipid phase within the cell. A second limitation of the direct cytochemical method is a lack of suitable chromogenic substrates for the majority of enzymes and the lack of specificity in the available chromogenic substrates.

Most of the attempts to determine the localization of enzymes in bacteria have been directed toward the demonstration of mitochondria. Although several staining procedures have been used, such as Janus green and the previously mentioned Nadi reagent, the commonest has been the treatment of the cells with an oxidizable substrate and one of the colorless tetrazolium salts. It has been assumed that bacterial mitochondria should contain the enzymes necessary to reduce the tetrazolium salts to insoluble, colored formazans, which, it has been further assumed, should accumulate within the postulated mitochondrion. Mudd and his co-workers[11, 12] have surveyed a number of aerobic bacteria and have reported a variety of granules visible by light microscopy after staining with tetrazolium salts, the Nadi reagent, Janus green B, or phospholipid stains. The granules which were found after staining with tetrazolium salts were distinct from the nucleus and were presumed to be mitochondria or at least the bacterial equivalent of mitochondria. Unstained cells examined by electron microscopy were found to contain prominent, spherical granules of diverse sizes, particularly in mycobacteria and in *Micrococcus cryophilus*. These granules are uniformly dense to the electron beam, but the interior subsequently volatilizes, leaving the appearance of a limiting membrane. Mudd has defined bacterial mitochondria as a summation of the properties suggested by cytochemical tests together with the properties of the dense bodies observed by electron microscopy.[13]

The interpretation of granules visible after staining with tetrazoles as structural elements of the cell was questioned by Weibull,[14] who followed the staining of *Bacillus megaterium* with triphenyltetrazolium by continuous microscopic observation. At first a large number of granules appeared near the periphery of the cell which later merged to form large granules. (A similar sequence was previously observed by Mudd et al.[12]) The cells were then disrupted and centrifuged to remove the colored granules; the supernatant liquid, free of microscopic granules, was found to produce more colored granules on incubation with triphenyltetrazolium and glucose. Similar observations were made of the staining of *Azotobacter agilis* with neotetrazolium and glucose. A preparation of empty cell envelopes, free of observable granules, was found to form granules which remained attached to the empty envelope after incubation with neotetrazolium and hydrogen.[15] It appears that the granules visible after staining with the tetrazolium salts are not preexisting structures but are precipitates of formazan.

The location of formazan in stained cells is not identical with the sites of the opaque granules in electron micrographs. Electron-dense granules have been recognized in many bacteria. Prior to their implication as mitochondria, these granules were frequently mistaken for nuclei. Several investigators have reported that the electron-dense granules are identical with volutin or metachromatic granules and that the granules owe their metachromasy to their content of polyphosphate.[16-18] The presence of these granules frequently results from any malnutrition which permits respiration and accumulation of phosphate from the medium but which slows or inhibits balanced synthesis. Not only do the number, size, and location within the cell of the electron-dense granules differ from the areas stained with tetrazoles, Nadi reagent, or Janus green B, but also observation of the *same* cell by both light and electron microscopy has shown that the stained areas are not congruent with the electron-dense granules.[17] In addition to dense granules *Caulobacter vibrioides* contains a basophilic vacuole that stains deeply with Sudan black B, triphenyltetrazolium chloride, or the Nadi reagent. One might conclude that this vacuole is the mitochondrion;

however, the vacuole of fixed cells is stained by formazan or indophenol blue.[18] These results illustrate that the site of accumulation of the dye is dependent on the greater solubility of the dye in the lipid phase and is not dependent on the location of enzymes.

The appearance of thin sections of bacteria by electron microscopy also argues against the presence of a mitochondrion. The mitochondria in thin sections of diverse plant and animal specimens have a characteristic fine structure, an osmophilic double membrane and a system of inner projections (either cristae or tubules) which appear to be folds or extentions of the inner membrane.[19] Protozoa,[20,21] fungi,[22,23] and yeast[24] have been shown to contain morphologically typical mitochondria. Nothing resembling the structure of a mitochondrion has been observed in thin sections of bacteria.[25-27] In fact, the structure of bacteria as revealed by electron microscopy of ultrathin sections bears little resemblance to the complex organization found in cells of higher forms. Not only are the mitochondria and similar chondriosomes lacking in bacteria, but also missing is the nearly universal system of membranes, variously referred to as the ergastoplasm or the endoplasmic reticulum, fragments of which constitute the bulk of the microsome fraction of liver cells.[28,29] Small, electron-dense granules about 200 A. in diameter, which are one feature of the endoplasmic reticulum, are not associated with a membranous structure in bacteria but rather are uniformly distributed through out the cytoplasm.[26] These and other remarkable differences in structure should serve as a caution to those who would apply nomenclature of the organelles of higher cells to the structures observed in or isolated from bacteria.

Any interpretation of the localization of enzymic function in bacteria must consider the unique cytology of the bacterium. Although broad similarities may exist between the enzymic architecture of bacteria and the cells of higher forms, it is not to be found in completely homologous organelles. The confusion resulting from the arbitrary application of the term mitochondrion to structures varying from metachromatic granules to formazan precipitates is exemplary.

III. Analytical Morphology

Some of the uncertainties of the direct cytochemical method can be eliminated if the structure in question is isolated in bulk and analyzed chemically and enzymically. This has proved to be a successful approach to several problems in microbial physiology. The most notable achievements have been the isolation of flagella and of cell walls, permitting an unambiguous determination of chemical composition. This method requires adequate techniques for the removal or release of the structure from the cell, a method of purification, and adequate criteria, both morphological and chemical, of the identity and purity of the final preparation.

A. Methods of Disrupting Bacterial Cells

Although several methods have been devised for the disruption of bacterial cells,[30] the methods most commonly used for the release of the contents of the cell for subsequent enzymological analysis are: (1) grinding with abrasive, (2) ballistic disintegration, (3) sonic treatment, and (4) osmotic shock. Unfortunately only a few studies have been made of the physical effects of the various methods of disruption. An ideal method of disrupting cells for the study of the localization of enzymes would merely open the wall and membrane without further damage to the subcellular structures. Most of the methods result in further degradation of some of the structural elements.

1. Abrasives

Grinding packed masses of bacteria with abrasives, such as powdered glass, alumina, or carborundum,[31, 32] has been successful in the preparation of enzymically active extracts from a wide variety of microorganisms. One obvious difficulty with this method is the failure to obtain quantitative recovery of all the input material from admixture with or adsorption to the abrasive. It is difficult to assess the fraction of cells disrupted and the extent of retention of large cell fragments. The recovery of various enzymes is based of necessity on the crude cell-free extract rather than on the starting material. The mechanism of disruption is apparently simple grinding. Although larger structures released by disruption may be comminuted by grinding, the submicroscopic macromolecules are too small to be appreciably comminuted.

2. Ballistic Disintegration

Bacteria, endospores, and yeast can be disrupted by agitation with glass reflector beads in the apparatus designed by Mickle[33] or the more powerful shaker designed by Nossal.[34] Lamanna and Mallette[35] have demonstrated that high-speed blendors, which alone are almost without effect on bacteria, become quite effective if glass beads are added to the suspension. The physical mechanism or mechanisms responsible for disruption of the cells is unknown. Cavitation occurs in the liquid during treatment but is probably not the major mechanism of disruption. Prolonged treatment results in inactivation of some enzymes and progressive comminution of some of the particulate material released from the cell.[36] Removal of the residual intact cells and larger subcellular fragments from the glass beads is considerably less difficult than extraction from powdered glass or alumina used in the abrasive methods; however, this fraction is contaminated to some extent with fragments of the glass beads and fragments of the container.

3. Sonic Treatment

Treatment of bacterial suspensions with high amplitude sound is another general method of extracting enzymes. Disruption results from the powerful secondary effects of resonating cavities in the liquid rather than from a primary effect of the sound field.[37] The frequencies which are most effective in producing cavitation of water (approximately 1 kc. to 1 mc.) would be expected to be most effective in the disruption of cells. Kinetics of disruption are first order and from 5000 to 300,000 acoustical watt-seconds are required for 50% disruption of various bacteria. The most susceptible are large Gram-negative rods, and the most resistant are micrococci. Sonic treatment has the obvious advantage of eliminating the abrasive; however, some of the subcellular structures are progressively disintegrated as in the ballistic method. The recovery of most enzymic activities after sonic treatment is quite good if precautions are taken to eliminate sonochemical oxidative inactivation. The cavitation of a liquid containing dissolved oxygen

TABLE I
Sonochemical Oxidation of *Escherichia coli* Glutamic Decarboxylase and Pyridoxal[a, b]

Atmosphere	Glutamic Decarboxylase (μM/hr./mg. N)	B$_6$ (μg./ml.)
H$_2$	245	0.009
Air	75	0.000
No treatment	—	0.010

[a] 50 ml. treated 20 minutes at 75 acoustical watts at 10 kc.
[b] Unpublished experiment of Mrs. Barbara Alexander.

results in the formation of hydrogen peroxide, probably from free radical precursors.[38] Sonic treatment in the presence of oxygen results in the inactivation of certain enzymes and even coenzymes. Table I illustrates the sonochemical oxidation of both the apoenzyme, glutamic decarboxylase, and the coenzyme, B$_6$. The treatment was comparable to that used to disrupt bacteria. The quantitative significance of sonochemical inactivation of enzymes has not been fully appreciated. Several investigators have attributed progressive loss of enzymic activity during sonic treatment to the loss of organized structures even though oxygen was not excluded to prevent oxidative inactivation.[39]

4. Osmotic Shock

Osmotic shock appears to be the most promising method of disrupting cells for the determination of the localization of enzymes. A cell is disrupted by a hydrostatic pressure exerted against the membrane. Once the membrane is destroyed, the force disappears; thus, subcellular fractions escape continued mechanical damage after release from the cell. Halophiles and halotolerant organisms have been shown to be in osmotic equilibrium with their environment. If the environment is suddenly diluted with water, the hydrostatic pressure developed against the membrane is sufficient to disrupt the cell.[40] *Azotobacter agilis, Rhodospirillum rubrum,* and *Serratia plymuthica* have been disrupted by osmotic shock after increasing the solute concentration of the cytoplasm by brief exposure to 1 M glycerol.[41] In most other bacteria the tensile strength of the wall prevents disruption by direct osmotic shock; however, the wall can be weakened or removed by specific enzymic digestion and the resulting protoplast can be lysed by osmotic shock.[42]

The technique used for disrupting the cell may be expected to affect the subcellular fractions isolated by subsequent centrifugal separation. In a few investigations attempts have been made to assess the magnitude of this effect. In the first comprehensive study of the macromolecular constituents in cell-free extracts of bacteria, Schachman *et al.*[43] compared the ultracentrifugal patterns of extracts obtained by several techniques of disruption. The macromolecules with sedimentation constants of 29 S and 40 S were present in approximately the same concentration regardless of the method used to disrupt the cell, which together with the discontinuous distribution of particle size suggests these macromolecules exist as such in the cell. By contrast most methods of disruption liberate deoxyribonucleic acid (DNA) as a relatively small molecule, usually recovered in the "soluble" fraction (sedimentation constant of 8 S). However, the disruption of cells by osmotic shock liberates DNA in a form which sediments at low centrifugal force in association with a structure of microscopic dimensions.[44] Because of the apparent organization of this structure (together with a preconception that DNA should be associated with a structure of microscopic dimensions), one concludes that the disruption of the cell by methods other than osmotic shock results in the concomitant mechanical destruction of the structure containing DNA. Obviously, any attempt to identify a fraction isolated by physical separation of extracts must consider the possibility that the material isolated comprises fragments of a larger structure.

In addition to primary mechanical disintegration of larger cell structures during disruption, a secondary destruction may result from opening the cell and exposing its contents to a new environment. The requirement of a suspending medium with moderately high osmotic pressure for the preservation of structure and certain enzymic activities of mammalian mitochondria[45] and the requirement of Ca^{++} and Mg^{++} to stabilize nucleoprotein particles of yeast[46] are good examples. Rotman[47] has described a secondary lysis of bacteria previously damaged by sonic treatment which superficially appears to be a secondary disruption of some large, integral structure. The phenomenon probably results from the gelation of DNA following the disruption of the cells in a medium of rather low pH, a condition necessary to demonstrate the effect, rather than from the disintegration of a structure existing *before* the mechanical damage to the cell.

If one measures the activity of a particular enzyme, it is obvious that the chemical and physical environment must be one which preserves that enzymic activity. In some instances this may require the use of different suspending fluids for different enzymes. For example, the use of unbuffered sucrose solutions as a medium for centrifugal fractionation will result in the complete loss of activity of an enzyme which requires a moderately high ionic strength.

B. Fractionation of the Extracts

After mechanical disruption of the bacterial cells in a suitable suspending fluid, the homogenate is usually fractionated by successive centrifugation at progressively higher centrifugal forces, sometimes using a density gradient to improve resolution. The components of the homogenate are thus separated according to their mass and density. Such a separation is obviously patterned after the fractionation of homogenates of animal tissue.

In the subsequent analysis of centrifugal fractions for enzymic activity, it is important to establish both the total activity recovered and the specific enzymic activity of a fraction. A fraction which contains the enzyme should have the bulk of the total enzymic activity of the homogenate together with a higher specific activity than the homogenate. The assay should, if possible, be based on a single step reaction. Frequently the assay of oxidative enzymes has been based on an over-all reaction resulting in consumption of oxygen. For example, the uptake of O_2 in the oxidation of ketoglutaric acid by bacterial extracts requires not only the specific dehydrogenase but the entire electron transport system as well. The dehydrogenase has usually been found in the soluble fraction but the electron transport system is contained in particles. Oxygen consumption by the soluble fraction would require a contamination with particles, and oxygen consumption by the particles requires a contamination by adsorbed or occluded dehydrogenase. Such an error in assay may be reflected in a failure to recover a large amount of the enzymic activity of the homogenate when the activities of the separate fractions are summed; however, the over-all reaction may still be catalyzed at a maximum rate if the contaminating enzyme is not rate-limiting.

Unfortunately, the homogenates of bacteria lack large structural elements, such as nuclei, mitochondria, or chloroplasts, which serve as a guide to the centrifugal separation of homogenates of plant and animal cells.

this necessitates blind centrifugal fractionation of the bacterial extracts and results in quite arbitrary fractions. A similar difficulty has been encountered in the "microsome fraction" of homogenates of animal tissue; the conventional method of preparation of microsomes results in a rather arbitrary collection of submicroscopic structural elements.[48] Several methods may be used to assess the validity of the centifugal fractionation. Electron microscopy of the fractions can serve both as an index of the homogeneity of the fraction and as a means of identifying the isolated structure. Electron microscopy has usually revealed a variety of particle types in the primary centrifugal fractions. Additional evidence of homogeneity (or the lack of it) can be obtained from independent methods of fractionation such as electrophoresis. Finally, indirect evidence of the validity of the fractionation may be obtained by determining the concentration (specific activity) of one or more enzymes in the fractions. If an enzyme is contained in a particle of discrete dimensions, this enzyme should be highly concentrated in adjacent fractions. An empirical approach would design the centrifugal fractionation to achieve a high concentration of this enzyme in a single fraction. If, however, the enzyme is contained in fragments of varying size, any protocol will result in a distribution of the enzymic activity over several fractions. The actual procedure used in fractionation of bacterial extracts has usually been based more on the convenience or whim of the investigator rather than on any of these considerations. This is reflected in the honest nomenclature proposed by Alexander[49] based only on the centrifugal force and time without implication of structures contained in the various fractions. For example, a sediment obtained by centrifugation at 20,000 g for 30 minutes would be designated $20p30$, the prefix indicating the average centrifugal force in 1,000 g and the suffix the time of centrifugation in minutes.

Despite the difficulties in designing a rational method of centrifugal separation of homogenates, it is possible to separate the homogenate into roughly three fractions: (1) structures sufficiently large to permit detection with the light microscope which sediment at moderately low centrifugal force (approximately 10,000 g for 15 minutes; (2) submicroscopic particles which are sedimented only by a high centrifugal force (approximately 100,000 g for 1 hour); and (3) the soluble fraction which is composed of those substances which remain in the supernatant liquid after removal of the submicroscopic particles. The composition of the various fractions will depend to a considerable extent on the method of disruption of the cells. The results obtained using sonic disruption, shaking with glass beads, or grinding with abrasives are, in general, sufficiently similar to consider them together. The more extensive comminution of small particles by sound or the difficulty in the loss of microscopic fragments from admixture with abrasive are minor by comparison with the difference between the results of any of these methods with that of osmotic disruption.

C. Properties of the Centrifugal Fractions

Because of the cytochemical evidence for "centers of oxidation-reduction" large enough to be visible by light microscopy and because of the identification of the mitochondrion as the site of many of the oxidative enzymes in the cells of higher plants and animals, the low-speed sediments which contain structures of microscopic dimensions have been carefully examined for oxidative enzymes. This fraction does contain oxidative enzymes under special conditions of disruption which will be considered later; however, if disruption is by the more common methods of grinding, ballistic, or sonic treatment, the low-speed centrifugal fractions do not contain an appreciable amount of oxidative enzymes. The low-speed fraction consists largely of electron-dense granules which contain polyphosphate and show metachromasy. The electron transport system is recovered in submicroscopic particles rather than in particles of microscopic dimensions.

The submicroscopic particles are rich in phospholipid and ribonucleic acid (RNA) and account for half of the total phosphorus and one-fourth to one-third of the nitrogen and dry weight of the homogenate. The soluble fraction contains most of the acid-soluble phosphorus and DNA. The chemical composition of the particulate fractions of *Azotobacter vinelandii*, *Escherichia coli*, *Micrococcus pyogenes*, and *Pseudomonas fluorescens* have been summarized by Alexander.[49]

The enzymic role of submicroscopic particles is illustrated by centrifuging homogenates at approximately 100,000 g for an hour or more which reduces the respiration of the resulting supernatant fluid to a small fraction of that of the homogenate.[50, 51] (An exception is *Bacillus megaterium* which yields a soluble fraction with a high rate of respiration.[52]) The enzymic activities found in the submicroscopic particles of a variety of aerobic bacteria are remarkably similar. These particles contain the electron transport system, oxidize a rather small number of substrates, and couple these oxidations to the phosphorylation of adenosine diphosphate (ADP).

1. Electron Transport System

The pellet from high-speed centrifugation of homogenates has a characteristic red color conferred by its content of cytochromes. In general the particle fraction shows the same alpha, beta, and Soret bands of cytochromes found in the intact cell. Table II gives a partial list of the alpha bands of a-, b-, and c-type cytochromes found in washed submicroscopic particles. The alpha and beta bands appear on the addition to the particles of an oxidizable substrate and disappear on shaking in air. The soluble fraction may have a weak cytochrome spectrum, particularly if the cells have

TABLE II
Cytochrome Components of Submicroscopic Particles

Organism	Cytochrome α maximum a	b	c	Reference
Azotobacter agilis	625	560	552–553	53
Pseudomonas fluorescens	None	560–562	552	54
Bacillus megaterium	600	558	None	55
Escherichia coli[a]	625, 590	556	None	56

[a] Low temperature spectrum.

been disrupted by sonic treatment; however, the soluble cytochrome is usually quantitatively insignificant. Submicroscopic particles also contain flavins, as evidenced by the decrease in light absorption in the region of 450 millimicrons.[54] Although the particles may contain bound pyridine nucleotides, the particles do not couple with external dehydrogenases unless diphosphopyridine nucleotide (DPN) or triphosphopyridine nucleotide (TPN) is added. The particles reduce a variety of oxidants other than oxygen, including nitrate, methylene blue, tetrazoles, 2,6-dichlorophenol-indophenol, and ferricyanide.

2. Substrates Oxidized by Particles

Particles obtained and washed by high-speed centrifugation oxidize only a few substrates by comparison with the broad spectrum of compounds oxidized by the whole homogenate. Table III illustrates the compounds oxidized by particles. Many other substrates are oxidized if the particles

TABLE III
Substrates Oxidized by Washed Particles

Organism	Substrates oxidized	References
Azotobacter agilis	Hydrogen, lactate, malate, succinate, DPNH, TPNH	57–60
Pseudomonas fluorescens	Gluconate, glucose, L-mandelate, malate, p-hydroxymandelate, succinate	54, 61
Escherichia coli	α-Glycerophosphate, formate, hydrogen, lactate, succinate, DPNH	56, 62
Proteus vulgaris	Formate, hydrogen, lactate, succinate, DPNH	63
Serratia marcescens	α-Glycerophosphate, α-ketoglutarate, formate, lactate, malate, succinate	51
Acetobacter suboxydans	Erythritol, ethanol, glucose, glycerol, lactate, mannitol, propanol, sorbitol	64

are supplemented with the soluble fraction which contains the majority of the substrate dehydrogenases. The most common substrates oxidized by washed particles are secondary alcohols (lactate, malate, mandelate, α-glycerophosphate, and polyhydric alcohols) and succinate. Hydrogenase of most aerobic bacteria is in the particles; however, one aerobic hydrogenomonad has a soluble hydrogenase.[65] Some substrates are oxidized by both soluble and particulate enzymes. *Azotobacter agilis* contains a soluble DPN-linked malic dehydrogenase in addition to the particulate enzyme,[58] and *Acetobacter suboxydans* contains a soluble glucose dehydrogenase as well as a particulate glucose "oxidase."[66]

Because of the limited range of substrates oxidized, the particles generally catalyze single-step reactions in contrast to the sequences catalyzed by whole homogenates of bacteria or by isolated mitochondria. For example, the oxidation of citrate by enzymes from *Azotobacter agilis* requires a soluble aconitase, isocitric dehydrogenase, and alpha-ketoglutaric dehydrogenase, a particulate succinoxidase, and either particulate malic oxidase or soluble malic dehydrogenase.[58] Perhaps the most significant catalytic activity of the particles is the reoxidation of reduced pyridine nucleotides which permits the coupling of the numerous soluble substrate dehydrogenases to oxygen as the ultimate oxidant.

3. Oxidative Phosphorylation

As one might anticipate from their content of respiratory enzymes, the submicroscopic particles contain a system for oxidative phosphorylation. Table IV is a summary of the measurements of oxidative phosphorylation by five aerobic bacteria. An obvious difference between the bacterial and mammalian systems is the much lower P:O ratio (atoms of inorganic phosphorus esterified per atom of oxygen consumed in respiration) which is usually less than 0.5 with bacterial extracts. A remarkable exception to the low P:O ratio found with most bacteria is the value of 1.78 reported for the oxidation of succinate by unfractionated homogenates of *Mycobacterium phlei*, which approaches the ratio for this reaction in mammalian mitochondria. The bacterial systems are unusually stable; particles from the azotobacter retain their ability for phosphorylation almost undiminished for at least two weeks.[67]

Several investigators have reported small to substantial increases in the P:O ratio if particles are supplemented with the soluble fraction. Oxidative phosphorylation by particles of *Azotobacter agilis* is only slightly increased by the soluble fraction.[60] The soluble fraction of this organism is known to contain myokinase, ATPase, and polynucleotide phosphorylase which are present in negligible amounts in the particles; these enzymes would be expected to decrease rather than increase apparent phosphorylation. Two

TABLE IV

Oxidative Phosphorylation by Bacterial Enzymes

Organism	Preparation	Substrate	Dinitrophenol sensitivity	P/O	Reference
Azotobacter agilis	Particles	H$_2$	—	0.21–0.52	59, 67
		DPNH[a]	—	0.27–0.48	60, 67
		TPNH	—	0.24–0.25	67
		Succinate	—	0.14–0.25	60
	Whole extract	Succinate	—	0.51	68
Alcaligenes faecalis	Whole extract	DPNH[b]	—	0.36	
		DPNH[a]	—	0.5	69
	Particles	DPNH		0.3	
Proteus vulgaris	Whole extract	Lactate	+[c]	0.42	63
	Particles	Lactate		0	
Aerobacter aerogenes	Whole extract	Lactate	—	0.33	63
	Particles	Lactate	—	0.16	
Mycobacterium phlei	Whole extract	Succinate	+[d]	1.78	70
	Particles	DPNH		0	

[a] Substrate levels.
[b] Generated with alcohol dehydrogenase.
[c] Approximately 50% inhibited by $1.3 \times 10^{-4} M$ DNP.
[d] $8 \times 10^{-5} M$ DNP reduced P/O to 0.05.

factors in addition to the particles have been reported to be required for phosphorylation by *Alcaligenes faecalis*.[69] Although the respiratory rate was increased by the addition of the two factors to the particles, the P:O ratio was as high with washed particles as in the reconstructed system. More recently the heat-stable factor has been identified as a polynucleotide which retains its activity after treatment with RNase; synthetic polynucleotides prepared by the action of polynucleotide phosphorylase on various nucleotide diphosphates stimulate phosphorylation in the presence of both particles and soluble fraction.[70] Oxidative phosphorylation by *Proteus vulgaris* shows an obligate requirement for soluble factors,[63] and the P:O ratio obtained with particles of *Mycobacterium phlei* is doubled by the addition of soluble fraction.[71] Menadione reductase from the soluble fraction of *M. phlei* was purified several fold and found to be as active as the whole soluble fraction in restoring high levels of phosphorylation.[72] Radiation of both particles and soluble fraction at a wavelength of 3600 A. diminished the respiratory rate and almost eliminated phosphorylation by the reconstituted system. Addition of flavin restored the respiratory rate but not phosphorylation; vitamin K_1 restored both.[73] Either flavins or naphthoquinones could have been photooxidized by this treatment.

Oxidative phosphorylation by bacterial enzymes is less sensitive to dinitrophenol than mammalian mitochondria (Table IV). If a concentration of dinitrophenol higher than 10^{-4} M is required to uncouple phosphorylation, the system has been designated as insensitive. Uncoupling by dinitrophenol parallels the requirement for soluble factors. Both *M. phlei* and *P. vulgaris* are sensitive to moderately low concentrations of dinitrophenol, and both require, or are stimulated by, the addition of soluble materials to the particles. It therefore appears that the action of dinitrophenol is indirect, requiring the participation of soluble factors to effect uncoupling.

D. Separation of Submicroscopic Particles

The submicroscopic particle fraction of bacterial extracts is heterogeneous. At least two types can be distinguished in electron micrographs of submicroscopic particles from bacteria; one is a spherical particle, opaque to electrons, about 200 A. in diameter; the second is irregular in size and shape, flattened, and less opaque to electrons than the spherical particle.[59, 68] Although the two types of particles are difficult to separate by centrifugation, they may be easily separated by electrophoresis. (Each of these particles has a counterpart in particles and fragments derived from mammalian tissue.)

1. RNA-Protein Particles

During electrophoresis of the submicroscopic particles on starch the RNA moves rapidly toward the anode as one or two bands. The particles containing RNA purified by electrophoresis have been clearly identified with the well-known 40 S and 29 S components of crude extracts. These particles contain 40 to 50% RNA; the remainder of the particle is protein.[59, 74] The RNA may be removed from the protein by treatment with high concentrations of urea, which suggests that hydrogen bonds are responsible for the linkage between the two moieties.[75] Because of the presence of the RNA-protein particles in the centrifugal fraction which also contains the oxidative enzymes, it has been assumed that these regular spherical particles were the site of oxidative enzymes; however, the purified ribonucleoprotein particles from *Azotobacter agilis* are devoid of oxidative enzymes.[59] The only enzymic activity demonstrated thus far is a ribonuclease, the activity of which increases on treatment with urea.[76] This endogenous ribonuclease liberates low molecular weight nucleotides, which may account for some reports of "soluble" RNA.[49] Similar ribonucleoprotein macromolecules have been obtained by fractionation of the microsomes from both plant and animal tissues. These particles contain approximately 40% RNA and have a sedimentation constant of approximately 50 S.[77-79] The most interesting biochemical activity of the plant and mammalian ribonucleoprotein particles is their rapid incorporation of labeled amino acids which suggests a role in protein synthesis.[78, 80]

2. Electron-Transport Particles

The particles containing the electron transport system can be separated from the RNA-protein particles by electrophoresis on starch[59] or by removing the RNA-protein particles by precipitation with 0.1 volume of ethanol.[53] The electron-transport particles have diverse sizes and shapes and fail to give a discrete boundary in the ultracentirfuge.[54, 59] Their random size, in contrast to the regular spherical shape of the RNA-protein particles, suggests that the electron-transport particles are derived from the fragmentation of a larger structure.

The electron-transport particles from bacteria are quite similar to submicroscopic fragments obtained by the disintegration of mammalian mitochondria. The electron-transport particle from *A. agilis* contains 26% lipid and amounts of flavin, copper, heme, and non-heme iron which are almost identical with the amounts found in small fragments of disrupted mitochondria.[53] The electron-transport particles from bacteria also resemble the mitochondrial fragments in their range of enzymic activity. The mitochondrial fragments oxidize succinate and DPNH but do not contain the numerous substrate dehydrogenases characteristic of whole mitochondria.[81] Carefully prepared mitochondrial fragments also show the low P:O ratios which are characteristic of the submicroscopic particles of bacteria.[82, 83] Because of these similarities it is tempting to suggest that the electron-transport particles from bacteria and the mitochondrial fragments are derived from homologous structures; however, the functional similarity need not reflect an architectural similarity in the parent structure of each.

The isolation of individual enzymes from the electron-transport particles has proved quite difficult, indicating that the proteins are firmly bound. Prolonged sonic treatment converts the particles to a form which no longer sediments in the high-speed centrifuge;[84] however, the component proteins are not released as individual molecules.[85] Treatment with deoxycholate also prevents subsequent centrifugal sedimentation.[86] It seems inappropriate to designate mere comminution as solubilization. A more meaningful and practical sense for the term "solubilization" is the solution of the individual proteins of the complex, the experimental test of which is the isolation and purification of a single protein from the parent complex. The successful solubilization of the particles is based on the properties of the lipid which serves as a cement. This lipid is mainly a phosphatidylethanolamine which is readily soluble in cold primary alcohols.[59] After extraction of the cementing phospholipid, the remaining protein is soluble in water or buffer. Tissieres[87] has used the two-phase butanol-water system[88] for the extraction of cytochromes from the particles of *A. agilis*. The proteins are more completely extracted by a two-step procedure devised by Mr. Jack Pangborn in this laboratory: the lipid is extracted from vacuum-dried particles with butanol; the residual butanol removed *in vacuo*, and the protein is dissolved in an appropriate buffer. This method results in a quantitative recovery of the c-type cytochromes and the hydrogenase from particles of *A. agilis*. The cytochromes have been highly purified after release by either method, which confirms the solubilization of the particles.

E. Location in the Cell

A knowledge of the chemical and enzymic properties of the centrifugal fractions of cell extracts does not, of itself, give evidence of the location of enzymes in the cell. It is necessary to identify the isolated fractions with the original topography of the cell.

1. RNA-Protein Particles

The ribonucleoprotein particles appear to be distributed randomly in the cytoplasm of the bacterium. Electron micrographs of sections of a variety of bacteria show electron-dense particles about 200 A. in diameter throughout the cytoplasm.[26] The individual particles resemble those which line the membranes of the endoplasmic reticulum of more highly differentiated cells. Thus far no similar system of membranes has been reported in bacteria.

The kinetics of release during the course of sonic disruption of cells suggests that the RNA-protein particles exist as such in the cytoplasm and are not derived from the comminution of a larger structure.[89] During sonic treatment the first event comminutes only a fraction of the envelope but liberates virtually all of the cytoplasm. If a substance is in solution or suspension in the cytoplasm, it should be released at a rate the same as that for destruction of cells. The RNA-protein particles fulfill this criterion. Also the RNA-protein particles are released quantitatively by osmotic shock which would not be expected to alter subcellular structure.[41] The ribonucleoprotein obtained by osmotic shock is contained in particles which have the same sedimentation constant and electrophoretic mobility as particles obtained by other methods of disruption.

2. Electron-Transport Particles

The origin of the submicroscopic particles which contain the respiratory enzymes is more difficult to establish. The spherical, electron-dense particles found in the cytoplasm, which have now been identified as the RNA-protein particles, have been erroneously assumed to house the respiratory apparatus.[26, 49] The only basis for this assumption was that centrifugal fractions which contained the respiratory enzymes also contained the spherical particles. Two independent methods suggest that the electron-transport particles are fragments of a larger structure produced by the method of disrupting the cell: (1) During the sonic disruption of the azotobacter the respiratory enzymes and phospholipid are released together at a lower rate than the disruption of cells, in contrast to the RNA-protein particles; (2) the irregular size of the electron-transport particles argues for their derivation from a larger structure.[89] Although it is relatively simple to prove that the electron transport system is not contained in preexisting submicroscopic particles, it is more difficult to establish which cellular structure is fragmented to produce them. A corollary to the delayed release of the respiratory enzymes during sonic treatment is that a microscopic structure must exist as a transient intermediate. The only distinct microscopic structure found during the sonic treatment of the azotobacter was an apparently empty cell envelope. It was purified by differential centrifugation and found to contain the same respiratory enzymes as the submicroscopic particles, including the capacity for oxidative phosphorylation.[59] This suggests that the respiratory enzymes are either part of, or attached to, the outer envelope of the azotobacter. Further sonic treatment of the purified hulls liberates submicroscopic particles identical with those obtained by the electrophoretic resolution previously mentioned. The usual mechanical methods of disruption are unsuited to the recovery of the native structure which houses the respiratory enzymes. Millman and Darter[90] were able to recover the bulk of the respiratory enzymes of *Mycobacterium tuberculosis* in a fraction sedimenting at low centrifugal force after the cells had been disrupted by grinding a thin paste with glass in a ball mill; electron micrographs of this fraction showed large granules (approximately 1000 A. in diameter) and apparently empty cell walls. Whether this method would yield similar results with other bacteria is uncertain. The most promising method for disrupting the cell with a minimum of damage to larger organelles is osmotic shock. After the removal of the cell wall of *Bacillus megaterium* with lysozyme, the resulting protoplast can be disrupted by the simple expedient of dilution with water. Low-speed centrifugation of the osmotically lysed cells sediments the cytochromes[55] and other oxidative enzymes.[52] Microscopic examination of this centrifugal fraction discloses large, empty spherical membranes or "ghosts" which are presumably the plasma membrane of the protoplast together with numerous free granules containing β-hydroxybutyrate. The respiratory activity is not reduced by removal of the granules, indicating that the membrane and not the free granules contains the respiratory enzymes.[52]

Militzer and his co-workers have obtained a "red fraction" rich in cytochromes by treatment of *B. stearothermophilus* with lysozyme.[91] The red fraction, which has been the object of numerous enzymological studies, has been found to contain cytochromes, malic and succinic oxidases, apyrase, and other enzymes.[92, 93] This red fraction was found by phase-contrast microscopy to consist of dense spheres approximately 1 micron in diameter. Since discrete areas of both cells and spheres stained with HCl-Giemsa (a conventional "nuclear" stain) it was implied that the chromatinic bodies of the cell developed into the larger spheres after release from the cell.[94] A subsequent paper which illustrated the staining of both cells and the spheres with triphenyltetrazolium chloride advanced a similar argument for the origin of the spheres from the areas in the cell containing formazan deposits.[95] A comparison of the published phase photomicrographs of the red fraction with photomicrographs of authentic protoplasts leaves little doubt that the red fraction consists of whole or lysed protoplasts.

More recently Georgi *et al.*[96] have published excellent electron micrographs of a well-washed red fraction. This fraction, which represented 20% of the dry weight of the original cells, contained most of the cytochromes, organic iron, lipid, and ATPase. Its appearance in electron micrographs was that of a membrane surrounding several dense granules about 850 A. in diameter. Weibull has reported an almost identical appearance of lysed protoplasts of *Bacillus* M.[97] Thus, the red fraction consists of "ghosts" of osmotically lysed protoplasts rather than whole protoplasts. It may be concluded that the location of the cytochromes is the same in *B. stearothermophilus* as in *B. megaterium*: either an integral part of, or attached to the cell membrane.

In the preceding studies either the membrane or the adhering granules could contain the respiratory enzymes. The most direct approach would require a physical separation of the two structures. Since this has not yet been accomplished, one must rely on less direct and less conclusive evidence. An extraction with fat solvents removed 10% of the dry weight of the red fraction and destroyed the membrane leaving the dense granules; this led to the conclusion[96] that the granules accounted for all of the residual 90% of the dry weight and that the cytochrome and ATPase must, therefore, be located in the dense granules. Neither of these conclusions is admissible. The membrane is probably lipoprotein; extraction of lipid destroys the membrane and releases but does not remove the protein. Although the only structures now resolvable by electron microscopy are the dense granules, individual protein molecules released from the membrane by extraction would still be present. Evidence for this is the fact that ATPase is released as a soluble protein by the solvent extraction. Also, as previously mentioned, solvent extraction is a preliminary to the isolation of cytochromes from lipoprotein complexes. The observation that the dense granules are *not* destroyed by solvent extraction, which is known to release the cytochromes, makes them unlikely candidates for the location of the respiratory enzymes. Thus, it appears that the respiratory enzymes of *Bacillus* are located in the membrane. The granules are digested by DNase and resemble a part of the nuclear material recently isolated from *B. megaterium* by osmotic lysis of protoplasts.

In the azotobacter, enterics, and possibly other Gram-negative bacteria, the wall and plasma membrane fail to separate on plasmolysis; rather, the wall collapses. Furthermore, the structure isolated by mechanical fragmentation of the cells of Gram-negative bacteria is chemically more complex than cell-wall preparations from Gram-positive microorganisms. Failure to demonstrate a separate membrane by plasmolysis together with the chemical complexity has prompted the use of intentionally vague and generic terms, such as "hull" or "Membran," to distinguish this structure from the isolated cell walls of the Gram-positive bacteria. Weidel and Primosigh[98] have postulated that the "Membran" of *Escherichia coli* consists of two layers; one is a lipopolysaccharide which is chemically similar to the isolated cell walls of Gram-positive microorganisms, and one is a lipoprotein. The lipoprotein layer may be the functional counterpart of, if not homologous with, the plasma membrane of Gram-positive bacteria. The previously mentioned evidence that respiratory enzymes are contained in, or firmly attached to, the hull of the azotobacter has recently been amplified by the application of osmotic shock as a method of disruption.[41, 99] Osmotic shock releases the soluble enzymes and the RNA-protein particles but leaves the DPNH oxidase, cytochrome, hydrogenase, and phospholipid in the empty envelope. It is tempting to assume that these substances are located in the layer of phospholipoprotein corresponding to the cell membrane; however, the determination of their exact location within the hull must await more refined methods of analysis.

3. OTHER STRUCTURES

Several other structural elements of the bacteria cell have been isolated by the techniques of analytical morphology. Flagella have been isolated and highly purified.[100, 101] Osmotic disruption has been found to release a large fraction of the DNA in a structure of microscopic dimension which has been purified from other cellular structures.[44] Granules rich in polymetaphosphate as well as lipid granules have been isolated from the cell. Thus far no enzymes have been found located specifically in these structures, although the examination for enzymes has been casual.

IV. Pigments of Photosynthetic Bacteria

For many years it was recognized that photosynthetic bacteria (and blue-green algae) differ from other photosynthetic organisms in that they lack microscopic plastids. It has been assumed that the photosynthetic pigments of these bacteria were homogeneously distributed throughout the cytoplasm. More recently it has become clear that the pigments are confined to submicroscopic plastids. The chlorophyll and carotenoids of mechanically disrupted *Rhodospirillum rubrum* can be sedimented by centrifugation for 1 hour at 25,000 g; these pigment-bearing particles were referred to by Schachman et al.[43] as chromatophores. Both electron microscopy and ultracentrifugation indicate a relatively homogeneous particle size. The purified chromatophores appear as disks approximately 1100 A. in diameter; the diameter of the equivalent sphere is about 600 A. From direct counts by electron microscopy of chromatophores in extracts it has been estimated that a cell of *R. rubrum* contains approximately 6000. The extracts of dark-grown cells do not contain comparable particles.

Similar chromatophores have been isolated from the sulfur purple bacterium, *Chromatium*.[102] They contain protein, carbohydrate, and phospholipid with a calculated content of 200 molecules of chlorophyll and a molar ratio of chlorophyll:protein:carotenoid of approximately 2:2:1. They also contain smaller amounts of cytochrome, flavin, and pyridine nucleotides. Similar chromatophores have not been isolated from the green sulfur bacterium, *Chlorobium thiosulphatophilum*.[103] The carotenoids of *Chlorobium* sediment at relatively low centrifugal force, but the chlorophyll remains in the supernatant liquid.

Sonic treatment and possibly other methods of disruption fragment the chromatophores to smaller particles. A brief sonic treatment of chromatophores isolated from *Chromatium* converted half of the chlorophyll, carotenoid, and cytochrome to a form which no longer sedimented as the original chromatophores at 25,000 g for 1 hour but sedimented at 100,000 g in 1.5 hours. The fragments accounted for all of the pigments which were lost from the chromatophore fraction as a result of sonic treatment. The fragments derived from the chromatophores had twofold higher amounts of phospholipid, chlorophyll, carotenoids, and cytochrome per milligram of protein and only a fraction of the initial carbohydrate.[102] Although it has been suggested that the carbohydrate in the original chromatophores is a structure responsible for its integrity, and that after sonic treatment a carbohydrate "shell" remains in the low-speed centrifugal fraction, it is possible that the carbohydrate represents a contamination of the isolated chromatophores with fragments of the cell wall which are more resistant to further comminution than are the chromatophores. Regardless of the interpretation the sonic conversion of the chromatophores to smaller particles offers a means of further analysis of the photosynthetic machinery. The chemical composition of the isolated small particles shares one feature with electron-transport particles obtained from bacteria: a high content of phospholipid. The similarity extends to the composition of the phospholipid, which contain ethanolamine as the only nitrogenous base.[59, 103] The concentration of this phospholipid in both particles is sufficient for the phospholipid to serve as a major structural element or cementing substance.

As one might anticipate from its content of chlorophyll, the chromatophore contains a major portion of the photosynthetic machinery. Both purified chromatophores and the smaller fragments phosphorylate ADP to ATP in the light without added reducing agents.[104-106] Photophosphorylation is moderately sensitive to dinitrophenol; $10^{-4}\ M$ dinitrophenol inhibits photophosphorylation by *Chromatium* 72%, which is comparable to the inhibition of oxidative phosphorylation in many bacterial systems.[104] The chromatophore fragments lose most of their ability for photophosphorylation during washing, suggesting a further similarity to oxidative phosphorylation by submicroscopic particles from nonphotosynthetic bacteria.[104, 106]

The CO_2-fixing enzymes are not present in the chromatophores but are present in the pigment-free, soluble fraction.[107] There is a functional similarity between the bacterial chromatophores and fragments of the chloroplasts of higher plants, from which the enzymes responsible for the fixation of CO_2 are released by mechanical disruptions while the photosynthetic pigments and enzymes for photophosphorylation are retained as part of the insoluble structure.[108]

Vatter and Wolfe have examined the location of the chromatophores in the cells of various photosynthetic bacteria by electron microscopy.[109] Thin sections of osmium-fixed cells of *R. rubrum* show ellipsoidal structures with a major axis of 500 A. distributed throughout the cytoplasm. In some sections they appear close-packed throughout the cytoplasm. The interior of these structures is more transparent to electrons than is the cytoplasm, while the outer membrane or cortex is more opaque. The most convincing evidence that these cytoplasmic granules are the chromatophores is that the granules are not present in dark-grown cells. Furthermore, granules of similar dimensions and morphology have never been found in sections of nonphotosynthetic bacteria. From the number of chromatophores appearing in a section, the number per cell was estimated to be 3000, which is in good agreement with the previous estimate of chromatophores released by mechanical disruption. Chromatophores could also be identified in *Rhodopseudomonas spheroides* and *Chromatium* strain D. *Chlorobium limicola* did not contain typical chromatophores; somewhat smaller, rather indistinct granules were present in the cytoplasm.

Indirect methods of determining the location of the photosynthetic pigments do not agree with the distribution of discrete chromatophores throughout the cytoplasm. During sonic disruption of *Chromatium*[102] and *R. rubrum*[110] the chlorophyll is released more slowly than the cells are disrupted, which suggests an attachment of the chromatophores to the cell wall or membrane. Osmotic shock of *R. rubrum* by diluting a suspension of these cells previously equilibrated with 1.5 M glycerol releases most of the nucleic acid but only 8% of the chlorophyll.[110] These results can be explained by the large dimensions of the chromatophores which may impede their release. However, the osmotic disruption of protoplasts of *R. rubrum* does release the larger "fat" globules yet fails to release a significant fraction of the photosynthetic pigments from the ghosts.[111] A possible explanation of these results is a network, possibly integral with the cell membrane, which physically connects the chromatophores, preventing their release.

V. Endospores

Since the development of criteria for biochemical purity of spores and of techniques for obtaining clean spore preparations, a variety of enzymes have been found characteristic of resting aerobic spores (reviewed by Murrell,[112] and Halvorson and Church[113]). Their discovery has prompted a few studies of the location within the spore of these enzymes and of other substances characteristic of the spore, perhaps with the hope of finding a clue to the enigma of germination.

Many of the enzymes of the resting spore are demonstrable in whole spores; these include alanine racemase,[114] adenosine deaminase,[115] ribosidase,[115] and the heat-resistant catalase.[112] Either these enzymes are located outside the main permeability barrier of the spore, or, less likely, the resting spore is permeable to the substrates of these enzymes. By contrast, the pyrophosphatase[116] and heat-labile catalase[112] are inactive until the spore is broken, suggesting that these two enzymes are inside the spore. Sonic disruption has given independent evidence of an external location of some enzymes.[117] The sonic disruption of the spores of *Bacillus cereus* occurs in two steps; the first removes the exosporium leaving a still viable and heat-resistant spore, and the second disrupts the spore proper. The release of the alanine racemase and adenosine deaminase during sonic disruption can be correlated with the loss of the exosporium rather than the disruption of the spore, which suggests that the racemase and deaminase are located in the exosporium. Isolated spore coats, obtained by ballistic disintegration, have been found to contain high specific activities of ribosidase and *p*-phenylenediamine oxidase. The kinetics of release of ribosidase agrees with a location somewhere in the spore body rather than in the exosporium.

The physical state of enzymes in extracts of spores appears to depend largely on the disrupting treatment. Since the spore is rather resistant to mechanical disruption, more extended treatment than that necessary for most bacterial cells is required to disrupt them. During this long treatment the further comminution of particulate material to progressively smaller structures no doubt occurs. The yield of spore coats appears to be greater if ballistic disintegration is used rather than sonic treatment or the Hughes press,[112] which suggests that empty spore coats are destroyed more readily by the latter techniques. During sonic treatment the particles which contain alanine racemase and adenosine deaminase are progressively converted to forms which fail to sediment in 1 hour at 100,000 g. Apparently both the exosporium and the spore body contribute to the submicroscopic particles in spore extracts.

Little of the measured enzymic activity of whole resting spores is destroyed by heat sufficient to kill the majority of the spores. Two theories have usually been advanced to explain the thermal resistance of these enzymes: (1) the catalysts are inherently heat-stable, and (2) the catalytic activity is not inherently heat-stable but is protected by some structural feature of the spore. The pyrophosphatase and heat-sensitive catalase are stable to heating only in whole spores; these enzymes are quite heat-labile if the spore is disrupted.[112, 116] From the fact that the enzymes do not act on external substrates one may tentatively conclude that they are inside the spore coat and may be protected by some general (but unknown) structural feature of an intact spore. The racemase and deaminase which are presumably outside the spore coat are still resistant to heat after the spore is disrupted. The thermal stability of these two enzymes depends on their physical state. As the particles containing these enzymes are disintegrated, the thermal stability, particularly of the deaminase, decreases. The enzymes in the isolated spore coat, ribosidase, and p-phenylenediamine oxidase are very resistant to heat.[112, 115]

References

[1] A. Claude, *Cold Spring Harbor Symposia Quant. Biol.* **9**, 263 (1941).

[2] A. Claude, *J. Exptl. Med.* **84**, 51 (1946).

[3] D. E. Green, *in* "Enzymes and Enzyme Systems" (J. T. Edsall, ed.), p. 15. Harvard Univ. Press, Cambridge, Massachusetts, 1951.

[4] G. H. Hogeboom, W. C. Schneider, and M. J. Striebich, *Cancer Research* **13**, 617 (1953).

[5] W. C. Schneider, *J. Histochem. and Cytochem.* **1**, 212 (1953).

[6] R. R. Bensley and N. Hoerr, *Anat. Record* **60**, 449 (1934).

[7] G. H. Hogeboom, W. C. Schneider, and G. E. Palade, *J. Biol. Chem.* **172**, 619 (1948).

[8] J. W. Harman, *Exptl. Cell Research* **1**, 382 (1950).

[9] G. E. Palade, *Anat. Record* **114**, 427 (1952).

[10] A. B. Novikoff, *Science* **124**, 969 (1956).

[11] S. Mudd, A. F. Brodie, L. C. Winterscheid, P. E. Hartman, E. H. Beutner, and R. A. McLean, *J. Bacteriol.* **62**, 729 (1951).

[12] S. Mudd, L. C. Winterscheid, E. D. DeLamater, and H. J. Henderson, *J. Bacteriol.* **62**, 459 (1951).

[13] S. Mudd, *J. Histochem. and Cytochem.* **1**, 248 (1953).

[14] C. Weibull, *J. Bacteriol.* **66,** 137 (1953).
[15] E. H. Cota-Robles, A. G. Marr, and E. H. Nilson, *J. Bacteriol.* **75,** 243 (1958).
[16] I. W. Smith, J. F. Wilkinson, and J. P. Duguid, *J. Bacteriol.* **68,** 450 (1954).
[17] A. M. Glauert and E. M. Brieger, *J. Gen. Microbiol.* **13,** 310 (1955).
[18] E. A. Grula and S. E. Hartsell, *J. Bacteriol.* **68,** 498 (1954).
[19] G. E. Palade, *J. Histochem. and Cytochem.* **1,** 188 (1953).
[20] A. W. Sedar and K. R. Porter, *J. Biophys. Biochem. Cytol.* **1,** 583 (1953).
[21] M. H. Greider, W. J. Kostir, and W. J. Frajola, *J. Protozool.* **5,** 139 (1958).
[22] G. Turian and E. Kellenberger, *Exptl. Cell Research* **11,** 417 (1956).
[23] S. Tsuda, *J. Bacteriol.* **71,** 450 (1956).
[24] H. D. Agar and H. C. Douglas, *J. Bacteriol.* **73,** 363 (1957).
[25] G. B. Chapman and J. Hillier, *J. Bacteriol.* **66,** 362 (1953).
[26] J. R. G. Bradfield, *in* "Bacterial Anatomy" (E. T. C. Spooner and B. A. D. Stocker, eds.), p. 296. Cambridge Univ. Press, London and New York, 1956.
[27] O. Maaløe and A. Birch-Andersen, *in* "Bacterial Anatomy" (E. T. C. Spooner and B. A. D. Stocker, eds.), p. 261. Cambridge Univ. Press, London and New York, 1956.
[28] G. E. Palade and K. R. Porter, *J. Exptl. Med.* **100,** 641 (1954).
[29] G. E. Palade, *J. Biophys. Biochem. Cytol.* **1,** 59 (1955).
[30] W. B. Hugo, *Bacteriol. Revs.* **18,** 87 (1954).
[31] W. P. Wiggert, M. Silverman, M. F. Utter, and C. H. Werkman, *Iowa State Coll. J. Sci.* **14,** 179 (1939).
[32] H. McIlwain, *J. Gen. Microbiol.* **2,** 288 (1948).
[33] H. Mickle, *J. Roy. Microscop. Soc.* **68,** 10 (1948).
[34] P. M. Nossal, *Australian J. Exptl. Biol. Med. Sci.* **31,** 583 (1953).
[35] C. Lamanna and M. F. Mallette, *J. Bacteriol.* **67,** 503 (1954).
[36] P. M. Nossal, *Biochem. J.* **57,** 62 (1954).
[37] E. N. Harvey, *Biol. Bull.* **59,** 306 (1930).
[38] O. Lindström, *J. Acoust. Soc. Am.* **27,** 654 (1953).
[39] A. F. Brodie and C. T. Gray, *Science* **125,** 534 (1957).
[40] M. Ingram, *in* "Microbial Ecology" (R. E. O. Williams and C. C. Spicer, eds.), p. 90. Cambridge Univ. Press, London and New York, 1957.
[41] S. A. Robrish and A. G. Marr, *Bacteriol. Proc. (Soc. Am. Bacteriologists)* **P70,** 130 (1957).
[42] C. Weibull, *J. Bacteriol.* **66,** 688 (1953).
[43] H. K. Schachman, A. B. Pardee, and R. Y. Stanier, *Arch. Biochem. Biophys.* **38,** 245 (1952).
[44] S. Spiegelman, A. I. Aronson, and P. C. Fitz-James, *J. Bacteriol.* **75,** 102 (1958).
[45] J. W. Harman and A. Kitiyakara, *Exptl. Cell Research* **8,** 411 (1955).
[46] F. C. Chao and H. K. Schachman, *Arch. Biochem. Biophys.* **61,** 220 (1956).
[47] B. Rotman, *J. Bacteriol.* **72,** 827 (1956).
[48] M. Takanami, *J. Histochem. and Cytochem.* **5,** 503 (1957).
[49] M. Alexander, *Bacteriol. Revs.* **20,** 67 (1956).
[50] A. Tissieres, *Nature* **174,** 183 (1954).
[51] A. W. Linnane and J. L. Still, *Biochim. et Biophys. Acta* **16,** 305 (1955).
[52] R. Storck and J. T. Wachsman, *J. Bacteriol.* **73,** 784 (1957).
[53] J. H. Bruemmer, P. W. Wilson, J. L. Glenn, and F. L. Crane, *J. Bacteriol.* **73,** 113 (1957).
[54] R. Y. Stanier, I. C. Gunsalus, and C. F. Gunsalus, *J. Bacteriol.* **66,** 537 (1953).
[55] C. Weibull, *J. Bacteriol.* **66,** 696 (1953).
[56] R. E. Asnis, V. G. Vely, and M. C. Glick, *J. Bacteriol.* **72,** 314 (1956).

[57] L. A. Hyndman, R. H. Burris, and P. W. Wilson, *J. Bacteriol.* **65**, 522 (1953).
[58] M. Alexander and P. W. Wilson, *J. Bacteriol.* **71**, 252 (1956).
[59] E. H. Cota-Robles, A. G. Marr, and E. H. Nilson, *J. Bacteriol.* **75**, 243 (1958).
[60] A. Tissieres, H. G. Hovenkamp, and E. C. Slater, *Biochim. et Biophys. Acta* **25**, 336 (1957).
[61] W. A. Wood and R. F. Schwert, *J. Biol. Chem.* **201**, 501 (1953).
[62] H. Gest, in "Phosphorus Metabolism" (W. D. McElroy and B. Glass, eds.), Vol. 2, p. 522. Johns Hopkins Press, Baltimore, Maryland, 1952.
[63] P. M. Nossal, D. B. Keech, and D. J. Morton, *Biochim. et Biophys. Acta* **22**, 412 (1956).
[64] C. Wismer, T. E. King, and V. H. Cheldelin, *J. Bacteriol.* **71**, 737 (1956).
[65] C. R. Bovell, Ph.D. Thesis, University of California, Davis, California, 1957.
[66] T. E. King and V. H. Cheldelin, *Biochem. J.* **68**, 31 P. (1958).
[67] I. A. Rose and S. Ochoa, *J. Biol. Chem.* **220**, 307 (1956).
[68] P. E. Hartman, A. F. Brodie, and C. T. Gray, *J. Bacteriol.* **74**, 319 (1957).
[69] G. B. Pinchot, *J. Biol. Chem.* **205**, 65 (1953).
[70] G. B. Pinchot, *J. Biol. Chem.* **229**, 1 (1957).
[71] A. F. Brodie and C. T. Gray, *J. Biol. Chem.* **219**, 853 (1956).
[72] A. F. Brodie and C. T. Gray, *Biochim. et Biophys. Acta* **19**, 384 (1956).
[73] A. F. Brodie, M. M. Weber, and C. T. Gray, *Biochim. et Biophys. Acta* **25**, 448 (1957).
[74] A. B. Pardee, K. Paigen, and L. S. Prestidge, *Biochim. et Biophys. Acta* **23**, 162 (1957).
[75] D. Elson, *Biochim. et Biophys. Acta* **27**, 207 (1958).
[76] D. Elson, *Biochim. et Biophys. Acta* **27**, 216 (1958).
[77] M. L. Petermann and M. G. Hamilton, *Cancer Research* **12**, 373 (1952).
[78] G. C. Webster, *Plant Physiol.* **32**, Suppl., 36 (1957).
[79] M. Takanami, *J. Histochem. and Cytochem.* **5**, 503 (1957).
[80] J. W. Littlefield, E. B. Keller, J. Gross, and P. C. Zamecnik, *J. Biol. Chem.* **217**, 111 (1955).
[81] F. L. Crane, J. L. Glenn, and D. E. Green, *Biochim. et Biophys. Acta* **22**, 475 (1956).
[82] C. Cooper and A. L. Lehninger, *J. Biol. Chem.* **219**, 489 (1956).
[83] W. W. Kielley and J. R. Bronk, *Biochim. et Biophys. Acta* **23**, 448 (1957).
[84] R. Repaske, *J. Bacteriol.* **68**, 555 (1954).
[85] J. Pangborn, unpublished. 1956.
[86] M. J. Wolin and H. C. Lichstein, *J. Bacteriol.* **72**, 762 (1956).
[87] A. Tissieres, *Biochem. J.* **64**, 582 (1956).
[88] R. K. Morton, *Nature* **166**, 1092 (1950).
[89] A. G. Marr and E. H. Cota-Robles, *J. Bacteriol.* **74**, 79 (1957).
[90] L. Millman and R. M. Darter, *Proc. Soc. Exptl. Biol. Med.* **91**, 271 (1956).
[91] W. Militzer, T. B. Sonderegger, L. C. Tuttle, and C. E. Georgi, *Arch. Biochem.* **24**, 75 (1949).
[92] W. E. Militzer, T. B. Sonderegger, and L. C. Tuttle, *Arch. Biochem.* **26**, 299 (1950).
[93] W. E. Militzer and C. Tuttle, *Arch. Biochem. Biophys.* **31**, 416 (1951).
[94] C. E. Georgi, W. E. Militzer, L. Burns, and J. Heotis, *Proc. Soc. Exptl. Biol. Med.* **76**, 598 (1951).
[95] L. Burns and W. E. Militzer, *Proc. Soc. Exptl. Biol. Med.* **82**, 411 (1953).
[96] C. E. Georgi, W. E. Militzer, and T. S. Decker, *J. Bacteriol.* **70**, 716 (1955).
[97] C. Weibull and K. G. Thorsson, in "Electron Microscopy" (F. S. Sjöstrand and J. Rhodin, eds.), p. 266. Academic Press, New York, 1956.

[98] W. Weidel and J. Primosigh, *J. Gen. Microbiol.* **18,** 513 (1958).
[99] A. G. Jose and R. M. Pengra, *Bacteriol. Proc. (Soc. Am. Bacteriologists)* **P76,** 132 (1957).
[100] C. Weibull, *Biochim. et Biophys. Acta* **2,** 351 (1948).
[101] C. Weibull, *Biochim. et Biophys. Acta* **3,** 378 (1949).
[102] J. W. Newton and G. A. Newton, *Arch. Biochem. Biophys.* **71,** 250 (1957).
[103] J. Gibson, *J. Gen. Microbiol.* **16,** ix (1957).
[104] J. W. Newton and M. D. Kamen, *Biochim. et Biophys. Acta* **25,** 462 (1957).
[105] A. W. Frenkel, *J. Biol. Chem.* **222,** 823 (1956).
[106] I. C. Anderson and R. C. Fuller, *Plant Physiol.* **32,** Suppl., 16 (1957).
[107] R. C. Fuller and I. C. Anderson, *Plant Physiol.* **32,** Suppl., 16 (1957).
[108] D. I. Arnon, F. R. Whatley, and M. B. Allen, *Science* **127,** 1026 (1958).
[109] A. E. Vatter and R. S. Wolfe, *J. Bacteriol.* **75,** 480 (1958).
[110] A. G. Marr, unpublished. 1957.
[111] G. Stanier, personal communication. 1958.
[112] W. G. Murrell, "The Bacterial Endospore" (a monograph). University of Sidney, Sidney, Australia, 1955.
[113] H. Halvorson and B. Church, *Bacteriol. Revs.* **21,** 112 (1957).
[114] B. T. Stewart and H. O. Halvorson, *J. Bacteriol.* **65,** 160 (1953).
[115] J. F. Powell and J. R. Hunter, *Biochem. J.* **62,** 381 (1956).
[116] H. S. Levinson, J. D. Sloan, Jr., and M. T. Hyatt, *J. Bacteriol.* **75,** 291 (1958).
[117] J. A. Berger and A. G. Marr, *J. Gen. Microbiol.* (in press).

AUTHOR INDEX

Numbers in parentheses are reference numbers and are included to assist in locating references when the authors' names are not mentioned in the text. Numbers in italics refer to the page on which the reference is listed.

A

Abelson, P. H., 13(29), 15(29), 18(29), 19(29), 20(29), *33*
Abraham, E. P., 27(87), *34*, 124(159, 159a), 142(159a, 267), *148*
Abraham, G., 159(36), *199*
Abrams, A., 116(108), 122(108), 142(159), *147*, 255, 264, 313, *357*
Abrams, R. Y., 366, *384*
Abramson, H. A., 295(127), *355*
Adams, J. N., 394(12), *412*
Adams, M. H., 99(9), 114, *144*, 388, 395(20), 399(26), *412*, *413*, 438, *441*
Adye, J., 165(76), *200*
Agar, H. D., 446(24), *466*
Albertsson, P., 117, 118, *147*
Alderton, G., 261(37, 38), *353*
Alexander, B., *448*
Alexander, H., 116, *147*
Alexander, H. E., 109(64), *145*
Alexander, M., 28(93), 29(93), *34*, 54, *95*, 451, 452, 453(58), 454(58), 456(49), 458(49), *466*, *467*
Ali-Cohen, C. H., 189(267), 190(267), 192(267), *204*
Allen, M. B., 463(108), *468*
Allen, P. Z., 428(45, 46), *440*
Allison, A. C., 417(6), *439*
Allison, V. D., 130(184), *148*
Almquist, E., 362, *383*, *384*
Altenbern, R. A., 275(221), *358*
Amano, T., 130, *148*, 269, 286, 289, 292, *354*, *355*
Ambler, R. P., *21*
Ames, B. N., 403(41), *413*
Amies, C. R., 376(92), *385*
Anderson, C. G., 113(96), 116(96), *146*
Anderson, E. S., 39, *94*
Anderson, I. C., 463(106, 107), *468*
Anderson, T. F., 115(104), *146*, 158(22), 163(51), 176(159, 160), 177(174), *199*, *202*, 346, *357*, 394(13), 395(14, 18, 19, 21, 22), *396*, *397*, 403(43), 408(50), 410(21, 63), *412*, *413*, *414*
Ando, K., 109, 113(60), *145*
Anfinsen, C. B., 4(4a), *32*
Angerer, K. v., 177(181), *202*
Angulo, J. J., 176(162), *202*
Angus, T. A., 243(77), *247*
Aninga, S., 219(35), 222, 223(35), *228*, 236(35), *246*
Arber, W., 404(48), *413*
Armstrong, J. J., 122, 125, 139(152), *148*, 254(186, 187), *357*
Arndt, W. F., 267, 269, 270, 289, 329(58), 331, 339, 345, 346, *345*, 381(103), *386*
Arnon, D. I., 463(108), *468*
Aronson, A. I., 18(46), 19(46), 28(46), *33*, 52(59), 70(59), 83(59), 90(59), *95*, 305(142), 309(142), *356*, 449(44), 461(44), *466*
Aschner, M., 98(4), *144*
Asnis, R. E., 453(56), *466*
Aspinall, G. O., 109(67, 68), *145*, *146*
Astbury, W. T., 158(23), 165(23, 66), 173(66), *199*, *200*
Aubert, J. P., 107, *145*
Auerbach, H., 171(110), *201*
Avery, O. T., 24(74), *34*, 99(8), 113, *144*

B

Babudieri, B., 176(152, 166), 177(175), *202*
Bach, F. W., 175, *201*
Bacon, G. A., 100, *144*
Baddiley, J., 116(110), 119, 122(110, 152), 125(152), 129, 139, *147*, *148*, *150*, 254, *357*
Badian, J., 66, 67, *95*, 234, 236, *247*
Baer, B. S., 134(211), *149*
Bahr, G. F., 90(135), *96*
Baker, R. F., 154(1), *198*, 238(68), *247*
Baracchini, O., 189(266), 190(266), 191, *204*
Baranowska, J., 144(282), *151*

Barbu, E., 19(51), *33*, 401(36), *413*
Barer, R., 37(13), *93*
Barkulis, S. S., 122, 128, *147*
Barlow, G. H., 173, *201*
Barlow, J. L., 403(42), 404(44), *413*, 438(68), *441*
Barner, H., 18(45), *33*
Barnett, L., 405, 406(48a), 407(48a), 408(48a), *413*
Baron, L. S., 270(213), *357*
Barrington, L. F., 118, 130(128), 135, 136(128), *147*, *149*
Barry, G. T., 127(172), *148*
Barry, P. J., 295(131), *355*
Bartmann, K., 369(58, 59, 60), *385*
Bateman, J. B., 163(55), *199*
Bauchop, T., 12(27a), *32*
Bauer, K., 133, *149*
Bauman, N., 140, *150*, 256(86), 276, *354*
Baumann-Grace, J. B., 223, *247*, 253, 258, 261, 306(16, 144), 307, *353*, *356*
Baur, E., 184(222), 185(222), *203*
Bayley, S. T., 121(180), 129(180), *148*
Bayne-Jones, S., 37(12a), *93*, 211(21), 212, *246*
Bazeley, P. L., 104(38), *145*
Beeson, P. B., 428(42), *440*
Beighton, E., 165(66), 173(66), *200*
Beiser, S. M., 110, *146*
Belar, K., 235(57), *247*
Beljanski, M., 263, 308, 311, 312, 323, *354*, *358*
Belozersky, A. N., 70(95), *95*
Bensley, R. R., 443, *465*
Benzer, S., 31(101), *34*, 405, 406(48a), 407(48a), 408(48a), *413*
Berger, J. A., 464(117), *468*
Berger, L. R., 131(198), *149*
Bergold, G. H., 219(42), *247*
Bergström, L., 301(241), *358*
Berliner, E., 243(80), *247*
Bernhard, W., 85(130), *96*
Bertani, G., 392(10), *412*
Beutner, E. H., 444(11), *465*
Beyerinck, M. W., 189(256), 190(265), 197, *204*
Billing, E., 216, *246*
Binkley, F., 110(78), *146*
Birch-Andersen, A., *54*, 57(70), 84(129),

85(64, 70), 89, *95*, *96*, 120, *147*, 251, *353*, 446(27), *466*
Bisset, K. A., 36, 77(105), 80(106), 81, 83(106), *93*, *96*, 163(244), 187, *204*, 209, 234(55), 236, *246*, *247*
Bister, F., 111(79), *146*
Black, S. H., *247*
Blanchard, M. L., 12(27), *32*
Blum, J. J., 173, *201*
Blum, L., 138(237), *150*
Bobasch, K., 111(86), *146*
Bocciarelli, D., 176(152), *202*
Bock, R. M., 28(93a), *34*
Böhme, H., 331, *358*, *359*
Boevé, J. J., 144(281), *151*
Bolle, A., 271, *357*
Boltjes, T. Y. Kingma, 165(80), 169(80, 99), *200*
Bolton, E. T. 13(29), 15(29), 18(29), 19(29), 20(29), 30(96a), *33*, *34*
Bonifas, V., 273(77), *354*, 366, 369(38), 381(97), *384*, *386*
Bordner, R. H., 21(60), *33*, 110(72), *146*
Boreck, B. A., 114(99), *146*
Bortels, H., 185, *203*
Bosco, G., 117, *147*
Boulanger, P., 125(163), *148*
Bovarnick, M., 105, 114(99), 125(45), *145*, *146*
Bovell, C. R., 454(65), *467*
Boyd, W. C., 417(7), 418, 419, 422(7), 428, *439*
Boyden, A. A., 424(15), *439*
Boyden, S. V., 416(2), *439*
Brachet, J., 5(8), *32*
Bradfield, J. R. G., 25(83), *34*, 42, 43, 53, 63(79), 85(28), 90(79), *94*, *95*, 175(142), 176(142), 177(142), *201*, 251(6), *353*, 446(26), 458(26), 459(26), *466*
Bradley, D. E., 219(39), 223, 228, *246*, *247*
Braun, H., 163(61), 165(63), *199*, *200*
Braun, W., 6(11), *32*
Breed, R. S., *10*, 180(193), *203*
Brenner, S., 258, 261(40, 42), 308, 309, 312, 324(40, 42), 325, 329(40, 42), 335(40, 42), *341*, *353*, *357*, 405, 406(48a), 407(48a), 408(48a), *413*
Brewer, C. R., 11(22), *32*
Bridoux, M., 308, 311, 312, *356*
Brieger, E. M., 57, *95*, 445(17), *466*

Briggs, S., 142(267), *150*
Bringmann, G., 141(257), *150*
Brinton, C. C., 163(57), *199*, 288(115), 345(115), *355*
Britt, E. M., 144(280), *151*, 286(109), 313, *355, 358*
Britten, R. J., 13(29), 15(29), 18(29), 19(29), 20(29), 30(96a), *33, 34*
Brodie, A. F., 29(95), *34*, 444(11), 448(39), 455(71, 72), 456(73), *465, 466, 467*
Bronk, J. R., 457(83), *467*
Bronn, H. G., 154(11), *198*
Broser, W., 174, *201*
Brown, A. D., 116(109), 126, *147*
Brown, D. D., 136(221), *149*, 336, *356*
Brown, D. H., 139(241), *150*
Brown, R., 103, *145*
Brown, T. McP., 362(13), *384*
Bruckner, V., 105, 106, 125(46, 49), *145*
Bruemmer, J. H., 453(53), 457(53), *466*
Brumfitt, W., 131, *149*, 255, *357*
Buchanan, J. G., 116(110), 119(110), 122(110, 152), 125(152), 129(110), 139(110, 152, 246, 247, 248), *147, 148, 150*, 254(185, 186, 187), *357*
Buder, J., 165(78), 195, *200, 205*
Burdon, K. L., 300(139), *356*
Burkholder, P. R., 180(202), 181(202), 184, *203*
Burns, L., 460(94, 95), *467*
Burris, R. H., 453(57), *467*
Burrous, J. W., 394(12), *412*
Burrows, T. W., 100, *144*
Burrows, W., 138, *149*
Butler, J. A. V., 320(153), 321, *356, 358*
Buttin, G., 322(157), *356*
Buzzell, A., 163(57), *199*, 288(115), 345(115), *355*

C

Calvin, M., 22(62), *33*
Cantacuzène, J., 175, *201*
Carey, W. F., 270, *357*
Carhart, S. R., 159(37), 160(37), *199*
Carne, H. R., 99(12), *144*
Carrère, L., 369(54), *385*
Carss, B., 116(110), 119(110), 122(110, 152), 125(152), 129(110), 139(110, 152, 247, 248), *147, 148, 150*, 254(185, 186, 187), *357*

Carter, C. E., 18(44), *33*
Cary, S. G., 376(91), *385*
Cassel, W. A., 66(86), 67(90), 70(86), 80(107), *95, 96*
Cater, D. B., 175(142), 177(142), *201*
Catlin, B. W., 100(16), *144*
Certes, A., 175(134), *201*
Chain, E., 131(197), *149*, 255(29), *353*
Challice, C. E., 135(216), *149*, 163(53), *199*
Chambers, L. A., 21(60), *33*, 115(104), *146*
Champe, S. P., 405, 406(48a), 407(48a), 408(48a), *413*
Chang, H., 377, *385*
Chao, F. C., 449(46), *466*
Chapman, G. B., 38(17), 43(30), 50(46), 54, 57(68), 83, 85(30, 63), *94, 95*, 223(45), 229(45), 236(45), 238(69), 241(69), *247*, 251, 252, *353*, 446(25), *466*
Chargaff, E., 70(96), *95*, 109(64), *145*, 336, 337, 338, 346, *356*
Chase, M., 19(51a), *33*, 409, *413*
Chatton, E., 209(13, 15), 210, *246*
Cheldelin, V. H., 453(64), 454(66), *467*
Chibnall, A. C., 106, 125(52, 53), *145*
Chung, S. F., 426, *440*
Church, B., 464, *468*
Ciak, J., 273, 274, 275(225), *354, 358*
Cifonelli, J. A., 139, *150*
Cinader, B., 430, *440*
Citri, N., 431, *440*
Clark, F. E., 107, *145*
Clark, W. R., 113, *146*
Claude, A., 443, *465*
Clayton, J., 184, 185(225), *203*
Clayton, R. K., 189(257), 190(272, 273, 274), 191, 192(274), 193, 195, 196, 197, 198, *204, 205*
Clegg, L. F. L., 207(4), 238(4), *246*
Cohen, G. N., 17(34), *33*, 322(154, 157), *356*
Cohen, S. S., 18(45), 19(47, 48), *33*, 399(28), 410(63), *413, 414*
Cohen-Bazire, G., 22(62), *33*
Cohn, F., 158(18), 180(195), 181(195), 182, 190, 192(18), *199, 203*
Cohn, M., 4(7), 6(7, 14), 23(14), *32*, 416(1), 418, 420(10), 421(10), 422(10), 423(10),

430(1, 48), 433(48, 61), *435*, *436*, 437 (62, 63, 64), *439*, *440*, *441*
Collin, B., 209(11), *246*
Colobert, L., 253, 263, 268, *353*, *357*
Colowick, S. P., 22(67), *33*
Conway, E. J., 144(284), *151*
Cook, R. P., 216, *246*
Coons, A. H., 416(4), *439*
Cooper, C., 457(82), *467*
Cooper, P. D., 116, 142(269), *147*, *150*, 271(72), *354*
Cori, C. F., 343(174), *356*
Cori, G. T., 343(174), *356*
Correns, C., 181, 182(209), 183(209), 184, *203*
Cota-Robles, E. H., 52(57), 54(57), *95*, 117, *147*, 445(15), 453(59), 455(59), 456(59), 457(59), 458(59, 89), 459(59, 89), 463(59), *466*, *467*
Coupin, H., 181, *203*
Cowie, D. B., 13(29), 15(29), 18(29), 19(29), 20(29), *33*
Cowles, P. B., 280(95), *355*
Crane, F. L., 453(53), 457(53, 81), *466*, *467*
Crathorn, A. R., 320, 321, *356*, *358*
Crawford, I. P., 433(59), *440*
Crawford, K., 142(267), *150*, 275(224), *311* *318*, *358*
Crick, F. H. C., 19(50), 24(50), *33*, 399, 401, *413*
Crosby, J. H., 188(251), 190(251), 191, *204*
Croson, M., 18(40), *33*, 48(44), *94*, 284(106), 305(106), *355*
Crozier, W. J., 180(198), 181(208), *203*
Crumpton, M. J., 127(171), *148*
Cuckow, F. W., 368(43, 44), 369, 377, 378, 381, *384*
Cummins, C. S., 27(86a), *34*, 116(114), 117(114), 118(114), 120(114), 121(114), 122(114, 144), 123(114, 144), 124(144, 154, 155, 156), 126(114, 144), 127(114, 144), 128(144, 154, 155), 129(154), 137, *147*, *148*, *149*, *241*, 254, 256(27), 343(26), *353*
Curnen, E. C., 428(41), *440*
Czekalowski, J. W., 177(178), *202*

D

Dagley, S., 18(39), 21(61), *33*
Dark, F. A., 122, 126(143), 129, 130(189),

133, *147*, *148*, *149*, 258(194), 265, *354*, *357*
Darter, R. M., 459, *467*
das Gupta, N. N., 163(54), *199*
Davidson, J. N., 18(43), *33*
Davies, D. A. L., 110, 111, 112, 113, 127(171), *146*, *148*
Davis, B. D., 19(55), *33*, 110, 140, *146*, *150*, 256 (86), 276, 277, *289*, *354*
Davis, D. L., 427, *440*
Dawes, E. A., 21(61), *33*
Dawson, I. M., 115(105), 116(105), 135(216, 217), *146*, *149*, 163(53), *199*
Day, N. K., 424(16), *439*
De, M. L., 163(54), *199*
de Bary, A., 238(65), *247*
Deborin, G. A., 313(245), *358*
Decker, T. S., 42(27), *94*, 460(96), *467*
De Gregorio, P., 366, *384*
de Harven, E., 85(130), *96*
DeLamater, E. D., 51(47), 65(81), 67(92), 77(81, 101, 102, 103, 104), 80(106a), 81(81, 102, 111), 83(102), 84(101, 103), *94*, *95*, *96*, 176, 177(158), *202*, 243(74), *247*, 444(12), 445(12), *465*
Delaporte, B., 18(40), *33*, 36(1, 6), 38, 46, 48(44), 50(6), 65(1, 6), 72(97), 76(1, 6), *93*, *94*, *96*, 209(12, 18), 210, 230(18), 243(73), *246*, *247*, 284(106), 305(106), *355*
Delbrück, M., 389, 390(7), 395(22), *412*, *413*
DelCampillo, A., 12(27), *32*
DeLey, J., 23(72), *33*
De Mars, R. I., 439(69), *441*
Demerec, M., 30(100), *34*, 389, *412*
Dempster, G., 163(58), 164(58), *199*, 288(116), *355*
Denbigh, K. G., 174, *201*
Denes, G., 106(49), 125(49), *145*
de Robertis, E., 165(64), 173, 174, *200*, *201*
De Sipin, M., 107(572), *145*
Desjardins, P. R., 275(96), 280, 281(96), *355*
Deubner, B., 176(169), *202*
Deutsch, H. F., 419(14), *439*
Dewy, D. L., 372, *385*
d'Herelle, F., 388, 392, *412*
Dienes, L., 26(86), *34*, 141(262), 143(262),

AUTHOR INDEX

150, 256, *357*, 363, 365, 366, 367(22), 368(45), 372(72), 373, *384*, *385*
DiGrado, C. J., 114(99), *146*
Dimitroff, W. T., 175(141), *201*
Distaso, A., 44, *94*
Dixon, J., 19(51a), *33*
Doak, B. W., 31(102), *34*
Dobell, C., 175, 176(132), 180, *201*
Dobell, C. C., 44, 65(34), 67, *94*
Dobson, M. J., 38(16), *94*
Doermann, A. H., 394(13), 410(61), *412*, *414*
Dolin, M., 12(28), *32*
Dolman, C. E., 377, *385*
Don, P. A., 135(216), *149*, 163(53), *199*
Dondero, N. C., 219(36), 222, 223(36), 224, 228, 236(36), *246*
Dorfman, A., 139, *150*
Douglas, H. C., 446(24), *466*
Douglas, S. R., 44, *94*
Dowding, E. S., 37(12), *93*
Drews, G., 59, *95*
Dreyer, W. J., 271, *357*
Dubin, D. T., 403(41), *413*
Dubos, R. J., 99(8), 113, *144*, *146*, 428, *440*
Dudman, W. F., 109, *145*, *146*
Duggan, P. F., 144(284), *151*
Duguid, J. P., 98(3, 5), 102(3), 109, 140, 141, *144*, *145*, *150*, 163(58), 164(58), *199*, 271(69), 288(116, 117), *354*, *355*, 365(30), *384*, 445(16), *466*
Dunn, D. B., 19(49), *33*
Dunn, M. S., *20*
Dutky, S. R., 243(75), *247*
Dyar, M. T., 99(11), *144*, 175(130), *201*

E

Eaves, G., 177(178), *202*
Eddy, A. A., 281(104), *355*
Eddy, J. J., 424(16), *439*
Edmunds, P. N., 109(66, 71), *145*, *146*, 288(116), *355*
Edward, D. G. FF., 373, 374, 375(87), 376(86), *385*
Edwards, P. N., 163(58), 164(58), *199*
Edwards, P. R., 109(70), 110(72), *146*
Ehrenberg, C. G., 158(17), *199*
Eisen, H. N., 419(13), *439*
Eisenstark, A., 369(53), *384*
Elek, S. D., 426, *440*

Elford, W. J., 135(216, 217), 141(256), *149*, *150*, 163(53), *199*, 377, 381, *385*
Ellis, E. L., 390(7), *412*
Elsden, S. R., 126, *148*
Emerson, M. R., 282, 345, *357*
Emerson, S., 282, 345, *357*
Enderlein, G., 362, *383*
Engelhardt, W. A., 4(6), *32*
Engelmann, T. W., 181, 188(252), 189, 190(258, 259, 261, 262), 194, 195, 197, *203*, *204*, *205*
Epstein, L. A., 131(197), *149*, 255(29), *353*

F

Fabricant, C., 47(40), *94*, 101(21), *144*
Falcone, G., 273, *358*
Fano, U., 389, *412*
Fantham, H. B., 175(138), *201*
Fauré-Fremiet, E., 60, *95*
Fechner, R., 180(205), 181, 182, 183(205), 184, *203*
Federighi, H., 181(208), *203*
Felix, A., 108, *145*, 411(65), *414*
Felsenfeld, O., 171(110), *201*
Fevold, H. L., 261(37, 38), *353*
Few, A. V., 129(179), *148*, 292, 293, 295(137), 300, 301(137), 302, 303(137), 304(137), 312, *355*, *356*, *358*
Fife, M. A., 109(70), *146*
Finland, M., 428(41), *440*
Fischer, A., 39(20), *94*, 158(20), *199*, 251, *353*
Fisher, H., 439(69), *441*
Fisher, H. W., 409(55a), *413*
Fitzgerald, P. L., 109(64), *145*
Fitzgerald, W. A., 373, *385*
Fitz-James, P. C., 18(46), 19(46), 28(46), *33*, 52(59), 58, 70(59, 94), 74(94), 82(115, 121), 83(59), 84(126), 90(59), *95*, *96*, 219, 223(49), *224*, 236(60, 63), *237*, 244(82), 245, *247*, *248*, 283, 290, 305(142), 309(142), 340, *356*, *358*, *359*, 449(44), 461(44), *466*
Flag, W., 369(55), *385*
Fleming, A., 130, *148*, 168, *169*, 171, *200*, 263, *354*
Flewett, T. H., 234(56), 235, 236, *247*
Flournoy, T., 176(156), *202*
Fogg, G. E., 180(204), *203*
Folkes, J. P., 117, *147*, 312, *356*
Forsyth, W. G. C., 107, *145*

Foster, J. W., 31(103), *34*, 214, *246*, 338 (168), *356*
Fowler, E. H., 243(76), 247
Fraenkel-Conrat, H., 24(80), *34*
Frajola, W. J., 446(21), *466*
Franchi, C. M., 165(64), 173, *199*, *201*
Francis, T., 104, *145*
Franklin, J. G., 219(39), 223(39), 228, *246*
Fraser, D., 116, 135(219), *147*, *149*, 257, 268, 284, 329(33, 61, 107), 330, *331*, 332, 333, 335, 336, *345*, *353*, *354*, *355*
Fraser, D. K., 394(12), 395(15), *412*, *413*
Fredericq, P., 392(11), *412*
Freeman, V. J., 411(66), *414*, *426*, *440*
French, C. S., 196, *205*
French, R. G., 410(59), *414*
Frenkel, A. W., 30(97), *34*, 463(105), *468*
Freundt, E. A., 373, 374, 376(86), *385*
Fritsch, F. E., 180(187, 190, 203), *202*, *203*
Fromme, I., 111(81, 83, 84), *146*
Fuerst, C. R., 130(194), *148*, 271, *357*, 403 (39a), *413*
Fuhrmann, F., 158(26), *199*
Fujikawa, K., 269(63), *354*
Fukui, K., 138(232), *149*, 242(72), *247*
Fukushi, K., 52(60), 59(77), *95*
Fukuya, I., 113(93), 138(232), *146*, *149*
Fuller, R. C., 463(106, 107), *468*
Fulthorpe, A. J., 425, *440*
Fulton, M., 159(37), 160(37), *199*

G

Gale, E. F., 17(36), 19(36), 24(36), *33*, 117, *147*, 271(70), 312, *354*, *356*
Gallant, D. L., 277(89, 90), *354*, *355*
Galloway, B., 104(35), *145*
Gamaleia, N., 362(9), *384*
Gamow, G., 400(32), *413*
Gardner, A. D., 271(68), *354*
Garnjobst, L., 185(228), *203*
Gebicki, J. M., 268, 275(217), *357*
Gelman, N. S., 313(245), *358*
Georgi, C. E., 42(27), *94*, 460, *467*
Gerhardt, P., 11(21), *32*, 137, 144(280), *149*, *151*, *247*, 258(194), 281, 286, 299, 301, 303, 306, 307, 313, 343, *355*, *356*, *357*, *358*, 427(32), *440*
Gest, H., 269(201), *357*, 453(62), *467*
Ghuysen, J. M., 125, 130(164), 131, 132, 137(165), *148*, *149*

Giaja, J., 281(103), *355*
Gibbons, I. R., 63(79), 90(79), *95*
Gibbons, N. E., 99(14, 15), 100, 116(15), 121(180), 129(180), 143(15), *144*, *148*, 281, *358*
Gibson, J., 462(103), 463(103), *468*
Gibson, T., 209(10), *246*
Gierer, A., 24(78), *34*, 334(164), *356*
Giesbrecht, P., 89(134), *96*
Gigger, R. P., 125(162), *148*
Gilby, A. R., 129(179), *148*, 253(22), 263, 265, 287(22), 288(22), 289, 290, 292, 293, 294, 295, 300, 301, 302, 303, 304, 312, *353*, *355*, *356*, *358*
Gilchrist, F. M. C., 126(170), *148*
Gillchriest, W. G., 28(93a), *34*
Gillies, R. R., 98(5), *144*, 288(117), *355*
Ginoza, H. S., 275(221), *358*
Ginsburg, B., *20*
Giuntini, J., 154(3), 176(165), 187(3), *198*, *202*
Gladstone, G. P., 142(267), *150*
Glaser, L., 139(140, 141), *150*
Glauert, A. M., *49*, 57, *95*, 445(17), *466*
Glendenning, O. M., 124(156), *148*
Glenn, J. L., 453(53), 457(53, 81), *466*, *467*
Glick, D., 424, *439*
Glick, M. C., 453(56), *466*
Goebel, W. F., 110, 111, *146*, 428(42), *440*
Gonder, R., 177(173), *202*
Good, R. A., 424(16), *439*
Gooder, H., 135(215), *149*, 264, *357*
Gordon, R. E., 107, *145*
Gorin, M. H., 295(127), *355*
Gottschalk, A., 128(177), *148*
Grabar, P., 425, *440*
Grace, J. B., 161(47), *199*
Graham, A. F., 410(59), *414*
Graham-Smith, G. S., 207(1), *246*
Grasset, E., 366, 369(38), *384*
Graudal, H., 171(111), *201*
Gray, C. T. 29(95), *34*, 448(39), 455(71, 72), 456(73), *466*, *467*
Gray, J., 154(14, 15), 155, 156, 157, *198*
Graziosi, F., 125, *148*, 273, *358*
Green, D. E., 117(118), *147*, 443(3), 457(81), *465*, *467*

AUTHOR INDEX

Green, M., 114(98), *146*
Green, T. D., 163(55), *199*
Greenberg, G. R., 254(187), *357*
Greenberg, L. J., 424(16), *439*
Greenberg, R. A., 133, *149*
Greenwood, C. T., 105(44), *145*
Greider, M. H., 446(21), *466*
Grelet, N., 208, 216(30), *246*
Griffiths, M., 22(62), *33*
Griffiths, R. M., 188(247), *204*
Groman, N. B., 426, *440*
Gross, J., 175(143), 176(143, 144), 177 (173), *201*, *202*, 457(80), *467*
Grossbard, E., 129, *148*
Grula, E. A., 132(201, 202), *149*, 263, *354*, 445(18), 446(18), *466*
Guex-Holzer, S., 102(27, 28), 107, 137, *145*, *149*, 253, 258, 306(143), *353*, *356*, 381 (100), *386*, 427(31), *440*
Guha, A., 163(54), *199*
Guillemin, M., *15*
Guilliermond, A., 66, *95*
Gunness, M., *20*
Gunsalus, C. F., 143(279), *151*, 453(54), 457(54), *466*
Gunsalus, I. C., 143(279), *151*, 453(54), 457(54), *466*
György, P., 107(57a), *145*

H

Haanes, M., 176(158), 177(158), *202*
Hackman, C. R., 369(63), *385*
Hadley, Ph., 362, *383*
Haendel, 177(182), *202*
Hagiwara, A., 346, *359*
Hahn, F. E., 273, 274, 275(225, 226), *354*, *358*
Hahnel, E., 131(196), *149*
Hale, C. M. F., 253, *357*
Halpern, P. E., 125(162), *148*
Halvorson, H. O., 133, *149*, 207(3), 238, *246*, 464(114), *468*
Ham, A. W., 51(50), *94*
Hamers, R., 437, *441*
Hamers-Casterman, C., 437, *441*
Hamilton, D. M., 109(62), *145*
Hamilton, M. G., 457(77), *467*
Hampp, E. G., 176(148), *202*
Hanby, W. E., 106, 107(50), 125(50), *145*

Hancock, G. J., 154(10, 14), 155, 156, 157, *198*
Hancock, R., 119, 142(131), *147*
Hannay, C. L., 31(104), *34*, 223(46), 243, 244(83), 245, *247*, *248*
Hannoun, C., 373, *385*
Hanotier, J., 308, 311, 312, *356*
Hansgirg, A., 181, *203*
Hardwick, W. A., 338(168), *356*
Harman, J. W., 443(8), 449(45), *465*, *466*
Harper, E. M., 104(35), *145*
Harrington, M. G., 271(66), 280, 288, *354*
Harris, H., 27(86a), *34*, 122(144), 123(144), 124(144, 154, 155, 156), 126(144), 127 (144), 128(144, 154, 155), 129(154), *147*, *148*, 254(26), 343(26), *353*
Harrison, J. A., 243(76), *247*
Hartman, P. E., 24(77), *34*, 44(35), 67(35), *94*, 444(11), 455(68), 456(68), *465*, *467*
Hartman, R. E., 163(55), *199*
Hartsell, S. E., 132, *149*, 263, *354*, 445(18), 446(18), *466*
Harvey, E. N., 448(37), *466*
Harvey, T. S., 107(57a), *145*
Hashimoto, T., 113(93), 138(232), *146*, *149*, 218(34), 219(34), 229(34), 230(34), 232 (34), 242(72), *246*, *247*
Hawthorne, J. R., 302, *356*
Hayes, W., 6(17), *32*
Hecht, E., 18(41), *33*
Hedén, C.-G., 37(14), *94*
Hehre, E. J., 109(62, 63), *145*
Heidelberger, M., 98(2), 102, 103, 104, *144*, *145*, 428(43), *440*
Heimpel, A. M., 243(77), *247*
Henderson, H. J., 51(47, 54), *94*, 444(12), 445(12), *465*
Henry, H., 45(37), *94*
Heotis, J., 460(94), *467*
Heppel, L. A., 24(81), *34*, 343(176), *357*
Herbert, D., 23(68, 69), *33*, 173, *201*
Herbst, E. J., 11(22), *32*
Herriott, R. M., 398(25), 404(44), *413*
Hershey, A. D., 19 (51a), *33*, 403, 409, *413*
Hess, G. E., 163(55), *199*
Hestrin, S., 98(4), *144*
Higashi, Y., 286(110), 289(110), 292(110), *355*

AUTHOR INDEX

Hijmans, W., 141(262), 143(262), *150*
Hill, A. V., 196, *205*
Hillier, J., 47(40), 54, 83, 85(63), *94 95*, 101(21), *144*, 223, 237, 238, *247*, 251, *353*, 446(25), *466*
Hilmoe, R. J., 343(176), *357*
Himmelspach, K., 111(84), *146*
Hirsch, J. G., 269, *354*
Hitchens, A. P., 180(193), *203*
Hoare, D. S., 126, 131(168), 140(215), *148*, *150*, 276(84), 277(84), *354*
Hobson, P. N., 105, *145*
Höpken, W., 369(58, 59, 60), *385*
Hoerr, N., 443, *465*
Hoffmann, E., 174(123), 177(123, 170), *201*, *202*
Hofmann, E., 174(122, 123), 177(122, 123), *201*
Hogeboom, G. H., 443(4, 7), *465*
Hogness, D. S., 437(62), *440*
Holbert, P. E., 219(36), 222, 223(36), 224, *228*, 236(36), *246*
Holden, J. T., 278, *358*
Holdsworth, E. S., 128, *148*
Hollande, A. C., 82, 93, *96*
Holman, J., 278, *358*
Horikoshi, K., 282, *357*
Horne, R. W., 115(106), 116, 122(106), 125(106), *147*, 161(42), *199*, 342(170), *347*, *356*, 405, 406(48a), 407(48a), *413*, 427(34), *440*
Horvath, S., 102, 103, 106(25, 54), 107, 125(54), *144*, *145*
Hosoda, J., 133, 134, *149*, 264, 282, 322(246, 247), *354*, *358*
Hosoi, A., 183(220), *203*
Hotchin, J. E., 135(216, 217), *149*, 163(53), *199*
Hotchkiss, R. D., 24(75), *34*
Houwink, A. L., 120, *147*, 161(46), 163(46, 52), *199*
Hovenkamp, H. G., 453(60), 454(60), 455(60), *467*
Howatson, A. F., 51(50), *94*
Huber, L., 52(61), *95*
Huertos, M. R., 369(55), *385*
Hughes, D. E., 116, *147*
Hughes, W. H., 168(94), *200*
Hugo, W. B., 28(91), *34*, 447(30), *466*
Human, M. L., 410(62), *414*
Humphrey, J. H., 417(6), *439*

Humphries, J. C., 114, *146*
Hund, A., 141(263, 265), 142(265), 143(263, 265), *150*, 298(240), *358*
Hunter, G. D., 320, 321, *356*, *358*
Hunter, J. R., 464(115), 465(115), *468*
Hunter, M. E., 81(111), *96*, 243(74), *247*
Hunter-Szybalska, M. E., 77(104), *96*
Huppert, J., 134(212, 213), *149*, 270, 271, *357*
Hurwitz, C., *27*, 323, *358*
Hutchinson, H. B., 184, 185(225), *203*
Hutchinson, W. G., 66(86), 70(86), 80(107), *95*, *96*, 368(48), *384*
Hyatt, M. T., 464(116), 465(116), *468*
Hyndman, L. A., 453(57), *467*

I

Ichikawa, S., 269(63), *354*
Iida, S., 282(207, 208), *357*
Ikawa, M., 122(146, 147), 123(147), 124(147), 125(146, 147), *147*
Ingram, M., 449(40), *466*
Ingram, V. M., 126, 127(168), *148*
Inoi, S., 138(232), *149*
Ishimoto, N., 139(245), *150*
Ito, E., 139, *150*
Ivanovics, G., 102, 103, 105, 106(25, 54), 107, 125(46, 54), 138, *144*, *145*, *149*
Izumi, Y., 242(72), *247*

J

Jacob, F., 6(17), *32*, *34*, 130(194), *148*, 271, 346, *357*, 395(18), *397*, *413*
Jacobs, S. E., 207(4), 238(4), *246*
Jacobsen, H. C., 185(233), *203*
Jaenisch, 177(182), 178(182), *202*
Jahn, E., 185, 186, *204*
Jakob, A., 176(164), *202*
James, A. M., 268, 275(217), 295(131, 132), 345(177), *356*, *357*
Jamieson, R. S. P., 109(68), *145*
Jeantet, P., 176(157), *202*
Jennings, H. S., 188(251), 190(251), 191, *204*
Jensen, H. L., 97(1), *144*
Jerne, K., 417(9), *439*
Jerrel, E. A., 268(62), *354*
Jesaitis, M., 111, *146*
Jeynes, M. H., 258(194), 275(98), 280, 281(98), 342, 345, 346, *355*, *357*
Johnson, A. R., 18(39), *33*

AUTHOR INDEX 477

Johnson, F. H., 154(1), 161(41), *198*, *199*
Johnson, M. J., 11(23, 24), *32*
Jones, M. F., 122, 128, *147*
Jones, S. A., 376(91), *385*
Jose, A. G., 323, *356*, 461(99), *468*
Joseph, N., 111(81), *146*

K

Kabat, E. A., 418, 422(11), 423(11), 428 (44, 45, 46), *439*, *440*
Kakutani, I., 113(93), 138(232), *146*, *149*
Kamen, M. D., 30(98), *34*, 463(104), *468*
Kandler, G., 274(79), *354*
Kandler, O., 141, 142(265), 143(263, 265, 266), *150*, 258(194), 274, 298, *354*, *357*, *358*, 372, 374, *385*
Kaplan, N. O., 22(67), *33*
Karunairatnam, M. C., 268, 269, *357*
Kashiba, S., 269(63), *354*
Kass, E. H., 104(37), *145*
Kater, J. C., 99(12), *144*
Kato, K., 130(186), *148*, 286(110), 289(110, 119), 292(110, 119), *355*
Kaufmann, W., 133, *149*
Kay, D., 135(218a), *149*
Kearney, E. B., 176(161), *202*
Keech, D. B., 453(63), 455(63), *467*
Keil, F., 180(197), *203*
Keilin, D., 304(141), *356*
Kekwick, R. A., 419, *439*
Kellenberger, E., 43(32), 51(51), 52, 57 (71), 72(98), 82(120), 83(122), 85(122) 86(71, 122), 89, *92*, *94*, *95*, *96*, 120, 129, 135, 141(261), 143(261), *147*, *150*, 236 (59), *247*, 252, 258(194), *270*, 271, 273, 274(75), 275(75), 284, 287, *331*, 342(75), *354*, *355*, *357*, 369(62), 381, *385*, *386*, 404(48), 409(52), 410(57), *413*, *414*, 446(22), *466*
Kellenberger, G., 287(113), *355*
Keller, E. B., 457(80), *467*
Kendall, F. E., 104, *145*
Kent, L. H., 127, *148*, 343(173), *356*
Kent, P. W., 103, 122(29), *145*
Kermorgant, Y., 176(157), *202*
Keysselitz, G., 175(136), *201*
Khorazo, D., 131(195), *149*
Kielley, W. W., 457(83), *467*
King, H. K., 116, *147*
King, T. E., 453(64), 454(66), *467*

Kitiyakara, A., 449(45), *466*
Kjeldgaard, N. O., 18(44a), *33*
Klebs, B. G., 180(184), *202*
Kleczkowski, A., 107, *145*
Klein, L., 238(66), *247*
Klieneberger-Nobel, E., 234, 236, *247*, 258 (194), *357*, 361, 362, 363(15, 21, 28), 367 (15, 21, 42), 368(1, 43, 44), 370, 371(15), 372, 373, 377(94), 377, 378, 381, *383*, *384*, *385*
Kluyver, A. J., 2(2), *32*
Knapp, R. E., 176(155), *202*
Knaysi, G., *10*, 36, 38(2, 18), 40(22), 46, 47(2, 40), 65(18), 81(110), 82, *93*, *94*, *96*, 101, 115, *144* *147*, 211 (23), 212(23), 216, 232(51), 237, 238, *246*, *247*
Kniep, H., 190(270), 191, 192, *204*
Knight, B. C. J. G., 430(49), *440*
Knoll, M. L., 154(2), 187(2), *198*
Kobayashi, T., 165(75), *200*
Koblmüller, L. O., 171(109), *201*
Koch, G., 111(86), 123(225), 136, *146*, *149*, 271, *357*, 409(53, 54), *413*
Koch, R., 158(19), *199*
Kodicek, E., *294*, *355*
Koffler, H., 165, 169(98), 170(104), *200*
Kolb, J. J., 116(116), 119(116), 120(132), 124(132), 125(132), 140(132), *147*, 260 (36), 267(36), 277(36, 92, 93), 278(36), *353*, *355*
Kolkwitz, R., 180(196), 181(196), *203*
Kono, A., 138(232), *149*
Korkes, S., 12(27), *32*
Kornberg, A., 23(73), *34*
Kornberg, H. L., 4(5, 5a), *32*
Kornberg, S. R., 22(63), *33*
Kostir, W. J., 446(21), *466*
Kottmann, V., 395, *413*
Kovacs, J., 106(49), 125(49), *145*
Kozloff, L. M., 16(32), *33*, 118, 130(128), 135, 136, *147*, *149*, 336, *356*, 404(47), 407(49), *413*
Kramer, I. R. H., 168(94), 171(94), *200*
Kream, J., 114(99), *146*
Krebs, E. G., 430, *440*
Krebs, H. A., 4(5, 5a), *32*
Krenner, J. A., 181, 182(214), 184, *203*
Kroll, A. J., 43(30), 57(68), 85(30), *94*, *95*, 252, *353*
Krueger, A. P., 134(211), *149*

Kühlwein, H., 186, *204*
Kufferath, H., 185(235), *204*
Kuhn, Ph., 361, *383*
Kuster, E., 369(55), *385*
Kvittingen, J., 168, *200*
Kyle, T. S., 369(53), *384*

L

Labaw, L. W., 102, 103, 120, *145*, *147*, 158 (24), 165, *199*, *200*
Lackman, D. B., 115(103), *146*
Lamanna, C., 31(102), *34*, 238, *247*, 447, *466*
Lancefield, R. C., 26(85), *34*, 98(6), 99 (10), 104, 105(36), 114, 137, *144*, *145*, *149*
Landau, J. V., *27*, 323(252), *358*
Landman, O. E., 261(43), 262(43), 275, 308, 317(43), 318, 319(43), *353*, *358*
Lanni, F., 399(27), 404(27), *413*, 438, *441*
Lanni, Y. T., 399(27), 404(27), *413*, 438, *441*
Lark, C., 280, 343, *358*, *359*
Lark, K. G., 275, 280, 342, 343, *358*, *359*, 399(26), *413*
Larson, W. P., *15*
Lauffer, M. A., 163(57), *199*, 288(115), 345 (115), *355*
Lauterborn, R., 182(217), *203*
Lautsch, W., 174, *201*
Lavillaureix, J., 366, 374, *384*, *385*
Lederberg, E. M., 84(127), *96*
Lederberg, J., 6(16), 7(18), 24(76), 31(76), *32*, *34*, 84(127), *96*, 141(260), 143(260, 274), *150*, 190(275), *204*, 256, 267(59), 273, 274, 275(59, 192), 276, 288, 322, 331, 337, 338, 342, 344, 346, *354*, *357*, 369 381, *384*, *386*, 412(68), *414*
Lee, Ki Yong, 19(51), *33*, 401(36), *413*
Lehninger, A. L., 457(82), *467*
Leidy, G., 109(64), *145*
Leifson, E., 159(35, 37), 160(37), 170, 176 (149), *199*, *201*, *202*
Leloir, L. F., 139(239), *150*
Lemoigne, M., 18(40), *33*, 48(44), *94*, 284 (106), 305, *355*
Lennox, E., 437(63), *441*
Lenoir, J., 253, *353*
Léon-Blanco, F., 176(162), *202*

Lerner, P., 432(55), *440*
Lesley, S. M., 410(59), *414*
Lester, R. L., 263, 308, 312, *354*
Levaditi, C., 176(163), *202*, 373(79), *385*
Levine, L., 403(42), *413*, 438(68), *441*
Levine, S., 21(60), *33*, 110, *146*
Levinson, H. S., 464(116), 465(116), *468*
Levinthal, C., 402, 409(55a), *413*, 439(69), *441*
Lewis, D., 126(170), *148*
Lewis, I. M., 65(80), *95*, 216, *246*
Lichtstein, H. C., 457(86), *467*
Lieb, M., 72(98), *96*
Lieberman, M., 134(211), *149*
Liebermeister, K., 141(261), 143(261), *150*, 273, 274(75), 275(75), 284, 287, 342(75), *354*, 369(62), 381(97), *385*, *386*
Lindberg, R. B., 376(91), *385*
Lindström, O., 448(38), *466*
Linnane, A. W., 452(51), 453(51), *466*
Lipardy, J., 45, 67, *94*
Littauer, U. Z., 23(73), *34*
Littlefield, J. W., 457(80), *467*
Lively, E. R., 84(127), *96*
Loebeck, M. E., 185(238), *204*
Löhnis, F., 362, *383*
Loewenthal, W., 174(125), *201*
Lofgren, R., 176(150, 151), *202*
Lorkiewicz, Z., 366, *384*
Lovett, S. L., *20*
Lowick, J. H. B., 295(132), 345(177), *356*, 357
Lüderitz, O., 111(79, 80, 81, 83, 84), *146*
Luria, S. E., 6(10), 31(105), 32(106), *32*, *34*, 389, 394(12), 395(19, 22), 410(62), *412*, *413*, *414*, 439(69), *441*
Lute, M., 16(32), *33*, 136, *149*, 404(47), *413*
Lwoff, A., 3(4), *32*, 327(161), *356*, 392(9), *412*

M

Maaløe, O., 18(44a), *33*, 54, 57(70), 85 (64, 70), 89(70, 133), *95*, *96*, 120, *147*, 251(10), *353*, 446(27), *466*
Maassen, M., 362(10), *384*
McCarty, M., 24(74), *34*, 104(31), 116(31), 118(31), 121(31), 130(188), 132, 133 (31), 137, *145*, *148*, *149*
McCullogh, W. G., 11(22), *32*
Macdonald, J. B., 154(2), 187, *198*
MacFarlane, M. G., 430(49), *440*

AUTHOR INDEX

McIlwain, H., 23(71), *33*, 447(32), *466*
McLaughlin, J., 113(94), *146*
McLean, R. A., 444(11), *465*
Maclennan, A. P., 111(88), *146*
MacLeod, C. M., 24(74), *34*
Macpherson, M. J., 105, *145*
McQuillen, K., 17(36), 19(36), 24(36), 27 (89), *33*, *34*, 129(179), 138(238), 140, 141(254), 142(272), 143(272), *148*, *150*, 253, 255(17), 256(31, 32), 257(32), 258, 259(31), 260, 261(17, 20, 39, 41, 44), 263, 270, 271(31, 71), 273, 274, 275, 276, 277, 278, 279, 280, 281(195), 284(31 32), 287, 288, *292*, *294*, 295, *296*, 297 (17), 298, 299(32), 300, 301(137), 302 (137), 303(137), 304(137), *311*, 312(20), 313, 314, 315, 316, 317, 318, 319, 320 (17), 321 (17, 243a), 322, 324(17, 39), 325, 326, 327, 328, 329(39), *331*, 335 (39), 338(17, 41), *339*, 340, 341, 342 (31, 32), *343*, *344*, 345 (71, 128), 346, 348, *353*, *354*, *355*, *356*, *357*, *358*, 381 (101), *386*
Macrae, R. M., 50(45), *94*
Maculla, E. S., 280(95), *355*
Madoff, S., 256, *357*, 365(31), 372(72), 373, *384*, *385*
Magasanik, B., 30(99), *34*
Magerstedt, C., 176(146), *201*
Mahler, H. R., 257, 268, 284(107), 329(33, 61, 107), 330, 331, 332, 333, 335(33, 107), 336(107), *353*, *354*, *355*
Mallet, G. E., 165(76, 77), 170(104), *200*
Mallette, M. F., 447, *466*
Mandel, P., 281, *358*
Mandelstam, J., 142, *150*
Mandin, J., 369(54), *385*
Manten, A., 188(253), 195(253), 196, *204*
Margalith, P., 114(100), *146*
Mark, H., 109(61), *145*
Marr, A. G., 52(57), 54(57), *95*, 117, *147*, 445(15), 449(41), 453(59), 455(59), 456 (59), 457(59), 458(41, 59, 89), 459(59, 89), 461(41), 463(59, 110), 464(110, 117), *466*, *467*, *468*
Martens, P., 180(191), *203*
Mason, D. J., 68, 77, *95*
Massart, J., 189, 190(276), 191, 192(265), 193, *204*
Mast, S. O., 193, *204*
Mathias, A. P., 139(246, 247), *150*

Mattes, O., 243(81), *247*
Maxon, W. D., 11(24), *32*
Maxted, W. R., 105, 114(40), 133(204), 134(214), 135(215), *145*, *149*, 264, *357*
May, J. W., 83(123), 85(123), *96*, 251, *353*
Mayall, B. H., 58(72), *95*, 218, 219(37), 220, 221, 222(37), 223(37), *226*, 229 (37), 238(37), 241(37), *246*
Mayer, M. M., 418, 422(11), 423(11), *439*
Mayers, V. L., 334(165), *356*
Meadow, P., 139, 140, *150*, 276(83, 84), *277*, *354*
Medill, M. A., 365, 368(48), *384*
Mellon, R. R., 362, *383*
Metzner, P., 165(79), 166(79), 190(79), 191, 192, 193, *200*
Meyer, A., 46, 48, *94*, 222, 229, *247*
Meyer, K., 131(195, 196), *149*
Meyer-Pietschmann, K. (Pietschmann, K.), 159(39), 185, 186, *199*, *203*, *204*
Mickle, H., 116, *147*, 447, *466*
Migula, W., 158(21), 165(21), 180(192), *199*, *203*
Mii, S., 343(175), *356*
Miles, A. A., 377(95), *385*
Militzer, W. E., 42(27), *94*, 460(96), *467*
Millman, L., 459, *467*
Mills, G. T., 104(35), *145*
Mills, R. C., 11(22), *32*
Minck, R., 372, 374, *385*
Mingioli, E. S., 277, *354*
Misra, S. S., 377(95), *385*
Mitchell, P., 17(35), *33*, 39, 40(24), *94*, 117 (125), 126(125), 127(125), 130(192), 134, 142, 143, *147*, *148*, *150*, 158, 184, *198*, 252, 253(21), 260, 263, 265, 266 (35, 56), 280(12), 288(21), 289, 290, 291, 300, 301, 304, 312(21), 322(155), *353*, *354*, *355*, *356*
Miyoshi, M., 189(269), 191, 192(269), *204*
Mölbert, E., 176(154, 168), 177(176), *202*
Molisch, H., 190(271), 192(271), *204*
Monod, J., 4(7), 6(7), 17(34), *32*, *33*, 322 (154, 157), *356*, 411(64), *414*, 437(62), *440*
Morgan, W. T. J., 110, *146*
Morioka, T., 269(63), *354*
Morse, M. L., 18(44), *33*
Morton, D. J., 453(63), 455(63), *467*
Morton, H. E., 176(159, 167), 177(174), *202*

Morton, R. K., 458(88), *467*
Mosimann, W., 110(77), *146*
Mosley, V. M., 102, 103, 120, *145*, *147*, 158 (24), 165, *199*, *200*
Moureau, M., 176(165), *202*
Moyer, L. S., 295(127), *355*
Moyle, J., 17(35), *33*, 39, 40(24), *94*, 117 (125), 126(125), 127(125), 130(192), 134, 142, 143, *147*, *148*, *150*, 252, 253 (21), 260, 263, 265, 266(35, 56), 280 (12), 288(21), 289, 290, 291, 300, 301, 304, 312(21), *353*, *354*, *355*, *356*
Mudd, S., 51(47, 48, 49, 54), 52, 54(65), 77(101), 80(106a), 81(111, 112), 84, *94*, *95*, *96*, 115(103, 104), *146*, 158(22), 176(160), *199*, *202*, 444, 445, *465*
Mühlens, P., 177(172), *202*
Mühlethaler, K., 21(58), *33*, 100(18), *144*
Müller, O., 180, *202*
Mulder, E. G., 16(33), *33*
Muller, F., 372(67), *385*
Muller, M. R., 368(49), *384*
Munoz, J., 424, 426, *439*
Murphy, J. S., 130(193), 135, *148*, 270, *357*
Murray, E. G. D., *10*, 180(193), *203*
Murray, R. G. E., 37(11), 40(23, 31), 42 (23), 43(31), 47(41), 54(125), 67(91), 74(99), 76(99), 77(99), 81(91), 82(91, 99, 117), 83(125), 85, *93*, *94*, *95*, *96*, *217*, *225*, 236(58), *247*, 251, 252, *353*
Murrell, W. G., 207(2), *246*, 464, 465(112), *468*
Muscatine, N. A., 397(23), *413*
Mutsaars, W., 324(159), *356*

N

Nakamura, O., 267, *354*
Nakamura, Y., 109, 113(60), *145*
Naylor, H. B., 218(34), 219(34), 229(34), 230(34), 232(34), *246*
Nazimoff, O., 373(79), *385*
Nečas, O., 281, 308, *355*, *358*
Nelles, A., 366, *384*
Neser, M. L., 159(36), *199*
Neumann, F., 159(38), 175(128), 176, *199*, *201*
Newton, B. A., 286(108), 287(108), 293, 338(108), 348, *355*
Newton, G. A., 462(102), 463(102), *468*

Newton, J. W., 30(98), *34*, 428(39), *440*, 462(102), 463(102, 104), *468*
Nickerson, W. J., 124(157), 128(157), *148*, 264(202), *357*
Nicol, C. S., 375(87), *385*
Nicolle, M., 133, *149*
Nijenhuis, L. E., 195(286), 196, *205*
Niklitschek, A., 181(211), 182(211), 183 (211), 184, *203*
Niklowitz, W., 59, *95*
Nilson, E. H., 52(57), 54(57), *95*, 445(15), 453(59), 455(59), 456(59), 457(59), 458(59), 459(59), 463(59), *466*, *467*
Nishimura, S., 322(246, 247), *358*
Nishino, K., 113(93), 138(232), *146*
Niven, C. F., 105(45), *145*
Noguchi, H., 174(121, 124), 177(121, 179), 178(179), *201*, *202*
Nomura, M., 130(190), 133, 134, *148*, *149*, 264, 282, 322, *354*, *358*
Norén, B., 184, *203*
Norris, J. R., 119, 130, 134, *147*, 311, *356*
Norris, L., 176(156), *202*
Norris, R. F., 107(57a), *145*
Northcote, D. H., 342(170), *356*
Nossal, P. M., 447, 453(63), 455(63), *466*, *467*
Novick, A., 6(15), *32*
Novikoff, A. B., 443(10), *465*
Novy, F. G., 176(155), *202*
Nunemaker, J. C., 362(13), *384*
Nunn, A. J., 159, *199*

O

Oakley, C. L., 425, *440*
O'Brien, P. J., 256, *357*
Ochoa, S., 12(27), 24(81), *32*, *34*, 323, 343 (175), *356*, *358*, 454(67), 455(67), *467*
Oelze, F. W., 177(171), *202*
Ørskov, J., 169(97), *200*
Ogiuti, K., 154(6), *198*
Okada, K., 286(110), 289(110), 292(110), *355*
O'Kane, D. J., 365, *384*
Olitzki, A. L., 426, *440*
Oltmanns, F., 180(186, 189), *202*
Oparin, A. I., 313, *358*
Ordal, E. J., 185(238), *203*
Osteux, R., 125(163), *148*

Otsuji, N., 322, *358*
Oudin, J., 416(5), 417(5), 424, 425, 426(5), *439*
Oxford, A. E., 184, *203*

P

Paigen, K., 456(74), *467*
Palade, G. E., 443(7, 9), 446(19, 28, 29), *465*, *466*
Palen, I., 159(35), *199*
Palmer, J. W., 131(195), *149*
Pangborn, J., 457(85), *458*, *467*
Panijel, J., 134(212, 213), *149*, 270, 271, *357*
Pappenheimer, A. M., 176(156), *202*
Pappenheimer, A. M., Jr., 16(31), *33*, 417(8), *439*
Pardee, A. B., 19(53), 28(92), *33*, *34*, 56(67), 58(67, 74), *95*, 142(268), 143(277), *150*, *151*, 271(67), 273, 280, *354*, *358*, 449(43), 456(74), 462(43), *466*, *467*
Park, B. H., 99(9), 114, *144*
Park, J. T., 19(56), 20, 27(56), *33*, 119, 122(148), 125, 126(148), 127(148), 128(148, 166, 174), 131(199), 138(148, 174), 139, 140, 142(131, 148), *147*, *148*, *149*, *150*, 255(189), 271, 272, 273, 275(219), 279, *354*, *355*, *357*, *358*, 382(105), *386*
Partridge, S. M., 110(76), *146*, 364(28), 372(28), *384*
Payne, J. I., 44(35), 67(35), *94*
Pease, P., 161(48), *199*, 280(97), *355*
Peluffo, C. A., 160(40), 173, 174, *199*, *201*
Pengra, R. M., 323, *356*, 461(99), *468*
Pérard, C., 209(13, 15), 210, *246*
Perkins, H. R., 128(176), *148*
Perlman, E., 110(78), *146*
Perrin, W. S., 175(135), *201*
Perry, J. J., 31(103), *34*, 214(25), *246*
Peshkoff, M. A., 36, 65(10), 82(118), *93*, *96*
Petermann, M. L., 457(77), *467*
Peterson, R. G., 132(200), *149*
Petrie, G. F., 426, *440*
Petrilli, A., 37(12a), *93*, 211(21), 212(21), *246*
Pfeffer, W., 188(249), 189, 190(255, 264), 191, 192, *204*
Phillips, O. P., 180(201), 181(201), *203*
Piéchaud, M., 67(87), *95*

Piekarski, G., 36, 65(4), 67, 76, 89(134), *93*, *95*, *96*
Pietschmann, K., *see* Meyer-Pietschmann, K.
Pijper, A., 158(27), 159, 160(36, 103), 165(81, 82, 83, 84), 167(27, 85), 168(27, 32, 33, 81, 85, 87, 88, 89, 90, 91, 92, 93, 103), 169, 170, 171, 172, 176(33), *199*, *200*, *201*
Pillemer, L., 138(236, 237), *150*
Pinchot, G. B., 455(69, 70), *467*
Pinsent, J., 23(68), *33*
Pirie, A., 324, *356*
Pitt, R. M., 108, *145*
Poetschke, G., 365(32), *384*
Polevitsky, K., 115(104), *146*, 176(160), *202*
Pollock, M. R., 366, *384*, 430, 431, *440*
Pontieri, G., 83(124), 89(124), *96*, 369(57), *385*
Porter, A., 175(139), 176(139), *201*
Porter, J. R., *13*
Porter, K. R., 3(3), *32*, 446(20, 28), *466*
Poulik, M. D., 425, *440*
Powell, E. O., 238, *247*
Powell, J. F., 22(65, 66), *33*, 124(160), *148*, 219, 241(70, 71), *246*, *247*, 343(171), *356*, 464(115), 465(115), *468*
Powelson, D. M., 68, 77, *95*
Prater, C. D., 410(60), *414*
Preer, J. R., 425, *440*
Pressman, D., 419(13), *439*
Prestidge, L. S., 19(53), *33*, 271(67), 279–280, *354*, 456(74), *467*
Preston, N. W., 174(118), 189(118), *201*
Preston, R. D., 129, *148*
Prévot, A. R., 154(3), 187, *198*, *204*
Primosigh, J., 19(52), 26(52), *33*, 111(91), 122(91, 150), 130(150), 135, 136, 141(91, 150), *146*, *147*, 256(30), 257(30), 269, 270, 279(65), *353*, *354*, 461, *468*
Pringsheim, E. G., 82(119), *96*, 180(194), 181(194), 187, 188, *203*, 209(14), *246*
Puck, T. T., 408(51), *413*
Pulvertaft, R. J. V., 369(51), *384*

Q

Quadling, C., 6(12, 13), *32*

R

Rachmeler, M., 433(58), *440*
Radin, N. S., 122(146), 125(146), *147*

Rake, G., 176(167), *202*
Ralston, D. T., 134(211), *149*
Rappaport, C., 397(23), *413*
Rashevsky, N., 196, *205*
Redfield, R. R., 4(4a), *32*
Rees, M. W., *21*, 106(52, 53), 125(52, 53), *145*
Reichert, K., 154(12), 158(12), 159(12), 165(12), 167(12), 173, *198*
Reiner, J. M., *27*, 323(252), *358*
Repaske, R., 267, 268, *354*, 381(102), *386*, 457(84), *467*
Rhuland, L. E., 27(88), *34*, 140, *150*, 276(85), *354*
Rich, A., 400(32), *413*
Richards, F. M., 106(53), 125(53), *145*
Richmond, M. H., 130, 132, 133, 134, *148*, 264, 265, *354*
Rickenberg, H. V., 322, *356*
Rinker, J. N., 169(98), 170(104), *200*
Roberts, I. Z., 321(243a), *358*
Roberts, R. B., 13, 15, 18(29), 19(29), 20, 30(96a), *33*, *34*, 316, 321(243a), *356*, *358*
Robinow, C. F., 36, 40(23), 42(23), 58(72), 65(5, 82, 83), 66(5, 83), 67, 70, 71(83), 72(82), 77(82, 83), 80(83, 108), 82(82, 116, 119), 84(82), *93*, *94*, *95*, *96*, 169(98), *200*, 209(16), 210, 218, 219(37, 38, 40, 42), 220, 221, 222(37), 223(37, 38) 229(37), 230(38), 232(38, 150), 236(38, 61, 62), 237, 238(37), 241(37), 243, *246*, *247*, 251, 283, *353*, *355*
Robrish, S. A., 449(41), 458(41), 461(41), *466*
Roessler, W. G., 11(22), *32*
Rogers, H. J., 104(39), 128(176), 142, *145*, *148*, *150*
Rogers, P., 322, *358*
Romano, A. H., 124(157, 158), 128(157, 158), *148*, 264(202), *357*
Rose, I. A., 454(67), 455(67), *467*
Rose, N. R., 176(167), *202*
Rosenthal, S. M., 403(41), *413*
Ross, K. F. A., 37(13), *93*, 216, *246*
Ross, O. A., 138(236), *150*
Rothert, W., 188(250), 189, 190(250), 191, 192(250), *204*
Rothstein, A., 144(283), *151*
Rotman, B., 449, *466*
Rouiller, C., 60, *95*
Roux, J., 369(54), *385*

Rowley, D., 138(235), *150*
Rubio-Huertos, M., 258(194), 275(96), 280, 281(96), *355*, *357*
Rudzinska, M. A., 51(52), *94*
Ruffilli, D., 8(20), *32*
Ruiter, M., 375, *385*
Rydon, H. N., 106, 107(50), 125(50), *145*
Ryter, A., 43(32), 57(71), 83(122), 85(122), 86(71, 122), 89(131, 132), *92*, *94*, *95*, *96*, 120, 129, *147*, 236(59), *247*, 252, *357*, 410(57), *414*

S

Saidel, H. F., 70(96), *95*
St. Clair, J., 143(274), *150*, 256, 275(192), 276, 331, 344, 346, *357*
Saito, M., 139(245), *150*
Sakaguchi, K., 282(206), *357*
Salton, M. R. J., 18(37), 21(59), 26(59), *33*, 98(7), 104(7, 32), 105(41), 110(74), 111, 115, 116, 117, 118, 120(74, 136), 121(32, 41, 74), 122(74, 106, 145, 149, 151), 123(74, 153), 124(41, 74), 125, 126, 127(74, 145, 149, 153, 168), 128(74, 90, 149, 175), 129(74), 130(149, 185), 131, 132(149, 165), 133(199b), 135(218), 136(149, 151, 218), 137, 140, 141(149, 151), 143, *144*, *145*, *146*, *147*, *148*, *149*, *150*, *151*, 161(42), *199*, 253(24), 254, 255(24), 256(25), 258(194), 259(24), 260, 261(39, 44), 267, 277, 287, 299, 324(39), 326, 327, 328, 329(39), 331, 335(39), 338, 339, *353*, *354*, *356*, *357*, *358*, 381(99), *386*, 427(34, 35, 36, 37), *440*
Sampson, J., 135(218a), *149*
Sanarelli, G., 154(7), *198*
Sanderson, A. R., 139(246), *150*
Santer, M., 15(30), *33*
Sato, K., 52(60), *95*
Schachman, H. K., 28(92), *34*, 56, 58(67, 74), *95*, 143(277), *151*, 449, 462, *466*
Schaechter, M., 18(44a), *33*
Schaudinn, F., 177(170), *202*
Schellack, C., 175(129, 137), 176, *201*
Schellenberg, H., 369(56), 371(64), *385*
Schenk, S. L., 193, *204*
Schewiakoff, W., 188(246), *204*
Schleich, F., 163(60), *199*
Schmerold, W., 176(169), *202*

Schmid, G., 180(206, 207), 181, 182, 183, 184(210), *203*
Schnauder, G., 373, *385*
Schneider, W. C., 443(4, 5, 7), *465*
Schoenberg, M. D., 138(237), *150*
Schramm, G., 24(78), *34*, 334(164), *356*
Schrammeck, J., 195, *205*
Schreil, W., 163(59, 60), *199*
Schroeder, B., 180(185), *202*
Schulman, H. M., 336(167), 337(167), 338(167), 346(167), *356*
Schulman, J. H., 293, *355*
Schultze, M., 182(219), 183(219), *203*
Schulz, G., 154(4), 180(4), 181, 182, 183(4), *198*
Schwert, R. F., 453(61), *467*
Scott, D. B., 176(148), *202*
Seastone, C. V., 104(37), *145*
Séchaud, J., *92*, 236(59), *247*
Sedar, A. W., 51(52), *94*, 446(20), *466*
Seki, Y., 269(63), *354*
Sekiguchi, M., 336, *359*
Sensenbrenner, M., 281(231), *358*
Senseney, C. A., 163(55), *199*
Sergent, E., 185(234), *203*
Sermonti, G., 7(19), *32*
Sertic, V., 134(210), *149*
Shafa, F., 122(151), 136(151), 138, 140, 141(151), *148*, *149*, 287, 299, *358*
Shapiro, H. S., 336(167), 337(167), 338(167), 346(167), *356*
Sharp, J. T., 141, 143(262), *150*, 363, 366, *384*
Shaskan, E., 16(31), *33*
Sheinin, R., 322, *358*
Sherman, J. M., 105(45), *145*
Sherris, J. C., 174, 189(118, 266), 190(266), 191, *201*, *204*
Shimizu, R., 130(186), *148*, 289(119), 292(119), *355*
Shinohara, C., 52(60), 59, *95*
Shockman, G. D., 116, 119, 120, 124(132), 125, 140, *147*, 260, 267, 277(36, 91, 92, 93), 278, *353*, *355*
Shoesmith, J. G., 174(118), 189(118), *201*
Shug, A. L., 284(107), 329(107), 331(107), 335(107), 336(107), *355*
Shugar, D., 144(282), *151*
Sia, R. H., 426, *440*
Siebold, C. T. v., 182(218), 183(218), *203*
Silberstein, J., 373, *385*

Silverman, M., 447(31), *466*
Singer, M. F., 343(176), *357*
Sinkovics, J., 345, 346, *357*, *358*
Sinsheimer, R. L., 395(16), 400(33), 401(34), 402(16), *413*
Sistrom, W. R., 22(62), *33*, 322, *358*
Sjöstrand, F. S., 54, 85(64), 89(133), *95*, *96*, 120(133), *147*, 251(10), *353*
Skuja, H., 183(221), *203*
Slade, H. D., 105(42), 114, 117, *145*, *147*, 264, *357*
Slamp, W. C., 264, *357*
Slater, E. C., 453(60), 454(60), 455(60), *467*
Sloan, J. D., Jr., 464(116), 465(116), *468*
Smiles, J., 38(16), *94*, *383*
Smiley, K. L., 105(45), *145*
Smith, A. G., 54(65), 76(100), 81(112), *95*, *96*
Smith, C. G., 11(23), *32*
Smith, E. E. B., 104, *145*
Smith, F. W., 117(118), *147*
Smith, I. W., 163(56, 58), 164(58), *199*, 288(116), *355*, 445(16), *466*
Smith, J. D., 19(49), *33*
Smith, N. R., *10*, 107, *145*
Smith, W. E., 368(45), *384*
Smithies, W. R., 99(14), 121(180), 129, *144*, *148*
Snell, E. E., 12(26), *32*, 122(146, 147), 123(147), 124(147), 125, *147*
Sobernheim, G., 174(125), *201*
Sohler, A., 124(158), 128(158), *148*, 264, *357*
Sonderegger, T. B., 460(91, 92), *467*
Soule, M. H., 176(150, 151), *202*
Spada-Sermonti, I., 7(19), *32*
Spiegelman, S., 18(46), 19(46), 24(79), 28(46), *33*, *34*, 52(59), 70(59), 83(59), 90, *95*, 261(43), 262(43), 305, 308, 309, 310, 317(43), 318, 319, 320, 321, 323, *331*, *338*, *353*, *356*, *358*, 437(63), *441*, 449(44), 461(44), *466*
Spilman, W., 270(213), *357*
Spizizen, J., 269(201), 331, 333, 334, 336, *356*, *357*, 410(58), *414*
Stacey, M., 45(37), *94*
Stadler, J., 433(60), *440*
Stähelin, H., 251, 253, 289, 290, 346, *353*
Stahmann, M. A., 114(98), *146*
Stanier, G., 464(111), *468*

Stanier, R. Y., 22(62), 28(92), *33*, *34*, 52(58), 56(67), 58(67, 74), *95*, 143(277, 279), *151*, 154(5), 184(223, 224), 185(224, 232, 236), 186, 187, *198*, *203*, *204*, *302*, 449(43), 453(54), 457(54), 462(43), *466*
Stapp, C., 185, *203*
Starr, M. P., 158(25), 165, *199*
Staub, A. M., 111(83), *146*
Stavitsky, A. B., 416(3), *439*
Stempen, H., 141(259), 142(259), *150*, 369, *384*
Stent, G. S., 261(40), 324(40), 325, 329(40), 335(40), *353*, 403(38, 39a), *413*, 439(70), *441*
Sternberg, K., *383*
Stevens, C. M., 125(162), *148*
Stevenson, H. J. R., 21(60), *33*, 110(72), *146*
Stewart, B. T., 464(114), *468*
Stier, T. J. B., 180(198), *203*
Still, J. L., 452(51), 453(51), *466*
Stocker, B. A. D., 6(12), 26(84), *32*, *34*, 170, 171, 172, *201*, 339, *356*, 412(67), *414*
Stokes, A. R., 399(30), *413*
Stokes, J. L., *20*
Storck, R., 29(96), *34*, 143, *151*, 161(45), *199*, 253(23), 255(23), 258(23), 287(23), 299, 305, 311(23), 317(23), 135, 136), 318(23), 338(23), *353*, *356*, 452(52), 459(52), *466*
Strange, R. E., 22(66), *33*, 122, 124(160), 126, 127, 129, 130(189), 133, *147*, *148*, *149*, 241(70, 71), *247*, 258(194), 265, 343(171, 172), *354*, *356*, *357*
Strangeways, W. I., 362(14), *384*
Streisinger G., 401(35), 404(45, 46), 405, 406(48a), 407(48a), 408(48a), *413*
Striebich, M. J., 443(4), *465*
Strominger, J. L., 19(56), 20, 27(56), *33*, 128(174), 138(174), 139(174), *148*, 272, *354*, 382(105), *386*
Stumpf, P. K., 117(118), *147*
Sulitzeanu, A., 426, *440*
Sundman, V., 369(63), *385*
Susich, G., 109(61), *145*
Suskind, S. R., 432(54), *440*
Sutton, R. M., 154(2), 187(2), *198*
Suzuki, J., 52(60), 59(77), *95*

Swain, R. H. A., 176(153), 177(153), *178*, *202*
Sweeny, P. R., 214(24), *246*
Swellengrebel, N. H., 175(140), *201*
Swift, H., 50(46a), *94*
Szybalski, W., 77(104), *96*

T

Tabor, E. C., 21(60), *33*
Takagi, Y. 322, *358*
Takahashi, I., 99(15), 100, 116(15), 143(15), *144*, 281, *358*
Takaishi, K., 113(93), 138(232), *146*, *149*, 242(72), *247*
Takanami, M., 451(48), 457(79), *466*, *467*
Takeya, K., 51(54), *94*
Tamatani, Y., 286(110), 289(110), 292(110), *355*
Tanaka, S., 113(93), 138(232), *146*, *149*, 242(72), *247*
Tani, I., 242(72), *247*
Tatum, E. L., 6(16), *32*
Taubeneck, U., 331, *358*, *359*, 368(49), *384*
Tawara, J., 161(49), *199*
Taylor, E. S., 271(70), *354*
Taylor, G., 154(9, 13), 155, 172, *198*
Taylor, J. H., 5(9), *32*, 84(128), *96*
Teece, G., 125, *148*
Ter Louw, A. L., 38, *94*, 211(22), *246*
Terranova, T., 366, *384*
Tessman, E. S., 403(38), *413*
Tessman, I., 403(38, 39), *413*
Thaxter, R., 185(230), *203*
Thimann, K. V., 154(8), *198*
Thomas, C. A., Jr., 284(107), 329(107), 331(107), 335(107), 336(107), *355*, 402, *413*
Thomas, J. B., 195(285, 286), 196, *205*
Thomas, R., 332, 336, *356*
Thompson, R., 131(195), *149*
Thompson, R. C., *17*
Thorne, C. B., 21(57), *33*, 105(47), 114, 121(47), 125(47), *145*
Thorsson, K. G., 43(29), *94*, 287, *355*, *358*, 460(97), *467*
Thouvenot, H., 154(3), 187(3), *198*
Tillett, W. S., 104, *145*
Tissieres, A., 452(50), 453(60), 454(60), 455(60), 458, *466*, *467*

AUTHOR INDEX

Tkaczyk, S., 37(13), *93*
Toennies, G., 116(116), 119(116), 120 (132), 124(132), 125(132), 140(132), *147*, 260(36), 267(36), 277, 278, *353*, *354*, *355*
Tokuyasu, Y., 252, *357*
Tomcsik, J., 25(82), *34*, 102, 103, 106, 107, 137, 141(258), 143(258), *144*, *145*, *149*, *150*, 223, *247*, 253, 258, 259, 261, 292, 306, 307, *353*, *355*, *356*, *357*, 381(100), *386*, 426(30), 427(31), *440*
Tomlin, S. G., 83(123), 85(123), *96*, 251, *353*
Torriani, A. M., 430(48), 433(48), 436(48), 437(64), *440*, *441*
Toumanoff, C., 245(85), *248*
Truant, J. P., *40*, 47(41), *48*, *94*
Trucco, R. E., 142(268), *150*, 273, *358*
Tsuda, S., 446(23), *466*
Tsuji, F. I., 427, *440*
Tulasne, R., 67, *95*, 363(23, 24), 367(23), 368(46), 372, 374, 376(23), *384*, *385*
Turian, G., 51(51), *94*, 446(22), *466*
Tuttle, L. C., 460(91, 92, 93), *467*
Twort, F. W., 388, *412*

U

Uhlela, V., 167, *200*
Ullrich, H., 180(199, 200), 181(200), 182, 183(199), *203*
Ungermann, E., 177(182, 183), *202*
Utter, M. F., 447(31), *466*

V

Vahle, K., 185, *204*
van den Ende, M., 135(216), *149*, 163(53), *199*
van den Hooff, A., 219(35), 222, 223(35), *228*, 236(35), *246*
Vanderwinkel, E., 161(45), *199*, 253(23), 255(23), 258(23), 287(23), 311(23), 317 (23), 318(23), 338(23), *353*
van Iterson, W., 100(17), *144*, 158(28), 159 (28), *161*, *162*, 163(46), 165(28), 172, 177(177), *199*, *202*
van Niel, C. B., 195(282), *205*
Van Rooyen, C. E., 410(59), *414*
van Thiel, Ph., 177(177), *202*
van Tieghem, P., 169, *200*

Van Vunakis, H., 403(42), *413*, 438(68), *441*
Varela, G., 176(162), *202*
Vatter, A. E., 59, *95*, 143(278), *151*, 252, *357*, 463, *468*
Vely, V. G., 453(56), *466*
Vendrely, R., 45, 67(36, 89), *94*, *95*, 372, *385*
Vennes, J. W., 137, 144(280), *149*, *151*, 286 (109), 299, 301, 303, 306, 307, 343, *355*, *356*, 427(32), *440*
Vetter, J. K., 117, *147*
Viggall, R. H., 176(158), 177(158), *202*
Vigoroux, J., 373(79), *385*
Vincendon, G., 281(231), *358*
Vincent, J. M., 108(59), *145*
Vinter, V., 216(29), *246*
Virieux, J., 188, *204*
Vishniac, H. S., 18(42), *33*
Vishniac, W., 15(30), *33*
Vogel, H. J., 4(7a), *32*
Volcani, B. E., 114(100), 126(170), *146*, *148*
von Klinckowstroem, A., 211, 212, 213, *246*
von Prittwitz und Gaffron, J., 368(47), 369(61), 371(65), 373, *384*, *385*
Voureka, A., 168(94), 171(94), *200*

W

Wachsman, J. T., 29(96), *34*, 143, *151*, 299, 305, 317(135, 136), *356*, 452(52), 459 (52), *466*
Wahl, R., 19(51), *33*, 401(36), *413*
Waley, S. G., 106, 125(51), *145*
Wallenfels, K., 23(70), *33*
Wamoscher, L., 115(102), *146*
Ward, B. H., 261(37), *353*
Ward, C. B., Jr., 369(53), *384*
Wardlaw, A. C., 131(199), *149*, 255(189), *357*
Warren, G., 161(41), *199*
Watson, J. D., 19(50), 24(50), *33*, 399, 401, *413*
Watson, J. H. L., 176(162), *202*
Watson, M. L., 54(62), *95*
Weber, M. M., 456(73), *467*
Webley, D. M., 107, *145*

Webster, G. C., 457(78), *467*
Webster, M. E., 113(94), *146*
Wedderburn, C. G., 176(162), *202*
Weibull, C., 18(38), 27(38), *27*, 27(90), *33*, *34*, 42(25, 26), 43(29), 51, 52, *94*, *95*, 99(13), 117(126), 141, 142(271), 143, *144*, *147*, *150*, 158(23, 29, 30), 159 (29, 34), 161(44), 162(50), 163(44, 50), 165(23, 65, 66, 67, 68, 69, 70, 71, 72, 73), 173, 174(67), *199*, *200*, *201*, 253, 256, 258(194), 261(13, 15), 262(13, 15), 263, 274, 283(13, 120), 287, 288, 289, 290(120), 297, 299, 300, 301, 304, 305, 311, 324(13), 338, 339, *353*, *354*, *355*, *356*, *357*, *358*, 363, 372, 381(98), *384*, *386*, 427, *440*, 445, 449(42), 453(55), 459(55), 460, 461(100, 101), *466*, *467*, *468*
Weidel, W., 19(52), 26(52), *33*, 111, 119, 120(139), 122(91, 150), 123(225), 128, 130(129, 150), 135, 136, 141(91, 150), *146*, *147*, *149*, 256(30), 257(30), 269, 270, 279(65), *353*, *354*, 409(52, 53, 54), *413*, 461, *468*
Weigle, J. J., 72(98), *96*, 401(35), *413*
Weinberger, H. J., 26(86), *34*, 363(22), 365(31), 367(22), 372(72), 373, *384*, *385*
Weiner, M., 6(15), *32*
Weiser, R. S., 131(198), *149*
Weiss, P., 185(237), *203*
Welsch, M., 130(187), 132, *148*, *149*, 275 (229), 281, *358*
Wensinck, F., 144(281), *151*
Wentholt, H. M. M., 375, *385*
Werkman, C. H., 447(31), *466*
West, G. S., 188(247), *204*
Westphal, O., 111, *146*
Whatley, F. R., 463(108), *468*
White, P. B., 171, *201*
Whitehouse, M. W., 103, 122(29), *145*
Whitfield, J. F., 37(11), 74(99), 76(99), 77(99), 82(99, 117), 85, *93*, *96*, 236(58), *247*
Wiame, J. M., 22(64), *33*, 161(45), *199*, 253(23), 255(23), 258, 287(23), 311, 317, 318, 338(23), *353*
Wickham, N., 99(12), *144*
Widra, A., 46, 51(38), 52, 70(38), *94*
Wierzchowski, P., 107, *145*

Wiggert, W. P., 447(31), *466*
Wile, U. J., 176(161), *202*
Wilkins, M. H. F., 399(30), *413*
Wilkinson, J. F., 50(45), *94*, 100, 109, *144*, *145*, *146*, 445(16), *466*
Williams, C. A., 425, *440*
Williams, R. C., 24(80), *34* 118, 120(136), 135(219), *147*, *149*, 158(25), 165, *199*, 395(15), *413*
Williamson, D. H., 281(104), *355*
Wilson, E. B., 2(1), *32*
Wilson, H. R., 399(30), *413*
Wilson, M. W., 425, *440*
Wilson, P. W., *323*, 453(53, 57, 58), 453 (58), 457(58), *466*, *467*
Winkler, A., 36, 48, 65(7), *93*, *94*, 363(25), 369, *384*
Winogradsky, S., 185, 189(268), *204*
Winterscheid, L. C., 51(47), *94*, 444(11, 12), 445(12), *465*
Wismer, C., 453(64), *467*
Wittler, R. G., 376(91), *385*
Wolfe, R. S., 59, *95*, 143(278), *151*, 252, *357*, 463, *468*
Wolfrom, M. L., 428(43), *440*
Wolin, M. J., 457(86), *467*
Wollman, E. L., 6(17), *32*, *34*, 346, *357*, 395 (18), *397*, 411(64), *413*, *414*
Wood, W. A., 453(61), *467*
Work, E., 19(54), *33*, 121(140), 122, 123 (140), 126, 131(168), 140, 141, *147*, *148*, *150*, 254(28), 256(28), 276, *277*, 343 (28), *353*, *354*, 372, *385*
Wright, R. R., 430, *440*
Wurz, L., 138(237), *150*
Wyatt, G. R., 19(48), *33*, 399(28, 29), *413*
Wyckoff, R. W. G., 38, *94*, 176(148), *202*, 211(22), *246*

Y

Yamada, E., 252, *357*
Yanofsky, C., 432(55, 56, 57), 433(58, 59, 60), *440*
Ycas, M., 400(32), *413*
Yoshida, N., 113(93), 138, *146*, *149*, 242 (72), 247
Young, I. E., 208, 212(9), 218(9), 219(9), 222(9), 223(9), 229(9), 230(9), 232, 234, 235, 236, 245, *246*, *248*

Z

Zamecnik, P. C., 365, *384*, 457(80), *467*
Zamenhof, S., 109, *145*
Zarnitz, M. L., 23(70), *33*
Zehender, C., 141, 142(265), 143(263, 265, 266), *150*, 274, 298, *354*, *358*, 372, 374, *385*
Zettnow, E., 176(147), *202*
Zilliken, F., 256, *357*
Zilliken, F. W., 107(572), *145*

Zinder, N. D., 24(76), 31(76), *34*, 84(127), *96*, 267, 269, 270, 289, 329(58), 331, 339, 345, 346, *354*, 381(103), *386*, 412(68, 69), *414*
Zuckerman, Avivah, 209(17), *246*
Zuelzer, M., 174(126), 175(127), 176(145), 177(180), *201*, *202*
Zukal, H., 181, *203*
Zworykin, K. A., 38(17), *94*, 238(69), 241(69), *247*
Zworykin, N., 161(41), *199*

SUBJECT INDEX

A

Abequose, lipopolysaccharides and, 111
Abrasives, cell disruption and, 447
"Accessory chromatic granules", 93
  structure of, 84–85
Acetate,
  incorporation of, 314–316
  lactobacilli and, 12
  oxidation, protoplasts and, 323
Acetobacter, surface of, 98, 109, 112
Acetobacter capsulatum, capsule of, 109, 112
Acetobacter suboxydans, particles, substrate oxidation and, 453, 454
Acetobacter viscosum, capsule of, 109
Acetobacter xylinum, cellulose and, 21, 100
Acetocarmine, chromatin bodies and, 66
Acetylamino sugars,
  lysozyme and, 130–131
  Streptomyces enzymes and, 133
N-Acetylfucosamine, cell walls and, 127
Acetylglucosamine
  cell walls and, 140
  group-specific polysaccharide and, 137
  lysozyme and, 131
  O antigens and, 110
Acetyl groups,
  amino sugars and, 127
  cell walls and, 116, 122
  C substance and, 104
  lysozyme and, 255
Acetylmuramic acid, lysozyme and, 131
N-Acetylneuraminic acid,
  cell walls and, 127
  function of, 128
Achromatium, movement and, 187–188
Achromatium oxaliferum, vacuoles of, 47
Achromobacter fischeri, antisera and, 427
Acridine orange, chromatin and, 39
Aconitase, cell fractions and, 454
Actinolysopeptidase,
  bacteriolysis and, 130
  receptors and, 137
Actinomyces bovis, size of, 9

Actinomycetes,
  cell walls of, 27
  lamellated structures in, 52
Actinophrys, gametes of, 235
Actomyosin, flagella and, 165
Adenosine deaminase,
  endospores and, 464
  heat resistance of, 465
Adenosine triphosphatase,
  biochemical unity and, 4
  cell fractions and, 454
  electron-transport particle and, 460
  flagella and, 26, 173
Adenosine triphosphate,
  bacterial growth and, 12
  bacteriophage and, 406–407
  flagella and, 173
  polyphosphates and, 22
Aerobacter,
  cell wall isolation and, 118
  chemotaxis and, 189
  polysaccharides and, 21
Aerobacter aerogenes,
  capsule of, 109, 112
  galactosidase, inactive protein and, 437
  oxidative phosphorylation in, 455
  penicillin and, 275
  protoplasts and, 268
  sedimentation constants and, 29
  vitamins in, 17
Aerotaxis,
  oxygen concentration and, 190–191
  Rhodospirillum rubrum and, 197
Agar,
  etching, myxobacteria and, 186
Agglutination, serological reactions and, 416
Agglutinins, normal sera and, 423
Aggregation, flagella and, 158–159
Agrobacterium tumefaciens, spherical forms of, 280
Alanine,
  bacteriolysins and, 132
  cell walls and, 122, 123, 125, 139
  incorporation and, 315

lysozyme and, 131
phage enzymes and, 136
protoplasts and, 278
Alanine racemase,
  endospores and, 464
  heat resistance of, 465
Albumin, protoplasts and, 289
Alcaligenes faecalis,
  cell membranes of, 323
  oxidative phosphorylation in, 455
  penicillin and, 275
  protoplasts of, 280–281, 342–343
Alcohols, protoplasts and, 294
Aldoheptose, surface components and, 111
Alkaloids, chemotaxis and, 192
Alum, antigen precipitation and, 420
Alumina, cell disruption and, 447
Amberlite, protoplasts and, 268
Amino acids,
  bacterial content of, 20
  bacteriolysins and, 132
  cell walls and, 123–126, 140, 254, 256
  incorporation, protoplasts and, 308, 312–316, 319–320
  chemotaxis and, 192
  chloramphenicol and, 142
  flagella and, 165
  L-forms and, 365
  protoplasts and, 259, 301
Amino sugars,
  capsules and, 103, 104
  cell walls and, 121, 126–128, 140
Amylase, protoplasts and, 322
Antibiotics, 14
  cell wall synthesis and, 140–142
  chromatin bodies and, 77
Antibodies,
  bacteriophage and, 404, 438–439
  enzyme inhibition and, 429–431
  reaction in gels, 424–425
  reaction in liquid, 423–424
Antigens,
  bacteriophage and, 399
  common, detection of, 426
  dosages of, 421–422
  injection routes and, 421
  injection schedules and, 421
  L-forms and, 372–373
  minor components and, 422
  particulate, preparation of, 419

protoplasts and, 306
soluble, preparation of, 420
somatic, receptors and, 136
Antireceptors, antibodies and, 136
Antisera,
  cross reacting, chemical basis of, 428–429
  genetic mutants and, 431–438
  L-forms and, 365, 373
  preparation,
    animals and, 418
    antigens and, 419–420
    bleedings and, 422–423
    collection and storage, 418
    fractionation and, 419
    injection and, 420–422
    testing of, 423
Antisera,
  useful properties of, 415
Apyrase, 460
Arabinose,
  cell walls and, 128
  protoplasts and, 291
  surface components and, 112
Arabokinase, protoplasts and, 317
Arginine,
  flagellar movement and, 174
  molar growth yield and, 12
Arlacel A, antigen preparation and, 420
Aromatic compounds, chemotaxis and, 192
Ascorbate, chemotaxis and, 192
Aspartic acid,
  cell walls and, 123, 124
  protoplasts and, 314, 315–316
Aspergillus oryzae, protoplasts and, 282
Autolysin,
  protoplasts and, 264
  specificity of, 133–134
Autolysis,
  cell wall isolation and, 119
  protoplasts and, 265–267
Azide,
  amino acid incorporation and, 313
  flagella and, 174
Azotobacter,
  chromatin bodies of, 72
  nitrogen fixation and, 16
  plasmolysis and, 253
  spores and, 209
  vacuoles of, 47

SUBJECT INDEX

Azotobacter agilis,
  cytochrome oxidase in, 52
  cytochromes of, 453, 458
  electron-transport particles and, 457, 459
  neotetrazolium and, 445
  osmotic disruption of, 449
  oxidative phosphorylation in, 454, 455
  particles, substrate oxidation and, 453, 454
  ribonucleoprotein of, 456
Azotobacter vinelandii,
  disintegration of, 117
  particles of, 452
  protoplasts, 268
    nitrogen fixation and, 323-324
    sedimentation constants and, 29
Azure,
  chromatin bodies and, 67
  L-forms and, 367

B

Bacillus,
  bacteriolysins and, 132, 133-134
  capsules, 21, 108
    enzymes and, 114, 115
  cell walls of, 125
  chemotaxis and, 189
  chromatin bodies of, 66, 74
  cytoplasmic membrane of, 43
  di-picolinic acid and, 22
  flagella of, 165
  spores, 209
    enzymes and, 265
Bacillus alesti, parasporal bodies and, 244-245
Bacillus anthracis,
  capsule, 102, 105, 108
    antigens of, 306
    enzymes and, 114, 115
  exosporium of, 223
  fusion cylinders in, 236
  plasmoptysis and, 251
  protoplasts of, 290, 346
  size of, 9
  spore chromatin in, 234, 235
  water content of, 13
Bacillus brevis, flagella of, 165
Bacillus cereus,
  amino sugars and, 126, 127
  autolysin and, 264

cell walls,
  amino acids and, 124
  isolation of, 119
  chromatin bodies of, 68, 83, 85
  cytoplasmic membrane and, 251, 252
  endospores of, 464
  endospore walls of, 129
  enzymes, protoplasts and, 311
  exosporium of, 223, 227
  extracellular polysaccharides of, 107
  fusion cylinders in, 235, 236
  lysozyme and, 261
  lytic enzymes of, 133
  nuclear membrane in, 81
  penicillinases of, 430-431
  spores, 208, 212, 216
    chromatin and, 231, 232, 234-237, 243
    cortex of, 222
    envelopes and, 229
    enzymes of, 265
    germination of, 238, 241-242
Bacillus circulans,
  capsule of, 107, 108
  enzymes, protoplasts and, 282
Bacillus coli, see *Escherichia coli*
Bacillus laterosporus, parasporal bodies and, 244-245
Bacillus licheniformis,
  polyglutamic acid of, 106
  spore cortex of, 219
Bacillus M,
  cytoplasmic membranes of, 42, 43
  electron-transport particle and, 460
Bacillus medusa, parasporal bodies and, 243-244
Bacillus megaterium,
  amino sugars and, 127
  antisera and, 427
  autolysin and, 264
  bacteriolysins and, 132
  bacteriophage,
    enzymes and, 135, 270
    receptors and, 135
  capsule of, 102, 106, 107, 108
  cell walls, 125, 255
    amino acids and, 126
    antigens and, 138
  chromatin bodies of, 70, 83, 84, 89
  cytochrome oxidase in, 52
  cytochromes of, 453
  cytoplasmic membranes of, 27

deoxyribonucleic acid of, 18
diaminopimelic acid and, 276
disintegration of, 117
endospore walls of, 129
exosporium of, 223–226, 228
extracellular polysaccharide and, 107
flagella of, 162, 163
granules in, 305
linoleic acid and, 294–295
lipid droplets in, 48–50
lysogenic, protoplasts of, 327–328
lysozyme and, 131, 253–255, 261–263
nuclear bodies of, 28, 461
osmotic properties of, 289–290
polysaccharides in, 51
protein in, 19
protoplasmic movements and, 37
protoplasts, 27, 40, 381
    appearance of, 258
    bacteriophage and, 260, 324–328, 346
    cell walls and, 259, 372
    chemical composition of, 259, 295–297
    detergents and, 292
    electron microscopy of, 287
    electron transport and, 459
    enzymes and, 308–311, 317–319
    fixation of, 286
    growth of, 324, 339–342
    immunochemical properties of, 137, 259, 306–307
    lysed, 320–321
    membrane and, 42, 299–301
    motility and, 338–339
    number per cell, 283
    osmotic properties of, 258–259, 289–290
    radioactive tracers and, 313–316
    respiration and, 311
    reversion of, 343–344
    spores and, 338
    stabilization of, 288, 289
    staining characteristics of, 259–260, 286–287
receptors of, 136–137
sedimentation constants and, 29
serological types of, 306
size of, 9
soluble fraction of, 452
spores, 208, 211, 216
    chromatin and, 232, 234–237, 243

coats of, 223
envelopes of, 218–222
enzymes and, 265
germination of, 238–241
triphenyltetrazolium and, 445
Bacillus mycoides,
  disintegration of, 117
  exosporium of, 223
  spore chromatin in, 234
  sporulation of, 338
Bacillus polymyxa,
  exosporium of, 224, 228
  extracellular polysaccharide and, 107, 108
  spores,
    chromatin and, 236
    coats of, 223
    cortex of, 222
    wall of, 219
Bacillus proteus, L-forms of, 363, 377
Bacillus pumilis, extracellular polysaccharides and, 107
Bacillus sphaericus,
  cell walls, 124
  isolation of, 119
Bacillus stearothermophilus,
  cell wall amino acids, 124
  electron-transport particles and, 460
Bacillus subtilis,
  autolysin and, 264
  capsules and, 105–106
  cell walls of, 121, 254, 255
  cytoplasmic membrane and, 252
  endospore walls of, 129
  enzymes, capsules and, 114, 115
  extracellular polysaccharides and, 107, 108
  flagella of, 158, 165
  lipids of, 18, 129
  lysozyme, 130, 132, 253–255
    production of, 264
  lytic enzymes of, 133–134
  membranous structures in, 60
  mitochondria and, 51
  protoplasts, 40
    appearance of, 258, 282
    electron microscopy of, 287
    enzymes and, 317, 321–322
    respiration of, 311
  size of, 9

SUBJECT INDEX

spores, 216
   chromatin and, 236–237, 243
   germination of, 238, 239
Bacillus thuringiensis,
   exosporium of, 223
   parasporal bodies and, 243–245
Bacitracin,
   action of, 142
   cell walls and, 124, 273
Bacteria,
   aberrant forms of, 361–362
   acid soluble fraction of, 16–17
   cell size and, 8–10
   chemical analysis of, 13–22
   classification, bacteriophage and, 411–412
   continuous flow cultures of, 11
   cytology, outline of, 25
   fractionation of, 25–28
   growth limitation and, 10–13
   intercellular space and, 13
   intramolecular organization and, 30–31
   osmotic pressure of, 17
   submicroscopic structure of, 28–31
   surface components of, 98–99
   unicellularity of, 5–6
   water content of, 13
Bacteriocines, protoplasts and, 260
Bacteriophage,
   adenosine triphosphate and, 406–407
   adsorption,
      cell wall digestion and, 409
      kinetics of, 408
      receptors and, 408–409
      sites of, 403–404
   antibodies and, 404, 438–439
   antisera and, 422
   bacterial classification and, 411–412
   bacterial typing and, 411
   bacteriolysis and, 130, 134–135
   capsule and, 105, 114
   cell wall interactions and, 135–137
   composition of, 31
   contractile properties of, 404–408
   damaged, protoplasts and, 333–336
   deoxyribonucleic acid, 400–403
      injection of, 409–410
   enzymes, protoplasts and, 269–271
   glucose and, 401
   host properties and, 410–411
   liquid cultures and, 392

mutants,
   host-range, 389–390
   plaque-type, 388
   osmotically shocked, 397–399
   infectivity of, 410
   phenotypic mixing of, 404
   protoplasts and, 324–336, 346–347, 410
   radioactive decay and, 402–403
   receptors and, 324
   resistant mutants and, 389
   "serum-blocking power" and, 439
   sources of, 388
   spherical forms and, 255
   structure,
      basic materials and, 403
      gross morphology and, 394–397
      head membrane and, 397–399
      nucleic acid and, 399–403
      tail protein and, 403–408
   suprainfection and, 410
   temperate,
      life cycles of, 392–394
      nucleic acid of, 401
      transduction and, 401
   tryptophan and, 403–404
   value of, 387–388
   virulent, life cycles of, 390–392
Bacterium, chemotaxis and, 189
Bacterium photometricum, phototaxis and, 194–195
"Bacterium termo", chemotaxis and, 190
Ballotini beads, cell wall isolation and, 116
Bartonella bacilliformis, size of, 9
Basophilia, spores and, 230, 232
Bayol F, antigen preparation and, 420
Beggiatoa,
   chemotaxis and, 189
   movement of, 180–184
Beggiatoa mirabilis, contractions of, 181–182
Biotin, bacteria and, 17
Bleeding, antisera and, 422–423
Blepharoplasts, bacterial flagella and, 161–162
Blood group substance, cross reactions of, 428–429
Blood platelets, lysins and, 130
Blue-green alga, sedimentation constants and, 29
Bordetella, chemotaxis and, 189

Bordetella pertussis, cell wall antigens and, 138
Borrelia, movement of, 176
Bouin's fixative, L-forms and, 367
Brilliant cresyl blue, vital staining and, 38
Brucella bronchiseptica, flagella of, 158
Brucella melitensis, size of, 9
Brucella suis, aeration and, 10–11
Buffers, bacterial growth and, 11–12

C

Cadmium,
  cyanide complex, bacteriophage and, 404
Calcium,
  bacterial content of, 15, 16
  bacteriophage and, 399
  chromatin bodies and, 89
  fusion cylinders and, 236
  spores and, 216
Capsules,
  characteristics of, 101–102
  composition of, 102–113
  growth conditions and, 100
  removal of, 113
  spore formation and, 214
  viability and, 99
Carbohydrates,
  bacterial content of, 21
  chemotaxis and, 192
  chromatophores and, 462
  flagella and, 165
  protoplasts and, 302
Carbon, bacterial content of, 14, 15
Carbon dioxide,
  fixation, chromatophores and, 463
Carborundum, cell disruption and, 447
Carotenoids,
  chromatophores and, 462
  phototaxis and, 195
  protoplasts and, 302
Caryophanon latum,
  chromatin bodies of, 82, 83, 89
  cytoplasmic membrane and, 252
  size of, 10
Catalase,
  bacterial content of, 23
  endospores and, 464
  heat resistance of, 465
Cations, chromatin bodies and, 74, 85–89

Caulobacter vibrioides, vacuoles in, 445–446
Cellobiose, protoplasts and, 261, 288
Cellobiuronic acid,
  capsules and, 103, 104
  enzymes and, 113
Cells,
  as factories, 7–8
  disruption,
    abrasives and, 447
    ballistic, 447
    osmotic, 448–450
    sonic, 448
  fractionation of, 450–452
  fractions,
    electron-transport and, 452–453
    localization of, 458–461
    oxidative phosphorylation and, 454–456
    properties of, 452–456
    submicroscopic particles and, 456–458
    substrate oxidation and, 453–454
  organization and, 1–2
  variations among, 4–5
Cell theory,
  biochemical unity and, 2
  structure and, 2–3
Cellulose, 139
  bacteria and, 21, 109
  bacterial surface and, 98, 100
Cell walls,
  amino acid incorporation in, 313
  antibiotics and, 275
  antigen preparation and, 419
  antigens of, 306–307
  antisera and, 426, 427–428
  bacteriophage and, 135–137, 324, 404, 408–409
  biosynthesis of, 138–142
  composition of, 21, 120–129, 254
  contamination of, 118
  definition of, 250–251
  digestion, bacteriophage and, 409
  electron microscopy and, 53
  electron-transport particle and, 459–461
  enzymes and, 129–135
  extraction of, 115–116
  function of, 142–144
  gram-negative bacteria and, 256

SUBJECT INDEX

isolated, behavior of, 259
isolation, 26–27, 115–116
  autolysis and, 119
  mechanical disintegration and, 116–119
  L-forms and, 369, 372
  physical properties of, 120
  pleuropneumonia-like organisms and, 374
  spore walls and, 229
  uridine nucleotides and, 271–273
Centrifugation, cell wall isolation and, 117
Centrioles, 93
  chromatin bodies and, 85
Centrosomes, 3
Cetyltrimethylammonium bromide, protoplasts and, 292
Charcoal, L-forms and, 366
Chemotaxis,
  definition of, 189
  Weber law and, 192–193
Chicken,
  antisera and, 418
  bleeding of, 422
Chitin, 139
  lysozyme and, 131
Chitinase, protoplasts and, 282
Chloramphenicol,
  action of, 142
  flagella and, 170
  teichoic acid and, 139
Chlorobium limicola,
  chromatophores of, 59, 463
  cytoplasmic membrane and, 252
Chlorobium thiosulphatophilum,
  amino sugars and, 127
  chromatophores of, 462
Chloroform, chemotaxis and, 192
Chlorophyll,
  bacteria and, 21–22
  chromatophores and, 462
Chlortetracycline, adaptation and, 319
Cholesterol,
  L-forms and, 364, 371, 372
  protoplasts and, 294–295
Chondrioids, bacteria and, 51–52
Christispira anodontae, movement of, 175–176
Christispira balbianii, movement of, 175–176

Chromatin,
  demonstration of, 44–45
  electron microscopy and, 53–54
  extrusion of, 234
  fusion cylinders and, 235–236
  phase microscopy and, 37
  protoplasts and, 283
  spore germination and, 243
  spores and, 216, 230–237
Chromatin bodies,
  arrangement of, 71–76
  demonstration of, 64–65
  division and, 76–80
  electron-transport particles and, 460
  staining of, 66–68, 70–71
  structure, 65
    chromatinic portion, 83–90
    nonchromatinic portion, 80–83
    total, 90–93
Chromatium,
  chemotaxis and, 189
  chromatophores and, 59, 462, 463
  movement of, 166
  photophosphorylation in, 463
  sonication of, 463–464
Chromatium okenii, movement of, 167
Chromatophores,
  antisera and, 427–428
  carbon dioxide fixation and, 463
  cell walls and, 143
  composition of, 462
  photosynthetic bacteria and, 58–59, 461–464
  sedimentation of, 30
Chromobacterium violaceum, polysaccharides of, 112
Chromosomes, chromatin bodies and, 92–93
Citrate,
  bacterial growth and, 11
  chemotaxis and, 192
  lactobacilli and, 12
  oxidation, cell fractions and, 454
  protoplasts and, 268, 288–289
Clones, differentiation in, 6–7
Clostridium,
  chemotaxis and, 189, 191
  di-picolinic acid in, 22
  polysaccharides in, 21, 50
  spores and, 209
  α-toxin of, 430

Clostridium butylicum, vitamins in, 17
Clostridium kluyveri,
  growth yields of, 12
  sedimentation constants and, 29
Clostridium pectinovorum,
  exosporium of, 225–229, 242
  spores,
    chromatin and, 232–234, 236
    envelopes of, 217
    formation of, 213–215
    germination of, 238, 242
Clostridium perfringens, bacteriolysins and, 132
Clostridium sporogenes,
  spores,
    cortex of, 219
    envelopes of, 229
Clostridium tertium, blood group substances and, 429
Clostridium tetani, spore chromatin in, 234
Clostridium welchii, bacteriolysins and, 132
Cobalt, bacteria and, 16
Coccus LC$_1$, amino sugars and, 126, 127
Colchicine, chromatin bodies and, 77
Colicines, 392
Complement, L-forms and, 365
Conjugation, protoplasts and, 346
Continuous flow, bacterial cultures and, 11
Copper sulfate, chemotaxis and, 192
Cortex, spores and, 219–223
Corynebacterium,
  cell wall sugars, 128
  chemotaxis and, 189
  cytoplasmic membrane and, 252
  L-forms of, 365
Corynebacterium diphtheriae,
  arabinose in, 128
  bacteriophage and, 411
  cell wall antigens of, 137
  diaminopimelic acid and, 276
  toxigenic strains of, 426
  water content of, 13
Corynebacterium ovis, antigens of, 137
Cross-reacting material,
  β-galactosidase and, 433–437
  tryptophan synthetase and, 432–433

Crystal violet,
  cytoplasmic membrane and, 42
  uridine nucleotides and, 139
Cultures, synchronized, 234, 235
Curare, chemotaxis and, 198
Cyanide, flagella and, 174
Cysteine, flagella and, 165
Cystine, bacteriophage and, 409
Cytidine diphosphate glycerol, isolation of, 139
Cytidine diphosphate ribitol, cell walls and, 139
Cytochemistry, bacteria and, 444–446
Cytochrome oxidase, bacterial mitochondria and, 52
Cytochromes,
  cell fractions and, 452–453
  chromatophores and, 462
  cytoplasmic membranes and, 27
  electron-transport particle and, 458, 460
  protoplasts and, 301, 304–305, 460
Cytology,
  bacterial, outline of, 25
  biochemical, 5
Cytophaga, agar etching and, 186–187
Cytoplasm,
  fine structure of, 52–58
  general features of, 44–45
  inclusions in, 46–52
Cytoplasmic membrane,
  antisera and, 427
  cytochrome oxidase and, 52
  demonstration of, 39–44, 251–253
  isolation of, 27
  ribonucleic acid and, 45

D

Dehydrogenases,
  cell fractions and, 454
  cytoplasmic membranes and, 27
Deoxyadenosine triphosphate, bacteriophage and, 406–407
Deoxycholate, electron-transport particles and, 457
Deoxymethyl pentoses, surface components and, 111
Deoxyribonuclease, 461
  bacteriophage and, 335
  chromatin bodies and, 90

SUBJECT INDEX

cytoplasmic granules and, 56–57
enzyme formation and, 319
protoplasts and, 299–300, 308–309, 310,
   311, 313, 325, 331–333, 343
suprainfection and, 410
Deoxyribonucleic acid,
bacterial, 84
bacterial surface and, 99–100
bacteriophage, 31, 135
   infection and, 334–336
   injection of, 409–410
   radioactive decay and, 402–403
   structure of, 399–400, 401–403
   temperate, 401
   T-even, 400–401
cations and, 86–89
cell disruption and, 449, 450
cell fractions and, 452
in vitro synthesis of, 23
L-forms and, 372
osmotic disruption and, 28, 320–321,
   322, 461
protein and, 70
protoplasts and, 283, 305, 340
spheroplasts and, 322
spores and, 232, 236–237
transformation and, 23, 24
Deoxyribose, protoplasts and, 301
Detergents, protoplasts and, 292
Dextran,
capsules and, 109
slime and, 105
α,ε-Diaminopimelic acid,
bacteria and, 19, 20
cell walls and, 122, 123–124, 140, 141,
   254
incorporation and, 314–315
isomers of, 126
L-forms and, 372
lysozyme and, 132
penicillin and, 274
phage enzymes and, 136
protoplasts and, 27, 259, 276–277, 297–
   299, 314, 343
spherical forms and, 255, 256
Diffusion coefficient, serological reactions and, 417
Digitonin, protoplasts and, 292
2,4-Dinitrophenol,
adaptation and, 319

amino acid incorporation and, 313, 316
bacterial enzymes and, 456
photophosphorylation and, 463
Dipicolinic acid,
bacteria and, 22
endospores and, 31
spores and, 219, 234
Dodecylamine·HCl, protoplasts and, 293
Dodecyltrimethyl ammonium bromide,
   protoplasts and, 293
Dyes, precipitins and, 424

E

Eberthella typhi, protoplasts and, 268
Egg-white, lysozyme of, 130–131
Elasticotaxis, movement and, 185
Electrokinetics, movement and, 158
Electron microscopy,
artifacts and, 85
bacterial mitochondria and, 446
capsules and, 101–102
cell fractionation and, 451
chromatophores and, 463
cytoplasm and, 52–58
cytoplasmic membrane and, 43–44
embedding damage and, 64
L-forms and, 368, 377–381
lipid droplets and, 50
metachromatic granules and, 48
pleuropneumonia-like organisms and,
   378–381
protoplasts and, 287–288
Electron transport, cell fractions and,
   452–453
Electron-transport particles,
cell fractions and, 457–458
localization of, 458–461
Electrophoresis, cell fractionation and,
   451, 456, 457
Endoplasmic reticulum, 3
bacteria and, 446
Endospores,
composition of, 31
enzymes in, 464–465
isolation of, 464
wall constituents of, 129
Energy,
biochemical unity and, 4
flagella and, 173–174
Enterobacteriaceae, cell walls of, 27

Enzymes,
  bacteriophage infection and, 411
  cell walls and, 129–135
  electron-transport particles and, 457
  flagella and, 174
  heat resistance of, 465
  identification, antisera and, 430
  induced, serology of, 434–437
  oxidative, cell fractions and, 452
  protoplasts and, 304–305, 308–311
  surface components and, 113–115
Equivalence point, galactosidase and, 433–434
Ergastoplasm, 3
  bacteria and, 446
Erythritol, protoplasts and, 261, 291
Erythrocytes, pneumococcus antiserum and, 428–429
Escherichia, chemotaxis and, 189
Escherichia coli,
  acridine orange and, 39
  aeration and, 10
  amino sugars and, 127
  autolysin and, 264
  bacteriophage,
    enzymes and, 135–136, 269–271
    receptors and, 135, 136
    resistance to, 389–390
  capsule of, 102, 110, 112
  cell walls, 118, 120, 121
    amino acids of, 123
    synthesis of, 139–140
  chromatin bodies of, 68, 70, 72, 84, 89
  cytochromes of, 453
  cytoplasmic membrane of, 43, 251–252
  deoxyribonucleic acid, 18
    synthesis of, 23
  diaminopimelic acid and, 274, 276–277
  disintegration of, 117, 119
  elementary analysis of, 14–16
  fimbriae and, 163–164
  flagella of, 165
  fusion cylinders in, 236
  β-galactosidase of, 23, 429, 430, 433–438
  D-glutamic acid in, 125–126
  growth, energy sources and, 11
  heptoses in, 128
  host-range mutants and, 389–390
  intercellular space and, 13
  leucozyme C and, 269
  L-forms of, 363
  lipids of, 18, 129
  lysine formation by, 19
  lysozyme and, 267–268
  mating in, 6–7
  membrane of, 461
  mucopolysaccharide and, 256–257
  osmotic pressure of, 17
  particulates, 452
    substrate oxidation and, 453
  penicillin and, 273
  phagocytin and, 269
  polysaccharides in, 51
  protein in, 19
  "protoplasts", 143, 255, 381
    autolysis and, 265
    bacteriophage and, 260, 329–336, 346
    chemical composition of, 297–299
    enzymes and, 316, 319–320, 323
    growth of, 342, 346
    immunochemical properties of, 259
    lysed, 322–323
    metabolic disturbance and, 279–280
    motility and, 339
    production of, 279–280
    reversion of, 344–345
  ribonucleic acid in, 18
  sedimentation constants and, 29
  size of, 9, 10
  surface components of, 111
  transduction of, 401
  tryptophan synthetase in, 432–433
  Vi antigen of, 113
  water content of, 13
Ethanol, chemotaxis and, 192
Ethanolamine, chromatophores and, 463
Ethylenediaminetetraacetic acid, protoplasts and, 267–269
Ethyl ether, chemotaxis and, 192
Exosporium,
  enzymes and, 464
  structure of, 223

F

Feedback, biosynthetic control and, 30
Fermentation, anaerobic, 4
Ferric chloride, Vi antigen and, 113
Feulgen reaction,
  accessory chromatic granules and, 84–85
  bacteria and, 38

chromatin bodies and, 44–45, 67, 70
  spores and, 230, 232, 234, 237
Filaments, bacterial movement and, 155–157
Filtration, L-forms and, 377
Fimbriae, motility and, 163
Fixation, L-forms and, 367
Flagella,
  antigens and, 306
  antisera and, 426
  blepharoplasts and, 161–162
  chemistry of, 165
  energy transport in, 173–174
  enzymes and, 174
  helical shape of, 172
  isolation of, 26, 461
  lateral, movement and, 167–169
  L-forms and, 372
  morphology of, 158–164
  mode of action of, 169–174
  origin of, 160–161
  propulsion mechanics and, 172–174
  protoplasts and, 286, 287, 338–339
  rate of growth of, 170
  removal of, 169–170
  spirochetes and, 178–179
  structure of, 165
  subfibrils and, 173
  synchrony and, 172–173
  terminal, movement and, 165–167
Flavins,
  cell fractions and, 453
  chromatophores and, 462
  oxidative phosphorylation and, 456
Flavobacterium, L-forms of, 363
Flavobacterium polyglutamicum, *Bacillus* capsules and, 114, 115
Fluorescence, serological reactions and, 416
Fluoride,
  chromatin extrusion and, 234
  flagella and, 174
Fluoroacetate, chemotaxis and, 192
Fluorodinitrobenzene,
  amino sugars and, 127
  cell walls and, 126
Folic acid, bacteria and, 17
Formalin, protoplasts and, 286
Formvar, L-forms and, 368
Fractionation, antisera and, 419
Fuchsin, cytoplasmic membrane and, 42

Fucose,
  blood group substance and, 429
  capsules and, 109
  surface components and, 111, 112
Fungi, cell walls of, 27
Fusiformis necrophorus, L-forms of, 362, 364, 371–372, 377
Fusobacteria, movement of, 187
Fusocillus girans, 187

G

Galactan, pneumococcus antiserum and, 428, 429
Galactosamine,
  capsules and, 107
  cell walls and, 122, 127
Galactose,
  capsules and, 104, 105, 107, 109
  cell walls and, 128
  O antigen and, 110
  surface components and, 112
β-Galactosidase,
  antibodies and, 429, 430
  bacterial content of, 23
  blood group substance and, 429
  inactive protein and, 434–437
  inhibited growth and, 437–438
  protoplasts and, 317–319, 322
  transduction of, 437
Galactosyl glucosamines, blood group substances and, 429
Galacturonic acid, capsules and, 105
L-Gala-D-mannoheptose, surface components and, 111
Gelatin, phase-contrast microscopy and, 69–70
Gelatinase, 311
Gels,
  precipitin reactions, 424–425
  applications of, 426
Gene loci, biosynthetic control and, 30–31
Geotaxis, spirilla and, 193–194
Giemsa stain,
  chromatin bodies and, 67, 68, 70
  L-forms and, 367
Glass,
  powdered, cell disruption and, 447
Glass beads, cell disruption and, 447
Glucosamine,
  capsules and, 107

cell walls and, 127
group-specific polysaccharide and, 137
C substance and, 104
O antigen and, 110
protoplasts and, 259
Glucose,
  bacterial growth and, 11
  bacteriophage and, 400–401
  capsules and, 103, 104, 107, 109
  cell walls and, 128
  incorporaion of, 314–315
  molar growth yield and, 12
  O antigens and, 110
  protoplasts and, 261, 301, 313
  surface components and, 112
Glucose dehydrogenase, cell fractions and, 454
Glucose "oxidase", cell fractions and, 454
Glucuronic acid, capsules and, 103, 104, 109
Glutamic acid,
  cell walls and, 122, 123, 125, 139
  incorporation and, 315
  muramic acid and, 125
  phage enzymes and, 136
Glutamic decarboxylase, sonic disruption and, 448
Glutathione, bacterial content of, 16, 17
Glycerol,
  cell disruption and, 449
  chemotaxis and, 192
  protoplasts and, 291
Glycerophosphate, washed particles and, 453–454
Glycine,
  bacteriolysins and, 132
  cell walls and, 123, 273
  incorporation of, 312–313, 314, 315, 317, 319–320
  L-forms and, 365
  protoplasts and, 280–281, 291
Glycogen, 39
  accumulation of, 50
  bacteria and, 21
Glyoxylate bypass, 4
Gram reaction,
  cell wall and, 144
  growth phase and, 44

Granulose, 50
  bacteria and, 21
Growth curve, bacteriophage and, 390–391
Gymnoplast, definition of, 253, 260

H

Haemophilus,
  L-forms of, 363
  transformation and, 24
Haemophilus aegyptius, antigens of, 426
Haemophilus influenzae,
  antigens of, 426
  capsule of, 109, 112
  L-forms of, 365, 366
Haemophilus pertussis, antigens of, 426
Halobacterium halobium, cell walls of, 120
Haustoria, myxobacteria and, 185
Helix pomatia,
  enzymes, protoplasts and, 281–282
Hematoxylin, chromatin bodies and, 66
Hemicellulase, protoplasts and, 282
Heptoses, surface components and, 111
Histidine, flagella and, 165
Histone, bacteria and, 70
Hyaloplasm, 2
Hyaluronic acid,
  capsules and, 104
  cell walls and, 139
Hyaluronidase, capsules and, 104–105, 114, 115
Hydrochloric acid-Azure A/SO_2 stain, spores and, 232
Hydrochloric acid-Giemsa stain, spores and, 232
Hydrodynamics, bacterial movement and, 154–157
Hydrogen, bacterial content of, 14, 15
Hydrogenase,
  cell fractions and, 454
  electron-transport particle and, 458
Hydrogen peroxide, bacteriophage and, 404–405
β-Hydroxybutyrate, 459
  granules of, 18, 27, 48, 305
Hydroxylamine, bacteriophage and, 404
5-Hydroxymethyl cytosine,
  bacteria and, 19
  bacteriophage and, 399, 400
Hydroxyproline, flagella and, 165

SUBJECT INDEX

I

Immersion oil, spores and, 216
Immunoelectrophoresis, 425
Immunology, cell structure and, 137–138
Indoleglycerophosphate, tryptophan and, 433
Inertia, bacterial movement and, 155
Infections, bacteriophage and, 392
Inositol, bacteria and, 17
Intine, spore envelopes and, 222
Iron, bacterial content of, 15, 16
Iron bacteria, growth and, 12–13
Isocitric dehydrogenase, cell fractions and, 454

J

Janus green B,
  bacteria and, 444
  mitochondria and, 51

K

α-Ketoglutaric dehydrogenase,
  cell fractionation and, 450
  cell fractions and, 454
  protoplasts and, 305
Kinetosomes, 3
Kjeldahl reaction, antibodies and, 423–424
Klebsiella,
  capsules of, 109–110
  chemotaxis and, 189
Klebsiella pneumoniae,
  capsule of, 98–99
  enzymes, capsules and, 114, 115
  L-forms of, 366
  water content of, 13

L

Lactase,
  β-galactosidase and, 430, 433
Lactate,
  washed particles and, 453–454
Lactic dehydrogenase, protoplasts and, 304–305
Lactobacilli,
  growth, buffers and, 12
Lactobacillus arabinosus,
  cell walls, amino acids of, 124, 125, 254
  cytidine nucleotides in, 139
  malate and, 12
  vitamin B_6 and, 278

Lactobacillus bifidus,
  cell walls of, 140
  polysaccharide of, 107, 108
Lactobacillus casei,
  cell wall amino acids of, 123, 124, 125
  unsaturated fatty acids and, 294–295
Lanthanum nitrate,
  chromatin bodies and, 89
  polysaccharide and, 63
  spores and, 220–222
Latent period, bacteriophage and, 390–391
Lead nitrate, chemotaxis and, 192
Lecithinase, 311
  antibodies and, 430
Leptospira,
  chemotaxis and, 189
  movement of, 177–178, 179
Leptospira canicola, axial filament of, 178
Leptospira icterohaemorrhagiae, oxygen and, 191
Leucine, incorporation of, 312
Leuconostoc mesenteroides, slime of, 105, 107, 108
Leucozyme C, protoplasts and, 269
Levan,
  extracellular, 107
  slime and, 105
L-Forms,
  composition of, 372
  definition of, 363
  discovery of, 362–363
  filtration of, 377
  growth,
    liquid media and, 365
    requirements for, 376
    solid media and, 363–364
  metabolism of, 371–372
  microscopic demonstration of, 367–368
  morphology of, 368–371
  pathogenicity of, 373–374
  pleuropneumonia-like organisms and, 363, 374–377
  production of, 365–367
  protoplasts and, 381–382
  serology of, 372–373
Lineola, size of, 10
Linkage groups, chromatin bodies and, 84

Linoleic acid, protoplasts and, 294–295
Lipase,
  enzyme formation and, 317
  nuclear bodies and, 28
  protoplasts and, 301, 308, 309–311
Lipids,
  bacterial content of, 17–18
  cell walls and, 121, 129
  cytoplasmic membrane and, 42
  flagella and, 165
  granules and, 461
  L-forms and, 372
  protoplast membrane and, 295, 301–302, 340
  spores and, 212
Lipoglycoprotein, receptor sites and, 409
Lipomucoprotein, cell walls and, 139–140
Lipopolysaccharide,
  cell walls and, 26, 256, 461
  pyrogenic substances and, 111
Lipoprotein,
  bacteria and, 19
  cell wall and, 26–27, 125, 256, 257
  cytoplasmic membranes and, 27, 461
  lysozyme and, 131–132
  penicillin and, 142
  protoplasts and, 300, 342
  removal of, 269
Listeria monocytogenes, lipids of, 50
Luciferase, antibodies and, 427
Lysergic acid, chemotaxis and, 198
Lysine,
  cell walls and, 123–124, 139–140, 277
  incorporation of, 313
Lysogeny, 392–394
  bacterial classification and, 412
Lysozyme,
  Bacillus megaterium and, 131, 253–255, 261–263
  bacterial, 264–265
  cell walls and, 127, 128, 259
  classification of, 130
  egg-white, 261
  flagella and, 161, 162
  gram-negative bacteria and, 255, 267–271
  gram-positive bacteria and, 253–255
  Micrococcus lysodeikticus and, 130, 131, 253–255, 263–264
  protoplasts and, 27, 42, 99
  receptors and, 136

Sarcina lutea and, 130, 131, 253–255, 263–264
Streptomyces and, 264
substrate of, 255

M

Macronucleus, chromatin bodies and, 80
Magnesium,
  bacterial content of, 15, 16
  protoplasts and, 288–289
Malate,
  bacterial growth and, 11
  chemotaxis and, 192, 198
  lactobacilli and, 12
  washed particles and, 453–454
Malic dehydrogenase,
  cell fractions and, 454
  protoplasts and, 305
Malic oxidase, 460
  cell fractions and, 454
Malleomyces mallei, water content of, 13
Malonate, chemotaxis and, 192
Mandelate, washed particles and, 453–454
Manganese, bacterial content of, 15, 16
Manno-D-galaheptose, cell walls and, 128
Mannose,
  capsules and, 107
  cell walls and, 128
  C substance and, 104
  protoplasts and, 302–303, 342
Meat extract, chemotaxis and, 192
Melibiose, protoplasts and, 261
Membranous systems, bacteria and, 58–62
Menadione reductase, oxidative phosphorylation and, 455
Merthiolate, antisera and, 418
Metabacterium polyspora, spores and, 209, 211
Metachromatic granules, nature of, 47–48
Methionine, L-forms and, 365
6-Methylaminopurine, bacteria and, 19
Methylcellulose, flagellar movement and, 167
5-Methyl cytosine, bacteria and, 19
Methylene blue,
  chromatin bodies and, 68
  L-forms and, 367

nuclear membrane and, 82
spores, 232
vital staining and, 38
volutin and, 48
Methyl green, chromatin bodies and, 66
ε-N-Methyllysine, 21
Microcapsules,
  definition of, 100
  nature of, 110–113
Micrococci, disruption of, 448
Micrococcus,
  chemotaxis and, 189
  cytoplasmic membrane and, 252
Micrococcus citreus, cell wall sugars, 129
Micrococcus cryophilus, granules in, 445
Micrococcus halodenitrificans,
  deoxyribonucleic acid and, 100
  lipids and, 129
  spheroplasts of, 281
Micrococcus lysodeikticus,
  autolysins and, 134, 264
  bacteriophage,
    enzymes and, 271
    receptors and, 135
  cell walls of, 121, 255, 259
  crystalline enzymes and, 23
  leucozyme C and, 269
  lysozyme and, 130, 131, 253–255, 263–264
  protoplasts, 381
    appearance of, 258
    autolysis and, 265
    detergents and, 292–293
    electron microscopy of, 287
    enzymes and, 308
    growth of, 342
    inhibitors and, 316
    membrane of, 300, 301–304
    osmotic properties of, 290–291
    radioactive tracers and, 312–313
    respiration of, 311
    stabilization of, 288
  Streptomyces enzymes and, 132, 133
Micrococcus pyogenes,
  particulates of, 452
  sedimentation constants and, 29
Micrococcus roseus,
  amino sugars and, 127
  cell wall amino acids, 124
Micrococcus urea, cell wall amino acids, 124

Micrococcus varians,
  cell walls, 126, 127
  sugars and, 129
Microsomes,
  bacteria and, 446
  enzyme activity of, 457
Mitochondria, 3
  bacteria and, 51–52, 60, 444–446, 452
Mitosis, chromatin bodies and, 77
Molybdenum, bacteria and, 16
Motility, *see* Movement
Movement,
  electrokinetic theory of, 158
  external stimuli and, 188–198
  fimbriae and, 163
  flagella and, 158–174
  fusobacteria and, 187
  gliding, 180–188
  hydrodynamic theory of, 154–157
  myxobacteria and, 184–187
  spirochetes and, 174–180
  thermokinetic theory and, 157–158
  torque and, 172
M Protein, cell walls and, 137
Mucocomplex,
  cell walls and, 122
  gram reaction and, 144
  lysozyme and, 130, 131–132
  penicillin and, 141
Mucopolysaccharide,
  antibiotics and, 275
  antigens and, 306
  bacteriophage enzyme and, 270
  cell walls and, 256–257
  gram-positive bacteria and, 254
  protoplasts and, 279
Muramic acid,
  cell walls and, 19–21, 26, 27, 122, 126, 127, 141
  cell wall synthesis and, 138–139
  D-glutamic acid and, 125
  phage enzymes and, 136
  protoplasts and, 259
  structure of, 127
Mycobacterium,
  cell wall sugars of, 128
  lipids and, 18
  membranous structures in, 59–60
  mitochondria and, 51
  vacuoles of, 47
Mycobacterium butyricum, 420

504 SUBJECT INDEX

Mycobacterium phlei, oxidative phosphorylation and, 454, 455, 456
Mycobacterium thamnopheos, tellurite reduction and, 51–52
Mycobacterium tuberculosis,
  antigen preparation and, 420
  diaminopimelic acid and, 276
  electron-transport particles and, 459
  lipids of, 50
  slime of, 101
  water content of, 13
Myokinase, cell fractions and, 454
Myxobacteria, movements of, 184–187
Myxococcus, pneumococcal capsules and, 113
Myxococcus rubescens, movement of, 186

N

Nadi reaction,
  bacteria and, 444
  mitochondria and, 51
Neotetrazolium, bacterial granules and, 445
Neurospora crassa,
  "protoplasts" and, 282, 345–346
  tryptophan synthetase and, 432–433
Neutral red, vital staining and, 38
Nicotinic acid, bacteria and, 17
Ninhydrin-Schiff reagent, chromatin bodies and, 70
Nitrogen, bacterial content of, 14, 15
o-Nitrophenylgalactoside, lactase and, 433–434
Nocardia,
  cell wall sugars, 128
  cytoplasmic membrane and, 252
  lysozyme and, 264
Nuclear bodies, isolation of, 28
Nuclear membrane, chromatin bodies and, 80–82, 93
Nuclei, L-forms and, 371
Nucleic acids,
  bacterial content of, 18–19
  bacteriophage and, 398
  biochemical unity and, 3–4
  cell walls and, 122
Nucleolus, chromatin bodies and, 93
Nucleoprotein, tubercle bacilli and, 70
Nucleosome, 93
Nucleotides, cell wall synthesis and, 139

O

O antigens,
  bacteriophage and, 135
  cell walls and, 110
"*Organite antapical*", structure of, 62
Ornithine transcarbamylase, spheroplasts and, 322–323
Oscillatoria, 209
  movement of, 180–184
Oscillospira batrachorum, spores and, 209
Oscillospira guilliermondi, spores and, 209, 210
Oscillatoria jenensis, contraction of, 182
Oscillatoria sancta,
  contraction of, 182
  rotation of, 181
Oscillatoria tenuis, contractions and, 182
Osmium,
  chromatin bodies and, 76, 85–89
  protoplasts and, 286
Osmotaxis, chemotaxis and, 191–192
Osmotic forces, movement and, 181
Osmotic pressure, bacteria and, 17
Osmotic shock,
  cell disruption and, 448–450
  chromatophores and, 464
  electron-transport particle and, 459, 461
  ribonucleoprotein particles and, 458
Oxaloacetic decarboxylase, bacterial content of, 23
Oxamycin, cell walls and, 142, 273
Oxidative phosphorylation,
  cell envelope and, 459
  cell fractions and, 454–456
  factor requirements of, 455–456
Oxygen,
  bacterial content of, 14
  bacterial growth and, 10, 12
  chemotaxis and, 190–191

P

Pantothenic acid, bacteria and, 17
Papain, cytoplasmic granules and, 56–57
Paracolobactrum ballerup, Vi antigen of, 113
Parasporal bodies, composition of, 31, 243–245
Paratose, surface components and, 111

SUBJECT INDEX

Pasteurella, chemotaxis and, 189
Pasteurella pestis, capsule of, 100
Pasteurella tularensis, size of, 9
Penicillin,
  cell walls and, 138–139, 140–142
  flagellar movement and, 168–169
  L-forms and, 363, 365, 366, 371, 376, 382
  magnesium and, 288
  protoplasts and, 27, 271–275, 298–299
  site of action, 142
  spherical forms and, 255, 256
Penicillinases,
  interconversion of, 431
  serological relationships of, 430–431
Pentose, protoplasts and, 301
Pentose cycle, 4
Peptone, chemotaxis and, 192
Periodic acid-Schiff reaction, polysaccharides and, 51
Permeability, cell wall and, 143
Permeases, "protoplasts" and, 322
pH, protoplasts and, 293–294
Phagocytin, protoplasts and, 269
Phase-contrast,
  capsules and, 102
  chromatin and, 37
  chromatin bodies and, 68–70, 77
  protoplasts and, 283–286
Phenylalanine, L-forms and, 365
p-Phenylenediamine oxidase,
  heat resistance of, 465
  spore coat and, 464
Phobotaxis, bacteria and, 188–189
Phosphatase, protoplasts and, 304
Phosphate, protoplasts and, 261
Phosphatidic acid, protoplasts and, 301, 302
Phosphatidylethanolamine, electron-transport particles and, 458
Phospholipid,
  cell fractions and, 452
  chromatophores and, 463
  electron-transport particle and, 458, 459
  mitochondria and, 51
Phospholipoprotein, protoplasts and, 340
Phosphorus, bacterial content of, 15
Phosphotungstic acid, polysaccharide and, 63

Photophosphorylation, isolated chromatophores and, 463
Photosynthesis, 4
  phototactic action spectrum and, 196
Phototaxis,
  nervous responses and, 196–197
  photosynthesis and, 196–197
  purple bacteria and, 194–195
  Weber law and, 195
Phytomonas, see *Agrobacterium*
Pigments, bacterial content of, 21–22
Plakin,
  bacteriolysis and, 130
  protoplasts and, 289, 292
Plaques, mechanism of formation, 391–392
Plastids, 3
Plasmolysis, cytoplasmic membrane and, 252–253
Plasmoptysis, definition of, 251
Pleuropneumonia-like organisms,
  electron microscopy of, 378–381
  filtration of, 377
  growth requirements of, 376
  L-forms and, 363, 374–377
  protoplasts of, 298
Pneumococci,
  antigens of, 105
  capsules, 98–99, 102–104, 108
    removal of, 113, 115
  C substance of, 104
  L-forms of, 256
Pneumococcus,
  antisera, red cells and, 428–429
  transformation and, 24
Polyethylene glycol,
  cell wall isolation and, 118
  protoplasts and, 261, 288, 290
Poly-D-glutamic acid,
  capsules and, 21, 102, 105–106
  enzymes and, 114, 115
  structure of, 105–107
Polyhydric alcohols, washed particles and, 453–454
Polymixin, protoplasts and, 286, 293
Polynucleotide phosphorylase, cell fractions and, 454
Polynucleotides, oxidative phosphorylation and, 455

Polypeptides,
  bacterial content of, 19–21
  spore formation and, 214
Polyphosphates,
  bacterial content of, 22
  bacterial granules and, 445, 461
  cell fractions and, 452
Polyribophosphate, 119, 122
  capsules and, 109
Polysaccharides,
  bacteriophage receptors and, 135, 136
  capsules and, 21, 98–99, 102, 107
  chromatophores and, 428
  cell walls and, 121, 128
  electron microscopy and, 63
  group-specific, 137
  pneumococcal components and, 103
Polyvinyl sulfate, lysed protoplasts and, 322
Potassium,
  bacterial content of, 15, 16
  chemotaxis and, 191–192
Potato juice, chemotaxis and, 192
Precipitins,
  galactosidase and, 433
  measurement of, 423
Prodigiosin, 21
Proflavine, bacteriophage protein and, 399
Proline,
  flagella and, 165
  incorporation and, 315
Properdin, immunity and, 138
Prophage, lysogeny and, 394
Propionibacterium pentosaceum, molar growth yield and, 12
Protamine, bacteria and, 70
Protease inhibitor, lysed protoplasts and, 322
Protein-polysaccharide-lipid,
  isolation of, 110
  surface antigens and, 107–109
Proteins,
  bacterial content of, 19
  bacteriophage and, 398, 399, 403–408, 410
  biochemical unity and, 3–4
  capsules and, 105
  cell walls and, 26
  chemotaxis and, 192
  deoxyribonucleic acid and, 70

flagella and, 165
formation, serology and, 433–439
lysed protoplasts and, 321–322, 323
measurement, precipitins and, 423–424
parasporal bodies and, 243–244
protoplast membrane and, 295, 302
spores and, 216, 223
streptococcal surface and, 98, 99
synthesis, antibiotics and, 275
Proteus,
  beta radiation and, 281
  chemotaxis and, 189–190
  L-forms of, 365, 366
Proteus mirabilis,
  flagella of, 160, 162
  penicillin and, 275
  spheroplasts of, 331
Proteus vulgaris,
  diaminopimelic acid and, 274
  flagella of, 158, 159, 165, 169, 173, 174
  heptoses and, 128
  L-forms of, 141, 364, 365, 367–369, 371–373
  oxidative phosphorylation in, 455, 456
  particulates, substrate oxidation and, 453
  penicillin and, 273
  "protoplasts" of, 284, 287, 297–298, 342, 382
  sedimentation constants and, 29
  size of, 9
Protoplasm,
  consistency of, 37
  homogeneity of, 36–37
Protoplasts,
  amino acid incorporation in, 308, 312–316
  antigenic properties of, 306–307
  antigens and, 137
  antimetabolites and, 279
  antisera and, 427
  bacteriophage and, 260, 324–336
  cell wall synthesis and, 343–344
  chemical composition of, 259, 295–311
  chemical damage and, 292–295
  chromatin bodies and, 82–83, 310–311
  conjugation of, 346
  constitutive enzymes and, 317
  damaged bacteriophage and, 333–336
  definition of, 253, 257, 260

diaminopimelic acid and, 276
disruption of, 449
division of, 341–342
electron microscopy of, 287–288
electron-transport particles and, 459–460
electrophoretic studies on, 295
enzymes and, 304–305, 308–311
fixation of, 286
flagella and, 161
formation, 39–42, 99
  autolysis and, 265–267
  bacteriophage and, 269–271
  common features of, 279–280
  fungi and, 281–282
  glycine and, 281
  leucozyme C and, 269
  lysozymes and, 261–265, 267–269
  metabolic disturbance and, 271–280
  phagocytin and, 269
  snail enzymes and, 281–282
  spore enzymes and, 265
gram-negative bacteria and, 255–257
granules in, 305
growth of, 339–343
immunochemical properties of, 259
inducible enzymes and, 317–319, 322–323
inhibitors and, 316–317
isolation of, 27
L-forms and, 381-382
membrane of, 299–305
  amino acid incorporation and, 321–322
microscopic appearance of, 258
motile species and, 338–339
natural occurrence of, 346–347
nitrogen fixation by, 323–324
nucleic acids and, 283
number per cell, 283
osmotically shocked, 319–322
osmotic properties of, 258–259, 289–291
phase contrast microscopy of, 283–286
physicochemical properties of, 288–295
respiration of, 311
reversion of, 343–345
solute penetration into, 290–291
spherical form of, 282–283
spore formation and, 338
staining characteristics of, 259–260, 286–287

survival of, 142–143
suspension media and, 261–263, 288–289
transformation of, 336–338
Pseudomonas,
  chemotaxis and, 189
  flagella of, 174
Pseudomonas aeruginosa,
  bacteriophage receptors and, 135
  cell walls of, 121
  protoplasts and, 268
  size of, 9
  water content of, 13
Pseudomonas fluorescens,
  cytochromes of, 453
  particulates, 452
    substrate oxidation and, 453
  sedimentation constants and, 29
  size of, 9
  vitamins in, 17
Pseudomonas salinaria, lipids and, 129
Purines, nitrogen determination and, 14
Puromycin, adaptation and, 319
Putrescine, 16
  bacteriophage and, 403
Pyocyanin, 21
Pyridine-2,6-dicarboxylic acid, spores and, 22
Pyridine nucleotides,
  cell fractions and, 453, 454
  chromatophores and, 462
Pyridoxin, bacteria and, 17
Pyrimidines, nitrogen determination and, 14
Pyrophosphatase,
  endospores and, 464
  heat resistance of, 465
Pyruvate, oxygen and, 12

R

Rabbits,
  antisera and, 418
  bleeding of, 422
Raffinose, protoplasts and, 261
Rana pipiens, bacteria of, 230
Receptors,
  bacteriophage and, 135, 408–409
  specificity of, 136
Redintegration, protoplasts and, 336–338

Respiration,
  bacteriophage infection and, 410–411
  protoplasts and, 311
Rhamnose,
  capsules and, 104, 105
  cell walls and, 128
  group-specific polysaccharide and, 137
  O antigen and, 110
  protoplasts and, 259, 281–282
  surface components and, 111, 112
Rhizobium,
  chemotaxis and, 190
  plasmolysis and, 253
Rhodobacillus palustris, pneumococcal capsules and, 113
Rhodopseudomonas spheroides,
  chromatophores of, 59, 463
Rhodospirillum, chemotaxis and, 190
Rhodospirillum rubrum,
  aerotaxis and, 197
  amino sugars and, 127
  cell walls of, 120, 121
  chemotaxis and, 198
  chromatophores of, 30, 58–59, 462, 463
  cytoplasmic membrane and, 252
  disintegration of, 117
  osmotic disruption of, 449
  phototaxis and, 195–198
  protoplasts and, 269
  sedimentation constants and, 29
  sonication of, 463–464
Ribitol phosphate, cell walls and, 254
Riboflavin, bacteria and, 17
Ribonuclease,
  amino acid incorporation and, 319
  bacterial chromatin and, 45
  cell fractions and, 456
  chromatin bodies and, 67, 68, 90
  cytoplasmic granules and, 54, 56–57
  enzyme formation and, 319
  protoplasts and, 301, 308–309, 310, 311, 312, 321, 325, 331–333, 340
  spores and, 236
Ribonucleic acid,
  cell fractions and, 452, 456–457
  chromatin bodies and, 67
  cytoplasmic membranes and, 27
  distribution of, 45
  functional integrity of, 24
  L-forms and, 372

  protoplasts and, 283, 321, 322, 343
  thiouracil and, 437–438
Ribonucleoprotein,
  chromatin bodies and, 83, 92
  cytoplasmic granules and, 54–56
  precipitation of, 457
Ribonucleoprotein particles, localization of, 458
Ribose, protoplasts and, 291
Ribosidase,
  endospores and, 464
  heat resistance of, 465
Ribosomes,
  enzymatic activity of, 29–30
  isolation of, 28–30
Rickettsia prowazekii, size of, 9
Romanovsky stain, 44
  chromatin bodies and, 66–67
Rough antigen, nature of, 112–113

S

Saccharomyces carlsbergensis, "protoplasts" of, 281–282
Saccharomyces cerevisiae,
  molar growth yields and, 12
  "protoplasts" of, 281–282
  sedimentation constants and, 29
Salmonella,
  chemotaxis and, 190
  chromatin bodies and, 67
  flagella of, 165
  L-forms of, 363, 364, 366
  methyllysine in, 21
  polysaccharides and, 21
  protoplasts and, 268
  ribonucleic acid of, 18
  spherical forms of, 342, 345
  surface components of, 111, 112
Salmonella breslau, L-forms of, 367
Salmonella enteritidis, bacteriophage enzymes and, 270
Salmonella gallinarum,
  amino sugars and, 127
  antigens of, 138
  protoplasts of, 299
  spherical transformation of, 141
Salmonella paratyphi,
  bacteriophage typing of, 411
  L-forms of, 367
  polysaccharides of, 111
  protoplasts of, 280

Salmonella pullorum, cell walls of, 121
Salmonella schottmülleri, flagella of, 160 168
  bacteriophage,
    receptors and, 135
    typing of, 411
  L-forms of, 365, 367
  spherical forms of, 280
Salmonella typhimurium,
  deoxyribonucleic acid of, 18
  L-forms of, 365, 373
  penicillin and, 274
  "protoplasts" of, 255, 280, 338
Salmonella typhosa, Vi antigen of, 113
Salts, L-forms and, 366
Saprospira, movement of, 176
Sarcina,
  bacteriolysins and, 132
  electron microscopy of, 58
  movement of, 171
Sarcina flava, lysozyme and, 253
Sarcina lutea,
  autolysin and, 264
  cell walls, 121, 259
    amino acids of, 124
  fragment regeneration and, 345
  lysozyme and, 130, 131, 253–255, 263–264
  protoplasts of, 258, 288, 290–291
  size of, 9
Sarcina ventriculi,
  chromatin bodies of, 72
  size of, 9
Serine,
  cell walls and, 123, 124
  incorporation and, 315
  spheroplasts and, 281
  tryptophan and, 433
Serological reactions,
  explorable problems and, 417
  liquid media and, 416
  solid media and, 416–417
Serratia, chemotaxis and, 190
Serratia marcescens,
  growth, energy source and, 11
  particulates, substrate oxidation and, 453
  water content of, 13
Serratia plymouthicum, disruption of, 449
Serum, L-forms and, 365–366

Shigella,
  chemotaxis and, 190
  L-forms of, 363
  surface components of, 111, 112
Shigella dysenteriae,
  chromatin bodies of, 72, 82
  galactosidase in, 437
  O antigen of, 110
  rough antigen of, 112–113
Shigella flexneri,
  galactosidase in, 437
  O antigen of, 110
Shigella sonnei,
  bacteriophage receptors of, 136
  galactosidase, inactive protein and, 437
  polysaccharides of, 111
Slime,
  characteristics of, 101–102
  composition of, 105, 108, 109–110, 112
  motility and, 182–184, 185–186, 187
  myxobacteria and, 185–187
Sodium, bacterial content of, 15, 16
Sodium chloride,
  chemotaxis and, 192
  L-forms and, 366
  protoplasts and, 290–291
Sodium dodecyl sulfate, protoplasts and, 293
Sodium dodecyl sulfonate, protoplasts and, 293
Sodium versonate, bacteriophage and, 407
Sonication,
  cell disruption and, 448
  cell wall isolation and, 117
  chromatophores and, 462, 463
  electron-transport particle and, 459
  endospores and, 464
  ribonucleoprotein particles and, 458
Sorbitol, protoplasts and, 291
Spectrum, *Bacterium photometricum* and, 194–195
Spermidine, 16
  bacteriophage and, 403
Spherical forms, gram-negative bacteria and, 255–257
Spheroplast, definition of, 258
Spirilloxanthin, phototaxis and, 195
Spirillum,
  chemotaxis and, 190, 191, 192

SUBJECT INDEX

cytoplasmic membrane of, 43
flagellum of, 163
geotaxis and, 193–194
thermotaxis and, 193
Spirillum serpens,
  amino sugars and, 127
  cell walls of, 120
  chromatin bodies of, 85
  cytoplasmic membrane and, 252
  heptoses and, 128
  movement of, 166
Spirillum undula,
  movement of, 166
  protoplasts of, 40
Spirillum volutans,
  movement of, 166
  size of, 9
Spirochaeta plicatilis, movements of, 175
Spirochetes,
  movements, 174–178
  mechanism of, 178–180
Sporangium, 229
Spore coat,
  enzymes and, 464
  structure of, 223
Spore cortex, germination and, 238–241
Spores,
  brightness of, 216
  chromatin and, 230–237
  development of, 211–215
  disintegration of, 464–465
  envelopes, 216–229
    development of, 229
  formation, protoplasts and, 338
  germination, 237–242
    chromatin and, 243
  granules and, 57–58
  interior of, 216
  stains and, 229–230
Spore wall, fate of, 218–219
Sporosarcina ureae,
  amino sugars and, 127
  cell walls,
    amino acids of, 124
    sugars of, 128
Sporulation, distribution of, 209–211
Stains,
  flagella and, 159
  spores and, 229–230
Staphylococci, bacteriophage and, 411
Staphylococcus, chemotaxis and, 190

Staphylococcus aureus,
  amino sugars and, 126, 127
  autolytic enzymes and, 134, 264
  bacteriolysis and, 130
  bacteriophage receptors and, 135
  cell walls, 116, 254
    sugars of, 128
  chloramphenicol and, 142
  disintegration of, 117
  lysozyme and, 255
  osmotic pressure and, 17, 142–143
  penicillin-resistant, 141–142
  penicillin-treated, 125, 128, 138–139, 271–273
  protoplasts of, 265, 290–201, 300–301, 304–305
  size of, 9
Staphylococcus citreus, amino sugars and, 127
Staphylococcus saprophyticus, amino sugars and, 127
Starch, bacteria and, 21
Steroids, bacteria and, 18
Streptobacillus moniliformis, L-forms of, 362–363, 364, 365, 371, 372, 373, 377, 378
Streptococcus (Streptococci),
  bacteriophage enzymes and, 134–135
  capsules of, 104–105, 108
  cell walls,
    antigens and, 137
    nucleotides and, 139
  cell wall sugars of, 128
  group-specific polysaccharide of, 137
  L-forms of, 143, 363, 364, 366
  M-protein and, 98, 99
  penicillin and, 141
  protoplasts and, 264
  Streptomyces enzymes and, 133
  surface, trypsin and, 114, 115
  transformation and, 24
Streptococcus bovis, slime of, 105, 108
Streptococcus faecalis, 267
  amino acid deficiency in, 140
  cell walls, 120, 121, 254, 255
    amino acids of, 123, 124, 125
  molar growth yields and, 12
  protoplasts and, 264, 277–279, 313
  size of, 9
Streptococcus lactis, size of, 9

SUBJECT INDEX

Streptococcus pneumoniae, enzymes, capsules and, 114
Streptococcus pyogenes,
  cell walls of, 121
  disintegration of, 117
  protoplasts and, 264
  rhamnose in, 128
  trypsin and, 114
Streptococcus salivarius, capsules of, 105, 108
Streptococcus viridans, capsules of, 105
Streptococcus viridans sanguis,
  L-forms, pathogenicity of, 373
Streptomyces,
  bacteriolysins and, 130, 132–133
  differentiation in, 7
  enzymes, protoplasts and, 282
  membranous structures in, 59–60
  mitochondria and, 51
  protoplasts of, 264
Streptomyces albus,
  cell walls and, 125
  protoplast production and, 264
  streptococci and, 133
Succinate,
  chemotaxis and, 192
  oxidative phosphorylation and, 454, 455
  washed particles and, 453–454
Succinic dehydrogenase, protoplasts and, 304–305, 317
Succinoxidase, 460
  cell fractions and, 454
Sucrose,
  bacteriophage and, 398–399
  protoplasts and, 261, 263, 288, 290, 291, 313
Sudan black, cytoplasmic membrane and, 43
Sugars, cell walls and, 128–129
Sulfhydryl compounds, chemotaxis and, 198
Sulfonamides, L-forms and, 371, 376
Sulfur, bacterial content of, 15
Suppressors, tryptophan synthetase and, 432
Suprainfection, deoxyribonuclease and, 410
Surface charge, antibiotics and, 275
Surface tension, movement and, 181

Suspension medium, cell disruption and, 449–450
Synchronization, chromatin bodies and, 77

T

Tartrate, chemotaxis and, 192
Teichoic acid,
  accumulation of, 139
  cell walls and, 122, 125, 254
Tellurite, reduction of, 52
Tetracyclines, chromatin bodies and, 82, 83
Tetrazoles, mitochondria and, 51
Tetrazolium salts, bacteria and, 444–445
Thermokinetics, movement and, 157–158
Thermotaxis, spirilla and, 193
Thiamine, bacteria and, 17
Thiobacillus thioparus, sulfur and, 15
Thiobacillus thiooxidans,
  sulfur and, 15
  vacuoles of, 47
Thiol esters, bacteriophage and, 406
Thiol groups, flagella and, 174
Thionine,
  chromatin bodies and, 67, 68
  ribonucleic acid and, 45
  spores and, 232
Thiorhodaceae, movements and, 188
Thiouracil, enzyme formation and, 437–438
Thiovulum, vacuoles of, 47
Thiovulum majus, cytoplasm of, 60–62
Threonine,
  cell walls and, 123, 124
  incorporation and, 315
Threshold, phobotaxis and, 189
Thymine, incorporation of, 314, 315–316
Tobacco mosaic virus, ribonucleic acid function and, 24
Toluene, cell wall isolation and, 119
Toluidine blue, volutin and, 48
Topotaxis, 188
  spirilla and, 194
α-Toxin, antibodies and, 430
Transduction,
  bacteriophage and, 401
  bacterial classification and, 412
Transformation,
  deoxyribonucleic acid and, 24
  protoplasts and, 336–338

Trehalose, protoplasts and, 261
Treponema pallidum, movement of, 176–177
Tricarboxylic acid cycle, 4
Trichloroacetic acid, cell wall isolation and, 116, 119
Trichomonas foetus, blood group substance and, 429
Triose phosphate dehydrogenase, antibodies against, 430
Triphenyltetrazolium,
  bacterial granules and, 445
  electron-transport particle and, 460
Trypsin,
  cell wall isolation and, 119
  enzyme formation and, 319
  protoplasts and, 308–309
  streptococcal surface and, 114, 115
Tryptophan,
  bacteriophage and, 403–404
  flagella and, 165
Tryptophanase, spheroplasts and, 322
Tryptophan synthetase,
  cross-reacting material and, 432–433
  fractionation of, 433
  reactions of, 433
Turbidimetry, precipitins and, 424
Tyvelose, lipopolysaccharides and, 111

U

Ultrasound, cell wall isolation and, 117
Ultraviolet,
  bacterial protoplasm and, 38
  volutin and, 48
Uncoupling, bacterial enzymes and, 456
Undulation, bacterial movement and, 155–157, 171
Uracil, incorporation of, 314, 315
Uranyl chloride, protoplasts and, 316
Urea,
  bacteriophage infection and, 335–336
  chemotaxis and, 192
  protoplasts and, 291
  ribonucleoprotein and, 456
Uridine pyrophosphate N-acetylmuramic acid, cell walls and, 139
Uridine diphosphate amino sugar peptides, penicillin and, 141

V

Vacuoles,
  nature of, 46–47
  protoplasts and, 284
  vital staining and, 38–39
Valine, cell wall synthesis and, 140
Vanadium, bacteria and, 16
Vi antigen,
  nature of, 113
  surface components and, 108
Vibrio,
  chemotaxis and, 190
  L-forms of, 374
  spherical forms of, 342, 345
Vibrio cholerae,
  antigens of, 138
  L-forms of, 372, 374
  spherical forms of, 280
Vibrio comma, water content of, 13
Vibrio costicolus,
  cell walls of, 121
  lipids of, 129
Vibrio metchnikovi,
  amino sugars and, 126, 127
  antigens of, 138
  cell walls of, 121
  flagella of, 159, 161, 165
  protoplasts of, 287, 299
  spherical transformation of, 141
Victoria blue, cytoplasmic membrane and, 42–43, 251
Violacein, 21
Virulence, capsules and, 114
Viscosity, bacterial movement and, 154–155, 167
Vital staining, significance of, 38
Vitamin B_6, sonic disruption and, 448
Vitamin D_2, cell permeability and, 294–295
Vitamin K_1, oxidative phosphorylation and, 456
Vitamins, bacterial content of, 17
Vitreoscillaceae, movements of, 188
Volutin, 39, 48
  bacterial granules and, 445
  formation of, 22
  Thiobacillus and, 47

W

Water, bacterial content of, 12
Wavelength, flagella and, 159–160

X

X-rays, flagella and, 165
Xylose, surface components and, 112

Y

Yeast,
  growth, energy source and, 11
  size of, 10

Z

Zinc,
  bacteria and, 16
  bacteriophage enzymes and, 136
  cyanide complex, bacteriophage and, 404
Zymosan, immunity and, 138

DATE DUE			
~~APR 0 4~~			
~~Feb~~			

Demco, Inc. 38-293